Methods of Experimental Physics

VOLUME 16

POLYMERS

PART A: Molecular Structure and Dynamics

METHODS OF EXPERIMENTAL PHYSICS:

L. Marton and C. Marton, *Editors-in-Chief*

Volume 16

Polymers

PART A: Molecular Structure and Dynamics

Edited by

R. A. FAVA

ARCO Polymers, Inc.
Monroeville, Pennsylvania

1980

ACADEMIC PRESS

A Subsidiary of Harcourt Brace Jovanovich, Publishers

New York   London   Toronto   Sydney   San Francisco

ACADEMIC PRESS, INC.
111 Fifth Avenue, New York, New York 10003

United Kingdom Edition published by
ACADEMIC PRESS, INC. (LONDON) LTD.
24/28 Oval Road, London NW1 7DX

Library of Congress Cataloging in Publication Data

Main entry under title:

Polymer physics.

    (Methods of experimental physics ; v. 16)
    Includes bibliographical references and index.
    CONTENTS: pt. A. Molecular structure and
dynamics, edited by R. A. Fava.
    1. Polymers and polymerization.
I. Fava, Ronald A. II. Series.
QD381.P612   547'.84   79-26343
ISBN 0-12-475916-5

PRINTED IN THE UNITED STATES OF AMERICA

80 81 82 83   9 8 7 6 5 4 3 2 1

CONTENTS

1. Introduction
by R. A. FAVA

2. Polymer Molecular Weights
by DOROTHY J. POLLOCK AND ROBERT F. KRATZ

CONTRIBUTORS

Numbers in parentheses indicate the pages on which the authors' contributions begin.

C. V. BERNEY, *Department of Nuclear Engineering, Massachusetts Institute of Technology, Cambridge, Massachusetts 02139* (205)

L. LAWRENCE CHAPOY, *Instituttet for Kemiindustri, The Technical University of Denmark, 2800 Lyngby, Denmark* (404)

DONALD B. DUPRÉ, *Department of Chemistry, University of Louisville, Louisville, Kentucky 40208* (404)

R. A. FAVA, *ARCO Polymers, Inc., 440 College Park Drive, Monroeville, Pennsylvania 15146* (1)

J. S. KING, *Nuclear Engineering Department, University of Michigan, Ann Arbor, Michigan 48109* (480)

ROBERT F. KRATZ, *ARCO Polymers, Inc., Research and Development Department, P.O. Box 208, Monaca, Pennsylvania 15061* (13)

PHILIP L. KUMLER, *Department of Chemistry, State University of New York, College of Fredonia, Fredonia, New York 14063* (442)

J. R. LYERLA, *IBM Research Laboratories, San Jose, California 95193* (241)

G. D. PATTERSON, *Bell Telephone Laboratories, Murray Hill, New Jersey 07974* (170)

DOROTHY J. POLLOCK, *ARCO Polymers, Inc., Research and Development Department, P.O. Box 208, Monaca, Pennsylvania 15061* (13)

R. G. SNYDER, *Western Regional Research Center, Science and Education Administration, U.S. Department of Agriculture, Berkeley, California 94710* (73)

J. R. STEVENS, *Department of Physics, University of Guelph, Guelph, Ontario, N1G 2W1, Canada* (371)

H. W. WHITE, *Department of Physics, University of Missouri, University Park, Columbia, Missouri 65211* (149)

T. WOLFRAM, *Department of Physics, University of Missouri, University Park, Columbia, Missouri 65211* (149)

SIDNEY YIP, *Department of Nuclear Engineering, Massachusetts Institute of Technology, Cambridge, Massachusetts 02139* (205)

FOREWORD

The thoroughness and dedication of Ronald Fava in preparing these volumes may be verified by this work's impressive scope and size. This is the first time *Methods of Experimental Physics* has utilized three volumes in the coverage of a subject area.

The volumes, in part, indicate the future development of this publication. Solid state physics was covered in Volumes 6A and 6B (edited by K. Lark-Horovitz and Vivian A. Johnson) in 1959. Rather than attempt a new edition of these volumes in a field that has experienced such rapid growth, we planned entirely new volumes, such as Volume 11 (edited by R. V. Coleman), published in 1974. We now appreciate the fact that future coverage of this area will require more specialized volumes, and *Polymer Physics* exemplifies this trend.

To the authors and the Editor of this work, our heartfelt thanks for a job well done.

L. Marton
C. Marton

PREFACE

A polymer must in many ways be treated as a separate state of matter on account of the unique properties of the long chain molecule. Therefore, although many of the experimental methods described in these three volumes may also be found in books on solid state and molecular physics, their application to polymers demands a special interpretation. The methods treated here range from classical, well-tried techniques such as X-ray diffraction and infrared spectroscopy to new and exciting applications such as those of small-angle neutron scattering and inelastic electron tunneling spectroscopy. It is convenient to present two types of chapters, those dealing with specific techniques and those in which all techniques applied in measuring specific polymer properties are collected. The presentation naturally divides into three parts: Part A describes ways of investigating the structure and dynamics of chain molecules, Part B more specifically deals with the crystallization of polymers and the structure and morphology of the crystals, while in Part C those techniques employed in the evaluation of mechanical and electrical properties are enumerated. It should be emphasized, however, that this is not a treatise on the properties of polymeric materials. The authors have introduced specific polymer properties only incidentally in order to illustrate a particular procedure being discussed. The reader is invited to search the Subject Index wherein such properties may be found listed under the polymer in question.

I have endeavored to arrange chapters in a logical and coherent order so that these volumes might read like an opera rather than a medley of songs. The authors are to be commended for finishing their contributions in timely fashion to help achieve this end. I also wish to acknowledge with thanks the support of ARCO Polymers, Inc. and the use of its facilities during the formative stages of the production.

R. A. FAVA

CONTENTS OF VOLUME 16, PARTS B AND C

CONTRIBUTORS TO VOLUME 16, PARTS B AND C

Part B

EDWARD S. CLARK, *Polymer Engineering, University of Tennessee, Knoxville, Tennessee 37916*

LOIS J. FROLEN, *National Measurement Laboratory, National Bureau of Standards, Washington, D.C. 20234*

IAN R. HARRISON, *College of Earth and Mineral Sciences, The Pennsylvania State University, University Park, Pennsylvania 16802*

G. N. PATEL, *Corporate Research Center, Allied Chemical Corporation, Morristown, New Jersey 07960*

GAYLON S. ROSS, *National Measurement Laboratory, National Bureau of Standards, Washington, D.C. 20234*

JAMES RUNT, *College of Earth and Materials Science, The Pennsylvania State University, University Park, Pennsylvania 16802*

JOSEPH E. SPRUIELL, *Polymer Engineering, University of Tennessee, Knoxville, Tennessee 37916*

RICHARD G. VADIMSKY, *Bell Telephone Laboratories, Murray Hill, New Jersey 07974*

JING-I WANG, *College of Earth and Mineral Sciences, The Pennsylvania State University, University Park, Pennsylvania 16802*

Part C

RICHARD H. BOYD, *Department of Materials Science and Engineering, University of Utah, Salt Lake City, Utah 84112*

NORMAN BROWN, *Department of Materials Science and Engineering, University of Pennsylvania, Philadelphia, Pennsylvania 19174*

D. KEITH DAVIES, *Electrical Research Association Limited, Leatherhead, Surrey, KT22 7SA, England*

R. M. FELDER, *Department of Chemical Engineering, North Carolina State University, Raleigh, North Carolina 27650*

BRUCE HARTMANN, *Naval Surface Weapons Center, White Oak, Silver Spring, Maryland 20910*

IAN L. HAY, *Celanese Research Company, Summit Laboratory, Summit, New Jersey 07901*

G. S. HUVARD, *Department of Chemical Engineering, North Carolina State University, Raleigh, North Carolina 27607*

TOSHIHIKO NAGAMURA,* *Department of Mechanical and Industrial Engineering, Department of Materials Science and Engineering, College of Engineering, University of Utah, Salt Lake City, Utah 84112*

DONALD J. PLAZEK, *Department of Metallurgical and Materials Engineering, University of Pittsburgh, Pittsburgh, Pennsylvania 15261*

J. ROOVERS, *Division of Chemistry, National Research Council of Canada, Ottawa, Ontario, K1A 0R9 Canada*

JOHN L. RUTHERFORD, *Kearfott Division, The Singer Company, Little Falls, New Jersey 07424*

B. R. VARLOW, *Electrical Engineering Laboratory, University of Manchester, Manchester M13 9PL, England*

R. W. WARFIELD, *Naval Surface Weapons Center, White Oak, Silver Spring, Maryland 20910*

* Present address: Department of Organic Synthesis, Faculty of Engineering, Kyushu University, Higashi-ku, Fukuoka 812, Japan.

METHODS OF EXPERIMENTAL PHYSICS

Editors-in-Chief
L. Marton   C. Marton

Methods of
Experimental Physics

VOLUME 16

POLYMERS

PART A: Molecular Structure and Dynamics

1. INTRODUCTION

By R. A. Fava

1.1. Historic Development

The possibility of joining small molecular units together to form long-chain macromolecules, or polymers, must have occurred to chemists long before polymer science became a subject in its own right. With the knowledge we have today we can, in fact, trace back through the literature many instances of polymerization reported unknowingly by various authors. For example, Regnault[1] in 1838 described the development of turbulence and the deposition of small quantities of a white noncrystalline powder when vinylidene chloride was stored in sunlight. In the following year, Simon[2] reported extracting a volatile oil from a commercial substance known as Storax. This oil thickened after several months to a viscous mass no longer soluble in alcohol or ether. Then, in 1872, Baumann[3] described the effect of sunlight on vinyl chloride. A white powder was formed with the same elemental analysis as the starting material. Baumann described this as an isomer of vinyl chloride. It is almost certain, however, that what he really saw was the polymer of vinyl chloride, i.e., poly(vinyl chloride) or PVC. Similarly, we can deduce that Regnault had probably synthesized poly(vinylidene chloride), another important commercial polymer, and Simon's volatile oil was styrene that he polymerized to polystyrene.

The experimental techniques were not then available to characterize such substances, nor had the theoretical understanding of chemical bonding been sufficiently well developed.

By the early twentieth century such curious reactions of some molecules to heat and light had become an important branch of organic chemistry. Standard procedures of organic chemistry such as molecular weights by osmometry and freezing-point depression, melting-point measure-

---

[1] M. V. Regnault, *Ann. Chim. Phys.* [2] **69**, 151 (1838).
[2] E. Simon, *Ann. Pharm.* (*Lemgo, Ger.*) **31**, 265 (1839).
[3] E. Baumann, *Ann. Chem. Pharm.* **163**, 308 (1872).

1

ments, crystal structure by X-ray diffraction, solution viscosity and other solution measurements were used to analyze the products. From observations of their colloidal nature it was concluded that the products were loose agglomerates (micelles) of small molecules held together by secondary valence forces. The forces were believed to be associated with the unsaturation in those molecules which contained covalent double bonds, for example:

$$\text{vinyl chloride:} \quad \begin{array}{c} \text{H} \\ \diagdown \\ \text{H} \diagup \end{array} \text{C}{=}\text{C} \begin{array}{c} \diagup \text{H} \\ \diagdown \text{Cl} \end{array}$$

$$\text{vinylidene chloride:} \quad \begin{array}{c} \text{H} \\ \diagdown \\ \text{H} \diagup \end{array} \text{C}{=}\text{C} \begin{array}{c} \diagup \text{Cl} \\ \diagdown \text{Cl} \end{array}$$

$$\text{styrene:} \quad \begin{array}{c} \text{H} \\ \diagdown \\ \text{H} \diagup \end{array} \text{C}{=}\text{C} \begin{array}{c} \diagup \text{H} \\ \diagdown \text{C}_6\text{H}_5 \end{array}$$

The micelle theory began to crumble after Staudinger and Fritschi[4] found that if natural rubber (polyisoprene) is hydrogenated the molecules still remain of colloidal size. The micelle theory would predict that the agglomerated isoprene molecules, saturated by the hydrogenation, would dissociate and form a highly volatile liquid.

A new theory, developed in classic publications of Staudinger,[5] postulated that polymers are made up of small molecules held together by primary valence forces (covalent bonding). The history of the development of this now-accepted model of the structure of polymers is well documented in Staudinger's collected works.[6]

Suffice to say, it was some time before these new ideas were fully accepted. One of the nagging drawbacks was the fact that polymers displayed crystallinity and the X-ray diffraction pattern suggested a unit cell similar in size to that of the small molecular unit. As early as 1926, however, it was hinted by Sponsler,[7] with reference to cellulose fibers, that the small molecules from which polymers are built form repeat units along the polymer chain and the unit cell is formed by this and the juxtaposition of adjacent parallel chains (see also Chapter 6.1, this volume, Part B).

[4] H. Staudinger and J. Fritschi, *Helv. Chim. Acta* **5,** 785 (1922).

[5] H. Staudinger, *Ber. Dtsch. Chem. Ges.* **53,** 1073, (1920); **59,** 3019 (1926).

[6] H. Staudinger, "From Organic Chemistry to Macromolecules," Wiley (Interscience), New York, 1961.

[7] O. L. Sponsler, *J. Gen. Physiol.* **9,** 677 (1926).

1.2. Definitions

The International Union of Pure and Applied Chemistry (IUPAC)[8] has defined internationally agreed nomenclature in the field of polymer science. The term *macromolecule* is an all-embracing term covering all large molecules, while a *high polymer* is restricted to structures that are at least approximately multiples of a low-molecular-weight unit (a *monomer*). A high polymer may be further specified as follows:

homopolymer: a single monomer type,

copolymer: two different monomer units in irregular sequence, e.g., poly(ethylene-*co*-propylene),

terpolymer (etc.): three (etc.) units in irregular sequence,

alternating copolymer: two units in alternating sequence, e.g., poly(ethylene-*alt*-carbon monoxide),

block copolymer: a sequence of one unit followed by a sequence of a second unit in the same chain, e.g., poly(styrene-*b*-butadiene), poly(styrene-*b*-isoprene-*b*-styrene).

Due to the statistical nature of the polymerization process the polymer chains are not all the same length and very often contain defects in the form of short or long branches. A special kind of branch containing a different molecular unit may be synthesized subsequent to the main-chain polymerization. Such a polymer is called a *graft copolymer*, e.g., poly(butadiene-*g*-styrene). If branches bridge two chains a *cross-linked* network is formed.

Another type of defect occurs in polymers built from monomer units that are asymmetric with respect to the chain axis. If the covalent bonds are not freely rotating by virtue of double bonds or steric hindrances, neighboring molecular units may be permanently joined in parallel or antiparallel sequences. This quality is called *tacticity*. If the sequence is random the polymer is called *atactic* and will not generally crystallize. Regular tacticity in a polymer chain is illustrated below with two monomer units of polypropylene:

$$\text{syndiotactic:} \quad -CH_2-CH-CH_2-\overset{\overset{\displaystyle CH_3}{|}}{CH}-$$
$$\underset{\displaystyle CH_3}{|}$$

$$\text{isotactic:} \quad -CH_2-\underset{\underset{\displaystyle CH_3}{|}}{CH}-CH_2-\underset{\underset{\displaystyle CH_3}{|}}{CH}-$$

[8] International Union of Pure and Applied Chemistry, *J. Polym. Sci.* **8**, 257 (1952).

Such chain details (configurations) are conveniently studied by high-resolution nuclear magnetic resonance (see Part 4, this volume).

1.3. Formation and Conformation

A polymer molecule may contain a thousand or more monomer units. To synthesize a polymer the monomer must be at least bifunctional, i.e., be able to react at two sites. The most common types of polymerization processes are addition polymerization and condensation polymerization. In the former, bifunctionality is effected by a double bond as in ethylene or the opening of a cyclic structure as in ethylene oxide,

$$CH_2{=}CH_2 \rightarrow {-}({-}CH_2{-}CH_2{-}){-}_n$$

$$H_2C \overset{O}{\underset{}{\diagup\diagdown}} CH_2 \rightarrow {-}({-}CH_2{-}CH_2{-}O{-}){-}_n .$$

Condensation polymerization, pioneered by Carothers et al.,[9] involves a chemical reaction between two types of bifunctional monomer generally with the elimination of water. Polyesters such as poly(ethylene terephthalate) are of this class. Phenol-formaldehyde resins, discovered by Baekeland[10] before the advent of polymer science, are also of this type. For authoritative discussions on the chemistry of polymer synthesis the reader is referred to the various standard texts.[11-13]

An unperturbed polymer molecule can be treated statistically using a model of a chain with freely rotating links performing a "random walk" on a lattice. The length of segments between the links generally corresponds to multiples of a monomer unit and depends also on the chain stiffness. Such models are treated by Flory[11] and Billmeyer.[12] The free molecular chain is found to take on randomly coiled conformations with an approximately spherical envelope. For polystyrene, as an example, if

[9] W. H. Carothers, *J. Am. Chem. Soc.* **51**, 2548 (1929); W. H. Carothers and J. A. Arven, *ibid.* p. 2560.

[10] L. H. Baekeland, *J. Ind. Eng. Chem.* **1**, 149 and 549 (1909).

[11] P. J. Flory, "Principles of Polymer Chemistry." Cornell Univ. Press, Ithaca, New York, 1953.

[12] F. W. Billmeyer, "Textbook of Polymer Science." Wiley, New York, 1971.

[13] P. E. M. Allen and C. R. Patrick, "Kinetics and Mechanisms of Polymerization Reactions." Halsted Press, New York, 1974.

the number of monomer units in the chain (the degree of polymerization) is 100, the chain length is about 2500 Å and the theoretical mean diameter of the freely coiled chain envelope is about 200 Å.

Unperturbed conformations can be approximated in dilute solution in a poor solvent (a theta solvent) and this allows information on molecular size, branching, etc., to be obtained from measurements such as solution viscosity and light scattering (see Part 2, this volume).

In more concentrated solutions the polymer molecules interfere with each other and in the limit become entangled. At this point the properties of polymers diverge greatly from those of low-molecular-weight substances. The entanglements endow polymers with a quality known as *viscoelasticity*, which is elasticity with a fading memory. When deformed, the entangled chains deform elastically but if held in the strained state they will relax by unraveling at a rate dependent on the temperature, chain flexibility, freedom, etc. Conversely, if polymers are forced to deform, the response will depend on the rate of deformation. The elastic character will be more pronounced at deformation rates exceeding the natural relaxation time of the molecules and may lead to phenomena such as melt instability and fracture, which are the headaches of plastics molders and processors. Measurements of viscoelasticity in polymer melts are treated in detail in Part 11 (this volume, Part C). The reader is also referred to standard texts such as that of Ferry.[14]

An extreme case of viscoelasticity is displayed in certain silicone polymers ("silly putty"), which are soft and puttylike when plied in the hand but may be shattered with a hammer or made to bounce like a ball. Some of the curious properties of these polymers are described by Busse and Starr.[15]

The exact conformation of molecules in a polymer melt has only relatively recently been solved by the technique of small-angle neutron scattering described in Chapter 5.4 (this volume). It is a remarkable fact that the molecules in a relaxed melt or amorphous solid are in a state approximating the unperturbed random coil. It is as though the molecules, although intimately entangled, are unaware of each other's existence. This lack of interaction is even more evident when one attempts to blend together two different types of polymer. Unlike the situation with small molecules, it is an exception to the rule to find compatibility between two polymers. This has been ascribed to the very small entropy contribution to the free energy of mixing. The subject is discussed fully in Part 16 (this

[14] J. D. Ferry, "Viscoelastic Properties of Polymers." Wiley, New York, 1970.
[15] W. F. Busse and F. C. Starr, *Am. J. Phys.* **28**, 19 (1960).

volume, Part C) and also in the treatise edited by Paul and Newman,[16] for instance.

1.4. The Solid State

A polymer in the solid state, in the absence of solvent, is a familiar material in the consumer market. It manifests itself for instance as plastic bags and buckets, appliance cabinets, children's toys, and increasingly in automobile construction. Each polymer has its own characteristic solid-state properties. It may be partially crystalline or totally amorphous, depending on the stereoregularity and mobility of the polymer chains. There is evidence that even amorphous polymers have a morphology, in the form of short-range order; see the overview written by Boyer.[17]

1.4.1. Amorphous Polymers

Amorphous polymers, such as atactic polystyrene and poly(methyl methacrylate), possess a glass transition temperature above which they are viscoelastic melts and below which they are glassy hard. When a polymer is heated through the glass transition temperature the specific heat, coefficient of thermal expansion, and dielectric constant undergo abrupt increases similar in kind to those accompanying a second-order thermodynamic phase change, although the "abrupt" change may be spread over a 10–20°C temperature interval. Furthermore, the temperature of the glass transition in polymers depends on the time scale of the experiment. Much of the phenomenon is therefore kinetic as opposed to thermodynamic in origin. Glass formation is not restricted to polymers but is also extant in those small-molecular substances that for some reason or other do not crystallize readily. Concepts such as metastability and frozen-in free volume[18,19] have been advocated to describe the process. The Old Testament has also been invoked by Reiner[20] as providing insight into the glass-forming mechanism: "The mountains flowed before the Lord," sang the prophetess Deborah (Judges 5:5), indicating in her wisdom the Lord's infinite time scale of observation. Reiner went on to suggest that solid and fluid states may be expressed in terms of the

[16] D. R. Paul and S. Newman, eds., "Polymer Blends." Academic Press, New York, 1978.

[17] R. F. Boyer, *J. Macromol. Sci., Phys.* **12**, 253 (1976).

[18] W. Kauzmann, *Chem. Rev.* **43**, 219 (1948).

[19] N. Hirai and H. Eyring, *J. Polym. Sci.* **37**, 51 (1959).

[20] M. Reiner, *Phys. Today* 62 (Jan 1964).

Deborah number, defined as the ratio of molecular relaxation time to time of observation. This number is unity at the point of glass formation, which in our own time scale of reference occurs when the viscosity is in the range $10^{13} - 10^{14}$ poise.[21]

Transitions in amorphous polymers are commonly studied by dynamic-mechanical loss measurements as a function of temperature. There is a peak in the loss (a measure of energy dissipated in the sample per cycle of oscillation) during relaxation processes such as a glass transition. Peaks are also found below the glass temperature and these are ascribed to the onset of more localized molecular motion such as that of branches, phenyl groups, or small sections of main chain. The reader is referred to the comprehensive source book of McCrum, Read, and Williams,[22] which also draws a parallel between dynamic-mechanical and dielectric loss measurements (see also Part 11 and Chapter 18.1, this volume, Part C).

The onset of molecular motion in polymers is detectable by many of the other methods described in this volume. It is discussed in connection with Brillouin scattering (Chapter 3.3), broadline NMR (Part 4), positron annihilation (Chapter 5.1), fluorescence depolarization (Chapter 5.2), paramagnetic probe techniques (Chapter 5.3), thermal analysis (Part 9), ultrasonic measurements (Chapter 12.1), high-pressure measurements (Chapter 12.2), permeability to gases and vapors (Chapter 17), dielectric loss (Chapter 18.1), and thermally stimulated currents (Chapter 18.2).

1.4.2. Crystalline Polymers

In polymers that possess stereoregularity, sections of the molecular chain may line up parallel to form crystalline regions detectable by X-ray diffraction. It is rare for a polymer to crystallize completely because of molecular entanglement and orientation restrictions, although under pressures exceeding 5000 atm crystallinities approaching 99% have been achieved. In general, a crystalline polymer will display both a glass transition temperature and a melting point. For a comprehensive review of crystallinity in polymers the reader is referred to Wunderlich's book[23] in addition to the discussions in Part B of this volume.

The original model of crystallinity in polymers was the fringed-micelle

[21] J. M. Stevels, in "Amorphous Materials" (R. W. Douglas and B. Ellis, eds.), p. 133. Wiley (Interscience), New York, 1972.

[22] N. G. McCrum, B. E. Read, and G. Williams, "Anelastic and Dielectric Effects in Polymeric Solids." Wiley, New York, 1967.

[23] B. Wunderlich, "Macromolecular Physics," Vols. 1 and 2. Academic Press, New York, 1973.

model introduced by Herrmann and Gerngross.[24]   The picture was one of completely random entangled chains among which there were regions where neighboring chain sections lined up to form crystalline micelles. One chain could therefore pass through several crystalline regions.

In 1957 three independent reports were made of the crystallization of linear polyethylene from very dilute solution.[25-27]   The single crystals were lozenge-shaped lamellae approximately 100 Å thick.   Some photographs of polymer single crystals are shown in Part 7 (this volume, Part B).   The molecular chains were aligned perpendicular to the surface in these lamellae, which meant that, since they were typically several thousand ångstroms long, they must have folded back and forth during crystallization along the sides of the lamellae.   This is a model quite unlike the fringed-micelle model.   The subject has been extensively reviewed.[28-30] The degree of regularity of the folds at the lamellar surface is a continuing debate and has been a fertile source of novel experimentation.   There is particular interest as to whether the chain-folding model can be applied to melt-crystallized polymers.   Many instances of lamellar morphology in such polymers have been reported.   The latest feeling, however, with evidence gleaned from the small-angle neutron-scattering experiment (see Chapter 5.4, this volume) and a comparison of the rapid crystallization rate to the relatively long molecular relaxation times, is that if there is chain folding in melt-crystallized polymers it is not of the adjacent-reentry type.[31]   For polyethylene chains, average reentry distances are in excess of 25 Å and one molecule may pass through many lamellar crystallites. The model is not too far removed from the original fringed-micelle model.

1.5. Orientation

Despite their tendency toward random coils and folding during crystallization, polymer molecules are highly orientable entities and have a natural fiber-forming tendency.   Polymers in nature exist abundantly as fibers in plants, hair, wool, horn, silk, etc.   Synthetic polymers often polymerize in the fibrous form but as soon as they are melted or dissolved they revert to the randomly coiled conformations and need to be deformed me-

[24] K. Herrmann and O. Gerngross, *Kautschuk* **8**, 181 (1932).
[25] P. H. Till, *J. Polym. Sci.* **24**, 301 (1957).
[26] A. Keller, *Philos. Mag.* [8] **2**, 1171 (1957).
[27] E. W. Fischer, *Z. Naturforsch. Teil A* **12**, 753 (1957).
[28] P. H. Geil, "Polymer Single Crystals." Wiley (Interscience), New York, 1963.
[29] A. Keller, *Rep. Prog. Phys.* **31**, Part II, 623 (1968).
[30] R. A. Fava, *J. Polym. Sci., Part D* **5**, 1 (1971).
[31] P. J. Flory and D. Y. Yoon, *Nature 1London)* **272**, 226 (1978).

chanically before orientation can be reintroduced. Orientation can occur in amorphous and crystalline regions of polymers and may be measured by a variety of techniques. Most of these are collected in Part 13 (this volume, Part C). The edited publication of Ward[32] is also to be recommended.

The desire to attain ultimate properties has led to some interesting work on oriented polymers. The theoretical mechanical strength and elastic modulus of an extended polymer chain is far in excess of that realized in a bulk polymer,[33] even if oriented by stretching. One way of approaching the ultraorientation needed to match the theory is to extrude a crystallizable polymer such as polyethylene, under high pressure and in the solid state, through a capillary. The elongational flow is claimed to yield highly transparent, ultraoriented filaments with high crystallinity. The subject has been reviewed by Perkins and Porter.[34]

Another curiosity of potential commercial value is the production, by stretching under carefully controlled conditions, of highly elastic fibers, the so-called hard-elastic fibers. This phenomenon has also been reviewed.[35]

1.6. Impurities

No discussion of experimental methods would be complete without an appreciation of the anomalous effects impurities in a sample can have on the various measurements. When dealing with the physics of polymers one is dealing with materials that are impure in many senses. In commercially obtained polymers, not only are there traces of catalyst residues from the polymerization process but there are also up to several percent of deliberate additives such as antioxidants, stabilizers against ultraviolet degradation, antistatic agents, nucleating agents, fire retardants, pigments, and internal lubricants. Details may be found in the Modern Plastics Encyclopaedia published annually by McGraw-Hill (New York). Additives are introduced into polymers to assist in the production of superior molded articles but do not help when one is interested in fundamental physical properties.

Some measurements are more sensitive than others to the presence of

[32] I. M. Ward, ed., "Structure and Properties of Oriented Polymers." Wiley, New York, 1975.

[33] R. N. Britton, R. Jakeways, and I. M. Ward, *J. Mater. Sci.* **11,** 2057 (1976).

[34] W. G. Perkins and R. S. Porter, *J. Mater. Sci.* **12,** 2355 (1977).

[35] S. L. Cannon, G. B. McKenna, and W. O. Statton, *J. Polym. Sci., Macromol. Rev.* **11,** 209 (1976).

impurities and additives. Most stringent purification is necessary for light scattering studies in polymer solutions and also in crystallization studies. A typical clean-up procedure for polyethylene is shown in Fig. 3 of Part 10 (this volume, Part B). Electrical properties are also sensitive to impurities. For instance, it has been shown that the electrical conductivity of polyethylene can be reduced 100-fold from 10^{-19} to 10^{-21} Ω^{-1} cm^{-1} by removing extractable impurities.[36]

Impurity in a polymer can also be of the same species as the polymer itself. Residual monomer, dimer, trimer, etc., are classed as impurity and, when measuring colligative properties, such as number-average molecular weights by osmometry, one must decide whether or not such very low-molecular-weight species should be included. A similar problem also exists at the other end of the molecular-weight scale and manifests itself in the form of gell-like and frequently cross-linked species. The problem of impurity of the same species as the polymer is, however, a general one since a linear polymer is an impure mixture of linear molecules with a broad range of molecular lengths and often with a significant degree of long-chain branching. Most sensitive to these variations are the viscoelastic properties of the polymer melt, so much so in fact that the melt properties have been used to monitor molecular weight and molecular-weight distribution.[37]

Another class of impurity includes those chemically attached to the polymer chain. In polyethylene, for instance, the infrared spectrum reveals fractions of a percent of —CO and —OH groups and also double bonds in the form of vinylidene, trans vinylene, and terminal vinyl. The oxygenated groups should not be present in an ideal polymerization of ethylene. Such polar impurities have been shown to be responsible for much of the already low dielectric loss of polyethylene.[38,39]

Finally, one should be aware of the presence of dissolved gases and water vapor absorbed into the polymer from the atmosphere. Molecular oxygen has, for example, been shown to complex with the benzene ring in aromatic polymers and yield anomalous nuclear magnetic resonance measurements of spin–lattice relaxation times.[40] Water vapor can be absorbed in quantities up to 0.2% in polystyrene and in even higher quantities in polymers such as cellulose, polycarbonates, and polyamides. Water's extremely high polarity causes it to affect considerably measure-

[36] R. H. Partridge, *J. Polym. Sci., Part B* **5**, 205 (1967).

[37] R. Sabia, *J. Appl. Polym. Sci.* **7**, 347 (1963).

[38] K. Nakagawa and S. Tsuru, *J. Polym. Sci., Polym. Phys. Ed.* **14**, 1755 (1976).

[39] O. Yano, K. Saiki, S. Tarucha, and Y. Wada, *J. Polym. Sci., Polym. Phys. Ed.* **15**, 43 (1977).

[40] M. F. Froix and A. O. Goedde, *Polymer* **17**, 758 (1976).

ments of dielectric constant and loss. It can also act as a plasticizer and loosen up the polymer molecules to the extent of introducing extra peaks in the dynamic-mechanical loss spectrum.

To round off these cautionary notes we recount without comment the interesting but short story of the discovery of polywater. In 1968, a new highly viscous form of water was announced.[41] This water was prepared in minute quantities by condensation in very fine fused quartz capillaries and the experimental results were confirmed by many workers. The infrared spectrum indicated an unusual type of bonding[42] and it was concluded that the new substance was a polymeric form of water. The problem of joining water molecules together to form a macromolecule generated a number of plausible theories on the nature of such chemical bonds.[42-44] At this point, however, evidence began creeping in from elemental analyses and microscopic observations to show that there was considerable impurity (especially sodium, chlorine, and sulfate residues) present in the polywater[45,46] and this led a year later to the following startling revelation.[47] The infrared spectrum of "polywater" was identical to that of human sweat.

[41] B. V. Deryagin and N. V. Churayev, *Priroda* (*Sofia*) **4,** 16 (1968).

[42] E. R. Lippincott, R. R. Stromberg, W. H. Grant, and G. L. Cessac, *Science* **164,** 1482 (1969).

[43] L. C. Allen and P. A. Kollman, *Science* **167,** 1443 (1970).

[44] J. W. Linnett, *Science* **167,** 1719 (1970).

[45] D. L. Rousseau and S. P. S. Porto, *Science* **167,** 1715 (1970).

[46] S. L. Kurtin, C. A. Mead, W. A. Mueller, B. C. Kurtin, and E. D. Wolf, *Science* **167,** 1722 (1970).

[47] D. L. Rousseau, *Science* **171,** 170 (1971).

2. POLYMER MOLECULAR WEIGHTS

By Dorothy J. Pollock and Robert F. Kratz

2.1. Definitions of Molecular Weight

2.1.1. Introduction

The formation of a polymer, be it synthetic or natural, is the result of a series of random events whereby the repeating monomeric units are joined together to form molecules, not all necessarily of the same size. The consequent collection of molecules ordinarily reflects a distribution of sizes. The distribution may be rather narrow for polymers formed through linear growth under homogeneous conditions; or quite broad, especially so if branched or formed in heterogeneous environments. Although biopolymers (e.g., proteins) may exhibit an amazing uniformity of size, the polymers employed in commercial plastics, especially thermoplastics, are rather long—hundreds to tens of thousands of monomer units—and may vary over this entire range of size in a single sample. The expressions used to describe the molecular size are therefore determined by the need to describe the shape of the distribution of sizes. The conventional definition of molecular weight—the weight in grams of 6.023×10^{23} (Avogadro's number) molecules—is inadequate for other than the most uniform polymeric materials. In the general case, a count of the molecules of *all* sizes is necessary to describe the distribution of molecular weight, as suggested by the curve shown in Fig. 1. If the ordinate at the shaded segment represents the number of molecules n_i of molecular weight M_i, it is apparent that the summation $\Sigma n_i M_i$ over all molecular species i can be represented by an equivalent moment NM_n, where N is the total number of molecules Σn_i, and M_n is a molecular weight defined in exactly the same way as the conventional definition of molecular weight mentioned above. In order to describe the distribution fully, additional moments equal in number to the number of molecular species present would be required. The form of the moments would be $\Sigma n_i M_i^a$, such that the exponent a represents the order of the moment. Usually, however, three or four such moments are sufficient in a practical sense to describe the distribution, since the distributions tend to follow a limited number of shapes.

13

METHODS OF EXPERIMENTAL PHYSICS, VOL. 16A

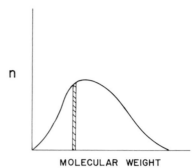

MOLECULAR WEIGHT

FIG. 1. Idealized molecular weight distribution.

2.1.2 The Moments of Molecular Weight

2.1.2.1. Number-Average Molecular Weight. As noted in Section 2.1.1, the number-average molecular weight (M_n) is defined as the weight in grams of Avogadro's number of molecules. The alternative expressions frequently used are simply equivalent relationships, substituting experimentally definable terms:

$$M_n = \text{weight of } 6.023 \times 10^{23} \text{ molecules} \qquad (2.1.1)$$

$$= \Sigma n_i M_i / \Sigma n_i \qquad (2.1.2)$$

$$= \Sigma w_i / \Sigma (w_i / M_i) \qquad (2.1.3)$$

$$= \Sigma c_i / \Sigma (c_i / M_i). \qquad (2.1.4)$$

Equation (2.1.1) illustrates the fact that the number-average molecular weight M_n represents the weight in grams (weight$_{Av}$) of the aforenoted Avogadro's number of molecules, since N_{Av} is the number of multiples of Avogadro's number of molecules considered. Alternative ways of expressing this same molecular weight parameter are given in Eqs. (2.1.2)–(2.1.4). In these relationships, as well as in those following in this section, n_i is the number of molecules of the ith fraction, which consists only of molecules having a single molecular weight M_i—being the weight in grams of Avogadro's number of such molecules. Correspondingly, w_i is the weight of the ith fraction and c_i is the weight concentration of the polymeric solute of the same fraction. Hence, each of the numerators in Eqs. (2.1.1)–(2.1.4) is in the same proportion to the weight of polymer as each denominator is to the number of moles of polymer. The choice of the relationship to use depends on the experimental procedure employed, as is discussed in the respective sections.

2.1.2.2. Weight-Average Molecular Weight. The second moment of molecular weight is found in the weight-average molecular weight M_w. The basic definition, which gives rise to its name, is the weighted average analogous to Eq. (2.1.2):

$$M_w = \Sigma w_i M_i / \Sigma w_i, \tag{2.1.5}$$

with corollary forms

$$M_w = \Sigma n_i M_i^2 / \Sigma n_i M_i \tag{2.1.6}$$

$$= \Sigma c_i M_i / \Sigma c_i. \tag{2.1.7}$$

The weight-average molecular weight is always greater than the number-average molecular weight for a polydisperse material, as is obvious from a comparison of Eqs. (2.1.2) and (2.1.6). It is only in the case of a material monodisperse in molecular weight (M_i is the same for all molecules and can be factored out of the summations) that M_n is identical to M_w. Moreover, the ratio of M_w / M_n is a frequently employed parameter describing the relative polydispersity of a polymer. This ratio may be as low as 1.05 for specially prepared (or fractionated) "monodisperse" polymers to higher than 20 for very broad distribution, commercial polyethylenes.

This ratio, M_w / M_n, is sometimes spoken of as "the" molecular weight distribution or MWD. Although this is not generally true, for a family of polymers whose distribution is fixed (e.g., binominal distribution, normal distribution, Zimm distribution) but with varying mean values of molecular weight, the ratio will indeed define the breadth of the distribution, while one or the other of these averages (M_n or M_w) will define the mean size of the molecules.

2.1.2.3. *Z* and Higher Averages of Molecular Weight. In analogous fashion, higher moments may be defined and calculated from fractionation data, to furnish better descriptions of the distributions. The first of these is generally called the Z-average molecular weight M_Z, and higher averages are named $Z + 1$ (Z plus one), $Z + 2$, $Z + 3$, etc. All these averages do retain a simple point of similarity in that they are the ratio of succeedingly higher moments divided by the next lower moment, so that they retain units of linear order of molecular weight. Hence, the definitions as ordinarily expressed are

$$M_Z = \Sigma n_i M_i^3 / \Sigma n_i M_i^2, \tag{2.1.8}$$

$$M_{Z+1} = \Sigma n_i M_i^4 / \Sigma n_i M_i^3, \tag{2.1.9}$$

$$M_{Z+2} = \Sigma n_i M_i^5 / \Sigma n_i M_i^4, \tag{2.1.10}$$

etc. Of course, alternative forms may be developed for each of these simply by substituting w_i or c_i for the $n_i M_i$ products, wherever found. But use of these lastly described molecular weight averages is limited to the more bizarre distributions, whence the information afforded is significant.

2.2. Intensive Properties of Polymers

2.2.1. Solution Viscosity

2.2.1.1. Background and Nomenclature. The viscosity of a dilute solution of a polymer is significantly greater than the viscosity of the solvent employed. The frictional properties of polymer molecules in solution are the basis of the theoretical dependence of the solution viscosity on the hydrodynamic volume and molecular weight. The utility of Staudinger's suggestion[1] to use the viscosity of a dilute solution as a measure of the dissolved polymer's molecular weight has, over nearly half a century, been confirmed empirically and justified theoretically. The currently practiced relationship is the Mark–Houwink equation:[2,3]

$$[\eta] = K M_v{}^a, \tag{2.2.1}$$

where $[\eta]$ is the limiting viscosity number, a measure of the viscosity adjusted for the concentration of polymer, K and a are constants, and M_v is the viscosity-average molecular weight.

The limiting viscosity number is a parameter derived from the flow of dilute solutions in specially designed capillary viscometers. The basic measurement is a comparison of the times (t and t_0) required for a specific volume of the solution and of the pure solvent, respectively, to flow through the same capillary at the same temperature. The evolution of several parameters of solution viscosity in terms of the flow times t and t_0 and of the polymer concentration c is given in Table I. The IUPAC recommendations,[4] although preferred, are far from being universally accepted; the former names occur in much of the literature including current publications.

The coefficient K and the exponent a are empirical constants generally useful over an extended, but nonetheless bounded range. Their nature and the meaning of the viscosity-average may be seen from the definitions

[1] H. Staudinger and W. Heuer, *Ber. Dtsch. Chem. Ges. B* **63**, 222 (1930).

[2] H. Mark, "Der Feste Koerper." Leipzig, 1938.

[3] R. Houwink, *J. Prakt. Chem.* **157**, 15 (1940).

[4] International Union of Pure and Applied Chemistry, *J. Polym. Sci.* **8**, 255 (1952).

TABLE I. Nomenclature for Solution Viscosity

IUPAC recommended nomenclature	Former nomenclature	Symbolic definition
Viscosity ratio	Relative viscosity	$\eta_r = t/t_0$
	Specific viscosity	$\eta_{sp} = \eta_r - 1$
Viscosity number	Reduced viscosity	$\eta_{red} = \eta_{sp}/c$
Logarithmic viscosity number	Inherent viscosity	$\eta_{inh} = (\ln \eta_r)/c$
Limiting viscosity number	Intrinsic viscosity	$[\eta] = \lim_{c \to 0}(\eta_{sp}/c)$
		$= \lim_{c \to 0}[(\ln \eta_r)/c]$

in Table I. In the case of a monodisperse polymer, which is a sample consisting of a single species of molecules all of identical molecular weight, the number-average, weight-average, Z-average, etc., molecular weights are all equal to this same "identical molecular weight." Equation (2.2.1) still holds for such a substance, and generally one can write

$$[\eta]_i = KM_i^a \qquad (2.2.2)$$

for any species i. The additive nature of the components of $[\eta]$, $[\eta]_i$, is evident from the definitions of Table I, since the limiting viscosity number is the limit of the viscosity number at infinite dilution, i.e., the point at which the molecules are independent entities. Hence, one can write

$$[\eta] = \Sigma c_i [\eta]_i = K\Sigma(c_i M_i^a), \qquad (2.2.3)$$

or

$$[\eta] = \Sigma w_i [\eta]_i / \Sigma w_i = K\Sigma(w_i M_i^a)/\Sigma w_i. \qquad (2.2.3a)$$

Since $w_i = n_i M_i$, combining Eqs. (2.11) and 2.13a) gives

$$KM_v^a = K\Sigma(n_i M_i^{1+a})/\Sigma(n_i M_i). \qquad (2.2.4)$$

Hence, the viscosity-average molecular weight becomes

$$M_v = [\Sigma(n_i M_i^{1+a})/\Sigma(n_i M_i)]^{1/a}. \qquad (2.2.5)$$

Since a varies from 0.5 in a theta-solvent (see below) to a maximum of nearly 1.0, the viscosity-average molecular weight always lies between M_n and M_w, is usually about 10–20% below M_w, and approaches M_w as a approaches unity.

The Mark–Houwink constants K and a are functions of both solvent and polymer type, and thus need to be determined experimentally for each polymer-solvent combination as needed. Although their usage is strictly valid only for linear polymers, compensating factors such as branches increasing in size or frequency monotonically with molecular weight often lead to practical applications not theoretically justified. The values for K and a are best obtained from double logarithmic plots of $[\eta]$ vs. the molecular weights of the polymer having M_n very close to M_w: fractionated polymer or polymer polymerized to yield homogeneous molecular weight. Since M_v is close to M_w, it is a useful, but less accurate approximation to substitute M_w in Eq. (2.2.1).

Kurata *et al.*[5] have compiled an extensive list of K and a for many polymer–solvent combinations; and new polymeric compositions are frequently described in terms of developed Mark–Houwink constants.

2.2.1.2. Hydrodynamic Volume. The size of a dissolved polymer molecule, that is to say, its effective hydrodynamic volume, depends on such parameters as molecular weight, the solvating power of the solvent, the stiffness and bulk of the backbone chain, and the branching of the molecule. For a linear polymer it is convenient to consider the hydrodynamic volume in terms of the root-mean-square end-to-end distance, $(\overline{r^2})^{1/2}$. This hydrodynamic volume is proportional to the cube of this end-to-end distance, $(\overline{r^2})^{3/2}$. It can be shown theoretically[6-8] that the limiting viscosity number is proportional to the ratio of the effective volume to the molecular weight:

$$[\eta] = \Phi(\overline{r^2})^{3/2}/M. \qquad (2.2.6)$$

Experimental verification shows Φ to be essentially constant, independent of polymer type or molecular weight, as well as of solvent.

To extend this result to the realm of practical use of the viscosity as a measure of molecular weight, it is helpful to employ the result that at sufficiently high chain length, the unperturbed end-to-end distance $(\overline{r^2})^{1/2}$ is proportional to the square root of the number of repeating segments, i.e., of the molecular weight. As long-range intramolecular interactions influence the molecular geometry, an expansion of the unperturbed dimension takes place. This is described by an expansion factor α, so that the actual dimension $(\overline{r^2})^{1/2}$ may be written as $\alpha(\overline{r_0^2})^{1/2}$. The expansion factor α is a function of the solvent–polymer interaction: a "good" solvent for a polymer leads to a value of α larger than 1.0, and a "poor" solvent produces a value of approaching unity. As the temperature is lowered, solvating power decreases and the conformation of the molecules approaches that of a random coil. The value of α approaches unity and becomes equal to unity at the θ-temperature where the solution is on the

[5] M. Kurata, Y. Tsunashima, M. Iwama, and K. Kamada, *in* "Polymer Handbook" (J. Brandrup and E. H. Immergut, eds.), 2nd ed., Chapter IV, pp. 1–30. Wiley, New York, 1975.

[6] P. Debye and A. M. Bueche, *J. Chem. Phys.* **16**, 573 (1948).

[7] J. G. Kirkwood and J. Riseman, *J. Chem. Phys.* **16**, 565 (1948).

[8] P. J. Flory and T. J. Fox, *J. Polym. Sci.* **5**, 745 (1950).

threshold of precipitation. In other words, the θ-temperature is the critical miscibility temperature for polymer of infinitely high molecular weight. In a θ-solvent, the molecular dimensions of the polymer equal the corresponding unperturbed dimensions $(\overline{r_0^2})^{1/2}$.

Now, substituting $\alpha(\overline{r_0^2})^{1/2}$ for $(\overline{r^2})^{1/2}$ in Eq. (2.2.6), the result is

$$[\eta] = \Phi(\overline{r_0^2}/M)^{3/2}M^{1/2}\alpha^3. \tag{2.2.7}$$

Noting that $\overline{r_0^2}/M$ and Φ are independent of solvent and molecular weight, an equivalent expression is

$$[\eta] = KM^{1/2}\alpha^3. \tag{2.2.8}$$

From the characteristics of α, it is clear that at the θ-temperature Eq. (2.2.8) reduces to Eq. (2.2.1) with the exponent a equal to $\frac{1}{2}$.

2.2.1.3. Experimental Procedures. The dilute solution viscosity is conveniently measured in one of many designs of capillary viscometers, that of Ubbelohde being especially advantageous in permitting a series of concentrations to be measured by successive dilution. Onyon[9] describes a number of capillary viscometers, along with details of their operation. Although rotational viscometers can be employed to measure viscosities of dilute solutions,[10] the usual laboratory practice employs a glass viscometer constructed with precision capillary tubing. In any case, the viscosity ratios (relative viscosities) of the solutions are determined, and extrapolated to zero concentration to obtain the limiting viscosity number $[\eta]$.

In the case of nonturbulent, laminar flow of a Newtonian fluid in a capillary, Poiseuille's law applies:

$$\eta = (\pi P r^4 t)/(8Vl) \tag{2.2.9}$$

where η is the viscosity, and P is the pressure driving the volume V of fluid through a capillary of radius r and length l in a period of time t. It is apparent, then, that the viscosity is proportional to the flow time t if the dimensions and driving force of the viscometer are maintained constant. In order to assure accuracy in the measurement of the flow times, apparently elaborate precautions must be taken. But since typical precision of $\pm 0.1\%$ is attainable with an ordinary stopwatch, and even better precision with automated optical timers, the care and attention are justified. Cleanliness is essential, so that to avoid erratic flow from clogged capillaries, chromic acid cleaning, high-temperature burn-out, and adequate

[9] P. F. Onyon, in "Techniques of Polymer Characterization" (P. W. Allen, ed.), Chapter VI. Butterworth, London, 1959.

[10] D. K. Carpenter and L. Westerman, in "Polymer Molecular Weights" (P. E. Slade, ed.), Part II, p. 441. Dekker, New York, 1975.

protection from dust are necessary. The last point carries over to the solvent and solutions used to the extent that they should be filtered just prior to introducing them into the viscometer. Temperature equilibrium and control are also needed to assure that the precision of the limiting viscosity number be no greater than about 1%. To explain, since the viscosities of the solvents employed are very sensitive to temperature changes (e.g., $\sim 3\%/°C$), control of the temperature to $\pm 0.05°C$ will allow a possible difference in viscosity or flow time of $\pm 0.15\%$. In turn, the viscosity ratio t/t_0 will have a precision of $\pm 0.3\%$. For $t/t_0 = 1.50$, the precision for both the specific viscosity ($t/t_0 - 1.0$) and the viscosity number would become $\pm 0.9\%$; with $t/t_0 = 1.25$, the result would be $\pm 1.5\%$. The extrapolations to limiting viscosity number (see below) would carry this precision with it, and so its expected precision based on temperature variation alone would be approximately $\pm 1\%$. Since the exponent a in Eq. (2.2.1) is usually $0.6-0.7$, the error introduced in the viscosity-average molecular weight is then limited to $\pm 1.5\%$.

Additional sources of error may also need to be taken care of for the most accurate work. Typical are the following: (1) non-Newtonian flow, whence extrapolation to zero shear rate is required; (2) the end effect, wherein the energy losses upon the confluence at the capillary's entrance must be accounted for if absolute values of viscosity are to be determined; (3) the kinetic energy, if a significant portion of the driving force or pressure is diverted to the kinetic energy of the flowing solution; (4) maintenance of concentration during preparation and measurement of the viscosity, which required attention to the possibility of evaporation or condensation; and (5) ensuring constant volume of flow, which may be affected by droplets of solution or solvent being retained within the viscometer body. Reference to appropriate texts[9,11,12] is recommended should experimental work be undertaken.

For determining the limiting viscosity number, the symbolic definition in Table I must be used, in either or both of its forms:

$$[\eta] = \lim_{c \to 0} (\eta_{sp}/c), \tag{2.2.10}$$

$$[\eta] = \lim_{c \to 0} [(\ln \eta_r)/c]. \tag{2.2.11}$$

To perform these extrapolations, no theoretical bases have been established, and one is led to empirical relationships. In order to limit random error and attain the desired precision in the viscosity measurement, it is

    [11] "Standard Test Method D 1601 for Dilute Solution Viscosity of Ethylene Polymers." Am. Soc. Test. Mater., Philadelphia, Pennsylvania, 1977.
    [12] "Standard Test Method D-2857 for Dilute Solution Viscosity of Polymers." Am. Soc. Test. Mater., Philadelphia, Pennsylvania, 1977.

recommended that the value of the solvent flow time t_0 be at least 100 sec, and that η_r be obtained at four or more concentrations such that η_r lies between 1.1 and 1.5. Then the extrapolation equations, which provide suitable linearity over as broad a concentration range as that being examined, may be employed. Among the more popular extrapolation equations are Huggins' equation[13]

$$\eta_{sp}/c = [\eta] + k' [\eta]^2 c, \tag{2.2.12}$$

where k' is constant for a given polymer–solvent system, and Kraemer's equation[14]

$$(\ln \eta_r)/c = [\eta] + k'' [\eta]^2 c, \tag{2.2.13}$$

where k'' is similar in properties to k', although generally of opposite sign. Moreover, since $\eta_r - 1 = \eta_{sp}$, $\ln \eta_r$ can be expanded in powers of η_{sp}:

$$\ln \eta_r = \ln(\eta_{sp} + 1) = \eta_{sp} - \eta_{sp}^2/2 + \cdots. \tag{2.2.14}$$

Considering only second-order terms, i.e., for very low concentrations, Eqs. (2.2.10) and (2.2.12)–(2.2.14) yield the result that

$$k' - k'' = 1/2. \tag{2.2.15}$$

With this result, the difference of Eqs. (2.2.12) and (2.2.13) results in the Solomon–Ciuta[15] equation:

$$[\eta] = 2(\eta_{sp} - \ln \eta_r)^{1/2}/c \tag{2.2.16}$$

which is reliable provided the polymer concentration is in the linear range *and* Eq. (2.2.15) is satisfied. Other empirical equations that have been recommended are Martin's equation[16]

$$\ln(\eta_{sp}/c) = \ln[\eta] + k''' [\eta]c, \tag{2.2.17}$$

which has an extended range of linearity, and the Schulz–Blaschke equation[17]

$$\eta_{sp}/c = [\eta] + k [\eta]\eta_{sp}, \tag{2.2.18}$$

which should provide good linearity, since η_{sp} may be represented by a power series in concentration. Each of these "name" equations can be the basis for the extrapolation to infinite dilution. Verification of their

[13] M. L. Huggins, *J. Am. Chem. Soc.* **64**, 2716 (1942).

[14] E. O. Kraemer, *Ind. Eng. Chem.* **30**, 1200 (1938).

[15] O. F. Solomon and I. Ciuta, *J. Appl. Polym. Sci.* **6**, 683 (1962).

[16] A. F. Martin, *Pap., 103rd Meet., Am. Chem. Soc., Div. Cellulose Chem., Sect. C,* Paper #1, (1942).

[17] G. V. Schulz and F. Blaschke, *J. Prakt. Chem.* **158**, 130 (1941).

applicability rests with confirmation of the extrapolated values obtained from Eqs. (2.2.12) and (2.2.13).

"Single-point" methods are frequently employed with these same equations, provided an evaluation of the empirical constants has been made for the polymer–solvent system in use. This does provide the advantage that better precision can be obtained from multiple readings at a concentration wherein the precision of the flow times can be optimized against the corrections that would be required at the extremes of the allowable range (i.e., $1.1 \leq \eta_r \leq 1.5$). The need to obtain the values of the constants also provides the means of assuring that the concentration employed does lie in the linear range implied by Eq. (2.2.12).

2.2.1.4. Polyelectrolytes. Polyelectrolytes are a class of polymers with ionizable groups along the chain. These materials in dilute solution will respond conventionally in nonionizing solvents, but in water and even wet solvents they will respond much differently. Polyacids, polybases, polyanhydrides as hydrolyzed, the copolymers, nucleic acids, and proteins are all examples of polyelectrolytes.

Essentially the ionization of the charged sites causes a chain expansion based on the mutual repulsion of the sites. An observed result is an increase in viscosity as the concentration decreases. Addition of a strong electrolyte to the water solution can produce a reversal such that the behavior again becomes normal and the usual relationships can be established. It is beyond the scope of this chapter to develop these concepts further, and this admonition is intended to point to the possibility of abnormalities occurring with polyelectrolytes.

2.2.2. Colligative Properties

2.2.2.1. Thermodynamic Basis. The colligative properties of polymer solutions, that is the physical properties such as osmotic pressure, vapor pressure lowering, freezing-point drop, and boiling-point rise, which are all sensitive to the concentration of the solute, are thermodynamic properties useful in the estimation of molecular weight. The analysis concerns itself with the equilibrium conditions existing between the solution and pure solvent: the reduction in vapor pressure (activity) caused by the solute and as reflected in the membrane-constrained pressure across a membrane, or in the measurable decrease in actual vapor pressure, as well as in the depression of the freezing point and the elevation of the boiling point. In the ideal solution, the free energy of dilution ΔG is expressed as

$$\Delta G = RT \ln(p/p^0), \qquad (2.2.19)$$

where the exerted pressure changes from a standard p^0 to an observed p. As Billmeyer[18] shows with the help of the Flory–Krigbaum theory of polymer solutions, the molar free energy of mixing becomes

$$\Delta G = RT[v_2/x + (\tfrac{1}{2} - x_1) v_2^2 + \cdot \cdot \cdot] \tag{2.2.20}$$

where RT is the product of the gas constant and the absolute temperature, v_2 is the volume fraction of polymer, x is the volume ratio of a polymer molecule to a solvent molecule, and x_1 is the interaction energy factor per solvent molecule.

The power series in v_2 implies a concentration dependence that is observed in practice for colligative properties. In terms of the moles of solvent n_1 and moles of polymer n_2, the definition of v_2 may be expressed as

$$v_2 = xn_2/(n_1 + n_2). \tag{2.2.21}$$

Then as the concentration of the polymeric solute decreases, v_2/x approaches n_2/n_1, since n_2 becomes insignificant in comparison to n_1. Moreover, n_2/n_1 is now the molar fraction cV/M, and higher terms in the power series may be ignored, whence a simple, linear response to concentration is attained:

$$\Delta G = RTcV/M, \tag{2.2.22}$$

where c is the concentration of solute, V the molar volume of the solvent, and M the molecular weight of the solute. The molar free energy may be represented by the product of the osmotic pressure π and the molar volume V of the solvent, whereby the limiting expression for the osmotic pressure–molecular weight relation is

$$\lim_{c \to 0} \pi/c = RT/M. \tag{2.2.23}$$

With the use of the Clausius–Clapeyron equation, the limiting expressions for the boiling point elevation ΔT_b and freezing point depression ΔT_f may be derived:

$$\lim_{c \to 0} \Delta T_b/c = (RT^2/\rho\Delta H_v)(1/M), \tag{2.2.24}$$

$$\lim_{c \to 0} \Delta T_f/c = (RT^2/\rho\Delta H_f)(1/M). \tag{2.2.25}$$

where ρ is the solvent's density, ΔH_v the latent heat of vaporization, and ΔH_f the latent heat of fusion.

Since Eqs. (2.2.23)–(2.2.25) require extrapolation to zero concentration, measurements at several concentrations are implied. Since a con-

[18] F. W. Billmeyer, "Textbook of Polymer Science," 2nd ed., p. 37. Wiley, New York, 1971.

centration dependence should be expected, it seems appropriate to cast these three equations into a modified form of Eq. (2.2.20). This is done, noting the correspondence of v_2 and c, and extending Eq. (2.2.20) in empirical form:

$$\Delta G/cV = (RT/M)(1 + Bc + Cc^2 + \cdots), \qquad (2.2.26)$$

which leads to the alternative forms of Eqs. (2.2.23)–(2.2.25):

$$\pi/c = (RT/M)(1 + Bc + Cc^2 + \cdots), \qquad (2.2.23a)$$

$$\Delta T_b/c = (RT^2/\rho M \Delta H_v)(1 + B'c + C'c^2 + \cdots), \qquad (2.2.24a)$$

$$\Delta T_f/c = (RT^2/\rho M \Delta H_f)(1 + B''c + C''c^2 + \cdots). \qquad (2.2.25a)$$

Experimentally, the term containing c^2 may often be neglected, thus utilizing a straight-line extrapolation. If curvature is evident, however, it may be convenient to rewrite Eq. (2.2.23a) as

$$\pi/c = (RT/M)(1 + kc)^2, \qquad (2.2.23b)$$

whereupon a plot of $(\pi/c)^{1/2}$ frequently becomes linear in c.

Generally the choice of a method for determining a colligative property depends on the solubility of a polymer in the available solvents, and the sensitivity of the (differential) measuring devices at hand. The choice, nonetheless, tends toward osmometry by virtue of the much greater response—provided membranes of adequate integrity, that is, selectively permeable to the solvent alone, can be found. Flory's[19] calculations for benzene solutions at various molecular weights of solute are summarized in Table II.

2.2.2.2. Osmometry. Although the principle of osmotic pressure is implicit in the foregoing description of the thermodynamics involved, it is useful to point out the operation of an osmometer. Basically, a polymer solution is separated from pure solvent by a membrane that is ideally per-

TABLE II. Theoretical Comparison of Colligative Properties of Solutes in Benzene

Molecular weight of solute	Property, extrapolated to $c = 0$		
	$\Delta T_b/c$ (°C-dl/g)	$\Delta T_f/c$ (°C-dl/g)	π/c (cm-dl/g)
10,000	0.0031	0.0058	28
50,000	0.0006	0.0012	5.7
100,000	0.0003	0.0006	2.8

[19] P. J. Flory, "Principles of Polymer Chemistry," p. 272. Cornell Univ. Press, Ithaca, New York, 1953.

meable to the solvent, but not to the solute. Since the chemical potential of the solvent is higher in the pure solvent, transfer of solvent molecules into the solution will take place. This thermodynamic drive continues until physical forces (e.g., pressure) or chemical forces (e.g., dilution) result in an equilibration of activity of the solvent molecules on the two sides of the membrane. In a "static" osmometer the transfer of solvent results in a rise of the liquid head in the solution chamber, and the osmotic pressure is the measure of this head. This type of osmometer requires hours to days to reach equilibrium and is not used much. Dynamic instruments, wherein the pressure is adjusted by servomechanisms coupled to pressure or flow sensors, can equilibrate in minutes. Commercial designs are available wherein the ease of operation, speed of cleaning, and simplicity of maintenance permit the complete determination including calibration and multiple concentrations to be performed within the time span formerly required for a single osmotic pressure reading by static instruments. There is an additional advantage attained by virtue of the speed of operation, which is particularly important since actual membranes are not absolutely impermeable to the solute: The short time needed for equilibration limits the degree of solute permeation, thus providing more reliable results than obtainable formerly.

The selection of a particular membrane is based on several criteria: (1) resistance to the solvent at the temperature in use; (2) retention of (impermeability to) the lower molecular weight species of the solute; and (3) rate of permeation of the solvent. In order to control these aspects of the membranes, they are supplied in various levels of compaction and thickness, and the treatment (swelling or conditioning) prior to use must be carefully attended to. Furthermore, various materials have been used, such as collodion, regenerated cellulose, gel cellophane, rubber, poly(vinyl alcohol), polyurethane, poly(vinyl butyral), and polychlorotrifluoroethylene. For more complete discussion of the properties and use of commercially available membranes, see the articles by Armstrong,[20] Chiang,[21] and Coll and Stross.[22]

The permeation of the low molecular weight species of the solute through the membrane is a persistent limitation of the method. Usually these permeating species are present in unfractionated polymers with M_n up to 50,000.[23] The theoretical treatment of this diffusion of the solute

[20] J. L. Armstrong, *N. A. S.—N.R.C., Publ.* **1573,** 51 (1968).

[21] R. Chiang, *in* "Newer Methods of Polymer Characterization" (B. Ke, ed.), Chapter 12. Wiley, New York, 1964.

[22] H. Coll and F. H. Stross, *N. A. S.—N.R.C. Publ.* **1573,** 10 (1968).

[23] F. W. Billmeyer, "Textbook of Polymer Science," 2nd ed., p. 72. Wiley, New York, 1971.

has been reviewed by Elias,[24] but no fully satisfactory means of overcoming the too low values of osmotic pressure have been developed. For substances of number-average molecular weight of 10,000–20,000, errors giving molecular weights two or three times the correct value might be observed, depending on the speed of the determination and the nature of the membrane.

2.2.2.3. Ebulliometry. The vapor pressure of a liquid is reduced, in response to Raoult's law, in proportion to the amount of solute dissolved. To raise the vapor pressure to that of pure solvent at the boiling point, the temperature of the solution must be raised. This dependence is indicated by Equation (2.2.24a), which also implies that extrapolation to zero concentration is appropriate. In order to measure the small differences in temperature of pure solvent and the solution in vapor-phase equilibrium with each other, multijunction thermocouples or matched thermistors are employed in a Wheatstone bridge circuit. Although theoretically the temperature difference could be employed to calculate directly the absolute number-average molecular weight from Eq. (2.2.24) or (2.2.24a), in practice a comparator (e.g., tristearin with molecular weight 891.5) is utilized. This avoids difficulties arising from impurity of the solvent and errors in the heat of vaporization.

There are only a few commercially available instruments suitable for the ebulliometric determination of the molecular weight of polymers, and moreover, problems develop through foaming and superheating. Thus, for experimental work, reference should be made to actual descriptions of operating procedures. Glover[25] provides a review of operation and calculations as most recently practiced, together with references to critiques of observed abnormalities. Generally, precision is seen in Glover's data[25] to rise from 5% at molecular weight 25,000 to about 15% for the most careful work on samples of molecular weights of 200,000. Thus, even though considerable care has been taken in ebulliometry, caution is necessary in respect to the interpretation of the results.

2.2.2.4. Cryoscopy. The depression of a solvent's freezing point by dissolved polymer provides another means of determining the *solute's* molecular weight from Eqs. (2.2.25) or (2.2.25a). The similarities of cryoscopy to ebulliometry, both theoretically and practically, abound from the use of a calibrating comparator to the methods for handling the data. The apparent simplicity of the equipment commends cryoscopy over ebulliometry. Indeed, the measurements employing thermistor bridge circuits similar to those used for ebulliometry make the method ef-

[24] H-G. Elias, *N. A. S.—N.R.C. Publ.* **1573**, 28 (1968).
[25] C. A. Glover, *in* "Polymer Molecular Weights" (P. E. Slade, ed.), Part I, p. 105. Dekker, New York, 1975.

fective for polymers of molecular weight up to about 30,000. The cautions to be taken in cryoscopic determinations include ensuring temperature equilibrium through stirring, initiating crystallization at some reproducible temperature by seeding, and ensuring that the solvent does not exhibit phase transitions near its freezing point nor interact with the polymeric solute. Any of these characteristics would introduce erratic or erroneous data. Glover[26] furnishes a critical review of the current state of the art.

2.2.3. Light Scattering

2.2.3.1. Principles and Definitions. Light-scattering methods make use of the intensity of the scattered light compared to the incident light to gain information concerning the size and conformation of macromolecules in solution. The current discussion is limited to the basic elements that relate to the determination of molecular weight. Billmeyer[27] provides a concise, practical exposition of the theory supporting the estimation of molecular weights of polymers by light scattering. The classical theory of the effect of electromagnetic radiation on matter is a reradiation of a portion of the incident energy in all directions away from the incident beam. Ideally, that is, for wavelengths well separated from absorption bands and much longer than the particles irradiated, the Rayleigh ratio R_θ (the ratio of the scattered intensity to the incident intensity divided by the volume traversed by the scattered beam) is employed with the fluctuation theory to obtain the net intensity of scattered light. The fluctuations, considered random, include the density, composition, and concentration brought on by the thermal and natural variations of the material and result in refractive index variations and additional scattering. The first step is to determine the scattering attributable to the dissolved polymer by taking the difference between the Rayleigh ratio for the solution and that for the solvent to give the excess scattering $\overline{R}_\theta$:

$$\overline{R}_\theta = R_\theta \text{ (soln)} - R_\theta(\text{solv}). \qquad (2.2.27)$$

This quantity is related to the molecular weight of the polymeric solute through the relationship

$$Kc/\overline{R}_\theta = 1/[MP(\theta)] + 2A_2c + \cdots, \qquad (2.2.28)$$

where $K = (2\pi^2n^2/N_0\lambda^4)(dn/dc)^2$, M is the molecular weight, $P(\theta)$ the

[26] C. A. Glover, in "Polymer Molecular Weights" (P. E. Slade, ed.), Part I, p. 125. Dekker, New York, 1975.

[27] F. W. Billmeyer, "Textbook of Polymer Science," 2n ed., p. 75. Wiley, New York, 1971.

particle-scattering factor, A_2 the second virial coefficient, n the refractive index of the solution, dn/dc the specific refractive index increment, N_0 Avogadro's number, λ the wavelength of the radiation in the solution, and θ the angle of the scattering beam.

The particle-scattering factor is a function of the product $(\sin \theta/2)(d/\lambda)$, where d is the diameter of the scattering particle, and is expressible in either form

$$P(\theta) = 1 - (16\pi^2/3)(d/\lambda)^2 \sin^2\theta/2,$$
$$P(\theta)^{-1} = 1 + (16\pi^2/3)(d/\lambda)^2 \sin^2\theta/2,$$

(2.2.29)

so that for θ approaching zero, $P(\theta)$ approaches unity. Hence for zero angle and zero concentration, Eq. (2.2.28) reduces to

$$Kc/\overline{R_\theta} = 1/M.$$

(2.2.28a)

Since Eq. (2.2.28) has the corollary result that the intensity of the scattered light is proportional to the square of the particle mass, the total intensity of the scattered light can be determined from a summation of the $\overline{R_\theta}$ for all particles:

$$\overline{R_\theta} = \Sigma\overline{R_{\theta i}} = K\Sigma c_i M_i = KcM_w.$$

(2.2.30)

A modest rearrangement of the rightmost equality gives

$$M_w = \Sigma c_i M_i/c = \Sigma c_i M_i/\Sigma c_i.$$

(2.1.7)

This is identical to the definition for the weight-average molecular weight as defined in Chapter 2.1. Thus the molecular weight obtained from the light-scattering analysis is the weight-average molecular weight.

Taking another look at Eq. (2.2.28),

$$Kc/\overline{R_\theta} = (1/M_w)P(\theta)^{-1} + 2A_2c,$$

(2.2.28)

and combining it with the second form of Eq. (2.2.29) results in the relationship (for a given polymer–solvent system)

$$Kc/\overline{R_\theta} = (1/M_w)(1 + k \sin^2\theta/2) + 2A_2c.$$

(2.2.31)

It is obvious that with a series of measurements at a number of angles for several concentrations, a multiple regression technique by computer would be very convenient to extract the intercept and M_w. Nonetheless, there are advantages to using Zimm's method[28] in which the left side of Eq. (2.2.31) is plotted against an abscissa of $\sin \theta/2 + Ac$, where A is an arbitrary constant. The plot is a mesh of parallel lines, permitting extrap-

---

[28] B. H. Zimm, *J. Chem. Phys.* **16**, 1093 (1948).

olation to both zero concentration and zero angle. This allows independent inspections of the fit of the points with each variable.

2.2.3.2. Applications. Experimental problems are extensive in light-scattering experiments. Small amounts of impurities have disproportionate effects on the scattering. Clarification of the solution by filtration through 0.2- to 0.5-μm membrane filters is usually satisfactory. Precleaning the cells and the solvent is also an aid to experimental work. Another source of difficulty is the need for precise measure of the refractive index increment, especially since it appears as its square in Eq. (2.2.28). Moreover, the value of dn/dc is about 0.1 for the usual polymer systems, so that with a 2% solution, a refractive index difference of 0.002 must be measured to $\pm 2 \times 10^{-5}$ to attain 1% precision. A differential refractometer is usually adequate to better the above precision. Other areas of concern have to do with the refraction at the liquid–glass interfaces; polarization corrections for the polarization of the scattering particles; spurious depolarization from optical aberrations; and absorption and fluorescence resulting in erroneous scattering intensity. This listing is intended not to erase the usefulness of the technique for known systems, but rather to point out that with unknown systems, caution is needed.

An area of interest concerns the analysis by light scattering of co-polymers. If the copolymers, for mechanistic or kinetic reasons, are known to have the same composition regardless of molecular weight (e.g., a copolymer consisting of alternating monomers A and B), the same principles hold as for homopolymers. If both the composition and molecular weight distribution are not defined, it is required to perform a number of measurements in solvents of different refractive indices to determine the value of M_w and information on the compositional heterogeneity.[29]

A recent development in light scattering techniques utilizes a laser source of radiation and low-angle detection.[30] The advantage is primarily in the simplification of Eq. (2.2.28) in that $P(\theta)$ becomes unity so that without the need for the angular dependence measurements, the experimental work can be speeded up considerably. There is, of course, a loss of information, in that the value of $P(\theta)$ provides information concerning the size of the solvated scattering particle d [cf. Eq. (2.2.29)]. An additional compensation might be brought out, however, which even further simplifies the analysis. Since the second virial coefficient A_2 is fixed for a given polymer–solvent system, if the concentration of the solute is known (measured), a single determination of the excess Rayleigh ratio

[29] E. F. Casassa and G. C. Berry, in "Polymer Molecular Weights" (P. E. Slade, ed.), Part I, Chapter 5. Dekker, New York, 1975.
[30] W. Kaye and A. J. Havlik, *Appl. Opt.* **12,** 541 (1973).

will suffice to calculate the molecular weight. This treatment has been used effectively by Ouano and Kaye[31] to improve the utilization of gel permeation analysis (Chapter 2.4) through concurrent measures of $\overline{R}_\theta$ and of the concentration from the refractive index, as the fractionated polymer solution elutes from the chromatograph.

2.3. Fractionation

2.3.1. Introduction

Since most polymers are heterogeneous with respect to molecular weight, their physical and mechanical properties, both in solution and in the solid state, are influenced by this molecular weight distribution. Fractionation is an experimental procedure by which the whole polymer is separated into fractions, each of which is more homogeneous than the original polymer. These fractionation procedures are of two types: analytical, in which the distribution curves can be produced and explored without isolating the fractions, and preparative, in which each individual fraction is isolated and characterized. Polymers can also be heterogeneous with respect to chemical composition and physical configuration of the molecules (linear, branched, amorphous, crystalline).

Most experimental techniques are developed to fractionate polymers according to molecular weight since more sophisticated procedures are required to deal with the complex variations of chemical composition and physical structure. Bello and Barrales-Rienda[32] present a comprehensive review of fractionation methods and authors. In a review such as this, space will not permit a detailed discussion of any of these methods; however, special emphasis will be given to fractional precipitation, column extraction, and gel permeation chromatography. Gel permeation chromatography (GPC) is an important, widely applicable technique for the determination of molecular weight distribution and for preparative fractionation. A complete review of GPC literature is beyond the scope of this review. However, cross-referencing with key references that are cited will provide an overview of the rapidly developing field.

2.3.2. Polydispersity and Fractionation

Linear polymers are mixtures of linear homologous molecules that vary in chain lengths and molecular weights. Branched polymers have poly-

[31] A. C. Ouano and W. Kaye, *J. Polym. Sci., Polym. Chem. Ed.* **12,** 1151 (1974).
[32] A. Bello and J. M. Barrales-Rienda, *in* "Polymer Handbook" (J. Brandrup and E. H. Immergut, eds.), 2nd ed., Chapter IV, p. 175. Wiley, New York, 1975.

disperse molecular structure as well as nonuniform chain length. Copolymers are polydisperse in chemical composition where the molecules consist of two or more chemically different types of structural units.

Fractionation reduces the polydispersity of high polymers, with an analytical procedure to determine the degree of polydispersity and with a preparative technique to isolate the narrow fractions for analysis. Different types of molecular weights (see Chapter 2.1 for definition of molecular weights—number-average M_n; weight-average M_w; Z-average M_z, and viscosity-average M_v) for the fractions are measured and/or calculated. Since the polydispersity index M_w/M_n does not sufficiently define the type of distribution of a polymer, fractionation, which leads to the construction of a distribution curve, does provide more detailed information concerning the molecular characteristics of the polymer.

Details of fractionation techniques are beyond the capacity of this review. However, details of solubility behavior and phase relationships of polymer/solvent systems are described by Allen,[33] Hall,[34] Scott,[35,36] Cragg and Hammerschlag,[37] Flory,[38] Tompa,[39] Kotera,[40] Billmeyer,[41] and Schneider.[42] In short, the solubility of polymer fractions decreases with increasing molecular weight. Therefore, fractions of decreasing molecular weight are separated by the addition of a nonsolvent to a polymer solution (fractional precipitation). On the other hand, extraction of polymer with progressively better solvents produces fractions of increasing molecular weight. When solubility is decreased either by loss of solvent power or by cooling below the critical solubility point, phase separation occurs and two layers form. Fractionation occurs because the ratio of the concentrations of high to low molecular weight species is greater in the precipitate than in the supernatant layer. Hall[34] describes how separation is improved by keeping the fraction size small and by working in dilute solution so as to keep R (the ratio of the volume of precipitate to supernatant

[33] P. W. Allen, ed., "Techniques of Polymer Characterization." Butterworth, London, 1959.

[34] R. W. Hall, in "Techniques of Polymer Characterization" (P. W. Allen, ed.), Chapter 2. Butterworth, London, 1959.

[35] R. L. Scott, Ind. Eng. Chem. 45, 2532 (1953).

[36] R. L. Scott, J. Chem. Phys. 13, 178 (1945).

[37] L. H. Cragg and H. Hammerschlag, Chem. Rev. 39, 79 (1946).

[38] P. J. Flory, "Principles of Polymer Chemistry," Chapter 13. Cornell Univ. Press, London and New York, 1953.

[39] H. Tompa, "Polymer Solutions." Butterworth, London, 1956.

[40] A. Kotera, in "Polymer Fractionation" (M. J. R. Cantow, ed.), p. 43. Academic Press, New York, 1967.

[41] F. W. Billmeyer, Jr., J. Polym. Sci., Part C 8, 161 (1965).

[42] N. S. Schneider, J. Polym. Sci., Part C 8, 179 (1965).

phase) small. Schulz[43] shows mathematically, by assuming values of R, the limitations of fractionation techniques due to head and tail overlapping of the fractions. Sharpness can subsequently be improved by refractionation of isolated fractions; however, total elimination of the "head" and "tail" in fractions is not possible.

2.3.3. Fractionation Techniques

2.3.3.1. Fractionation by Solubility. Many techniques for the fractionation of polymers by solubility differences are reviewed in detail by Kotera,[40] Hall,[34] and Guzman.[44] Of these, the methods of fractional precipitation, fractional solution, and chromatographic fractionation are briefly discussed here. The choice of technique depends on equipment available, the size and speed of fractionation desired, and the stability and solubility of the polymer. Hall[34] and Kotera[40] described in detail the various methods along with supplementary references.

Fractional precipitation techniques are fairly simple and require less sophisticated equipment than other methods demand. Small-scale (2 gm fractions) or large-scale (150 gm fractions) experiments are possible. The three most widely used procedures are (1) addition of a nonsolvent; (2) elimination of solvent by evaporation; (3) lowering the solution temperature. Addition of nonsolvent is the most useful and adaptable of the three procedures. Precipitation by cooling is limited to situations where precipitation of the polymer from solution can be achieved at an experimentally convenient temperature. If it is difficult to assign a convenient solvent/nonsolvent system, the solvent evaporation technique becomes a viable alternative.

In fractional solution, this direct extraction technique provides a means of isolating a predetermined low molecular weight portion of the polymer without completing the fractionation. However, equilibration problems between the extracting solvent and the polymer tend to broaden the molecular weight limits of the individual fractions. Consequently, more sophisticated equipment and procedures are needed for extraction, gradient elution, and film extraction procedures than is required for bulk fractional precipitation.

Turbidimetric titration provides a rapid, qualitative method for obtaining a first approximation of the distribution curve.

2.3.3.1.1. FRACTIONAL PRECIPITATION. 2.3.3.1.1.1. Addition of Nonsolvent. Fractional precipitation in bulk requires the addition of non-

[43] G. V. Schulz, Z. Phys. Chem. B **47**, 155 (1940).
[44] G. M. Guzman, in "Progress in High Polymers" (J. C. Robb and F. W. Peaker, eds.), p. 113. Academic Press, New York, 1961.

solvent to a dilute solution of a polymer with vigorous stirring until a slight turbidity is evident at the controlled temperature of the fractionation. Since equilibrium is necessary, the mixture is stirred for several hours after final turbidity adjustments are made, after which the temperature is first raised about 5° to permit dissolution and then cooled slowly to the controlled temperature. The precipitated phase, which settles in a layer, is separated from the supernatant phase either by decantation or syphoning of the latter or by removal of the precipitate from the bottom of the flask. The fractions are isolated in order of decreasing molecular weight. The precipitate that has been isolated from the original solution is recovered by a separate dissolution, precipitation, and drying procedure. The supernatant liquid is treated with the next increment of nonsolvent and the procedure is repeated until a large proportion of nonsolvent is needed to obtain a fraction. At this point, the remaining polymer is isolated by increasing the concentration by evaporation and adding more nonsolvent. The final fraction is obtained by evaporating the supernatant liquid to dryness. Details of this technique and apparatus used are described by Hall.[34]

The choice of the solvent/nonsolvent pair determines the physical characteristics of the precipitate and hence its ease of handling. Hall[34] and Flory[38] show that the efficiency of the fractionation increases as R [the ratio of the volume of the precipitate phase (V^1) to the volume of the supernatant phase (V)] decreases. In other words, in practice, R must be kept low by working in dilute solution and by keeping the fractions small. Therefore, efficiency increases as the concentration of the polymer decreases. The molecular weight of the polymer to be fractionated dictates the limits of the concentration of the polymer solution. That is, for a molecular weight ($\overline{M}_w$) of the polymer of 10^6 to 10^4, the initial concentration could range from 0.25 to 2.5%.

Each primary fraction can be refractionated to give more homogeneous fractions. Flory[45] suggests that a large fraction precipitated from a solution whose concentration is 0.5–1.0% can be redissolved at 0.1% and a new fraction isolated. The dilute phase can then be returned to the main solution after which a second large fraction can be precipitated. This technique combines features of refractionation and the dilute solution theory.

2.3.3.1.1.2. Solvent Volatilization. Kotera[40] describes in detail the procedure of precipitating a polymer from solution by the preferential removal of solvent by evaporation. The main advantages of this technique are (1) the experimental initial volume can be scaled up because it de-

[45] P. J. Flory, *J. Am. Chem. Soc.* **65,** 372 (1943).

creases as the fractionation proceeds; (2) the solvent/nonsolvent proportion changes continuously; (3) local precipitation on the surface of the liquid is unlikely.

Solvent, which must be more volatile than the nonsolvent, is selectively evaporated with stirring to the point of turbidity. After equilibration has been reached, the fraction is isolated and processed as described above for fractional precipitation.

2.3.3.1.1.3. Lowering the Temperature. Fractionation can be accomplished by gradually controlled cooling of the solvent or solvent/nonsolvent mixture to decrease the solvent power. Advantages of this technique, as detailed by Hall[34] and Kotera[40] are (1) a single-solvent system can be used; (2) the system volume is constant throughout the duration of the fractionation; (3) chance of local precipitation is minimized; (4) lack of homogeneity within the mixture during fractionation is minimized. Disadvantages are (1) the cooling technique is not effective for all polymers; (2) limitations exist on solvent/nonsolvent systems; (3) polymer degradation at elevated temperatures can occur. The cooling process must be slow and fraction size depends on the temperature change. Often a nonsolvent has to be used since few single-solvent systems will satisfactorily complete the fractionation.

2.3.3.1.2. TURBIDIMETRIC TITRATION. In this analytical technique as described by Hall,[34] Giesekus,[46] and Afifi-Effat and Hay,[47] a precipitant is added slowly to a very dilute polymer solution causing turbidity (precipitation), which is measured by an increase in intensity of scattered light or increase in optical density of the solution. Selection of the solvent/nonsolvent system is critical. Molecular weight distribution information is limited; however, the method is useful for control purposes and for the exploration of the effects of chemical composition in copolymer systems.

2.3.3.1.3. SUMMATIVE PRECIPITATION. Billmeyer and Stockmayer[48] and Battista[49] describe this technique where part of the polymer in each of a series of polymer solutions is precipitated to a specified and varied extent. Weight fractions w_x and molecular weights M_x are obtained. The product $w_x M_x$ relates to the integral of the cumulative distribution curve. Limited information about the molecular weight distribution is then obtained from the position of the maximum and breadth of the molecular weight distribution curve.

[46] H. Giesekus, *in* "Polymer Fractionation" (M. J. R. Cantow, ed.), p. 191. Academic Press, New York, 1967.

[47] A. M. Afifi-Effat and J. N. Hay, *Br. Polym. J.* **8**, 91 (1976).

[48] F. W. Billmeyer and W. H. Stockmayer, *J. Polym. Sci.* **5**, 121 (1950).

[49] A. O. Battista, *in* "Polymer Fractionation" (M. J. R. Cantow, ed.), p. 307. Academic Press, New York, 1967.

2.3.3.1.4. FRACTIONAL SOLUTION. The techniques of fractional solution are described in detail by Hall[34] and Elliott.[50] Here a polymer is in contact with a solvent or solvent mixture of specified solvent power. The low molecular weight fractions are isolated first. As the solvent power is increased, subsequent fractions of higher molecular weight are obtained. Hall[34] proposes both advantages and disadvantages of these methods. The advantages are (1) the higher molecular weight fractions should be only slightly broadened in distribution due to contamination from low molecular weight material; in fact, the solution method produces fractions with lower maximum molecular weight than those isolated by fractional precipitation; (2) the procedure is easily automated; (3) small quantities of polymer, less than 1 gm, can be fractionated. The disadvantages are that it is difficult to (1) attain true equilibrium between polymer and liquid; (2) scale up to obtain large fractions; (3) control fraction size. Also, large volumes of solvents are necessary.

The success of any fractionation can be evaluated as follows: (1) the sum of the weights of the fractions should match the weight of the original polymer; (2) the weight-average limiting viscosity number of the fractions should equal that of the original polymer, as in the equation

$$[\eta] = \sum_i \omega_i [\eta]_i, \tag{2.3.1}$$

where ω_i is the weight of the ith fraction with a limiting viscosity number $[\eta]_i$.

It is relevant to recognize the true range of molecular weight species in any fraction regardless of the polydispersity index (M_w/M_n) obtained from the measurements of the weight-average (M_w) and number-average (M_n) molecular weights. Elliott[50] demonstrates that a fraction with a log-normal or Wesslau[51] distribution whose M_w/M_n value is 1.05 ($M_w = 102,000$) will contain 5% by weight of polymer with molecular weight below 70,000 and 5% above 143,000. Kenyon and Salyer[52] found a range of molecular weight of 38,000–740,000 in a polystyrene sample with a polydispersity index of 1.17 ($M_w = 326,000$ and $M_n = 278,000$).

2.3.3.1.4.1. *Direct Sequential Extraction.* Procedural details are given by Hall[34] and Elliott.[50] In general, the polymer is dissolved in a predetermined solvent/nonsolvent mixture after which the supernatant is removed and isolated as the fraction. The precipitate then undergoes subsequent extractions with progressively solvent richer solvent/nonsolvent

[50] J. H. Elliott, *in* "Polymer Fractionation" (M. J. R. Cantow, ed.), p. 67. Academic Press, New York, 1967.
[51] H. Wesslau, *Makromol. Chem.* **20,** 111 (1956).
[52] A. S. Kenyon and I. O. Salyer, *J. Polym. Sci.* **43,** 427 (1960).

mixtures until all the polymer is accounted for. The only way of estimating fraction size is by isolation and recovery. Therefore, much trial and error can be avoided if a preliminary solubility curve of the polymer as a 1% solution in a series of solvent/nonsolvent combinations is determined before fractionation; from these data the proper solvent/nonsolvent combination can be selected.

2.3.3.1.4.2. Film Extraction. Either a continuous or batch operation can be used in the film extraction procedure for polymer fractionation. In the continuous operation, a slow-moving belt is thinly coated with polymer solution after which the solvent is evaporated to dryness. The fractions are isolated when the thin film on the belt is extracted continuously in a series of tubes containing solvent/nonsolvent mixtures of increasing solvent power.

Fuchs[53-55] describes the batch operation of coating metal foil with a thin film of polymer. A series of fractions of increasing molecular weight is obtained by extracting the coated foil with progressively richer solvent/nonsolvent mixtures. Hall,[34] Elliott,[50] and Fuchs[53] describe the procedures. The advantages of this technique lie in the simplicity of apparatus and rapidity of completion of the fractionation. Hall[34] states that 0.7 gm can be separated into 14 fractions of 50 mg each in 2.5 hours. The disadvantages are (1) the coating of the foil with polymer must be a developed art; (2) the film can detach from the foil if high molecular weight polymer swells in the solvent mixture; (3) the effects of nonequilibrium extraction can seriously distort the overall fractionation efficiency. Because of the inherent difficulties with this technique, it has been replaced in large part by column elution techniques.

2.3.3.1.4.3. Coacervation Extraction. The coacervation method is the reverse of fractional precipitation since the coacervate is the polymer-rich phase as a result of a liquid-phase separation. Nonsolvent is added to the solution until nearly all (about 90%) of the polymer becomes a part of the coacervate. The dilute polymer solution, which retains the lowest molecular weight fraction, is removed and isolated. The remaining coacervate is then extracted with a richer solvent/nonsolvent mixture, and this same procedure is repeated until the fractionation is complete. Details of this procedure are described by Hall[34] and Elliott.[50] Equilibrium is readily attained because the extraction occurs in the liquid coacervate phase rather than from the solid polymer. Even though this method is not widely used, its efficiency is comparable with that of fractional precipitation at intermediate and low molecular weight ranges.

[53] O. Fuchs, *Makromol. Chem.* **5,** 245 (1950).
[54] O. Fuchs, *Makromol. Chem.* **7,** 259 (1951).
[55] O. Fuchs, *Z. Electrochem.* **60,** 229 (1956).

2.3.3.1.4.4. Column Elution. Hall,[34] Elliott,[50] Schneider,[42] Kenyon *et al.*,[56] Kokle and Billmeyer,[57] and Anderson[58] present detailed descriptions of various uses and techniques of column elution fractionation. The work of Desreux[59] 25 years ago was a major breakthrough in the field of fractionation when he developed a solution method whereby column packing having a large surface area was coated with polymer, after which the polymer was fractionated by elution. This method, which has been used successfully with many polymer types,[60] provides speed, versatility, efficiency, automation, and ease of scale-up. Through study of the variables that are important to successful fractionation, a model was developed for isotactic polypropylene by Shyluk.[61] In column elution fractionation, fractions are eluted from the column with progressively better solvents. The changing solvent medium can be either a nonsolvent containing increasing proportions of solvent or with a single solvent or solvent/nonsolvent mixture by progressively raising the column temperature. The classic type of column extraction apparatus was developed by Francis *et al.*[62] and is described by Hall[34] and Elliott.[50] Each fractionation demands optimum loading conditions, proper column design, temperature control, and proper experimental techniques, which have been studied by Henry,[63] Baker and Williams,[64] Cantow *et al.*,[65,66] and Francis.[62] Criteria for solvent/nonsolvent selection are described by Elliott,[50] Gernert *et al.*,[67] and Guillet *et al.*[68] Details of polymer support types, techniques of polymer deposition, rate of elution, degradation, and methods of fraction recovery are described by Elliott,[50] Hall,[34] and Schneider.[42] Desreux and Spiegels[59] describe the variation of the extraction procedure where a polymer is completely precipitated from a hot solution by cooling. A single solvent is the extraction medium at a series of progres-

[56] A. S. Kenyon, I. O. Salyer, J. E. Kurz, and D. R. Brown, *J. Polym. Sci., Part C* **8**, 205 (1965).

[57] V. Kokle and F. W. Billmeyer, Jr., *J. Polym. Sci., Part C* **8**, 217 (1965).

[58] F. R. Anderson, *J. Polym. Sci., Part C* **8**, 275 (1965).

[59] V. Desreux and M. C. Spiegels, *Bull. Soc. Chim. Belg.* **59**, 476 (1950).

[60] M. J. R. Cantow, *in* "Polymer Fractionation" (M. J. R. Cantow, ed.), p. 461. Academic Press, New York, 1967.

[61] S. Shyluk, *J. Polym. Sci.* **62**, 317 (1962).

[62] P. S. Francis, R. C. Cooke, and J. H. Elliott, *J. Polym. Sci.* **31**, 453 (1958).

[63] P. M. Henry, *J. Polym. Sci.* **36**, 3 (1959).

[64] C. A. Baker and R. J. P. Williams, *J. Chem. Soc.* p. 2352 (1956).

[65] M. J. R. Cantow, R. S. Porter, and J. F. Johnson, *Nature (London)* **192**, 752 (1961).

[66] M. J. R. Cantow, R. S. Porter, and J. F. Johnson, *J. Polym. Sci., Part C* **1**, 187 (1963).

[67] J. F. Gernert, M. J. R. Cantow, R. S. Porter, and J. F. Johnson, *J. Polym. Sci., Part C* **1**, 195 (1963).

[68] J. E. Guillet, R. L. Combs, D. F. Slonaker, and H. W. Coover, Jr., *J. Polym. Sci.* **47**, 307 (1960).

sively increasing temperatures until all the polymer is dissolved. The operating temperature must be above the crystalline melting point of the polymer.

2.3.3.1.5. COMPARISON OF FRACTIONAL SOLUTION METHODS. In comparing fractional solution methods, Elliott[50] states that column gradient elution is the most flexible and efficient method. Direct extraction is best for preliminary separations. The Fuchs[53] film extraction is most effective for low molecular weight polymer. The coacervation method is most effective for preparative separations. Nasini and Mussa[69] prefer coacervation extraction over the Desreux or precipitation methods for polyethylene fractionation. Wijga et al.[70] prefer column elution for crystalline polypropylene. Davis and Tobias[71] found less degradation with column elution than with precipitation and coacervate extraction. Schneider et al.[72] obtained similar results in fractionating polystyrene using the Baker–Williams and gradient elution methods. In addition, Schneider concluded that the gradient elution and chromatographic methods give similar results even though they possess differing degrees of multistage characteristics.

2.3.3.1.6. EFFECT OF POLYMER STRUCTURE ON SOLUBILITY-BASED FRACTIONATION. Billmeyer[41] summarizes the effects of polymer structure on fractionation techniques. If heterogeneity of composition exists, fractionation by composition will dominate fractionation by molecular weight. Schneider et al.[73] suggest that branching increases solubility. Hence, a branched fraction will consist of a mixture of species that are low molecular weight with little branching along with more extensively branched higher molecular weight species. Consequently, the effect of the two variables must be considered separately in the fractionation of branched polymers.

2.3.3.1.7. CONSTRUCTION OF MOLECULAR WEIGHT DISTRIBUTION CURVES. A summarization of the methods of constructing molecular weight distribution curves as described by Hall[34] can elucidate the vital function of the fractionation process in characterizing polymers. If a fractionation is sharp, the M_w/M_n ratio is near unity and all molecular weight methods give similar molecular weight values for each fraction. However, if $M_w \neq M_n$, the shape of the distribution curve will depend on

[69] A. Nasini and C. Mussa, Makromol. Chem. 22, 59 (1957).

[70] P. W. O. Wijga, J. van Schooten, and J. Boerma, Makromol. Chem. 36, 115 (1960).

[71] T. E. Davis and R. L. Tobias, J. Polym. Sci. 50, 227 (1961).

[72] N. S. Schneider, J. D. Loconto, and L. G. Holmes, J. Appl. Polym. Sci. 5, 354 (1961).

[73] N. S. Schneider, R. T. T. Traskos, and A. S. Hoffman, J. Appl. Polym. Sci. 12, 1567 (1968).

the molecular weight parameter used and the instrumental technique for its determination.

The integral weight function W_x is

$$W_x = \sum_{x=1}^{x} w_x, \qquad (2.3.2)$$

where w_x is the weight fraction of molecules of chain length x. Although number-average molecular weight and limiting viscosity number can also be used to describe molecular weight distribution, weight-average molecular weight M_w is the most accurate parameter. Hall[34] illustrates tabulated and graphical fractionation data. Variations in treatment of the data to produce an accurate representation of the molecular weight distribution are discussed also by Mark and Raff.[74] The overlap of the fractions is defined in the construction of the integral curve. The differential weight distribution curve

$$dW_x/dx = w_x \qquad (2.3.3)$$

is obtained by graphical differentiation of the integral curve and a subsequent plot of slopes of tangents to the integral curve (dW_x/dM_x) vs. the corresponding M_x values.

The extent of loss of polymer during fractionation can be determined by Eq. (2.3.1). If the accumulated calculated limiting viscosity number exceeds the original, low molecular weight material has been lost. Then, too, degradation of the polymer will lower the calculated summative value of limiting viscosity number below the initial experimental value.

As with any data analysis technique, this procedure has limitations, which are described by Hall.[34] Tung[75] and Goodrich[76] present a detailed exposition concerning the various methods of deriving molecular weight distribution from data obtained from conventional fractionation experiments.

2.3.3.2. Fractionation by Column Chromatography. Baker and Williams[64] developed a chromatographic procedure to fractionate polymers using a column packed with support material, solvent gradient, heaters to produce a linear temperature gradient in the column, and a device for collecting the fractions. The polymer is coated on a small amount of packing

[74] H. Mark and R. Raff, "High Polymers," Vol. III. Wiley (Interscience), New York, 1941.

[75] L. H. Tung, in "Polymer Fractionation" (M. J. R. Cantow, ed.), p. 379. Academic Press, New York, 1967.

[76] F. C. Goodrich, in "Polymer Fractionation" (M. J. R. Cantow, ed.), p. 415. Academic Press, New York, 1967.

material (glass beads) by evaporation from a good solvent. The coated beads are then added to the top of the column as a slurry in a poor solvent. The solvent gradient ranges from 100% poor to 100% good solvent. Temperature gradient is highest at the top of the column. Molecular weight distribution curves are constructed from the collective data of the isolated fractions. This technique is a multistage fractional precipitation process employing continual reequilibration progressively throughout the column length until polymer emerges from the end of the column as a saturated solution at the lower column temperature.

Column chromatography has been used to fractionate many polymers—these are tabulated by Porter and Johnson.[77] Apparatus, experimental technique, and results are discussed by Porter and Johnson,[77] Hall,[34] Kokle and Billmeyer,[57] and Schneider.[78] Column construction, temperature control, column support materials, solvent gradient, flow control, and type of fraction collector are variables to be carefully selected and administered.

The use of a thermal gradient can under some circumstances improve fractionation efficiency. Porter and Johnson[77] and Schneider[72] present references that both support and refute its worth. Booth[79] finds agreement within limits of precision of fractionation of polystyrene by the Baker–Williams method and successive precipitation fractionation. The overall success of any fractionation depends on the choice and control of the variables involved.

2.3.3.3. **Miscellaneous Methods of Fractionation.** 2.3.3.3.1. THERMAL DIFFUSION. Fractionation by thermal diffusion gives molecular weight distributions similar to those from fractional precipitation. A high temperature gradient is set up between two surfaces containing a polymer solution. The solution is connected to an upper and lower reservoir. Because of the temperature gradient, the molecules circulate by thermal agitation and the large molecular polymer species separate and migrate toward the lower reservoir. The state of the art is reviewed by Emery.[80]

2.3.3.3.2. ISOTHERMAL DIFFUSION. The heterogeneity of a polymer mixture may be determined from isothermal diffusion. This measurement uses the frictional properties exhibited by macromolecules in solu-

[77] R. S. Porter and J. F. Johnson, in "Polymer Fractionation" (M. J. R. Cantow, ed.), p. 95. Academic Press, New York, 1967.

[78] N. S. Schneider, *Anal. Chem.* **33**, 1829 (1961).

[79] C. Booth, *J. Polym. Sci.* **45**, 443 (1960).

[80] A. H. Emery, Jr., in "Polymer Fractionation" (M. J. R. Cantow, ed.), p. 181. Academic Press, New York, 1967.

tion when subjected to ultracentrifugation. Buchard and Cantow[81] describe the theory, experimental methods, and critique of this method.

2.3.3.3.3. FRACTIONATION OF CHEMICAL INHOMOGENEITY. The complexity of the fractionation process for chemically heterogeneous polymers is discussed in detail by Molau[82] and Fuchs and Schmieder.[83] Solubility is determined by molecular weight as well as by the chemical structure of the molecules. Hence, there is a contributory cumulative effect of these two characteristics on the fractionation. In cases of extreme chemical heterogeneity, molecular weight has little influence on the separation of fractions. The varying chemical composition diminishes the influence of molecular weight upon solubility. Hence, both properties change discontinuously during the fractionation. Fuchs and Schmieder[83] describe the techniques used for heterogeneous polymers and present a literature survey of the recommended methods for a wide variety of copolymers.

2.3.3.3.4. GENERAL SUMMARY OF ADDITIONAL METHODS. Cantow[60] presents various methods of polymer fractionation not yet widely used. In addition, he includes a tabulated appendix that suggests methods as well as solvent/nonsolvent systems for a large variety of polymers. This can serve as an excellent reference library to provide preliminary information useful for formulating fractionation schemes.

2.4. Gel Permeation Chromatography

2.4.1. Introduction

Gel permeation chromatography (GPC) is a versatile, relatively new column fractionation technique for characterizing the molecular weights and distributions of polymers. For many years, time-consuming bulk fractionation was often necessary if a complete molecular weight distribution of a polymer was desired. Since 1961 when Moore[84] introduced gel permeation chromatography, the method has gained worldwide acceptance and has grown dramatically. It is rapid, convenient, reproducible, and adaptable to both analytical and preparative separations. Its many applications include quality control, guidance in blending and polymer

[81] W. Burchard and H. J. Cantow, *in* "Polymer Fractionation" (M. J. R. Cantow, ed.), p. 285. Academic Press, New York, 1967.

[82] G. E. Molau, *N. A. S. —N.R.C. Publ.* **1573**, 245 (1968).

[83] O. Fuchs and W. Schmieder, *in* "Polymer Fractionation" (M. J. R. Cantow, ed.), p. 341. Academic Press, New York, 1967.

[84] J. C. Moore, *J. Polym. Sci., Part A* **2**, 835 (1964).

synthesis, and the establishment of physical and structural property relationships.

Moore[84] used gel filtration as the basis for his technique. Gel filtration
had been developed in the biochemistry field for the separation of large,
water-soluble molecules, that is, proteins. Moore successfully synthesized crosslinked polystyrene of various porosities with which he demonstrated the separation of monomers and polymers. An important catalyst
to the growth of GPC was the manufacture by Waters Associates, Inc. of
commercial gel permeation chromatographs, which, through contract
with Moore and the Dow Chemical Company, utilized these gels. The instrument, its use, and potential were initially described in 1965 by
Maley.[85]

GPC is a special form of solid/liquid chromatography that sorts molecules according to their size in solution. It, like gas chromatography, is a
separation process. However, gas chromatography separations are
based on vapor pressure and liquid chromatography separations are based
on solubility. Chromatography basically involves separations due to a
difference in the equilibrium distribution of sample components between
two phases, namely, the moving or mobile phase and the stationary
phase. In liquid chromatography, the mobile phase is noncompressible.
Because of low diffusion in the mobile phase, high-speed liquid chromatography is possible. Since liquid chromatography is a solubility-based
phenomenon, factors such as temperature and vapor pressure affect the
separation only by their effect on solubility and mobile phase viscosity.
The sample components migrate through the chromatographic system
only when they are in the mobile phase. The separation occurs as the solute molecules in the solvent mobile phase percolate through a packed
column of porous particles (stationary phase) and thereby diffuse into and
out of the pores of the packing, which itself has a pore size distribution.
Different molecular size species permeate the gel particles to a different
extent. Molecules that are too large to enter the gel pores pass through
the column by way of the interstitial (or void) volume V_0, which is the
space occupied by the solvent outside the porous packing. Therefore,
the large molecules exit first, followed by progressively smaller molecules, which travel a longer path as they travel through the column. In
this liquid–liquid partitioning, the stationary phase is the portion of the
solvent that is inside the gel pores and the moving phase is the solvent
outside the packing particles. Because of this sorting process, the molecular size distribution of polymers can be measured by GPC. Figure 2
demonstrates the principle of the separation of the molecules according to

[85] L. E. Maley, *J. Polym. Sci., Part C* **8**, 253 (1965).

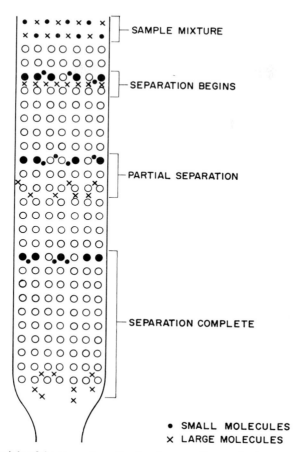

FIG. 2. Principle of the separation of molecules according to size by gel permeation chromatography.[90]

size. The molecules coming out of the column are detected by either a bulk property (differential refractometer) or solute property (ultraviolet absorption) type detector and a chromatogram is traced on a recorder chart, which plots the detector response vs. the volume eluted. This plot is a record of the molecular size distribution of the sample.

Calibration data from which size or molecular weight distribution of an eluted sample can be calculated are constructed by using standards whose molecular weights are independently known. The two most commonly used calibration schemes plot either log molecular weight or log hydrodynamic volume (intrinsic viscosity × molecular weight) vs. the retention volume, the latter of which is the peak position of the distribution curve of the calibration standards in units of volume.

No other single parameter provides as much information about the physical and processing properties of a polymer as does the molecular weight distribution. Subtle differences in the shapes of the molecular weight distribution curves may indicate gross differences in the processing characteristics of the polymers. In industry, GPC is useful for quality control of both raw materials and finished products, process control, new product development, and monitoring of product stability. The entire molecular weight distribution of a polymer is important for prediction of its end-use performance. For example, the two polymers shown in Fig. 3 may have identical average molecular weights but they have different distributions. They may have identical melt indexes but would mold to give different products because of the difference in the distributions. Therefore GPC fills the need for characterization of molecular weights and distributions of polymers better than any other single conventional technique.

The published literature concerning GPC is so extensive from the last 12 years that it is impossible to cover all aspects of the techniques in one review chapter such as this. References 84–116 are a compilation of se-

[86] American Society for Testing and Materials, "Bibliography on Liquid Exclusion Chromatography (GPC)," At. Mol. Data Ser. AMD 40, Sect. D-20.70.04. ASTM, Philadelphia, Pennsylvania, 1974.

[87] V. F. Gaylor, H. L. James, and H. H. Weetaal, *Anal. Chem., Fundam. Rev.* **48** (5), p. 44R (1976).

[88] J. G. Cobler and C. D. Chow, *Anal. Chem., Appl. Rev.* **49** (5), p. 159R (1977).

[89] Waters Associates, Inc., "Bibliography of Gel Permeation Chromatography Publications," Tech. Bull. No. 19907. Waters Assoc., Milford, Mass., 1969.

[90] J. Cazes, *J. Chem. Educ.* **43**, Part I, p. A567; Part II, p. A625 (1966).

[91] J. Cazes, *J. Chem. Educ.* **47**, Part I, p. A461; Part II, p. A505 (1970).

[92] K. J. Bombaugh, in "Modern Practice of Liquid Chromatography" (J. J. Kirkland, ed.), p. 237. Wiley (Interscience), New York, 1971.

[93] K. H. Altgelt and J. C. Moore, in "Polymer Fractionation" (M. J. R. Cantow, ed.), p. 123. Academic Press, New York, 1967.

[94] F. W. Billmeyer, Jr., "Textbook of Polymer Science," p. 53. Wiley (Interscience), New York, 1971.

[95] N. Nakajima, *Adv. Chem. Ser.* **125**, 98 (1973).

[96] J. H. Ross, Jr. and R. L. Shank, *Adv. Chem. Ser.* **125**, 108 (1973).

[97] T. Provder, J. C. Woodbrey, J. H. Clark, and E. E. Drott, *Adv. Chem. Ser.* **125**, 117 (1973).

[98] E. F. Casassa, *N. A. S.—N.R.C. Publ.* **1573**, 285 (1968).

[99] L. D. Moore, Jr. and J. I. Adcock, *N. A. S.—N.R.C. Publ.* **1573**, 289 (1968).

[100] D. D. Bly, *Phys. Methods Macromol. Chem.* **2**, 1 (1972).

[101] H. Determan, "Gel Chromatography." Springer-Verlag, Berlin and New York, 1968.

lected articles, books, and comprehensive reviews that cover to some degree all phases of the development of GPC. The current proposals for recommended terms for GPC are described by Bly.[111] Of practical value is the ASTM Standard Test Method[113] for Molecular Weight Averages and Molecular Weight Distribution of Polystyrene by Liquid Exclusion Chromatography (GPC). A current bibliography update for liquid chromatography by Attebery *et al.*[117] and a treatment of the role of GPC in the characterization of polymers by Abbott[118] contribute to an update of the fast-growing field.

Each reference in itself is extensively referenced so that the potential for a serious study of any phase of the gel permeation technique can be based on the above selected bibliography. Of special note are the three reviews[86-88] of the subject through 1972, 1975, and 1976, respectively; the AMD-40[86] report reviews liquid exclusion chromatography literature from 1964–1971; supplement AMD-40S1 reviews 1972–1975; supplement AMD-40S2 of the bibliographical series is soon to be published. Therefore, background information is available to elucidate both the practical and theoretical profundities of gel permeation chromatography.

[102] J. Cazes, "Gel Permeation Chromatography," Audio Course. Am. Chem. Soc., Washington, D. C., 1971.

[103] A. C. Ouano, E. M. Barrell, and J. F. Johnson, *Tech. Methods Polym. Eval.* **4**, 287 (1975).

[104] J. C. Moore and J. G. Hendrickson, *J. Polym. Sci., Part C* **8**, 233 (1965).

[105] A. E. Hamielec, G. Walther, and J. D. Wright, *Adv. Chem. Ser.* **125**, 138 (1973).

[106] E. Nichols, *Adv. Chem. Ser.* **125**, 148 (1973).

[107] F. S. C. Chang, *Adv. Chem. Ser.* **125**, 154 (1973).

[108] B. S. Ehrlich and W. V. Smith, *Adv. Chem. Ser.* **125**, 164 (1973).

[109] L. H. Tung, *N. A. S.—N.R.C. Publ.* **1573**, 261 (1968).

[110] J. C. Moore, *N. A. S.—N.R.C. Publ.* **1573**, 273 (1968).

[111] D. D. Bly, K. A. Boni, M. J. R. Cantow, J. Cazes, D. J. Harmon, J. N. Little, and E. D. Weir, *J. Polym. Sci., Part B* **9**, 401 (1971).

[112] B. L. Karger, *in* "Modern Practice of Liquid Chromatography" (J. J. Kirkland, ed.), p. 3. Wiley (Interscience), New York, 1971.

[113] ASTM Committee, "Standard Test Method for Molecular Weight Averages and Molecular Weight Distribution of Polystyrene by Liquid Exclusion Chromatography (Gel Permeation Chromatography-GPC)," ANSI/ASTM D 3536-76, Sect. D 20.70. ASTM, Philadelphia, Pennsylvania, 1976.

[114] D. J. Harmon, *J. Appl. Polym. Sci.* **11**, 1333 (1967).

[115] J. Cazes, "Liquid Chromatography of Polymers and Related Materials," Chromatogr. Sci. Ser., Vol. 8. Dekker, New York, 1977.

[116] J. F. Johnson and R. S. Porter, *J. Polym. Sci., Part C* **21**, iii (1968).

[117] J. M. Attebery, R. Yost, and H. W. Major, *Am. Lab.* **9**, 79 (1977).

[118] S. D. Abbott, *Am. Lab.* **9**, 41 (1977).

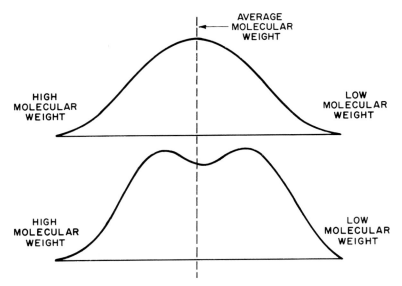

FIG. 3. Molecular weight distribution differences singularly shown by gel permeation chromatography.

2.4.2. Theory of Separation

2.4.2.1. Mechanism of Separation. Bombaugh,[92] Cazes,[90,91] and Dawkins and Hemming[119] have described the theory and mechanism of GPC. Altgelt[120] and Yau et al.[121] have recently reviewed and studied the mechanism of the separation process of GPC. The three types of mechanism suggested are steric exclusion, restricted diffusion, and thermodynamic effects. Yau et al.[121] present evidence that one or more of the above mechanisms can operate, depending on the experimental conditions. Yau and Malone[122] have studied the role of diffusion in separation. Casassa[123] studied the separation with a uniform gel pore size but with varying polymer coil size in solution. Carmichael,[124] and Cantow et al.[125–127] agree with Moore and Arrington's[128] proposal that during the size

[119] J. V. Dawkins and M. Hemming, *Makromol. Chem.* **176**, 1795 (1975).

[120] K. H. Altgelt, *Adv. Chromatogr.* **7**, 3 (1968).

[121] W. W. Yau, C. P. Malone, and S. W. Fleming, *J. Polym. Sci., Part B* **6**, 803 (1968).

[122] W. W. Yau and C. P. Malone, *J. Polym. Sci., Part B* **5**, 663 (1967).

[123] E. F. Casassa, *J. Polym. Sci., Part B* **5**, 773 (1967).

[124] J. B. Carmichael, *Macromolecules* **1**, 526 (1968).

[125] M. J. R. Cantow and J. F. Johnson, *J. Polym. Sci., Part A-1* **5**, 2835 (1967).

[126] M. J. R. Cantow, R. S. Porter, and J. F. Johnson, *J. Polym. Sci., Part A-1* **5**, 987 (1967).

sorting in GPC, the entire range of possible dimensions of the macromolecule coils must be considered.

In addition to the role of diffusion in separation, Yau *et al.*[121] and Casassa[129] have studied the role of exclusion in GPC. They conclude that the solute molecule is excluded from part of the pores of the gel where solvent can permeate and that the extent of exclusion increases with molecular weight. The general conclusion seems to be that size exclusion is the dominant factor in GPC separation even though other effects such as diffusion, lack of equilibrium, or adsorption may distort the operation.

2.4.2.2. Components of Separation Performance. The efficiency of a gel permeation column is expressed in terms of the number of theoretical plates it contains. In chemical engineering terms, during fractional distillation, a theoretical plate refers to a discrete distillation stage constituting a simple distillation in which complete equilibrium is established between the liquid and vapor phases.[90] Since the solvent in the interstitial volume and the solvent inside the gel pores of a GPC column never achieve true equilibrium, the significance of the theoretical plate concept as described by a chemical engineering concept is diminished. However, it is a convenient way of comparing the relative efficiencies of different columns. The theoretical plate is a function of the column, operating conditions, and sample species used. The procedure for the calculation of a column plate count is described by Cazes.[90]

Resolution describes the degree to which adjoining chromatographic peaks are separated during fractionation. Acceptable baseline resolution is observed when the distance between the two peak volumes is at least 1.5 times the base width of one peak. Karger[112] and Cazes[90] present a detailed study of the calculations and variables involved. Additional separation terms including selectivity, capacity, and random dispersion are also described in the above references. Each column or set of columns should be evaluated for separation performance before it is used for a chromatographic analysis.

There are many variables that affect the resolution process in GPC. The effects of these variables are described by many authors: a general description of variables of efficiency of separation[130–133]; the effects of so-

[127] M. J. R. Cantow and J. F. Johnson, *Polymer* **8**, 487 (1967).

[128] J. C. Moore and M. C. Arrington, *Int. Symp. Macromol. Chem., Prepr.* Vol. VI, p. 107 (1966).

[129] E. F. Casassa, *Macromolecules* **9**, No. 1, 182 (1976).

[130] M. R. Ambler, L. J. Fetters, and Y. Kesten, *J. Appl. Polym. Sci.* **21**, 2439 (1977).

[131] J. Y. Chuang, A. R. Cooper, and J. F. Johnson, *J. Polym. Sci., Part C* **43**, 291 (1973).

[132] A. R. Cooper, *J. Polym. Sci., Polym. Phys. Ed.* **12**, 1969 (1974).

[133] J. Y. Chuang, *Diss. Abstr. Int. B* **35**, 166 (1974).

lution concentration[134-136] and corrections for overloading[137]; sample size[138]; flow rate[139-142]; column packing pore size and distribution[143,144] and temperature.[145] Bly[100] tabulates and references the static and dynamic variables affecting separation.

2.4.3. Instrumentation and Operation

2.4.3.1. The Basic Chromatograph. Figure 4 shows a schematic drawing of the solvent flow system in a gel permeation chromatograph such as that manufactured by Waters Associates, Inc. Solvent from the reservoir passes through the heated, nitrogen-purged degasser and then on to the pump and filters, where it divides into two streams. The sample stream of solvent passes through the closed sample injection valve, the columns, and one side of the differential refractometer. The four-way sample valve permits sample injection through a coil previously filled with solution into the solvent flow line. When the sample eluent empties into the 5-ml syphon, it interrupts a light beam that in turn marks the recorder chart and thereby permanently monitors the flow rate. The reference side of the solvent stream passes through a reference column, which creates back pressure for flow control, and then through the other side of the differential refractometer. A switching valve connected into the system with two selected column banks permits a rapid change in column choice. The chromatogram that is traced represents the continual response by the detector to the presence of solute in the sample-side solvent stream.

The ASTM Committee D-20.70.04, which is responsible for standardizing analytical methods (specifically gel permeation chromatography) for plastics has assembled a standard test method for obtaining molecular weight averages and molecular weight distribution of polystyrene by liquid exclusion chromatography (GPC). This method, namely, ASTM

[134] D. Berek, D. Bakos, L. Soltes, and T. Bleha, *J. Polym. Sci., Polym. Lett. Ed.* **12**, 277 (1974).

[135] Y. Kato and T. Hashimoto, *Kobunshi Kagaku* **30**, No. 334, 107 (1973).

[136] Y. Kato and T. Hashimoto, *J. Polym. Sci., Polym. Phys. Ed.* **12**, 813 (1974).

[137] A. Lambert, *Polymer* **10**, 213 (1969).

[138] B. E. Bowen and S. P. Cram, *J. Chromatogr. Sci.* **12**, 579 (1974).

[139] A. R. Cooper, J. F. Johnson, and A. R. Bruzzone, *Eur. Polym. J.* **9**, 1381 (1973).

[140] A. R. Cooper, J. F. Johnson, and A. R. Bruzzone, *Eur. Polym. J.* **9**, 1393 (1973).

[141] A. R. Cooper, *Br. Polym. J.* **5**, 109 (1973).

[142] H. A. Swenson, H. M. Kaustinen, and K. E. Almin, *J. Polym. Sci., Part B* **9**, 261 (1971).

[143] A. R. Cooper and I. Kiss, Jr., *Br. Polym. J.* **5**, 433 (1973).

[144] A. R. Cooper and J. F. Johnson, *J. Appl. Polym. Sci.* **15**, 2293 (1971).

[145] A. R. Cooper and A. R. Bruzzone, *J. Polym. Sci., Polym. Phys. Ed.* **11**, 1423 (1973).

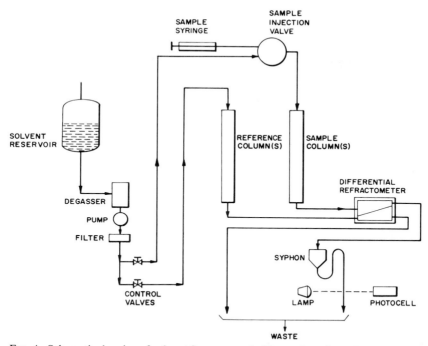

FIG. 4. Schematic drawing of solvent flow system in Waters Associates, Inc., gel permeation chromatograph.

D3536-76,[113] presents a basic, practical approach to an understanding of the instrumentation, experimental techniques, and the acquisition, calculation, and interpretation of data.

2.4.3.2. Variation in Assemblies. Many options exist either in sources of basic instrumentation or in components that can be utilized in any gel permeation operation. Cazes[90,91] and Bly[100] provide some references to equipment that is on the market. Assemblies representing all levels of sophistication are described by Bly.[100] Numerous detectors are used—the main criterion is that it be able to sense the presence of solute in the solvent with accuracy. Principles of detection are described by Keller.[146] Janik[147] suggests accurate techniques for interpreting the detector signal. The differential refractometer, which is the most useful detector, unfortunately is sensitive to temperature and pressure changes. Infrared detectors have poor sensitivity but do provide reproducibility of operation[96]

[146] R. A. Keller, *J. Chromatogr. Sci.* **11**, 223 (1973).
[147] A. Janik, *J. Chromatogr. Sci.* **13**, 93 (1975).

and selectivity in the analysis of composition.[148,149] Ultraviolet absorption has high sensitivity. A variable wavelength ultraviolet photometer is described by Rodgers.[150] The use of flame ionization has been reported by Coll *et al.*[151] Karasek[152] presents a general review of detectors used for liquid chromatography. Viscometric detectors have been described by Revillon *et al.*,[153] Ouano *et al.*,[154,155] Gallot *et al.*,[156,157] Goedhart and Opschoor,[158] Park and Graessley,[159,160] Scheinert,[161] Constantin,[162] Kato *et al.*,[163] Spatorico and Coutter,[164] and Servotte and DeBruille.[165] Ouano and Kaye[31] have successfully monitored GPC eluent by low-angle laser light scattering. Wise and May[166] review the current progress in the field of detectors for liquid chromatography.

Types of pumps, columns, and packings are reviewed by Bly,[100] Cazes[90,91] and Bombaugh.[92] Leading manufacturers of chromatographic equipment use the controlled volume "minipumps," which have low volume and high accuracy. The effects of pulses on the system can be minimized by inserting bellows between the pump and the columns.

Columns and packings have been reviewed by Altgelt and Moore,[93] Snyder,[167] and Spatorico and Beyer,[168,169] and in other chromatography textbooks where the details of requirements and techniques can be found. The two most widely accepted packings for GPC are crosslinked poly-

[148] S. L. Terry and F. Rodriguez, *J. Polym. Sci., Part C* **21**, 191 (1968).

[149] J. V. Dawkins and M. Hemming, *J. Appl. Polym. Sci.* **19**, 3107 (1975).

[150] D. H. Rodgers, *Am. Lab.* **9** (2), 133 (1977).

[151] H. Coll, H. W. Johnson, Jr., A. G. Polgar, E. E. Seibert, and F. H. Stross, *J. Chromatogr. Sci.* **7**, 30 (1969).

[152] F. W. Karasek, *Res. Dev.* **26**, 34 (1975).

[153] A. Revillon, B. Dumont, and A. Guyot, *J. Polym. Sci., Polym. Chem. Ed.* **14**, 2263 (1976).

[154] A. C. Ouano, *J. Polym. Sci., Part C* **43**, 229 (1973).

[155] A. C. Ouano, D. L. Horne, and A. R. Gregges, *J. Polym. Sci.* **12**, 307 (1974).

[156] Z. Gallot, L. Marais, and H. Benoit, *J. Chromatogr. Sci.* **83**, 363 (1973).

[157] Z. Grubisic-Gallot and H. Benoit, *Polym. Prepr., Am. Chem. Soc., Div. Polym. Chem.* **18**(2), 217 (1977).

[158] D. Goedhart and A. Opschoor, *J. Polym. Sci., Part A-2* **8**, 1227 (1970).

[159] W. S. Park and W. W. Graessley, *J. Polym. Sci., Polym. Phys. Ed.* **15**, 71 (1977).

[160] W. S. Park and W. W. Graessley, *J. Polym. Sci., Polym. Phys. Ed.* **15**, 85 (1977).

[161] W. Scheinert, *Angew. Makromol. Chem.* **63**, 117 (1977).

[162] D. Constantin, *Eur. Polym. J.* **13**, 907 (1977).

[163] Y. Kato, T. Takamatsu, M. Fukutomi, M. Fukuda, and T. Hashimoto, *J. Appl. Polym. Sci.* **21**, 577 (1977).

[164] A. L. Spatorico and B. Coutter, *J. Polym. Sci., Polym. Phys. Ed.* **11**, 1139 (1973).

[165] A. Servotte and R. DeBruille, *Makromol. Chem.* **176**, 203 (1975).

[166] S. A. Wise and W. E. May, *Res. Dev.* **28**, 54 (1977).

[167] L. R. Snyder, *Anal. Chem.* **39**, 698 (1967).

[168] A. L. Spatorico, *J. Appl. Polym. Sci.* **19**, 1601 (1975).

[169] A. L. Spatorico and G. L. Beyer, *J. Appl. Polym. Sci.* **19**, 2933 (1975).

styrene and porous glass. The former was introduced by Moore[84] and the latter by Haller.[170] Styragel, the crosslinked polystyrene, is available from Waters Associates, Inc., Milford, Massachusetts. The porous glass Porasil is available from Bio-Rad Laboratories, Richmond, California, and has been evaluated by Cantow and Johnson.[171] Column parts and packed columns are available from vendors listed above and are listed in tables published by Bombaugh,[92] Cazes,[90,91] and in aforementioned reviews.[87,88] Baker and DeStefano[172] suggest a performance criterion for high-performance liquid chromatography columns as a guideline for acceptable operation.

2.4.3.3. Preparatory Scale Units. The operation and use of the Anaprep units where larger columns, larger solvent volumes, and larger sample sizes are employed to produce preparative sized fractions are described by Bombaugh et al.,[173] Cooper et al.,[174] Montague and Peaker,[175] Kato et al.,[176,177] Peyrouset et al.,[178] Vaughan and Francis,[179] and Maley et al.[180] Because of the magnitude of operation, preparative GPC presents experimental difficulties. However, the advantages of isolating fractions by this technique rather than by conventional cumbersome bulk fractionation techniques in most cases outnumber the disadvantages.

2.4.3.4. High Resolution Using Recycle. Resolution can be increased by recycling the solute through the columns. This permits an increase in effective column length without adding columns that increase the pressure drop. Bombaugh et al.[181] and Cazes[91] describe the experimental details of recycle operation. Grubusic-Gallot et al.[182] used recycle to study narrow distribution fractions. The effects of recycle on peak width, sensitivity of the detector, and resolution are considered by Bombaugh.[92]

[170] W. Haller, *Nature (London)* **206,** 693 (1965).

[171] M. J. R. Cantow and J. F. Johnson, *J. Appl. Polym. Sci.* **11,** 1851 (1967).

[172] D. R. Baker and J. J. DeStefano, *Pittsburgh Anal. Conf. Presentation, 1977* p. 317 (1977).

[173] K. J. Bombaugh, W. A. Dark, and R. N. King, *Int. Semin. Gel Permeation Chromatogr., 4th, 1967* p. 25 (1967).

[174] A. R. Cooper, A. J. Hughes, and J. F. Johnson, *J. Appl. Polym. Sci.* **19,** 435 (1975).

[175] P. G. Montague and F. W. Peaker, *J. Polym. Sci., Part C* **43,** 277 (1973).

[176] Y. Kato, K. Sakane, K. Furukawa, and T. Hashimoto, *Kobunshi Kagaku* **30,** 558 (1973).

[177] Y. Kato, T. Kamitani, K. Furukawa, and T. Hashimoto, *J. Polym. Sci.* **13,** 1695 (1975).

[178] A. Peyrouset, R. Prechner, R. Panaris, and H. Benoit, *J. Appl. Polym. Sci.* **19,** 1363 (1975).

[179] M. F. Vaughan and M. A. Francis, *J. Appl. Polym. Sci.* **21,** 2409 (1977).

[180] L. E. Maley, W. B. Richman, and K. J. Bombaugh, *Polym. Prepr., Am. Chem. Soc., Div. Polym. Chem.* **8,** No. 2, 1250 (1967).

[181] K. J. Bombaugh, W. A. Dark, and R. F. Levangie, *J. Chromatogr. Sci.* **7,** 42 (1969).

[182] Z. Grubisic-Gallot, L. Marais, and H. Benoit, *J. Polym. Sci.* **14,** 959 (1976).

2.4.3.5. High-Temperature Operation. Many commercial polymers, such as polyethylene and polypropylene, are insoluble at room temperature and hence must be characterized at temperatures near 150°C. For this purpose, the injection port, columns, and detector must be heated. Temperature regulation is particularly critical for the detector.[103] Temperature controllers for the circulated-air oven are commercially available. Porous glass columns are by nature more durable at high temperature and pressure; however, Styragel columns from Waters Associates, Inc. have been used satisfactorily over extended periods of time for high-temperature GPC. Safety precautions should be exercised to control oven fire in case flammable solvent should leak from a faulty connection. High-temperature GPC studies of polyethylene have been reported by many authors including Nakajima,[183-185] Salovey and Hellman,[186] Ward *et al.*,[187,188] and for polypropylene by Coll and Gilding,[189] Ogawa *et al.*,[190] Ouano and Mercier,[191] and Atkinson and Dietz.[192]

2.4.3.6. High-Performance GPC. The availability of column packings that are microparticles has led to the development of high-performance or high-speed exclusion chromatography (GPC). Separations that require several hours with the larger size packings can now be done in minutes. Separations can now be more selective. Some practical aspects of high-performance chromatography theory are discussed by Know.[193] The most serious problem with the high-speed operation is that because of the much smaller retention times, flow rate variations can cause serious errors in molecular weight analysis. Bly *et al.*[194,195] have studied the effects of this error. Recent studies of high-performance GPC have been made by Kato *et al.*,[177,196,197] Dark *et al.*,[198] Sage *et al.*,[199] McNair and

---

[183] N. Nakajima, *J. Polym. Sci., Part C* **21**, 153 (1968).

[184] N. Nakajima, *J. Polym. Sci., Part A-2* **5**, 101 (1966).

[185] N. Nakajima, *Sep. Sci.* **6**, 275 (1971).

[186] R. Salovey and M. Y. Hellman, *J. Polym. Sci., Part A-2* **5**, 333 (1967).

[187] I. M. Ward and T. Williams, *J. Macromol. Sci., B* **5**, 693 (1971).

[188] T. Williams, Y. Udagawa, A. Keller, and I. M. Ward, *J. Polym. Sci., Part A-2* **8**, 35 (1970).

[189] H. Coll and D. K. Gilding, *J. Polym. Sci., Part A-2* **8**, 89 (1970).

[190] T. Ogawa, Y. Suzuki, and T. Inaba, *J. Polym. Sci., Part A-1* **10**, 737 (1972).

[191] A. C. Ouano and P. L. Mercier, *J. Polym. Sci., Part C* **21**, 309 (1968).

[192] C. M. L. Atkinson and R. Dietz, *Makromol. Chem.* **177**, 213 (1976).

[193] J. H. Know, *J. Chromatogr. Sci.* **15**, 352 (1977).

[194] D. D. Bly, H. J. Stoklosa, J. J. Kirkland, and W. W. Yau, *Anal. Chem.* **47**, 1810 (1975).

[195] D. D. Bly, H. J. Stoklosa, J. J. Kirkland, and W. W. Yau, *Anal. Chem.* **47**, 2328 (1975).

[196] Y. Kato, T. Kametani, and T. Hashimoto, *J. Polym. Sci., Polym. Phys. Ed.* **14**, 2105 (1976).

[197] Y. Kato, S. Kido, M. Yamamoto, and T. Hashimoto, *J. Polym. Sci., Part A-2* **12**, 1339 (1974).

[198] W. A. Dark, R. J. Limpert, and J. D. Carter, *Polym. Eng. Sci.* **15**, 831 (1975).

Chandler,[200] Baker *et al.*,[201] Majors and MacDonald,[202-204] Martin *et al.*,[205] Bly,[206] Dawkins and Yeadon,[207] Kirkland *et al.*,[208,209] Mori,[210] and Snyder.[211] By using monodisperse polystyrene standards as test cases, the molecular weight distribution of a polymer mixture was determined in less than 10 minutes as accurately as with conventional GPC. Effects of packing particle size, solvent flow rate, column length, solution concentration, injection volume, and fraction volume were studied. New instrumentation and column packings are continually being marketed as the field of high-performance chromatography grows to accommodate the current industrial needs for the characterization of polymers.

2.4.4. Molecular Weight Evaluations

2.4.4.1. Qualitative Information. The GPC analysis produces a recorded chromatogram picture of the size distribution of the polymer. Figure 5 is a typical gel permeation chromatogram. The vertical axis is the detector response (ΔRI for the differential refractometer) and the horizontal axis is the retention volume (V_r) of the sample component which represents the number of milliliters of mobile phase that must be passed through the column to elute the sample. The curve peak volume is inversely proportional to log M (molecular weight). Complete fractionation is indicated when the curve completely traces between volume limits that are free from influence of both the void or interstitial volume (V_0) and the impurity peaks. The proper choice of pore size packing in the columns leads to a complete separation. Bly[100] describes the types of information that the raw curve itself reveals to the chromatographer. Visual comparisons of sample traces generated under identical experimental conditions provide a rapid means of monitoring polymerization variables, impurities, and additives and observing the overall breadth of molecular

[199] D. Sage, P. Berticat, and G. Vallet, *Angew. Makromol. Chem.* **54**, 167 (1976).

[200] H. M. McNair and C. D. Chandler, *J. Chromatogr. Sci.* **14**, 477 (1976).

[201] D. R. Baker, R. C. Williams, and J. C. Steichen, *J. Chromatogr. Sci.* **12**, 499 (1974).

[202] R. E. Majors, *Am. Lab.* **7**, 13 (1975).

[203] R. E. Majors, *J. Chromatogr. Sci.* **15**, 334 (1977).

[204] R. E. Majors and F. R. MacDonald, *J. Chromatogr. Sci.* **83**, 167 (1973).

[205] M. Martin, G. Blu, C. Eon, and G. Guiochon, *J. Chromatogr. Sci.* **12**, 438 (1974).

[206] D. D. Bly, *Polym. Prepr., Am. Chem. Soc., Div. Polym. Chem.* **18**(2), 173 (1977).

[207] J. V. Dawkins and G. Yeadon, *Polym. Prepr., Am. Chem. Soc., Div. Polym. Chem.* **18**(2), 227 (1977).

[208] J. J. Kirkland and P. E. Antle, *J. Chromatogr. Sci.* **15**, 137 (1977).

[209] J. J. Kirkland, W. W. Yau, H. J. Stoklosa, and C. H. Dilks, Jr., *J. Chromatogr. Sci.* **15**, 303 (1977).

[210] S. Mori, *J. Appl. Polym. Sci.* **21**, 1921 (1977).

[211] L. R. Snyder, *J. Chromatogr. Sci.* **15**, 441 (1977).

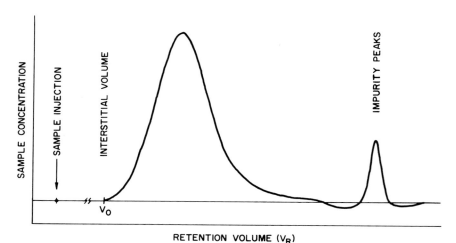

FIG. 5. A typical gel permeation chromatogram.

weight distributions of the samples. However, such qualitative evalua-
tions are useful primarily for short-range studies.

2.4.4.2. Quantitative Information. 2.4.4.2.1. COMPARATIVE EVALU-
ATION. Simple comparative techniques are often used to obtain reliable
estimates of molecular weight distribution (MWD) of polymers. Bly[100]
describes a method for calculating the polydispersity index M_w/M_n. By
using appropriate standards for comparison, if the entire chromatograms
fall in the linear region of the GPC calibration, the width of the curve represents the MWD. In this case, W/d is a constant [W is the curve width
and d is M_w/M_n (polydispersity index)]. Therefore d can be calculated
by comparing the width of curve a for the standard with the width of curve
b for the unknown sample:

$$W_a/d_a = W_b/d_b. \qquad (2.4.1)$$

This presents a rapid way of visually and semiquantitatively calculating
the polydispersity of an unknown sample provided that experimental conditions are the same for both the standard and the sample.

2.4.4.2.2. CALIBRATION FOR MOLECULAR WEIGHT CALCULATIONS.
Well-characterized standards for several polymer types are commercially
available from the National Bureau of Standards,[212] Phillips Petroleum
Company, Pressure Chemical Company (Pittsburgh, Pennsylvania) and
Arro Laboratories (Joliet, Illinois). Some of these standards are listed by
Cazes,[91] Ouano et al.,[103] and Hellman.[213] Unfortunately, the lack of such

[212] H. L. Wagner, Adv. Chem. Ser. 125, 17 (1973).
[213] M. Y. Hellman, in "Liquid Chromatography of Polymers and Related Materials" (J.
Cazes, ed.), p. 29. Dekker, New York, 1977.

calibration standards for the most polymer types does complicate the procedures for obtaining quantitative data.

The simplest direct calibration technique requires the chromatographing of several narrow molecular weight distribution standards of the specific polymer type to be analyzed and of varying molecular weights that cover the required range. The peak retention volume of each standard is then plotted graphically against the known molecular weight average. If the standards are sufficiently narrow, the peak molecular weight might be M_w[84] or M_0 [which is equal to $(M_w \times M_n)^{1/2}$]. Cazes,[90,91] Bly,[100] Ram and Meltz,[214] and the ASTM Committee[113] present general details of calibration procedures. The accuracy of the molecular weights to be calculated depends on the number and quality of the standards used, the selection of peak molecular weight values of the standards, and the correction techniques applied, as discussed later. Figure 6 shows a general GPC calibration where the effective limits of selective permeation are represented by the portion of the curve that is linear in log M vs. V_r (retention volume). The linearity of the plot is represented by the equation

$$\log(\text{molecular weight}) = a + b \cdot V \text{ (appearance volume).} \quad (2.4.2)$$

The molecular weight of the unknown is then determined from the calibration plot and the retention volume profile of the sample.

In the case of actual polymers, instrumental inefficiencies cause band broadening, which results in a chromatogram broader than the actual molecular weight distribution. This error becomes significant if M_w/M_n is less than 2.0. Balke and Hamielec,[215] Hess and Kratz,[216] Pickett et al.[217] and Tung[218] present methods whereby corrections for this band broadening can be applied to the chromatographic data for the narrow distribution standards so that a corrected calibration curve can be constructed by which molecular weights and distributions of the unknown samples can be calculated. This correction, though sometimes significant, has a minor effect on the final molecular weights compared with errors caused by poor selection of columns and standards, sample preparation, and inaccurate data acquisition and reduction techniques. The subject of dispersion corrections will be further discussed in Section 2.4.4.2.4.

A universal calibration accurately relates molecular size in a given solvent, independent of the polymer type. Reviews[86,87,100] of proposed uni-

[214] A. Ram and J. Meltz, *Polym.-Plast. Technol. Eng.* **4**, 23 (1975).
[215] S. T. Balke and A. E. Hamielec, *J. Appl. Polym. Sci.* **13**, 1381 (1969).
[216] M. Hess and R. F. Kratz, *J. Polym. Sci., Part A-2* **4**, 731 (1966).
[217] H. E. Pickett, M. J. R. Cantow, and J. F. Johnson, *J. Polym. Sci., Part C* **21**, 67 (1968).
[218] L. Tung, *J. Appl. Polym. Sci.* **13**, 775 (1969).

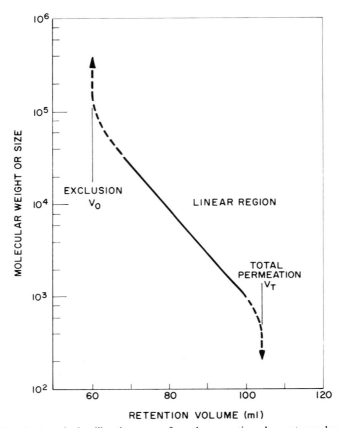

FIG. 6. A typical calibration curve for gel permeation chromatography.

versal calibration techniques summarize published work through November 1975. ASTM D20.70.04 Method D-3593-77 presents a practical technique for the application of this calibration procedure. Grubisic *et al.*[219] present the most widely used type of universal calibration, which employs a plot of the hydrodynamic volume of the molecule ($\log[\eta]M$) vs. retention volume as in the equation

$$\log[\eta]M = A + B \times \text{volume.} \tag{2.4.3}$$

By using $[\eta]M$ as a measure of hydrodynamic volume, a nearly linear calibration line results when $[\eta]M$ is plotted against retention volume for a wide variety of polymers as shown in Fig. 7. Such polymers listed in the reviews include polystyrene,[219] star-shaped polystyrenes,[219] polypro-

[219] Z. Grubisic, P. Rempp, and H. Benoit, *J. Polym. Sci., Polym. Lett. Ed.* **5**, 753 (1967).

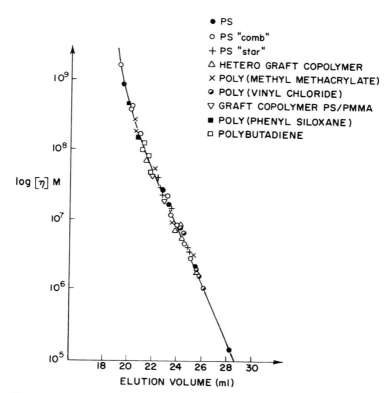

FIG. 7. A calibration curve for gel permeation chromatography based on hydrodynamic volume.[219]

pylene,[189,192,220] block and graft copolymers of styrene and methyl methacrylate,[219] poly(vinyl chloride),[219,221] poly(methyl methacrylate),[219,222] polybutadiene,[219,221,223,224] polyethylene,[220,224,225] ethylene-propylene copolymers,[220] poly(α-methyl styrene),[189] poly(dimethyl siloxane),[222,226] branched and graft copolymers.[219] Ambler[221] suggests that $[\eta]M$ serve as a universal calibration parameter when the molecular geometries of all the samples involved are similar and that the universality breaks down when geometry and rigidity differences are great. Conviction of the validity of

[220] P. Crouzet, A. Martens, and P. Mangin, J. Chromatogr. Sci. 9, 525 (1971).
[221] M. R. Ambler, J. Polym. Sci., Polym. Chem. Ed. 11, 191 (1973).
[222] J. V. Dawkins, J. Macromol. Sci. B 2, 623 (1968).
[223] B. L. Funt and V. Hornof, J. Appl. Polym. Sci. 15, 2439 (1971).
[224] K. A. Boni, F. A. Sleimers, and P. B. Stickney, J. Polym. Sci., Part A-2 6, 1579 (1968).
[225] J. V. Dawkins and J. W. Maddock, Eur. Polym. J. 7, 1537 (1971).
[226] J. V. Dawkins, J. W. Maddock, and D. Coupe, J. Polym. Sci., Part A-2 8, 1803 (1970).

these limitations is not shared by other authors. It is recognized, however, that applicability of a universal calibration technique to all polymers is not uniform. Errors caused by the variables of molecular weight range, Mark–Houwink constants, and polydispersity should be considered for each type of polymer to be characterized.

Once the calibration has been established using the available standards, the limiting viscosity number must be measured and/or the Mark–Houwink constants for the unknown polymer must be known in order to calculate the molecular weight from the gel permeation data. Dawkins[222,227] has shown that the use of unperturbed dimensions is in some cases superior to $[\eta]M$ as a calibration parameter. Bly[100] presents a review of published work on the two methods along with the limitations of each.

Calibrations can be developed from a single standard,[228] broad polydispersity standards,[229-235] and narrow polydispersity standards.[223,236] The use of internal standards[237-239] is a time-saving technique. Many authors have substantiated the application of universal calibration techniques.[220,240-247] Nonlinear calibration treatment is described by Cardenas.[248] The development of the Mark–Houwink constants as a corollary of the universal calibration is described by Ambler[221] and Boni.[224]

[227] J. V. Dawkins, *Eur. Polym. J.* **6**, 831 (1970).

[228] B. R. Loy, *J. Polym. Sci., Polym. Chem. Ed.* **14**, 2321 (1976).

[229] M. J. R. Cantow, R. S. Porter, and J. F. Johnson, *J. Polym. Sci., Part A-1* **5**, 1391 (1967).

[230] F. L. McCrackin, *J. Appl. Polym. Sci.* **21**, 191 (1977).

[231] A. R. Weiss and E. Cohn-Ginsberg, *J. Polym. Sci., Part A-2* **8**, 148 (1970).

[232] F. C. Frank, I. M. Ward, and T. Williams, *J. Polym. Sci., Part A-2* **6**, 1357 (1968).

[233] B. A. Whitehouse, *Macromolecules* **4**, 463 (1971).

[234] J. R. Purdon, Jr. and R. D. Mate, *J. Polym. Sci., Part A-1* **6**, 243 (1968).

[235] H. K. Makabadi and K. F. O'Driscoll, *J. Appl. Polym. Sci.* **21**, 1283 (1977).

[236] T. Williams and I. M. Ward, *J. Polym. Sci., Polym. Lett. Ed.* **6**, 621 (1968).

[237] R. C. Williams, J. A. Schmit, and H. L. Suchan, *J. Polym. Sci., Polym. Lett. Ed.* **9**, 413 (1971).

[238] G. N. Patel and J. Stejny, *J. Appl. Polym. Sci.* **18**, 2069 (1974).

[239] G. N. Patel, *J. Appl. Polym. Sci.* **18**, 3537 (1974).

[240] M. R. Ambler and D. McIntyre, *J. Polym. Sci., Polym. Lett. Ed.* **13**, 589 (1975).

[241] A. Rudin and H. L. W. Hoegy, *J. Polym. Sci., Part A-1* **10**, 217 (1972).

[242] A. R. Weiss and E. Cohn-Ginsberg, *J. Polym. Sci., Polym. Lett. Ed.* **7**, 379 (1969).

[243] J. V. Dawkins and M. Hemming, *Makromol. Chem.* **176**, 1777 (1975).

[244] E. P. Otocka and M. Y. Hellman, *J. Polym. Sci., Polym. Lett. Ed.* **12**, 331 (1974).

[245] G. R. Williamson and A. Cervenka, *Eur. Polym. J.* **8**, 1009 (1972).

[246] E. Nichols, *Adv. Chem. Ser.* **125**, 148 (1973).

[247] J. V. Dawkins, *Br. Polym. J.* **4**, 87 (1972).

[248] J. N. Cardenas and K. F. O'Driscoll, *J. Polym. Sci., Polym. Lett. Ed.* **13**, 657 (1975).

Numerous comparisons have been published of the various universal calibration methods.[189,219,222,223,225,227,249–252] A recent application of the use of internal standards for calibration in high-speed GPC was reported by Kohn.[253]

2.4.4.2.3. DATA ACQUISITION. Because of continual developments in the field of automatic data acquisition, the manual procedure for converting the chromatogram data to molecular weight values is seldom used. Data points used are the chromatographic heights at selected values of appearance volume. The incremental volumes are selected either manually or automatically for subsequent computations so as to reproduce an accurate chromatogram. Details of the means by which the various parameters such as M_n, M_w, M_Z, and $[\eta]$ are calculated from GPC data are given by Ouano et al.,[103] Hellman,[213] and Bly.[100]

2.4.4.2.4. CALCULATIONS. Many of the approaches that have been used to solve the calibration/molecular weight calculation problem are documented by ASTM.[86] The Hamielec method[254] permits calibration with one broad molecular weight distribution (MWD) standard ($M_w/M_n \sim 2$ or greater) of the polymer type of interest and has been widely used. It corrects for chromatographic dispersion; however, it is useful primarily when the sample and standard have similar MWDs and MWs. Yau et al.[255] have described an improved version of the Hamielec method that permits the use of very dissimilar MWD standards and samples of the same polymer type and at the same time corrects for chromatographic dispersion. Both methods use the linear calibration curve approximation.

The parameters most commonly calculated from GPC data are M_n, M_w, and M_Z.

Viscosity-average molecular weight M_v can be calculated if the Mark–Houwink exponent a is known for the specific solvent–polymer system from Eq. (2.2.5), and the limiting viscosity number can be estimated from Eq. (2.2.1) if the coefficient K is also known.

The equations by which M_n and M_w are calculated from a log-linear

[249] H. Coll and L. R. Prusinowski, J. Polym. Sci., Polym. Lett. Ed. 5, 1153 (1967).

[250] J. V. Dawkins and M. Hemming, Polymer 16, 554 (1975).

[251] M. Iwama, M. Abe, and T. Homma, Kogyo Kagaku Zasshi 72, 931 (1969).

[252] J. V. Dawkins, Eur. Polym. J. 13, 837 (1977).

[253] E. Kohn and R. W. Ashcraft, in "Liquid Chromatography of Polymers and Related Materials" (J. Cazes, ed.), Chromatogr. Sci. Ser. No. 8, p. 105. Dekker, New York, 1977.

[254] S. T. Balke, A. E. Hamielec, B. P. LeClair, and S. L. Pearce, Ind. Eng. Chem., Prod. Res. Dev. 8, 54 (1969).

[255] W. W. Yau, H. J. Stoklosa, and D. D. Bly, J. Appl. Polym. Sci. 21, 1911 (1977).

calibration of molecular weight and elution volume are

$$M_w = \sum_V W(V)D_1 \exp(-D_2 V), \qquad (2.4.4)$$

$$M_n = 1 \bigg/ \sum_V [W(V)/D_1 \exp(-D_2 V)], \qquad (2.4.5)$$

where $W(V)$ represents the theoretical chromatogram of the sample and $D_1 \exp(-D_2 V)$ the calibration, both in terms of the elution volume V; D_1 is the exponential of the intercept and D_2 the slope of the calibration lines.

The Hamielec method[254] uses these equations but substitutes $F(V)$ to approximate $W(V)$, where $F(V)$ is the observed or experimental chromatogram, which has been broadened by the chromatographic column dispersion process. Both $W(V)$ and $F(V)$ are normalized: $\Sigma_V W(V) = \Sigma_V F(V) = 1$. $F(V)$ is determined by a chromatographic analysis of the sample. Bly[100] gives details of data handling. In Hamielec's method, the coefficients D_1' and D_2' are determined for the calibration lines from the standard sample with known M_n and M_w. Using these primed calibration coefficients and the data for the chromatographed unknown samples, the $F(V)$ for the samples are inserted into the following equations and the molecular weights are calculated:

$$M_w = \sum_V F(V)D_1' \exp(-D_2' V), \qquad (2.4.6)$$

$$M_n = 1 \bigg/ \sum_V [F(V)/D_1' \exp(-D_2' V)]. \qquad (2.4.7)$$

The effective linear calibration $D_1' \exp(-D_2' V)$ corrects for the GPC curve broadening and yields correct M_n and M_w values when the standard and samples are similar. However it does not give accurate molecular weights when the standard and sample differ, nor does it represent the theoretical peak position calibration of narrow MWD standards.

The method of Yau et al.[255] employs an adjusting parameter σ the deviation caused by column dispersion that is measured to a first approximation as the experimental dispersion of a very narrow MWD polystyrene standard. (At $\sigma = 0$, Yau's method, which is called GPCV2, reduces to the Hamielec method.) The value of σ is determined from a separate chromatogram of a NMWD standard from the relation

$$\sigma = (A/H)(2\pi)^{1/2}, \qquad (2.1.8)$$

where H is peak height and A is peak area. The equations developed by Yau[255] are

$$M_w = \exp[-\tfrac{1}{2}(D_2\sigma)^2] \sum_V [F(V)D_1 \exp(-D_2 V)], \qquad (2.4.9)$$

$$M_n = \exp[\tfrac{1}{2}(D_2\sigma)^2] \Big/ \sum_V \frac{F(V)}{D_1\exp(-D_2V)}. \qquad (2.4.10)$$

Hamielec[256] and Provder and Rosen[257] have also developed these equations using a different mathematical approach. Results show that Yau's method reduces the difference in calculated molecular weights when using the Hamielec and the peak position calibration curves.

2.4.5. Applications

2.4.5.1. Introduction. Gel permeation chromatography has provided a precise means of determining the molecular weight distribution of polymers without the laborious classic fractionation and characterization of fractions that was necessary for many years. As a consequence, studies of the correlation of physical properties with molecular weight distribution have been significantly simplified. Examination of numerous reviews of the gel permeation field[87–90,100,103,116] will demonstrate how extensively this technique is used either as a direct independent characterization tool or in partnership with other investigative procedures for the study of polymer properties and structure.

Two of these many applications will be described briefly, namely, the determination of long-chain branching and the study of the changes of chemical composition of copolymers as a function of molecular weight distribution.

2.4.5.2. Branching. The effects of branching on the mechanical and melt properties of polymers is well known qualitatively, although less well documented quantitatively. Since branching changes the size of the dissolved molecule as compared with an unbranched molecule of the same molecular weight, the gel permeation analysis of molecular weight moments and distributions, which is based on a size (or hydrodynamic volume) separation, will be in error unless branching is accounted for. Short-chain branching (SCB) reduces the limiting viscosity number ($[\eta]$) of a branched polymer in comparison with a linear polymer of the same molecular weight but to the extent of only 1% of the effect of long-chain branching (LCB).

Four recent reviews[87,88,100,103] present a general summary of studies to determine long-chain branching in polymers. Most of the work on long-chain branching has been done on polyethylene. Wild and Guliana[258] characterized branched and linear fractions of polyethylene by viscosity and molecular weight measurements. They found that calibra-

[256] A. E. Hamielec, *J. Appl. Polym. Sci.* **14**, 1519 (1970).
[257] T. Provder and E. M. Rosen, *Sep. Sci.* **5**, 437 (1970).
[258] L. Wild and R. Guliana, *J. Polym. Sci., Part A-2* **5**, 1087 (1967).

tion by hydrodynamic volume ($[\eta]M$) instead of molecular weight (M) was satisfactory for both linear and branched polymers. Tung[259,260] combined sedimentation-velocity with GPC to determine the distribution of branching in styrene–divinylbenzene copolymers. Drott and Mendelson[261–263] attempted to quantify long-chain branching by developing a technique based on a GPC separation that is a function of hydrodynamic volume of the polymer. The decrease in the limiting viscosity number of the branched polymer with the same molecular weight as the linear polymer represented the degree of branching and hence a branching index. Assumptions, such as that branching distribution is independent of molecular weight, have been challenged, but the method provides a means of obtaining a quantitative estimate of long-chain branching from experimental viscosity and GPC data. Details of the method are available in the original paper.[262] Ram and Miltz,[264] Wild *et al.*,[265] Cote and Shida,[266] Miltz and Ram,[267] Wagner and McCracken,[268] Servotte and De-Bruille,[165] Nakano and Goto,[269] and Shultz[270] have studied long-chain branching in polyethylene by using variations of the hydrodynamic theory. The method of Drott has been evaluated by Cervenka *et al.*[271–273] Wild *et al.*[274] correlate the effects of long-chain branching with viscoelastic properties. The branching characteristics of diene polymers and copolymers have been investigated by Kraus and Stacy[275,276] and Ambler *et al.*[277,278] Scholte and Meijerink[279] combine viscosity and light scatter-

[259] L. H. Tung, *J. Polym. Sci., Part A-2* **7**, 47 (1969).

[260] L. H. Tung, *J. Polym. Sci., Part A-2* **9**, 759 (1971).

[261] R. A. Mendelson and E. E. Drott, *J. Polym. Sci., Polym. Lett. Ed.* **6**, 795 (1968).

[262] E. E. Drott and R. A. Mendelson, *J. Polym. Sci., Part A-2* **8**, 1361 (1970).

[263] E. E. Drott and R. A. Mendelson, *J. Polym. Sci., Part A-2* **8**, 1373 (1970).

[264] A. Ram and J. Miltz, *J. Appl. Polym. Sci.* **15**, 2639 (1971).

[265] L. Wild, R. Ranganath, and T. Ryle, *J. Polym. Sci., Part A-2* **9**, 2137 (1971).

[266] J. A. Cote and M. Shida, *J. Polym. Sci., Part A-2* **9**, 421 (1971).

[267] J. Miltz and A. Ram, *Polymer* **12**, 685 (1971).

[268] H. L. Wagner and F. L. McCracken, *J. Appl. Polym. Sci.* **21**, 2833 (1977).

[269] S. Nakano and Y. Goto, *J. Appl. Polym. Sci.* **20**, 3313 (1976).

[270] A. R. Shultz, *J. Polym. Sci.* **10**, 983 (1972).

[271] A. Cervenka and T. W. Bates, *J. Chromatogr.* **53**, 85 (1970).

[272] A. Cervenka and G. R. Williamson, *Eur. Polym. J.* **10**, 305 (1974).

[273] G. R. Williamson and A. Cervenka, *Eur. Polym. J.* **10**, 295 (1974).

[274] L. Wild, R. Ranganath, and D. C. Knobeloch, *Polym. Eng. Sci.* **16**, 811 (1976).

[275] G. Kraus and C. J. Stacy, *J. Polym. Sci., Part A-2* **10**, 657 (1972).

[276] G. Kraus and C. J. Stacy, *J. Polym. Sci., Polym. Symp.* **43**, 329 (1973).

[277] M. R. Ambler, R. D. Mate, and J. R. Purdon, Jr., *J. Polym. Sci., Polym. Chem. Ed.* **12**, 1771 (1974).

[278] M. R. Ambler, R. D. Mate, and J. R. Purdon, Jr., *J. Polym. Sci., Polym. Chem. Ed.* **12**, 1759 (1974).

[279] T. G. Scholte and N. L. J. Meijerink, *Br. Polym. J.* **9**, 133 (1977).

ing to provide additional experimental data to study branching. Kato *et al.*[280] separate polymers of differing branching structure by using high-efficiency columns. Otocka *et al.*[281] combine infrared analysis with viscosity and GPC experimental data. Ambler[282] evaluates different branching functions that can be used in the determination of random branching by GPC. This brief review of the variations on the general approaches to the determination of long-chain branching in polymers can serve to provide a background of awareness of the current state of the art. A more complete review of publications from 1966 to 1976 was recently presented by Drott.[283]

2.4.5.3. Chemical Composition Analysis. The chemical properties of a polymer depend on the nature and molecular weight of the functional groups in the monomer from which the polymer was made, as well as branching and heterogeneity within the polymer. Copolymers, which are formed by the union of two distinct monomer types, assume specific properties depending on the proportion of each type of monomeric unit and the distribution of each along the polymer chain. Consequently, the characterization of copolymers requires more sophisticated techniques than are demanded by homogeneous polymers.

In the GPC analysis of copolymers, a differential refractometer detector is sensitive to compositional changes, and so quantitative data for polymers with varying composition are difficult to analyze.

Terry and Rodriguez[148] used infrared detection monitored at different frequencies for successive runs of the polymer to detect specific functional groups of the copolymer. Multiple detectors, specifically a differential refractometer and ultraviolet spectrophotometer, in series were used by Runyon *et al.*[284] to measure simultaneously the amount of each component in styrene–butadiene copolymer. The general technique is to quantitatively calibrate the refractive index response to styrene and butadiene individually and the ultraviolet response to styrene at a wavelength at which butadiene is not absorbed. Knowing the relative response of each component, the distribution curves for the butadiene component, styrene component, the sum of the two, and the weight percent butadiene

[280] Y. Kato, T. Hashimoto, T. Fujimoto, and M. Nagasawa, *J. Polym. Sci., Part A-2* **13**, 1849 (1975).

[281] E. P. Otocka, R. J. Roe, M. Y. Hellman, and P. M. Muglia, *Macromolecules* **4**, 507 (1971).

[282] M. R. Ambler, *J. Appl. Polym. Sci.* **21**, 1655 (1977).

[283] E. E. Drott, *in* "Liquid Chromatography of Polymers and Related Materials" (J. Cazes, ed.), p. 161. Dekker, New York, 1977.

[284] J. R. Runyon, D. E. Barnes, J. F. Rudd, and L. H. Tung, *J. Appl. Polym. Sci.* **13**, 2359 (1969).

on styrene can be computed. Molau,[82] Ouano et al.,[103] Ceresa,[285] and aforementioned reviews[86-88] provide background for compositional analysis of copolymers. Coupling of GPC with automatic viscometry permits the characterization of copolymers with respect to both composition and molecular weight heterogeneity.[286] Styrene–butadiene copolymers have been studied by Cantow et al.[287,288] and Adams.[289] Styrene–butadiene star blocks,[290] block copolymers,[291] graft copolymers,[292] polystyrene[293] and α-methyl styrene star-branched polymers, and styrene–maleic anhydride copolymers[294] have been characterized by using GPC with variations of detectors and calibration techniques. Corbin and Prud'Homme,[295,296] Cramond et al.,[297] Ho-Duc and Prud'Homme,[298] and Urwin and Cramond[299] have used GPC techniques to determine composition and/or molecular weight distributions of styrene–isoprene block copolymers. Menin and Roux[300] used GPC to determine the composition of polymer mixtures of polybutadiene and polyisobutylene. Ethylene–propylene copolymers[301] and vinyl chloride–vinyl acetate copolymers[302] have been characterized by GPC. The above examples represent only a few examples of the variety of copolymers whose composition and molecular weights have been investigated by using modifications of GPC analysis.

2.4.6. Conclusions

The field of gel permeation analysis is growing so rapidly that no single text can either cover the subject comprehensively or provide up-to-date information. Instrumentation and calibration development have domi-

[285] R. J. Ceresa, in "Techniques of Polymer Characterization" (P. W. Allen, ed.), Chapter 8. Academic Press, New York, 1959.

[286] Z. G.-Gallot, M. Picot, P. Gramain, and H. Benoit, J. Appl. Polym. Sci. 16, 2931 (1972).

[287] H. J. Cantow, J. Probst, and C. Stojanov, Kautsch. Gummi, Kunstst. 21, 609 (1968).

[288] H. J. Cantow, J. Probst, and C. Stojanov, Rev. Gen. Caoutch. Plast. 45, 1253 (1968).

[289] H. E. Adams, Sep. Sci. 6, 259 (1971).

[290] L.-K. Bi and L. J. Fetters, Macromolecules 9, 732 (1976).

[291] A. Dondos, P. Rempp, and H. Benoit, Makromol. Chem. 175, 1659 (1974).

[292] J. P. Kennedy, J. Polym. Sci., Polym. Chem. Ed. 13, 2213 (1975).

[293] T. Kato, A. Itsubo, Y. Yamamoto, and T. Fujimoto, Polym. J. (Jpn.) 7, 123 (1975).

[294] C. D. Chow, J. Appl. Polym. Sci. 20, 1619 (1976).

[295] N. Corbin and J. Prud'Homme, J. Polym. Sci., Polym. Chem. Ed. 14, 1645 (1976).

[296] N. Corbin and J. Prud'Homme, J. Polym. Sci., Polym. Phys. Ed. 15, 1937 (1977).

[297] D. N. Cramond, J. M. Hammond, and J. R. Urwin, Eur. Polym. J. 4, 451 (1968).

[298] N. Ho-Duc and J. Prud'Homme, Int. J. Polym. Mater. 4, 303 (1976).

[299] J. R. Urwin and D. N. Cramond, Aust. J. Chem. 22, 543 (1969).

[300] J. P. Menin and R. Roux, J. Chromatogr. 64, 49 (1972).

[301] T. Ogawa and T. Inaba, J. Appl. Polym. Sci. 21, 2979 (1977).

[302] J. Janca and M. Kolinsky, J. Appl. Polym. Sci. 21, 83 (1977).

nated the literature during the past 13 years. Certainly the development of new detectors, on-line data acquisition, and computer data reduction have simplified the problems of converting a raw chromatogram into a molecular weight distribution of a polymer. In spite of progress to date, future development in the sophistication of instruments, increased automation, and new column-packing materials will add to the value of GPC as a versatile and accurate means of determining molecular weight distributions of polymers. However, skillful implementation of theoretical and experimental details by those who utilize this characterization tool is essential to obtain useful and accurate results.

2.5. Miscellaneous Methods

2.5.1. Sedimentation Analysis

2.5.1.1. Principles and Instrumentation. Ultracentrifugation, or sedimentation analysis, is an intricate, uncommon, specialized technique for determining the molecular weight and distribution of high molecular weight materials. Historically, it has found major application in the field of biochemistry for compact protein molecules rather than for random-coil polymers, where mechanical entanglement of the chains causes deviation from ideality and consequent disobedience of the laws of thermodynamics of polymer solutions.[303] However, in recent years, it has become a valuable aid for characterizing synthetic polymers. Even though the standards for instrumental and optical components of the ultracentrifuge are high, the variables affecting the experimental procedures are readily controlled and evaluation of the experimental results is direct and reliable.

The ultracentrifuge consists of an aluminum rotor several inches in diameter, which is rotated at high speed in a vacuum chamber (to minimize thermal effects). A small cell placed within the rotor holds in a cavity the solution being centrifuged. The rotor is driven electrically or by air or oil turbine. The cell has two windows, one on either side of the cavity. The concentration of polymer along the length of the cell is measured by optical methods based on refractive index or absorption. Billmeyer[304] and Scholte[305] each show a diagram of a commercial analytical ultracentrifuge. Scholte lists the names of firms who market analytical

[303] F. W. Billmeyer, Jr., "Textbook of Polymer Science," p. 32. Wiley (Interscience), New York, 1971.

[304] F. W. Billmeyer, Jr., "Textbook of Polymer Science," p. 90. Wiley (Interscience), New York, 1971.

[305] T. G. Scholte, in "Polymer Molecular Weights," (P. E. Slade, Jr., ed.), Part II, p. 501. Dekker, New York, 1975.

ultracentrifuges. Speeds range from 800 to 80,000 rpm and operation temperature can approach 100°C.

Solvents must have a refractive index and density different from the polymer being analyzed to ensure measurement and sedimentation, respectively. Concentration gradients in the cell are determined optically by using a beam of light traveling in a direction parallel to the axis of rotation and perpendicular to the cell, the beam being intercepted by the cell at each rotation. Refraction of rays of the beam passing through different portions of the cell may be determined either by the Lamm scale method[306] or by use of schlieren optics.[306] In both cases, the measured refractions are converted to concentration gradients along the direction of sedimentation within the cell. The progress of the solute in this centrifugal field is observed in determining sedimentation velocity, and the distribution of solute in a lower centrifugal field is traced in the equilibrium method. Experimental data are obtained as a photographed curve of dn/dr vs. r, where n is refractive index and r is the distance from the center of rotation to the point of observation in the cell. These data are converted to a curve of dc/dr vs. r if the specific refractive increment dn/dc is known. A synthetic boundary cell, in which pure solvent is layered onto a polymer solution during ultracentrifugation to form a sharp boundary, can be used to calibrate the optics of the ultracentrifuge or to determine dn/dc after calibration.[304]

Details of theory, experimental procedures, and applications of ultracentrifugation methods are presented by Flory,[307] Billmeyer,[304] Scholte,[305] Adams et al.,[308,309] Provencher and Gobush,[310] Gehatia and Wiff,[311] Pekar et al.,[312] Williams et al.,[313,314] Chervenka,[315] Bowen,[316] Mc-

[306] T. Svedberg and K. O. Pederson, "The Ultracentrifuge." Oxford Univ. Press (Clarendon), London and New York, 1940 (Johnson Reprint Corp., New York, 1959).

[307] P. J. Flory, "Principles of Polymer Chemistry," p. 303. Cornell Univ. Press, Ithaca, New York, 1953.

[308] E. T. Adams, Jr., P. J. Wan, D. A. Soucek, and G. H. Barlow, Adv. Chem. Ser. 125, 235 (1973).

[309] E. T. Adams, N. A. S.—N.R.C. Publ. 1573, 84 (1968).

[310] S. W. Provencher and W. Gobush, N. A. S.—N.R.C. Publ. 1573, 143 (1968).

[311] M. Gehatia and D. R. Wiff, Adv. Chem. Ser. 125, 216 (1973).

[312] A. H. Pekar, P. J. Wan, and E. T. Adams, Jr., Adv. Chem. Ser. 125, 260 (1973).

[313] J. W. Williams, K. E. van Holde, R. L. Baldwin, and H. Fujita, Chem. Rev. 58, 715 (1958).

[314] J. W. Williams, "Ultracentrifugal Analysis in Theory and Experiment." Academic Press, New York, 1963.

[315] C. H. Chervenka, "A Manual of Methods for the Analytical Ultracentrifuge," Spinco Division of Beckman Instruments, Inc., Palo Alto, California, 1969.

[316] T. J. Bowen, "An Introduction to Ultracentrifugation." Wiley (Interscience) New York, 1970.

Cormick,[317] and Blair.[318] It is the purpose of this review to briefly summarize the primary types of sedimentation analysis.

2.5.1.2. Sedimentation–Diffusion Equilibrium. In the sedimentation equilibrium technique, the ultracentrifuge is operated at low rotation speed for perhaps 1 to 2 weeks under constant conditions. Thermodynamic equilibrium is reached and the polymer is distributed in the cell according to its molecular weight and distribution. At equilibrium, the sedimentation force on each species is just balanced by its tendency to diffuse back down the concentration gradient resulting from its position in the centrifugal field. The sedimentation force on a polymer particle of mass m at a distance r from the axis of rotation is $\omega^2 r(1 - \bar{v}\rho)m$, where ω is the angular velocity of rotation, $\bar{v}$ the partial specific volume[319] of the polymer, and ρ the density of the solution. For ideal solutions[319] of heterogeneous polymers, a weight-average molecular weight can be calculated:

$$M = \frac{2RT \ln(c_2/c_1)}{(1 - \bar{v}\rho)\omega^2(r_2^2 - r_1^2)}, \qquad (2.5.1)$$

where RT is the thermal energy per mole and c_1 and c_2 are the concentrations at two points r_1 and r_2 in the cell. If the data are treated in terms of the refractive increment n, the difference between the refractive indices of solution and solvent, then

$$M_Z = \frac{RT}{(1 - \bar{v}\rho)\omega_2} \left[\left(\frac{1}{r}\frac{dn}{dr}\right)_2 - \left(\frac{1}{r}\frac{dn}{dr}\right)_1 \right] (\eta_2 - \eta_1)^{-1}, \qquad (2.5.2)$$

where M_Z is the Z-average molecular weight [see Eq. 2.1.8)]. In both cases above, positions 1 and 2 must be at the ends of the cell to include all molecular species. In nonideal solutions, experimentation at θ-temperature[320] reduces the dependence of molecular weight on concentration as expressed by the second virial coefficient.[321]

The sedimentation equilibrium experiment takes several days to reach equilibrium even with short cells. The distribution of molecular weights, M_n, M_w, M_z, and M_{z+1}, can be obtained as described by Scholte.[322–324]

[317] H. W. McCormick, in "Polymer Fractionation" (M. J. R. Cantow, ed.), p. 251. Academic Press, New York, 1967.

[318] J. E. Blair, J. Polym. Sci., Part C **8**, 287 (1965).

[319] P. J. Flory, "Principles of Polymer Chemistry," p. 495. Cornell Univ. Press, Ithaca, New York, 1953.

[320] F. W. Billmeyer, Jr., "Textbook of Polymer Science," p. 31. Wiley (Interscience), New York, 1971.

[321] F. W. Billmeyer, Jr., "Textbook of Polymer Science," p. 73. Wiley (Interscience), New York, 1971.

[322] T. G. Scholte, J. Polym. Sci., Part A-2 **6**, 91 (1968).

[323] T. G. Scholte, J. Polym. Sci., Part A-2 **6**, 111 (1968).

[324] T. G. Scholte, Eur. Polym. J. **6**, 51 (1970).

The accuracies with which average molecular weights have been determined are equal to 5% for M_w, 10% for M_n and M_z, and 25% for M_{z+1}.[305]

2.5.1.3. Equilibrium Sedimentation in a Density Gradient. Meselson et al.[325] have developed a sedimentation–diffusion equilibrium technique for a polymer in a solvent with a density gradient. The solvent is a mixture of a light and a heavy liquid to produce a density gradient during rotation. At equilibrium, a polymer dissolved in this solvent will collect at a point in the gradient where the density is equal to its own. Since the polymer will diffuse, it will not be packed in a sharp layer, but will assemble in a band whose width depends on the molecular weight of the polymer, the rotor speed, and the density gradient.[305] The band is Gaussian in shape[321] with respect to solute concentration, the half-width being inversely proportional to the solute molecular weight. The method is sensitive to small differences in effective density among the solute species. Density-gradient ultracentrifugation was used by Morawetz[326] to separate atactic and isotactic polystyrene and to obtain composition distributions in copolymers.

Archibald[327] showed that measurement of c and dc/dr at cell boundaries allows molecular weight to be determined at any stage in the equilibrium process. If measurements are made about 10–60 minutes after the start of the experiment, that is, before the molecular species have time to redistribute in the cell, the weight-average molecular weight and second virial coefficient can be evaluated. This method is applicable for random-coil polymers even though sedimentation equilibrium with short cells is often preferred.

2.5.1.4. Sedimentation Transport. In the sedimentation transport or velocity experiment, the centrifuge is operated at high speed so that the solute is moved to the bottom of the cell. Billmeyer[304] and Scholte[305] present details of the theory for this method. The number of approximations and the complex data treatment required in this procedure limit its use to qualitative observations of molecular weight distribution. Results are valuable because qualitative information about the nature of the distribution of species present is directly obtained.

2.5.1.5. Preparative Separations. The ultracentrifuge is useful in the preparation of samples. Preparative ultracentrifuges cannot only fractionate polymer samples but also efficiently remove easily sedimented impurities.

[325] M. Meselson, F. W. Stahl, and J. Vinograd, *Proc. Natl. Acad. Sci. U.S.A.* **43**, 581 (1957).

[326] H. Morawetz, "Macromolecules in Solution." Wiley (Interscience), New York, 1965.

[327] W. J. Archibald, *J. Phys. Colloid Chem.* **51**, 1204 (1947).

2.5.2. End-Group Analysis

2.5.2.1. Principles and Applications.

For many years, end-group analysis has been an accepted method for measuring number-average molecular weight, M_n. It is beyond the scope of this chapter to review with any detail the literature covering nearly 50 years of development and usage of the technique. Nor is it possible to present the experimental methods used to analyze the various types of synthetic polymers.

However, two comprehensive and somewhat detailed reviews of the subject of end-group analysis have been written by Price[328] in 1959 and Garmon[329] in 1975. This section is primarily a general summary of the salient principles and applications from these two earlier reviews, both of which have extensive reference background.

Historically, end-group analysis has been most successful with condensation polymers, which, because of their method of polymerization, have reactive end groups. Vinyl polymers do not show as close a correlation between end-group concentration and molecular weight. Since the concentration of end groups varies inversely with molecular weight, the method loses accuracy at high molecular weights, the limit of which depends on the sensitivity and accuracy of the method used. Instrumental methods based on nuclear magnetic resonance (NMR) and infrared spectroscopy (IR) are often less sensitive than chemical functional group methods. In addition, chemical methods are usually less sensitive than some ultraviolet and visible photometric methods (elemental analysis). The practical upper limit of molecular weight for chemical end-group analysis is close to 50,000. No limit can be set on radioisotope and elemental analysis since polymer solubility and quality of purification and separation are the limiting factors. The unique advantage of end-group analysis is that it can measure molecular weights of condensation polymers below 5000, which for most physical methods is difficult.

The total number of end groups of all types in a linear polymer is twice the number of polymer molecules. If each molecule has one end group of a particular kind, then the number of end groups of this kind is equal to the number of molecules. Such theory provides the basis for determining number-average molecular weight of linear polymers by the end-group analysis method. For accuracy of results, the nature and quantity of the end groups must be known.

End-group methods for determining molecular weight are most successful with condensation polymers because of their functional end

[328] G. F. Price, *in* "Techniques of Polymer Characterization" (P. W. Allen, ed.), p. 207. Butterworth, London, 1959.

[329] R. G. Garmon, *Tech. Methods Polym. Eval.* **4**, Part I, 31 (1975).

groups. The end groups are often acidic or basic in type, as with car-boxyl groups of polyesters or the amine groups of polyamides. Both groups can be estimated by titration. For example, if each molecule has one carboxyl group, it is sufficient to estimate the carboxylic content to determine the molecular weight. If each molecule possesses one amine and one carboxyl end group for each chain, molecular weight is obtained by analyzing for either group. If a polymer is derived from two difunc-tional monomers, such as dicarboxylic acid and a glycol, it will possess molecules with three possible combinations of end groups: both carboxyl, both hydroxyl, or one of each kind. Only if the original monomer mix-ture has equivalent numbers of each type of functional group will the end groups of each type be equal in number to each other and to the number of molecules present. Usually, one group will be dominant; therefore, the content of each type of end group must be determined and subsequently the total end-group content in order to permit a molecular weight calcula-tion. The epoxide ring of propylene oxide can open in either of two ways, to yield a primary or a secondary hydroxyl end group. Polypropyl-ene glycols and mixed polypropylene–polyethylene glycols contain both kinds of end groups. It is important to the industrial chemist to differen-tiate between primary and secondary hydroxyl groups in the production of urethane foam and fibers.

Therefore, the end-group method of molecular weight determination is used mainly for linear polymers prepared from difunctional monomers, for which there is a simple ratio between number of end groups and number of molecules. The use of polyfunctional monomer creates branched or cross-linked polymers with variable numbers of end groups per molecule. End-group estimation of molecular weight is not possible except in special cases.[330]

End-group analysis can also provide a measure of the degree of branching. Estimation of the total end-group content by analysis and of the number-average molecular weight by osmometry permits the calcula-tion of the average number of end groups per molecule, which is a mea-sure of the degree of branching.

End-group analysis is less applicable to addition polymers for the fol-lowing reasons: first, the molecular weights are too high for experimental accuracy; second, the nonfunctional end groups are difficult to analyze; third, the free radical polymerization process does not offer a clear-cut mechanism by which two polymer radicals interact. An exception is a tailor-made addition polymer with an end that can be analyzed.[331]

[330] J. R. Schaefgen and P. J. Flory, *J. Am. Chem. Soc.* **70**, 2709 (1948).
[331] C. C. Price and B. E. Tate, *J. Am. Chem. Soc.* **65**, 517 (1943).

End-group analysis requires that the chemical nature of the polymer and its end groups be known. Solvents must be chosen that dissolve the polymer but do not react with it. Colorimetry and infrared absorption are among the analytical techniques used to detect end groups. Low concentration of end groups may require microchemical analytical techniques. Impurities in the solvent or polymer cause errors in the determined end group.

Price[328] and Garmon[329] describe in detail the experimental procedures for end-group analysis for many polymers including aliphatic polyamides, aromatic polyamides, polyesters, polyurethanes, cellulose, and vinyl polymers.

2.5.2.2. Calculations. End-group concentrations are usually expressed in units of equivalents per 10^6 gm or in the identical units, microequivalents per gm (μeq/gm). Number-average molecular weight can be calculated:

$$M_n = 10^6 \, n/\mu eq/gm, \text{ end groups}, \qquad (2.5.3)$$

where n is 1 or 2 depending on whether the end groups measured are on one or both ends of the polymer molecule.

If the linear condensation polymers contain more than two types of end groups (where there is addition of an impurity to control molecular weight), all of the end groups must be measured:

$$M_n = 2 \times 10^6/\mu eq/gm, \text{ total end groups}. \qquad (2.5.4)$$

Shaefgan and Flory[330] describe such methods for nonlinear polymers. In this case, if the quantity of multifunctional monomers present is known, the number-average molecular weight can be calculated. Or, if M_n and total end-group concentrations are known, the quantity of branch points can be determined.

Infrared absorption data provide an assortment of long-chain methyl [$H_3C-(CH_2)_n$, with n greater than 2], ethyl, and unsaturated groups in a sample. Each molecule is assumed to show at least one long-chain methyl group, and possibly ethyl and/or unsaturated groups. Two ends of a linear polyethylene molecule must be either two methyl groups or one methyl and one vinyl ($H_2C{=}CH-$) group. For example, since Phillips-type polyethylene (Marlex) contains exactly one double bond per molecule, there are two ways of determining the number of molecules: (1) each double bond counts as one molecule; and (2) the sum of the long-chain methyl groups and the terminal vinyl groups is twice the number of molecules. Since infrared absorption provides these structural groups per unit of weight (per 1000 CH_2 groups, which represent a molecular weight of 14,000), the number-average can be calculated:

(1) $M_n = 14{,}000/(\text{double bonds}/1000\ CH_2s)$,   (2.5.5)

(2) $M_n = 14{,}000/(\tfrac{1}{2})$ (long chain methyl groups/100 CH_2s
    + terminal vinyl groups/1000 CH_2s).   (2.5.6)

In any case, even though both calculative procedures apply to the Marlex polyethylene, the second approach applies to any linear polyethylene.

3. SPECTROSCOPIC METHODS

3.1. Infrared and Raman Spectra of Polymers†

By R. G. Snyder

3.1.1. Introduction

Of the physical methods used to analyze molecular structure, infrared and Raman spectroscopy are among the oldest. In the field of polymers, infrared has long been used for identification and compositional analysis, and it still maintains a prominent position here in spite of the availability of newer techniques. The role of vibrational spectroscopy as a method of studying polymer structure is less well known and less well documented.

In recent years the area of vibrational spectroscopy concerned with the structure of polymers has undergone a change of emphasis away from the analysis of the normal vibrations of idealized periodic chain models toward the analysis of real polymer systems. That is not to say that in the earlier preoccupation with the interpretation of spectra based on periodic chains, applications were purposely neglected, or even not forthcoming, but rather that a certain critical degree of understanding of the vibrational dynamics of the regular chain was needed before it became possible to develop effective methods for investigating those deviations from structural ideality that influence the properties of a polymer. From the polymer scientist's point of view, recent developments in vibrational spectroscopy must appear more relevant. Studies on the structure of polymer crystals in the solid state, and chain conformation in both the solid and liquid states, are appearing at an increasing rate, and are tending more to complement studies by other techniques. It seems that after a long and slow start, the potential of this spectroscopy is beginning to be realized.

The relationship between vibrational spectra and the structure of polymers first became a subject of interest in the late 1940s. At that time vibrational spectroscopy was well developed: the theory of normal vibra-

---

† See also Vols. 13A and B (Spectroscopy) in this series, particularly Chapter 4.1 in Vol. 13B; as well as Vol. 3A (second ed.), Chapter 2.2.

73

tions had been cast in its present formalism,[1,2] and the infrared instrumentation available could provide spectra of good quality even by today's standards.[3] Development of the field was, however, impeded by the poor characterization of these complex systems.

Some of the early work on polymers involved determining the polarization properties of bands and their structural implications.[4,5] For the most part, however, interpretation of spectra tended to concern the identification of chemical groups through their characteristic frequencies in the same manner used for any large molecule.[6]

Impetus to the development of equations describing the vibrations of a regular polymer chain resulted from the determination of the structures of a few of the simpler polymers. Foremost among these was polyethylene. The structure found for this polymer could hardly have been surprising since it turned out to be, as might be expected, essentially the same as that of the *n*-paraffins worked out 10 years earlier.[7] The impact, rather, had to do with the fact that the polymer was highly ordered. This polymer chain, consisting entirely of identical triatomic entities (methylene groups), was found[8] to have a planar zigzag skeleton with a translational repeat unit containing two methylene units; the crystal unit cell was found to be traversed by two chains and to contain four methylene groups. It is among the simplest of known polymers and among the most important commercially. Consequently polyethylene and its oligomers, the *n*-paraffins, were recognized as ideal systems of chain molecules upon which to test relationships between theory and experiment. As will be seen here frequently, this polymer is also particularly well suited as an example to illustrate many aspects of the vibrations of a chain molecule.

After the structure of polyethylene became known, and its vibrational dynamics began to be understood,[9,10] interest was furthered by the availability of other simple regular polymers, a result of the discovery of techniques for stereospecific polymerization. A general method for the analysis of the vibrational dynamics of periodic helical polymer chains was

[1] E. B. Wilson, Jr., *J. Chem. Phys.* **7**, 1047 (1939); **9**, 76 (1941).

[2] E. B. Wilson, Jr., J. C. Decius, and P. C. Cross, "Molecular Vibrations." McGraw-Hill, New York, 1955.

[3] G. Herzberg, "Molecular Spectra and Molecular Structure," Vol. II. Van Nostrand-Reinhold, Princeton, New Jersey, 1945.

[4] A. Elliott, E. J. Ambrose, and R. B. Temple, *J. Chem. Phys.* **16**, 877 (1948).

[5] E. J. Ambrose and R. B. Temple, *Proc. R. Soc. London, Ser. A* **199**, 183 (1949).

[6] G. B. B. M. Sutherland, *Discuss. Faraday Soc.* **9**, 274 (1950).

[7] A. Müller, *Proc. R. Soc. London, Ser A* **120**, 437 (1928).

[8] C. W. Bunn, *Trans. Faraday Soc.* **35**, 482 (1939).

[9] J. G. Kirkwood, *J. Chem. Phys.* **7**, 506 (1939).

[10] T. Simanouti, *J. Chem. Phys.* **17**, 734 (1949).

worked out.[11] This method, and subsequent extensions of it, combined the theory of harmonic vibrations of small molecules with the theory of crystal vibrations. Shortly afterward there was a period of rapid advance in the use of digital computers for the normal coordinate analyses of small molecules,[12,13] and this had ramifications of great importance to the analysis of polymer spectra. To set up the vibrational equations of a polymer chain, numerical values of force constants were needed, and it now became possible to obtain them from the vibrational analyses of small molecules. With the aid of computers it also became relatively easy to solve these equations and to calculate dispersion curves and normal coordinates for polymer chains. Within the last decade highly satisfactory vibrational models have been established for periodic chain molecules, and the rectitude maintained in this pursuit is now apparent from the number of structural applications that are possible as a result of these developments.

One structural problem of special importance concerns the detection and analysis of chain disorder, normally conformational disorder in the liquid and solid states. In some ways vibrational spectroscopy is particularly well suited to this problem: its effectiveness is not dependent on the existence of structural periodicity (cf. diffraction methods); and vibrations occur during a time scale so fast ($< 10^{-11}$ seconds) as to be virtually unaffected by transitions between conformational states, or by the motion of chain segments (cf. NMR, Part 4, this volume).

It has long been recognized that even in the most highly crystalline polymers, there are significant, often gross, deviations from perfect order. Features that manifest this disorder appear in the spectra of all polymers, and recently it has become much easier to distinguish these from features associated with the periodic chain. To the methods used for the investigation of the structure of crystal lamellae, long the nearly exclusive domain of microscopy and of X-ray and electron diffraction methods, can now be added low-frequency Raman spectroscopy. This technique provides structural data different from that attainable by other methods. The Raman bands studied appear in the spectrum as a result of a special kind of deviation from perfect order: the straight-chain segments (which constitute the crystal and tend to be conformationally ordered) are finite rather than infinite in length. In addition, other types of disorder characteristic of solids can be investigated using spectroscopic methods. These include disorder associated with conformation, chain packing, and chain orientation.

[11] P. W. Higgs, *Proc. R. Soc. London, Ser. A* **220**, 472 (1953).
[12] J. Overend and J. R. Scherer, *J. Chem. Phys.* **32**, 1289 (1960).
[13] J. H. Schachtschneider and R. G. Snyder, *Spectrochim. Acta* **19**, 117 (1963).

There is usually a higher degree of conformational disorder in the liquid state than in the solid. For analyzing the spectra of disordered systems there is, unfortunately, no method that is the equivalent to statistical mechanics. Nevertheless, even though the spectra of liquids are impossible to understand to any degree of completeness approaching what is possible for solids, a limited interpretation often leads to unique insight into conformational structure.

The potential of vibrational spectroscopy for the quantitative measurement of conformation is important not only for structural investigations but also for testing and establishing models upon which to base calculations involving statistical mechanics of polymer chains. In some instances incentives such as these have been so strong as to inspire conformational interpretation of spectra that is characterized more by exigency than by rigor.

Different approaches are needed for the interpretation of the spectra of polymer chains depending on whether order or disorder is dominant. This difference has been used as a basis for organizing the sections following.

When principles and general methods are subordinate to specific techniques and to specific systems, a different kind of organization is appropriate. There are reviews that emphasize the infrared spectra of synthetic[14] and biological polymers,[15] and the Raman spectra of synthetic[16] and biological polymers.[17] Other reviews and books concerning various aspects of these topics are available.[18-24]

As mentioned earlier, there exists a vast area of vibrational spectroscopy, mostly infrared, which deals with techniques for identification,

[14] S. Krimm, *Fortschr. Hochpolym.-Forsch.* **2**, 51 (1960).

[15] H. Susi, *Biol. Macromol.* **2**, 676 (1969).

[16] J. L. Koenig, *Appl. Spectrosc. Rev.* **4**, 233 (1971).

[17] J. L. Koenig, *J. Polym. Sci., Part D* **6**, 59 (1972).

[18] F. J. Boerio and J. L. Koenig, *J. Macromol. Sci., Rev. Macromol. Chem.* **7**, 209 (1972).

[19] W. L. Peticolas, *Adv. Polym. Sci.* **9**, 285 (1972).

[20] I. W. Shepherd, *Rep. Prog. Phys.* **38**, 565 (1975).

[21] G. Zerbi, *Appl. Spectrosc. Rev.* **2**, 193 (1969).

[22] R. Zbinden, "Infrared Spectroscopy of High Polymers." Academic Press, New York, 1964.

[23] A. Elliott, "Infra-red Spectra and Structure of Organic Long-Chain Polymers." St. Martin's Press, New York, 1969.

[24] V. Fawcett and D. A. Long, *Mol. Spectrosc.* **1**, 352 (1973).

[25] J. C. Henniker, "Infrared Spectrometry of Industrial Polymers." Academic Press, New York, 1967.

[26] M. Tryon and E. Horowitz, *in* "Analytical Chemistry of Polymers" (G. M. Kline, ed.), Part 2, Chapter 8. Wiley (Interscience), New York, 1962.

[27] D. O. Hummel, "Infrared Spectra of Polymers in the Medium and Long Wavelength Region." Wiley (Interscience), New York, 1966.

characterization, and determination of composition.[25-28] These methods are discussed in the context of general analytical infrared spectroscopy in standard references[29,30] and are generally applicable to polymers. The infrared and Raman spectra of polymers show bands whose positions and intensities are associated with certain chemical groups, and these may be identified by reference to existing compilations of characteristic frequencies.[31,32] In a similar vein the composition of a copolymer may be determined once the absorptivities of those bands associated with its polymer components are known. The list of applications of this kind is long and varied, and is in large part responsible for the continued widespread interest in this field. The focus of this presentation, however, will be on those properties of polymers that, as reflected by their vibrational spectra, distinguish them from smaller molecules.

3.1.2. Experimental

3.1.2.1. Introduction. Although the two spectroscopies, infrared and Raman, are related in a fundamental way in that they both measure frequency differences between energy levels, the kinds of events that induce a molecule to undergo a transition are completely different in the two cases, as are the experimental techniques for measuring the spectra.

In the infrared the event is the absorption of a photon by a molecule. A spectrum consists of a plot of the frequency ν of the photon, against some measure of the probability of its being absorbed.† The range of frequencies is approximately 4000 to 10 cm^{-1}, although the frequency region below 300 cm^{-1} presents special experimental difficulties and is not routinely measured. The intensive ordinate of the spectrum is expressed as some function of the ratio of the intensity I of the radiation transmitted to the intensity I_0 of the incident radiation. For a sample having a uniform thickness b and a concentration c, the ratio I_0/I can be related to the absorptivity a_ν at frequency v through the Beer–Lambert equation.

$$\log_{10} I_0(\nu)/I(\nu) = a_\nu bc, \qquad (3.1.1)$$

which is much used in quantitative infrared analysis.

In the Raman the event is absorption–emission: a photon of frequency

---

† Frequency will be indicated by ν and expressed in terms of wavenumbers, cm^{-1}.

[28] I. Kössler, *Encycl. Polym. Sci. Technol.* **7**, 620 (1967).

[29] A. L. Smith, *Treatise Anal. Chem.* **6**, Part 1, 3535 (1965).

[30] W. J. Potts, Jr., "Chemical Infrared Spectroscopy." Wiley, New York, 1963.

[31] L. J. Bellamy, "The Infrared Spectra of Complex Molecules." Wiley, New York, 1975.

[32] N. B. Colthup, L. H. Daly, and S. E. Wiberley, "Introduction to Infrared and Raman Spectroscopy," 2nd ed. Academic Press, New York, 1975.

ν_0 from the excitation source (usually a visible or UV laser) interacts with, and is inelastically scattered by, a molecule. The scattered photon has a frequency $\nu_0 \pm \nu$, where $hc\nu$ is the vibrational energy absorbed or given up by the molecule. A Raman spectrum normally consists of a plot of the intensity of emission against the frequency displacement ν, measured on the low-frequency side of the excitation line. The frequency region covered is essentially the same as that of an infrared spectrum. For solids, but not liquids, it is possible to make measurements very close to the exciting line, perhaps to within 2 cm^{-1}.

It should be noted that in addition to the $\nu_0 - \nu$ spectrum on the low-frequency side of ν_0 (the Stokes spectrum) there is a complementary spectrum $\nu_0 + \nu$ on the high-frequency side (the anti-Stokes spectrum). The latter is seldom measured because it is normally much less intense than the Stokes spectrum.

Intensities are measured on a relative scale because it is extremely difficult to measure the ratio of incident radiation to scattered Raman radiation, and also because the ratio itself is dependent on ν_0.

In general, Raman spectra are more complex than infrared spectra. The relative intensities of the bands in the spectrum of a given sample are dependent on the frequency of excitation. This dependence usually becomes more significant as ν_0 approaches UV frequencies. In any case the intensity and polarization of the Raman radiation scattered in a give direction are related to the direction and polarization of the exciting radiation, and this relationship varies with ν. A Raman spectrum should therefore always be accompanied by a designation of the excitation frequency and a description of the geometry used in the measurement.

These properties, along with the fact that Raman spectra normally must be measured at a lower signal to noise ratio than infrared spectra, tend to discourage the use of Raman for routine quantitative analysis. There are exceptional cases where Raman is preferred, as in the analysis of aqueous solutions, since such samples are virtually opaque to infrared radiation. In general, because of its relative simplicity and greater sensitivity, infrared overshadows Raman in the analytical field.

For structural analysis the situation is quite different. The very factors that complicate the use of Raman for quantitative analysis can be used to advantage to relate structure and spectra: the dependence of the spectrum on polarization helps in assigning vibrational modes[2,3,33]; the dependence of the spectrum on the frequency of excitation can be exploited to relate molecular and electronic structure.[34] Thus, although the Raman experi-

---

[33] J. A. Koningstein, "Introduction to the Theory of the Raman Effect." Reidel Publ., Dordrecht, Netherlands, 1972.

[34] J. Behringer, *Mol. Spectrosc.* **3**, 163 (1975).

ment is somewhat more difficult than the infrared, in some ways it yields more information. More important, however, is that the two spectroscopies give different kinds of information, and serve in a complementary way in structural analysis.

Infrared spectrometers are usually obtained from commercial sources as integrated instruments. This is less often the case for Raman spectrometers, where it is advantageous to keep the source, monochromator, detector, and data processing devices as separate modules for the purpose of flexibility, and to facilitate the replacement of individual units in taking advantage of new technical developments. There are many companies that manufacture infrared and Raman spectrometers and their components; needless to say, there is much literature available about their products.

The specifications of a spectrometer needed for the study of polymers differ little from those needed for the study of other forms of matter in the condensed state. Although the bands in polymer spectra tend to be broader than those of small molecules, the advantages of good resolution (< 1 cm^{-1}) cannot be disregarded. It is occasionally necessary to resolve closely spaced bands, and more commonly to measure the shape of a band. Accurate band shape measurement requires good resolution even though the band might not be considered narrow.

There is a special need for accuracy in the measurement of intensities. Polymer spectra are complex, and the features of interest are often weak. It is useful to be able to compare two spectra in a quantitative manner. For such purposes the Fourier transform infrared spectrometer, with its data acquisition and processing capabilities, is widely used.[35] Similar capabilities can be incorporated into a Raman spectrometer.[36]

3.1.2.2. Infrared. 3.1.2.2.1. SPECTROMETERS. There are two types now in use, dispersive and interferometric, named according to the method employed to analyze the radiation from a continuous infrared source after it has passed through the sample. A third type of spectrometer, which consists of an infrared tunable diode laser source and a detector, has recently become commercially available.[37] Although in its present form this spectrometer appears useful only for certain specialized applications [where extremely high resolution is needed over a relatively narrow frequency interval (< 300 cm^{-1})], future development may open the possibility of polymer applications.

The design, operation, and use of dispersion spectrometers have been

[35] J. L. Koenig, *Appl. Spectrosc.* **29**, 293 (1975).
[36] S. Kint, R. H. Elsken, and J. R. Scherer, *Appl. Spectrosc.* **30**, 281 (1976).
[37] E. D. Hinkley, K. W. Nill, and F. A. Blum, *Top. Appl. Phys.* **2**, 125 (1976).

well described in the literature[30,38] and this accounts for the brevity of the description given here and summarized in Fig. 1.

Infrared radiation from a continuous source, usually a heated wire, or a silicon carbide or rare earth oxide rod, is collected and focused on the entrance slit of a monochromator. The sample is normally positioned in front of this slit. After entering the monochromator the light is collimated, dispersed by a grating (usually in combination with filters or a prism), and refocused to pass through an exit slit and onto a detector, usually a thermocouple, Golay cell, or bolometer. The signal from the detector is amplified and plotted against frequency. For the purpose of electronic amplification, the light beam is interrupted with a mechanical chopper to produce an AC signal. Nearly all spectrometers are designed to divide the light from the source into two separate but parallel beams. One beam passes through the sample while the other is used as a reference. After leaving the sample compartment, the separated beams are, in effect, combined to follow a common path through the spectrometer. The ratio of the intensities of the beams, I/I_0, is recorded, usually in terms of percentage transmission.

Although a Fourier transform spectrometer was first used some 30 years ago to obtain a spectrum in the visible region, its application to the infrared had to await recent developments in technology and computational methods to become competitive with the dispersive spectrometer.[39] The transform spectrometer surpasses its dispersive counterpart in that spectra can be measured more quickly or more accurately, or if desired, smaller samples can be used.

The optical component of a Fourier transform spectrometer is a Michelson interferometer. Consequently what is actually measured is an interferogram, which must subsequently be converted into a transmission spectrum.

The interferometer is shown schematically in Fig. 1. Collimated light from an infrared source impinges on a semireflecting surface, the beam splitter, which transmits half the light to a movable plane mirror and reflects the other half to a fixed plane mirror. The beams reflected from these two mirrors are recombined at the beam splitter. The intensity of the recombined beams is measured as a function of the displacement of the movable mirror, and the resulting function is the interferogram. If the distances of the mirrors to the beam splitter are equal, all wavelengths combine constructively to produce a large signal, which is used as a reference point for the movable mirror. If there is a sample between the source and the beam splitter, the reference signal appears as before, but in

[38] R. P. Bauman, "Absorption Spectroscopy." Wiley, New York, 1962.

GRATING MONOCHROMATOR

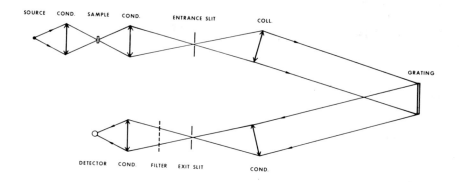

INTERFEROMETER

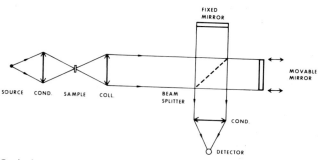

FIG. 1. Optical components of an infrared dispersion and interferometric spectrometer (cond., condenser; coll., collimator).

addition there will be other maxima in the interferogram resulting from selective absorption of energy by the sample.

The principal advantages[39] of the interferometer over the dispersive spectometer are the absence of slits and a greater throughput efficiency. Combined, these provide a very large gain over the dispersive spectrometer in terms of signal to noise. The S/N advantage is given approximately by $K(\Delta\nu/\Delta\nu_{1/2})^{1/2}$, where $\Delta\nu$ is the width of the region to be measured at resolution $\Delta\nu_{1/2}$ and where the value of K is estimated to be between 10 and 50. Direct comparisons between spectrometers are made difficult by differences in detectors, etc. However, using $K = 10$ a calculated advantage of 548 for a scan over a 3000 cm^{-1} frequency interval at 1 cm^{-1} resolution is impressive in any case.

[39] P. R. Griffiths, "Chemical Infrared Fourier Transform Spectroscopy." Wiley, New York, 1975.

Conversion of the resulting interferogram to an absorption spectrum entails the use of a digital computer. Improvements in computer design and in the mathematical technique for performing the required Fourier transformation have reduced the time for this calculation in an ordinary situation to a matter of a few seconds or minutes.

Because it is necessary to collect, store, and transform the interferogram, most systems employ a dedicated minicomputer. The developers of these systems have capitalized on the attendant capability to add interferograms, to process the data in any way desired (such as adding or subtracting weighted spectra), to display spectra on a CRT, and to store data for future retrieval. These latter capabilities are not unique to the Fourier transform spectrometer, but are often associated with it because of the ancillary presence of a computer. Systems for collecting and processing data have also been developed for dispersion spectrometers so as to enhance their performance.

The advantages and disadvantages of dispersive and interferometric spectrometers are subject to their intended use. In the mid infrared, the dispersion spectrometer may be adequate for most purposes. A research quality spectrometer of this type is capable of measuring a spectrum over a narrow-frequency interval at high resolution in a time comparable to that of a transform spectrometer, since the latter ordinarily measures the entire spectrum, needed or not needed. Even with the inclusion of a minicomputer, the cost of a dispersion spectrometer is less.

The advantage of the Fourier transform spectrometer is its superior signal to noise ratio; this can be exploited, as previously indicated, in a number of ways. The high accuracy attainable in measuring a spectrum, combined with the data processing capabilities, offers advantages of particular significance to polymer spectroscopy since these systems are always complex and in a sense multicomponent. Standard analytical methods for multicomponent analysis can be used to isolate the spectra of the individual components, provided a sufficient number of independent spectra have been accurately measured. In a simple case the bands due to the solvent may be removed from the spectrum of a polymer in solution. In a more complex application the spectra of the crystalline and the amorphous fractions of a polymer may be separated.[40] Many such applications are reported.[35]

3.1.2.2.2. MEASUREMENT OF SPECTRA. It is probably true that regardless of the constitution or physical form of a polymeric material, there is always a way to measure its infrared spectrum. There is a great variety of techniques for preparing samples, and the choice depends on

[40] M. M. Coleman, P. C. Painter, D. L. Tabb, and J. L. Koenig, *J. Polym. Sci., Polym. Lett. Ed.* **12**, 577 (1974).

the nature of the sample and on the quality of the spectrum desired.

There are two different methods of measuring the spectrum of a specimen. The commonest and simplest is by direct transmission through a thin layer. The second involves internal reflection of the incident beam from the interior surface of a transparent prism of high refractive index, which is in optical contact with the surface of the sample. With this method a geometry is employed to direct the incident beam in such a way that, if the sample were not present, total internal reflection would occur. With the sample present, internal reflection still occurs, but it is no longer total, since there is attenuation as a result of some shallow penetration of the infrared light into the sample.[41] The spectrum obtained using the ATR (attenuated total reflection) method much resembles, but is not identical to, the transmission spectrum.[42] This technique is particularly suitable for surface coatings whose spectra may be measured *in situ,* and for substances such as rubbery polymers, which are difficult to prepare as thin films for transmission measurements.

There is a variety of methods for preparing polymer samples for transmission measurements.[23,25] For a dispersive spectrometer the lateral dimensions of the specimen usually need to be no larger than 1.5×0.5 cm, and for an Fourier transform spectrometer, considerably less. This sample area can be greatly reduced if a good spectrometer is available or if beam-condensing optics are used. Thicknesses are of course determined by the absorptivities in the spectral regions of interest, normally falling in the range of 1 to 1000 μm. Methods for the bulk solid state include solvent casting, compression molding, and sectioning with a microtome. Special methods are available for fibers, and if the sample is a fine powder, the usual mulling or KBr pressed-disk techniques may be used.[32] A sample can sometimes be dissolved and measured in solution, but this is usually not a preferred method because of the interference from solvent bands. For quantitative work it is necessary that samples have uniform thickness, and in the case of solids, preferred orientation of the molecules is to be avoided since otherwise the intensities of bands will be discriminantly affected.

Polarized infrared spectra provide data useful for assigning bands and determining structure, and such spectra may be measured provided the sample can be oriented. High orientation of semicrystalline polymers can often be achieved through mechanical processing. Common methods include drawing (a film or fiber), rolling, or extruding. In the case of drawing, approximate uniaxial orientation results, while rolling produces a more complex, sometimes biaxial form of alignment. Polar-

[41] J. Fahrenfort, *Spectrochim. Acta* **17,** 698 (1961).
[42] N. J. Harrick, "Internal Reflection Spectroscopy." Wiley, New York, 1967.

ization of the incident infrared radiation is accomplished in the mid infrared by using transparent sheets having a high index of refraction, such
as silver chloride, which are placed in the beam and tilted to it at Brewster's
angle. Suitable polarizers are available for all wavelengths, and several
alternative polarizing arrangements have been described.[23]

A simple quantitative measure of the polarization of a band, applicable
to the uniaxial case, is the dichroic ratio R, defined here as

$$R = \int A_\sigma \, d\nu / \int A_\pi \, d\nu, \tag{3.1.2}$$

where A_σ and A_π are the absorptivities associated with incident radiation
polarized, respectively, perpendicular and parallel to the direction of the
sample axis and where integration is over the band. Normally, R is not
determined from integrated intensities, but rather approximated by using
values of A_σ and A_π measured at the frequency where maximum absorption occurs. For making vibrational assignments, approximate values of
R are usually sufficient.

Quantitative values of dichroic ratios are useful for estimating the orientation of chemical groups relative to the chain direction, and for determining the degree of alignment of polymer chains of known structure. In
these cases accurate values of R are needed, and special care must be exercised in determining them.[14,15] Incomplete polarization of the incident
radiation, inaccurate alignment, and complications resulting from biaxial
samples may result in serious errors. In addition, account must be taken
of the fact that an infrared spectrometer does not respond with equal sensitivity to parallel and perpendicular polarized radiation, and that this difference in sensitivity varies with frequency.

3.1.2.3. Raman. The usefulness of Raman spectroscopy for the study
of polymer structure was known and had been demonstrated[43] before the
laser era, although it is only recently its use has become so common that it
can be considered on an equal footing with infrared as a research tool.

3.1.2.3.1. SPECTROMETERS. The components and arrangement of components for measuring Raman spectra scattered at 90° are shown in Fig. 2.
The source is usually a laser since it provides an intense, stable, and parallel beam of highly monochromatic excitation. New varieties of lasers
continue to be developed, and at present there is available a sizable number of intense lines in the visible and ultraviolet suitable for use in Raman
spectroscopy. Tunable lasers also exist, and while they are well suited for
use in studying resonance Raman phenomena, the energy available is less
and band widths are greater than for fixed frequency gas lasers. Most
polymer spectroscopy is done with the 488.0 or the 514.5 nm lines of the

[43] J. R. Nielsen, *J. Polym. Sci., Part C* **7**, 19 (1964).

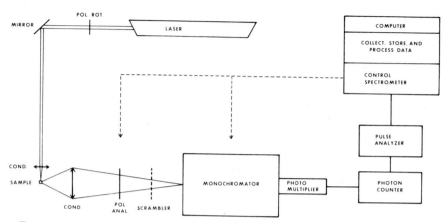

FIG. 2. Experimental arrangement for measuring Raman spectra (cond., condenser; coll., collimator; pol. rot. and anal., polarization rotator and analyzer).

argon ion laser, although, in cases where the sample fluoresces, the red line from a krypton or helium–neon laser may be preferred.

Before being focused on the sample, the laser light may pass through a polarization rotator, whereby the plane of polarization of the excitation can be adjusted. Scattered light from the sample is collected and passes through a polarization analyzer and a polarization scrambler. The latter is necessary to circumvent the problem of (frequency dependent) polarization discrimination by the monochromator.

The scattered radiation entering the monochromator consists of an extremely intense unshifted component, referred to as the Rayleigh and Tyndall component, and much weaker, frequency-shifted Raman components. The ratio of the intensities of Raman light to Rayleigh–Tyndall light entering the monochromator is approximately in the range 10^{-5} (for very clean, low molecular weight liquids) to 10^{-13} (for particulate solids). Thus an important problem in designing monochromators for Raman spectroscopy is to prevent the intense unshifted excitation from reaching the detector by scattering within the monochromater since it can thereby contribute to the background at all frequencies. This problem is of course most severe in the region nearest the exciting frequency (< 30 cm^{-1}), and it is just this region that has come to have special importance in the study of crystalline polymers. To reduce stray light, monochromators are used in tandem: double monochromators are standard but recently triple monochrometers have been used. Holographically replicated gratings also serve to alleviate this problem, since scattering and lines from ghosts are less intense for this type than for ruled gratings.

Although the following technique for reducing stray light does not pertain to monochromator design, it is appropriate to mention it here. In this method[44] a heated cell containing iodine vapor is placed between the sample and the monochromator, and the intense 514.5 nm line of the argon ion laser is used for excitation. It happens there is an iodine absorption line very near (514.4 nm) the laser line. An etalon within the laser cavity is tuned to select the particular longitudinal laser mode that superimposes on the iodine absorption band. The resultant absorption reduces the intensity of the Rayleigh–Tyndall light reaching the monochromator by a factor of at least 10^4 and thus greatly improves the performance of a double monochromator at low frequencies. The disadvantages are the necessity of being fixed to the 514.5 nm line, and the possible interference in the Raman spectrum by other absorption lines of the iodine vapor. The latter problem can be greatly alleviated through a ratioing technique.[45]

The scattered light is usually detected with a photomultiplier whose gain is sufficiently high to respond to single photons. The signal from photon-counting systems is plotted directly or stored.

The use of data acquisition and processing methods is less common in Raman spectroscopy than in infrared, but the advantages are at least as great. Since Raman signals are weak, accuracy and sensitivity can be improved by repetitive or slow scanning in conjunction with signal averaging. Also, since spectrometer throughput efficiency and detector response are nonlinear functions of frequency, it is desirable to correct the observed spectrum for these effects. Efficient and flexible systems have been described,[36] and are becoming commercially available.

Fluorescence from impurities in the sample is a frequent problem in Raman spectroscopy because the cross section for such emission can be several orders of magnitude larger than for Raman emission. The problem is especially troublesome in measuring the spectra of macromolecules since it is often very difficult in these cases to reduce the concentration of the offending species to an acceptable level. Since the noise in the amplified signal is proportional to the intensity of the signal, background radiation increases noise levels and weaker bands are obscured. Consequently the problem is not solved by merely readjusting the base line.

The intensity of fluorescence is often found to decrease with time as the sample is irradiated with the laser excitation. Presumably this occurs because the fluorescing species are in some way chemically modified by the

[44] G. E. Devlin, J. L. Davis, L. Chase, and S. Geschwind, *Appl. Phys. Lett.* **19**, 138 (1971).

[45] J. R. Scherer, *Appl. Opt.* **17**, 1621 (1978).

radiation. Fluorescence may also be reduced by using a laser line that is less effective in exciting it. Usually this means going to lower frequencies (to red lines), but since Raman intensities have an approximate fourth power dependence [Eq. (3.1.16)] on the excitation frequency, the reduction in fluorescence is accompanied by a reduction in the Raman signal as well.

Recently there have been increasingly successful attempts to discriminate against fluorescent emission in favor of Raman emission by capitalizing on the great difference in the time for these events to occur after initial excitation. Raman emission occurs, in effect, simultaneously with irradiation, while fluorescent emission occurs later, within nanoseconds in the case of solutions. With a pulsed argon ion laser and a gated amplifier, Raman to fluorescence enhancements of about two orders of magnitude have been reported for solutions.[46] This technique should be of considerable value in its application to measuring the spectra of macromolecules.

3.1.2.3.2. MEASUREMENT OF SPECTRA. Of the various sampling arrangements possible, the most common is right-angle collection shown in Figs. 2 and 3. As indicated in Fig. 3 the laser excitation is linearly polarized, while the scattered light has in general two components: one is perpendicular and one is parallel to the polarization direction of the excitation; their intensities are designated as $I(\perp)$ and $I(\|)$, respectively. The ratio of these intensities for a band is the depolarization ratio ρ, defined as

$$\rho = \int I(\perp) \, d\nu / \int I(\|) \, d\nu. \qquad (3.1.3)$$

Here the integration is over the band, although normally ρ is approximated by using the ratio of peak heights. The expression for ρ is analogous to that for the dichroic ratio R defined in Eq. (3.1.2). Unlike R in the infrared case, ρ can have different values for different Raman bands even when the sample is not oriented. Its value depends solely on the symmetry of the vibration in some cases, and on the nature of the vibration in others. This is one aspect of the statement made earlier that the Raman experiment can provide more information than the infrared.

Other orthogonal scattering geometries are possible, and it has become a convention to designate these symbolically as $G(gg')G'$ where the letters represent space fixed axes: G and g refer respectively to the direction and polarization of the incident beam, while G' and g' refer to the

[46] J. M. Harris, R. W. Chrisman, F. E. Lytle, and R. S. Tobias, *Anal. Chem.* **48**, 1937 (1976).

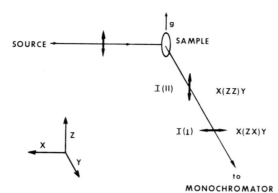

FIG. 3. Raman polarization measurements: arrangement and convention.

same quantities for the scattered radiation. In terms of this convention, the depolarization ratio defined in Eq. (3.1.3) may be expressed as

$$\rho = I[X(ZX)Y]/I[X(ZZ)Y], \tag{3.1.4}$$

where the axes are designated in Fig. 3.

Sampling is simplest for liquids. These are usually contained in a cylindrical glass vessel, such as a capillary tube, and oriented in a way to maximize intensity and preserve the polarization of the incident and scattered light. Microsampling techniques have been developed enabling high quality Raman spectra to be measured for sample volumes less than 0.1 μl.[47]

For solids the situation is less well defined. If the solid is transparent, as may be the case for an extruded or drawn polymer, there will probably be some degree of molecular alignment, and the measured spectrum will depend on the polarization direction of the incident and scattered light relative to the sample orientation. As in the case of the infrared, the polarized spectrum of an oriented sample is helpful in determining vibrational assignments and structure. On the other hand the interpretation of polarized spectra is less intuitively obvious in the Raman. This will be discussed in Section 3.1.3.4.1.

If a polymer is opaque or translucent, a sampling technique that provides a suitable Raman spectrum is usually found through a trial and error process. Depending on the physical characteristics of the sample, techniques have been described in the literature that involve shaped samples, pressed powders, inclined surfaces, back-scattering geometries, etc., so that generalizations seem inappropriate. The polarization of the excita-

[47] G. F. Bailey, S. Kint, and J. R. Scherer, *Anal. Chem.* **39,** 1040 (1967).

tion and of the Raman radiation is partially or entirely lost as a result of scattering by inhomogeneous samples, and this of course makes polarization measurements inaccurate or impossible. Qualitative values of depolarization ratios can sometimes be determined for powders by first immersing them in a liquid of a matching index of refraction.[48]

3.1.3. Vibrational Theory for Molecules

3.1.3.1. Introduction. It is appropriate to begin a discussion of normal vibrations of a polymer from the standpoint of the theory developed for small molecules. The vibrational dynamics and intensities of small molecules and of polymer chains are, of course, described formally by the same equations. Thus, in so far as it is possible to treat a polymer chain as a large molecule, the vibrational theory outlined in this section is directly applicable. There is in practice, however, a basic and obvious problem in the case of a polymer molecule: its sheer size precludes a calculation of all the vibrational modes. In view of this, some comment on the relevance of this section is needed.

In the solid state, polymer chains often assume structures that are periodic. If the approximation is made that such a chain is infinitely long, and perfect in its periodicity, the methods outlined in Section 3.1.4.3.2.1 can be used to simplify the vibrational equations so that they are no larger and only slightly more complex than those for a finite molecule. If the chain is disordered so that these methods are inapplicable, a vibrational analysis may be based on calculations performed on finite segments of the chain, so that again the situation is similar to that of a finite molecule. Finally, all such calculations assume the availability of values of force constants, and these are derived from the analysis of the spectra of small molecules. Since many vibrational calculations have been made on polymers, it is useful to know from where the force constants came, and the approximations assumed in evaluating them.

3.1.3.2. Calculations of Frequencies and Normal Coordinates. The vibrational equations of a molecule are derived within the framework of classical mechanics. Except for some low-frequency modes, the displacement a molecule undergoes as it vibrates represents only a small deviation from its equilibrium structure. It is because of small displacements that the equations of motion can be written in a simple form. Thus, when the potential and kinetic energies of a molecule are expanded in terms of internal displacement coordinates, terms higher than quadratic are ignored; when the dipole moment and polarization are similarly expanded, terms higher than linear are ignored. This approximation is supported by the fact that, except in rare cases, the main features of an ob-

---

[48] B. J. Bulkin, *J. Opt. Soc. Am.* **59**, 1387 (1969).

served infrared or Raman spectrum can be explained, often quantitatively, without invoking the effects of the higher-order terms.

The number of normal or fundamental vibrations for a nonlinear molecule having N atoms is $3N - 6$, and for a linear molecule, $3N - 5$. The kth normal mode is characterized by a frequency ν_k and by a normal coordinate Q_k. Since the energy of the molecule can be expressed in a quadratic form, this frequency is proportional to the spacing between energy levels of the harmonic oscillator, which is given by

$$E = hc\nu_k(n + \tfrac{1}{2}), \qquad n = 0, 1, 2, 3, \ldots \qquad (3.1.5)$$

Fundamental modes involve transitions between adjacent energy levels and therefore involve changes in energy corresponding to a frequency ν_k. The normal coordinate Q_k describes the vibration and may be expressed in terms of intramolecular coordinates. The latter coordinates measure the displacements, or relative displacements, that atoms undergo during a vibration.

The best sources of vibrational frequencies of a molecule are its infrared spectrum and its Raman spectrum. An analysis begins with a band assignment. This involves associating observed frequencies with a more or less qualitative description of the kinds of motion thought to be involved. Another facet of band assignment involves relating the symmetry of the vibration to the symmetry of the molecule. A normal coordinate calculation takes the analysis a step further and leads to a quantitative description of the form of the vibrational mode.

3.1.3.2.1. INTERNAL COORDINATES. A vibrating molecule undergoes a periodic distortion away from its equilibrium structure. The distorted structure may be described in terms of one of several kinds of coordinates, and it can be shown[2] that internal coordinates can be used for this purpose. These coordinates are referenced within the molecule, in contrast to external coordinates, which are referenced to axes fixed in space.

Various kinds of internal coordinates are used. Valence coordinates measure the changes in the lengths of bonds and the changes in angles or dihedral angles defined by bonds. These coordinates, which will be designated by r_i, are especially suitable for expressing the potential energy of the molecule since the force constants associated with them can often be related to the chemical properties of bonds. For example, a bond-stretching force constant can be related to the strength of a bond. Another useful and similar type of internal coordinate is a "group" coordinate. It consists of a linear combination of valence coordinates, appropriately combined so as to describe the motion of a group of atoms or some localized motion occuring within the group. Group coordinates are used because a vibration can often be described succinctly and vividly in

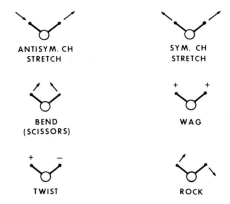

FIG. 4. Group coordinates for methylene vibrations.

terms such as "methyl rocking," or "methylene scissoring," such as shown in Fig. 4 for the methylene group. Another type often used is the Cartesian displacement coordinate. This coordinate, which will be designated by x_i, measures the displacement of an atom in terms of Cartesian axes fixed at its equilibrium position.

The vibrational problem can be set up and solved in terms of any of these (or other suitable) types of coordinates, as seems appropriate. Although valence and group coordinates are physically appealing, the use of Cartesian displacement coordinates simplifies the expression for the kinetic energy of the molecule, and thereby offers some computational advantages. In addition, since Cartesian displacement coordinates measure atom displacements they are useful in depicting the form of the normal vibrations.

The transformation between these two kinds of internal coordinates is of course linear, and may be written

$$\mathbf{r} = \mathbf{Bx}, \tag{3.1.6}$$

where $\mathbf{r}$ and $\mathbf{x}$ are column matrices of the internal coordinates and $\mathbf{B}$ is a matrix readily found from the structural geometry of the molecule. In the following, valence coordinates will be used as a basis for expressing the potential and kinetic energies.

3.1.3.2.2. VIBRATIONAL EQUATIONS. Equations will be derived leading to vibrational frequencies ν_k and normal coordinates Q_k expressed in terms of a set of internal coordinates $\mathbf{r}$. These equations follow from those properties which define[49] normal coordinates, namely, that when normal coordinates are used as a basis for expressing potential and kinetic

[49] E. T. Whittaker, "Analytical Dynamics," Chapter 7. Cambridge Univ. Press, London and New York, 1937.

energies, there are no cross terms:

$$2V = \mathbf{Q'\Lambda Q},\qquad(3.1.7)$$

$$2T = \mathbf{\dot{Q}'E\dot{Q}},\qquad(3.1.8)$$

where $\mathbf{Q}$ is the column matrix of the Q_k; $\mathbf{\Lambda}$ is a diagonal matrix whose elements are the frequency parameters λ_k, where $\lambda_k = 4\pi^2 c^2 \nu_k{}^2$; $\mathbf{\dot{Q}}$ is the time derivative of $\mathbf{Q}$; $\mathbf{E}$ is the unit matrix; and the prime indicates a transposed matrix.

The potential energy written in terms of internal coordinates is

$$2V = \mathbf{r'Fr},\qquad(3.1.9)$$

where $\mathbf{F}$ is the matrix of valence force constants, whose elements are related to V by

$$F_{ij} = \partial^2 V/\partial r_i\, \partial r_j.$$

The kinetic energy similarly expressed is

$$2T = \mathbf{\dot{r}'G^{-1}\dot{r}},\qquad(3.1.10)$$

where $\mathbf{G}^{-1}$ is the inverse of a matrix $\mathbf{G}$ whose elements may be written in closed form in terms of the atomic masses and structural parameters of the molecule. The elements of $\mathbf{G}$ are in consequence structurally dependent, and this accounts in large part for the dependence of the frequency of a normal vibration on the conformation of a molecule.

The vibrational equations are derived through a transformation of coordinates

$$\mathbf{r = LQ},\qquad(3.1.11)$$

which, when applied to Eqs. (3.1.9) and (3.1.10), leads[2] to a set of eigenvalue equations

$$\mathbf{GFL = L\Lambda},\qquad(3.1.12)$$

from which $\mathbf{L}$ and $\mathbf{\Lambda}$ may be calculated. There are other ways of expressing this equation, some more suitable for computer computation.

In the event that both the force constants and the molecular structure are known, it is a straightforward matter to calculate the normal coordinates and the frequencies. A more common situation is that the vibrational frequencies and the molecular structure are known, but the force constants are unknown. It is not a simple matter to evaluate force constants from frequency and structural data.[50] We shall consider this problem in the next section.

[50] P. Gans, "Vibrating Molecules," Chapter 6. Chapman & Hall, London, 1971.

Force constants play an important role in polymer physics. If the vibrations of polymer chains are to be analyzed quantitatively for application to structural problems, it is necessary to have available accurate values of force constants derived from the spectra of small molecules. Beyond their use in vibrational analysis, force constants relate to other properties of polymers. For example, force constants may be used to calculate the elastic moduli of chains,[51] or even of polymer crystals[52] if intermolecular force constants are also available. Force constants are used indirectly in conformational analysis[53] and directly in mapping energy surfaces[54] associated with the various chain defects postulated to account for such phenomena as the relaxation loss peaks observed in dynamic–mechanical spectra of polymer solids.

3.1.3.3. Determination of Force Constants. The principal difficulty in determining the force constants of a molecule from its observed frequencies and known structure is that unless the molecule is diatomic, or else very small and very symmetric, the problem is badly underdetermined in that there are far too many force constants relative to the number of known frequencies. In the worst case, where the (nonlinear) molecule has no elements of symmetry, the ratio of the number of independent force constants to the number of frequencies is $(3N - 5)/2$, where N is the number of atoms. If the molecule has some symmetry, this ratio will be smaller, but the situation is still hopeless unless additional independent frequency data are included, and approximations are made to reduce the number of independent force constant parameters. The problem is usually handled as follows.

The number of independent frequency data can in effect be increased by using observed frequencies from one or more isotopically substituted (usually deuterated or partially deuterated) molecules and by using observed frequencies from molecules that are chemically very similar. These frequencies are used together to evaluate a set of force constant parameters in a calculation in which the fundamental assumption is made that many of these parameters have the same value in different molecules. If the independent force constant parameters are designated by ϕ_i, and ϕ represents a column matrix of them, their relation to the F-matrix of molecule m may be expressed[12] through a Z-matrix defined as

$$\mathbf{F}^{(m)} = \mathbf{Z}^{(m)}\boldsymbol{\phi}. \qquad (3.1.13)$$

[51] T. Shimanouchi, M. Asahina, and S. Enomoto, *J. Polym. Sci.* **59**, 93 (1962).

[52] Y. Shiro and T. Miyazawa, *Bull. Chem. Soc. Jpn.* **44**, 2371 (1971).

[53] P. J. Flory, "Statistical Mechanics of Chain Molecules." Wiley (Interscience), New York, 1969.

[54] R. H. Boyd and S. M. Breitling, *Macromolecules* **7**, 855 (1974).

It is desirable that the number of ϕ_i be much less than the total number of $F_{ij}^{(m)}$ summed over all molecules in the calculation.

If the molecules are carefully chosen, the assumption of equivalence can prove to be a good one, and as a result, the force constants are transferable to other systems. This property of transferability is of great importance in the analysis of polymer spectra.

The number of parameters can be further reduced if force constant equivalence is assumed on the basis of local symmetry or on chemical grounds. For example, it is usually assumed that interaction constants between coordinates distant from one another can be neglected. It should be emphasized that there is an unavoidable element of arbitrariness in this manner of reducing the number of independent parameters, and so it is unrealistic to expect their computed values will necessarily be quantitatively meaningful. Nevertheless, unless the assumptions made have a sound physical basis, the property of transferability will suffer.

In favorable cases the number of observed frequencies from a set of related molecules may exceed the number of force constant parameters by a factor approaching 10. The parameters are evaluated through an iterative adjustment procedure[50] that uses a weighted least squares criterion to minimize differences between the observed and calculated frequencies. Using the best determined force fields, the error in calculated frequencies may average about 10 cm^{-1}.

Measuring and assembling reliable frequency data, and setting up and solving the equations, are formidable tasks, and for this reason relatively few systems have been analyzed sufficiently well to provide reliable force constants. Among the sets of parameters derived from small molecules that have proven to be transferable and have found application to polymer systems are those for unstrained saturated hydrocarbons,[55,56] primary[57] and secondary[58] saturated alkyl chlorides, aliphatic ethers,[59] alkyl benzenes,[60] and cyclo-olefins.[61] In addition, a force field useful for biopolymers has been derived from selected amides and polyamides.[62]

3.1.3.4. Vibrational Intensities. Relative to vibrational frequencies, integrated infrared and Raman intensities are more difficult to measure accurately, and more difficult to interpret in terms of transferable parame-

[55] R. G. Snyder and J. H. Schachtschneider, *Spectrochim. Acta* **21,** 169 (1965).
[56] R. G. Snyder, *J. Chem. Phys.* **47,** 1316 (1967).
[57] R. G. Snyder and J. H. Schachtschneider, *J. Mol. Spectrosc.* **30,** 290 (1969).
[58] W. H. Moore and S. Krimm, *Spectrochim. Acta, Part A* **29,** 2025 (1973).
[59] R. G. Snyder and G. Zerbi, *Spectrochim. Acta, Part A* **23,** 391 (1967).
[60] C. La Lau and R. G. Snyder, *Spectrochim. Acta, Part A* **27,** 2073 (1971).
[61] N. Neto, C. DiLauro, and S. Califano, *Spectrochim. Acta, Part A* **24,** 385 (1968).
[62] W. H. Moore and S. Krimm, *Biopolymers* **15,** 2439 and 2465 (1976).

ters. Parametric models for calculating intensities have been developed and tested on small molecules. Applications to polymers are limited, and at present the most useful aspect of the relationship between intensities and polymer structure concerns selection rules and polarization properties.

The symmetry of a molecule (or polymer chain) determines which normal vibrations are allowed to appear in a spectrum, and influences the polarization of the allowed transitions. Those symmetry operations (axes of rotation, planes of symmetry, etc.) that, when applied to the molecule, leave it unchanged, form a set that defines the symmetry of the molecule. This set will be found to correspond to one on the point groups listed in standard references.[2] After a molecule is identified with one of the point groups, the number of its vibrations that belong to each symmetry species can be determined by a simple procedure.[2] Assigning a vibration to a symmetry species is equivalent to finding the relationship between the symmetry of the vibration and the symmetry of the molecule. Vibrations within a given species have the same activity and polarization properties, but these properties are different for different species. In cases where the symmetry of a molecule is high, normal vibrations are usually distributed among a large number of species, and this makes polarization measurements especially helpful in determining assignments. Of course, the process can be reversed, and the observed polarization can be used to determine molecular symmetry, or something about the orientation of the molecules.

3.1.3.4.1. INFRARED. The kth normal vibration will be infrared active if the derivative of the dipole moment of the molecule with respect to the normal coordinate is not zero. If this vector derivative is designated as $\mathbf{M}'_k$ the integrated infrared intensity averaged over all orientations of the molecule is

$$A_k = \frac{\pi}{3c^2} \mid \mathbf{M}'_k \mid^2 \qquad\qquad (3.1.14)$$

for the kth fundamental mode. On the other hand, if the molecule is fixed in space, the observed intensity will be proportional to the square of the dot product $\mathbf{E} \cdot \mathbf{M}'_k$, where $\mathbf{E}$ is the electric vector associated with the incident radiation. The direction of $\mathbf{M}'_k$ can be ascertained for a molecule or a crystal of known orientation by measuring the intensity of the band as a function of the direction of $\mathbf{E}$. Conversely, if the direction of $\mathbf{M}'_k$ is determined by molecular symmetry, polarization measurements enable the band to be assigned to a symmetry species; when the band has already been assigned, polarization measurements may indicate the orientation of the molecules.

Molecular symmetry may or may not determine the direction of M_k'. For helical polymers having a threefold or higher (screw) rotation axis, M_k' is either parallel or perpendicular to the chain, so that measured values of the dichroic ratio defined in Eq. (3.1.2) are sufficient to assign all infrared active vibrations. For systems of lower symmetry, the direction of M_k' tends to be more dependent on the nature of the vibration. The transition moment of a highly localized vibration mode, such as an N–H stretching mode of a polypeptide, for example, may be assumed to be along the N–H bond. If the system has low symmetry, it may be possible to determine the direction of the bond from polarization measurements.[15]

Orientation is usually imperfect for polymers. Expressions for dichroic ratios have been calculated for the situation where chains are parallel, but are oriented randomly about their long axes.[23] Other situation can be envisioned including partial orientation. For some of these cases theoretical expressions for dichroic ratios have been worked out.[22]

3.1.3.4.2. RAMAN. Equations for Raman intensity are more complex. For our purposes these will be expressed in a simplified form that emphasizes certain factors. Given the orthogonal scattering geometry shown in Fig. 3, the Raman intensity I_k of normal mode k is proportional to the intensity of the excitation I_0 and to a scattering activity S_k, and may be written

$$I_k(g') = K_k S_k(g, g') I_0(g), \qquad (3.1.15)$$

where g and g' refer to the polarization states of the incident and scattered radiation, and where K_k is a proportionality constant.

For Stokes bands, K_k may be written

$$K_k = \frac{a(\nu_0 - \nu_k)^4}{\nu_k[1 - \exp(-hc\nu_k/kT)]}, \qquad (3.1.16)$$

where a is a constant and T is the temperature of the sample.

3.1.3.4.2.1. *Normal Raman Scattering.* The scattering activity is related to the electric moment induced by the exciting radiation; the electric moment is in turn related to the polarizability tensor α of the molecule. A great simplification in Eq. (3.1.15) results if $S_k(g, g')$ is expressed in terms of the elements of the derivative of α with respect to the normal coordinate, i.e., expressed in terms of $\partial\alpha/\partial Q_k$, which will be written α_k'. It is possible[63] to express Raman intensities in terms of the elements of α_k' when the excitation frequency ν_0 is well separated both from the frequencies associated with electronic transitions and from the vibrational frequency ν_k. These conditions are well met in most situations and result in the follow-

[63] G. Placzek, "Marx Handbuch der Radiologie," 2nd ed., Vol. 6, Part 2, p. 209. Akad. Verlagsges, Leipzig, 1934.

ing simplifications: α'_k is a symmetric tensor; the values of its elements are independent of ν_0; higher order derivatives of α do not contribute to Raman intensities. In this case Raman scattering is said to result from the "normal" Raman effect.

The form of the scattering activities then depends on the elements of the derived tensor α'_k, the polarization directions g and g', and the orientation of the molecules. The symmetry species of the vibration determines the symmetry of α'_k and which of its elements are zero.

When the principal axes of the polarizability tensor of the molecule are aligned parallel to the X, Y, and Z axes (Fig. 3), the scattering activities are simply

$$S_k(g, g') = (\alpha'_k)^2_{gg'}, \qquad (3.1.17)$$

so that in this case it is possible to measure the value (or at least the relative value) of the square of each element of α'_k for a sample so oriented.

If the molecules are randomly oriented, as in a liquid, $S_k(g, g')$ represents an average that is conveniently expressed in terms of two of the invariants of α'_k: the spherical part,

$$\overline{\alpha}'_k = (\alpha'_1 + \alpha'_2 + \alpha'_3)/3, \qquad (3.1.18)$$

and the anisotropic part,

$$\overline{\beta}'^2_k = \tfrac{1}{2}[(\alpha'_1 - \alpha'_2)^2 + (\alpha'_2 - \alpha'_3)^2 + (\alpha'_3 - \alpha'_1)^2], \qquad (3.1.19)$$

where the subscripts of the elements of α'_k now refer to the principal axes of the polarizability tensor.

$$S_k(ZX) = (3/45)\overline{\beta}'^2_k, \qquad (3.1.20)$$

$$S_k(ZZ) = \overline{\alpha}'^2_k + (4/45)\overline{\beta}'^2_k, \qquad (3.1.21)$$

corresponding respectively to the $I(\perp)$ and $I(\|)$ used in Eq. (3.1.3) to define ρ. Thus ρ_k can be related to $\overline{\alpha}'_k$ and $\overline{\beta}'_k$ as

$$\rho_k = 3\overline{\beta}'^2_k /(45\overline{\alpha}'^2_k + 4\overline{\beta}'^2_k) . \qquad (3.1.22)$$

As an alternative to displaying spectra as $I(\perp)$ and $I(\|)$, and using ρ to relate them as is commonly done, it is more desirable in some ways to display the spectra directly in terms of $\overline{\alpha}'^2$ and $\overline{\beta}'^2$.[64]

For a powdered or otherwise inhomogeneous sample in which the excitation and Raman radiation are scattered randomly, the scattering activity is $\overline{\alpha}'^2_k + (2/9)\overline{\beta}'^2_k$.

Between the limits of fixed and completely random orientation are intermediate cases such as uniaxial orientation, which occurs in fibers or

[64] J. R. Scherer, S. Kint, and G. F. Bailey, *J. Mol. Spectrosc.* **39**, 146 (1971).

extrudates, and uniplanar orientation in which case the long axes of the chains are confined to a plane. The scattering activities that result from averaging may or may not be different for different symmetry species. These have been evaluated in terms of the elements of α'_k for all the common point groups.[65] A more complex and less tractable case arises if there is a continuous distribution of molecular orientation. If one axis is unique and θ is the angle between this axis and a chain segment, the distribution of orientations may be expanded in terms of spherical harmonics. It has been shown that the Raman experiment can be used to determine both $\overline{\cos^2\theta}$ and $\overline{\cos^4\theta}$.[66]

   3.1.3.4.2.2. Resonance Raman Scattering. If ν_0 overlaps or falls near electronic absorption bands, resonance Raman scattering occurs. Its effect is to greatly enhance the Raman intensities of certain vibrational modes that are coupled with the electronic transitions.[33,34] The Raman bands enhanced are few in number, and are usually associated with modes involving a chromophore. These characteristics make resonance Raman spectroscopy especially suitable for applications to biological macromolecules since the intensity enhancement simplifies spectra and brings to prominence certain bands that may be associated with a biologically active site. For example, since the spectra are dependent on the nature of the chromophore, the binding associated with the heme has been studied in heme proteins using this technique.[67] Other aspects unique to resonance Raman spectroscopy are that α'_k is a function of ν_0, and that α'_k is not necessarily symmetric. The field is complex and not fully understood, but is promising and being actively pursued, especially in its biological applications.[68,69]

3.1.4. Vibrations of a Periodic Chain Molecule

   3.1.4.1. Introduction. In its highest state of order, an isolated polymer chain is periodic, infinitely long, and free of defects. The equations of motion of such a chain can be expressed in closed form and solved exactly. The basic equations for the harmonic vibrations of periodic lattices were first derived in connection with the calculation of heat capacities of crystals,[70,71] and in a sense the equations for a polymer chain

[65] R. G. Snyder, *J. Mol. Spectrosc.* **37**, 353 (1971).
[66] D. I. Bower, *J. Polym. Sci., Polym. Phys. Ed.* **10**, 2135 (1972).
[67] T. G. Spiro, *Biochim. Biophys. Acta* **416**, 169 (1975).
[68] A. Lewis and J. Spoonhower, *in* "Spectroscopy in Biology and Chemistry" (S.-H. Chen and S. Yip, eds.), Chapter 11. Academic Press, New York, 1974.
[69] T. G. Spiro and B. P. Gaber, *Annu. Rev. Biochem.* **46**, 553 (1977).
[70] M. Born and T. von Kármán, *Phys. Z.* **13**, 297 (1912).
[71] L. Brillouin, "Wave Propagation in Periodic Structures." Dover, New York, 1953.

represented a special case of these in which waves propagate in only one dimension. It remained to recast them in a form in keeping with the conventions used in the vibrational analysis of the normal modes of finite molecules.

3.1.4.1.1. RELATION TO ISOLATED POLYMER CHAINS. Although, in reality, polymer chains are neither infinitely long nor free of defects, an idealized model is a useful approximation in many situations as, for example, in providing a basis for the interpretation of the infrared and Raman spectra of a crystalline polymer. Most features of the spectra of a highly crystalline polymer can be accounted for on this basis. The effects of intermolecular interaction are usually small in the higher frequency region. At lower frequencies (< 500 cm^{-1}) these effects become more significant, and if there is specific interaction between chains, such as result from the hydrogen bonds between chains in the β-form of polyamides, the intermolecular effects must be taken into account.

3.1.4.1.2. GENERAL SOLUTION. The vibrational equations of a one-dimensional periodic infinite chain have general solutions

$$u_l{}^m = u_0{}^m \exp[-2\pi i(\nu ct - Kal)], \qquad (3.1.23)$$

where $u_l{}^m$ is a displacement coordinate for atom m in repeat unit† l, a is the length of the repeat unit, and $u_0{}^m$ is an amplitude factor. The periodicity of the wave in time is the frequency ν, and the periodicity in space is the wavevector K (equal to $1/\lambda$, where λ is the wavelength). Since the polymer chain is one dimensional, K may be written as a scalar quantity, its sign indicating the direction of propagation of the wave. A phonon refers to a particlelike entity whose energy is $hc\nu$ and whose momentum is hK. Momentum plays a significant role in some spectroscopic phenomena. Spectra resulting from vibrational transitions that involve single phonons are sometimes referred to as one-phonon spectra and may be thought of as analogous to the spectra that involve transitions between adjacent vibrational levels of a finite molecule.

It is convenient to use a parameter ϕ, defined as the phase angle difference between the motion of adjacent repeat units, or alteratively between the motion of the subunits within the repeat unit. If ϕ refers to adjacent repeat units then

$$\phi = 2\pi aK, \qquad (3.1.24)$$

so that the wavevector conditions $K = 0$ and $K = 1/2a$ are equivalent, respectively, to $\phi = 0$ (all repeat units in phase) and to $\phi = \pi$ (adjacent repeat units out of phase).

---

† The translational repeat unit of the chain (the equivalent of a one-dimensional unit cell) will be referred to as the "repeat unit." It may contain two or more identical "subunits."

Infrared and Raman bands are observed only for those vibrations of the infinite chain for which $K = 0$. All other vibrations ($K \neq 0$) are inactive since for these the total dipole moment (infrared) or induced dipole moment (Raman) associated with a periodic motion cancels to zero when summed over the infinite chain. The $K = 0$ condition greatly limits the amount of vibrational data available directly from infrared and Raman spectra. We shall return to this point after the modes of the periodic chain are discussed in more detail.

3.1.4.2. Description of Vibrations. 3.1.4.2.1. DISPERSION CURVES. The relation between ν and K (or ϕ) is described by a dispersion curve of the form

$$\nu = f(K), \qquad \text{or} \qquad \nu = f(\phi). \qquad (3.1.25)$$

A chain whose repeat unit contains n atoms has $3n$ dispersion curves (also referred to as branches) and their shape, relative frequency disposition, and identification with specific kinds of motion provide a basis for understanding spectra. Even a crude estimate of the form of these curves such as may be attained from a calculation based on approximate force constants is often informative.

In addition to dispersion curves, another useful and related function is the frequency distribution or density of vibrational states $g(\nu)$. It is defined as the number of states per unit interval of frequency and is equal to $d\phi/d\nu$ since the vibrational modes are distributed uniformly along ϕ. Only in the very simplest model systems, such as will be used here for illustration, is it possible to derive the dispersion curves analytically. Normally $g(\nu)$ is expressed as a histogram computed from calculated frequency data through

$$g(\nu_i) = CN_i/\Delta\nu, \qquad (3.1.26)$$

where C is a normalization factor. In this expression, $\Delta\nu$ is a constant frequency interval whose midpoint is ν_i, and N_i is the number of modes between $\nu_i + \Delta\nu/2$ and $\nu_i - \Delta\nu/2$. If the dispersion curve is in graphical form, N_i is proportional to the interval $\Delta\phi_i$, where $\Delta\phi_i = \phi(\nu_i + \Delta\nu/2) - \phi(\nu_i - \Delta\nu/2)$. Each density of states curve has at least one singularity, called a critical point, where $g(\nu)$ becomes infinite for $\Delta\nu = 0$. Critical points usually, but not always, occur at frequencies for which $K = 0$, and they have special importance in the interpretation of multiphonon spectra. Density of vibrational states are used in the calculation of heat capacities.

3.1.4.2.2. ACOUSTIC AND OPTIC MODES. A few general comments on dispersion curves are appropriate. The region of the dispersion curve from $K = -1/2a$ to $K = 1/2a$ is called the "first" Brillouin zone, and $K = 0$ is referred to as its center. Since higher order zones are redundant in this case and since the frequency of the wave is the same in either direction of propagation, only the region $K = 0$ to $K = 1/2a$ need be considered. Dispersion curves are classified as either acoustic or optic. Acoustic vibrations involve a momentum change of the whole repeat unit. The momentum change may be translational in a direction along the chain axis or perpendicular to it, or rotational, involving a change in the angular momentum about the chain axis. The zone center frequency of an acoustic branch is always zero since the in-phase motion of all repeat units results in an unconstrained translation or rotation of the rigid chain. There are four acoustic branches unless the chain consists of a colinear row of atoms. In that case one branch is missing since rotation about the chain axis cannot be defined.

The longitudinal vibrations of a collinear chain will serve to illustrate some properties of an acoustic branch. In the example, shown in Fig. 5, the chain is monatomic and linear, and its atoms are constrained to move

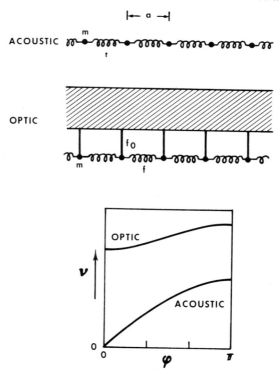

FIG. 5. Model monatomic lattices and their associated acoustic and optic dispersion curves.

along the chain axis. Consequently there is only one dispersion curve, and it is acoustic. Actually this model has more than illustrative significance since its vibrations are closely related in their form to the so-called longitudinal acoustic modes observed at low frequencies in the Raman spectra of crystalline polymers. These Raman active modes will be discussed in Section 3.1.5.3.2 in relation to their use in the study of crystal morphology.

The collinear chain consists of atoms of mass m, separated at a distance a. Adjacent atoms are joined by massless springs, whose force constant is f. The dispersion curve in the interval $\phi = 0$ to $\phi = \pi$ is given by

$$\nu = \frac{1}{\pi c} \left(\frac{f}{m}\right)^{1/2} \sin \frac{\phi}{2} \qquad (3.1.27)$$

and is shown in Fig. 5. Of special interest is the region where ϕ is very small and where, as a consequence, the dispersion is nearly linear:

$$\nu = \frac{1}{2\pi c} \left(\frac{f}{m}\right)^{1/2} \phi \qquad (3.1.28)$$

In this region the vibrations have wavelengths very long relative to the interatomic spacing and are compressional in character, so that they are associated with sound waves. For this reason the dispersion curve is referred to as an acoustic branch.

Most of the dispersion curves of a polymer chain are optic. Unlike acoustic modes, optic modes do not involve a change of momentum of the repeat unit, and therefore the zone center frequencies are not zero. A consequence of this is that the slope of an optic branch is zero at the zone center because the Brillouin zone is continuous through and symmetric about $K = 0$.

A model system that has an optic dispersion curve is also depicted in Fig. 5. This system resembles the monatomic lattice used to illustrate acoustic vibrations in that it consists of masses m, coupled by springs having a force constant f. In addition, however, the masses are fixed to the ends of flexible (massless) rods attached to a supporting structure of infinite mass. As before, only motion along the chain is allowed. The dispersion curve shown in Fig. 5 is given by

$$\nu^2 = \nu_0^2 + \frac{f}{\pi^2 c^2 m} \sin^2 \frac{\phi}{2} \qquad (3.1.29)$$

where ν_0, the frequency of a mass uncoupled from its neighbors, is given by $\nu_0 = (1/2\pi c)(f_0/m)^{1/2}$, where f_0 is the force constant associated with

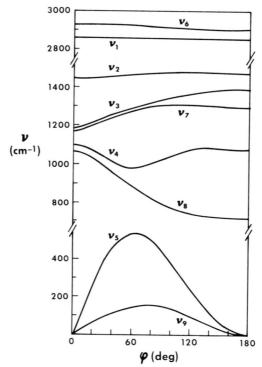

FIG. 6. Dispersion curves for an isolated extended polymethylene chain. The zone center modes are assigned in Table I.

bending the rods.   Near the zone center the curve assumes a parabolic shape:

$$\nu = \nu_0 \left(1 + \frac{f}{2f_0} \, \phi^2 \right).$$
(3.1.30)

Proceeding from model systems to a real polymer we show in Fig. 6 the calculated dispersion curves[72,73] of an extended polyethylene chain.   The phase angle difference refers to adjacent methylenes, so that the zone center is both at 0 and at π.   Two branches, ν_5 and ν_9, are acoustic and represent respectively in-plane and out-of-plane modes, where the plane is defined by the carbon skeleton.   The vibrations of most of the branches

[72] R. G. Snyder, *J. Mol. Spectrosc.* **31**, 464 (1969).
[73] M. Tasumi, T. Shimanouchi, and T. Miyazawa, *J. Mol. Spectrosc.* **9**, 261 (1962).

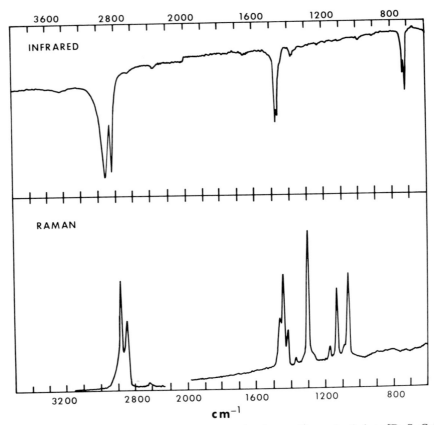

FIG. 7. Infrared and Raman spectra of a sample of crystalline polyethylene [D. S. Cain and A. B. Harvey, *Dev. Appl. Spectrosc., Part B* 7, 94 (1970)].

can be associated with motions such as those depicted in Fig. 4 for the methylene group. In some cases the character of the motion changes in moving across the Brillouin zone. For example, $\nu_7(0)$ and $\nu_8(\pi)$ are rocking modes, whereas $\nu_7(\pi)$ and $\nu_8(0)$ are twisting modes. From the density of vibrational states derived from dispersion curves such as these, the heat capacity of the crystalline polymer has been calculated.[74]

3.1.4.3. Vibrational Analysis. 3.1.4.3.1. FACTOR-GROUP MODES. The infrared and Raman spectra of a sample of crystalline polyethylene are shown in Fig. 7. Both spectra are relatively simple and they are quite different from each other. To explain them, selection rules must be considered.

[74] B. Wunderlich, *J. Chem. Phys.* **37**, 1207 (1962).

As in the case for the finite molecule, infrared activity occurs when $\partial \mathbf{M}/\partial Q_i\ (K) \neq 0$, and Raman activity when $\partial \boldsymbol{\alpha}/\partial Q_i\ (K) \neq 0$, where $Q_i(K)$ refers to the ith branch. For the infinite chain both of these derivatives are zero for $K \neq 0$. If $K = 0$, all repeat units vibrate in phase, and these special vibrations are referred to as factor-group modes. Their number is $3n - 4$, where n is the number of atoms in a repeat unit. They alone are allowed to appear in a spectrum.

The factor-group modes have the same periodicity as the chain in its equilibrium state, and therefore the symmetry of these modes can be related to the symmetry of the chain. The latter is designated in terms of a one-dimensional space group, referred to as a "line group." The line group is comprised of the kinds of symmetry operations associated with finite molecules and the kinds associated with an infinite system, i.e., operations involving translations of the chain, namely screw axes and glide planes. The resulting collection of symmetry elements constitutes a factor group, which is analogous to and isomorphous with a point group.

A point-group classification of the factor-group vibrations is possible because for these vibrations one end of every repeat unit is exactly in phase with the other end so that the vibrational system is reduced, in effect, to that of a single isolated repeat unit. As a result, the frequencies of the factor-group modes can be calculated using a slight modification of the procedures described earlier for a finite molecule, in which the finite molecule is replaced by the repeat unit.[13]

The classification of factor-group modes into the symmetry species of an (isomorphic) point group leads to selection rules and polarization properties just as for finite molecules. The procedure for determining the symmetry of the polymer chain and for classifying its factor-group modes is to be found in the literature[11,75] and has been repeatedly illustrated for the extended polyethylene chain.[22,75-77] Only the results of this classification will be discussed.

The isolated extended polyethylene chain has the highest symmetry of any well-characterized polymer. Its factor-group is isomorphous with point group D_{2h}, and its 14 ($= 3 \times 6 - 4$) factor-group modes are divided among the eight symmetry species of this point group as follows:

$$3A_g + 2B_{1g} + 2B_{2g} + B_{3g} + A_u + B_{1u} + 2B_{2u} + 2B_{3u}.$$

All g species imply Raman activity, and all u species infrared activity except for the A_u species, which is inactive in either. Since there are only a

[75] M. C. Tobin, *J. Chem. Phys.* **23**, 891 (1955).

[76] G. Turrell, "Infrared and Raman Spectra of Crystals." Academic Press, New York, 1972.

[77] S. Krimm, C. Y. Liang, and G. B. B. M. Sutherland, *J. Chem. Phys.* **25**, 549 (1956).

few vibrations within each species, polarization measurements in the infrared and Raman have facilitated assignments.

Extensive spectroscopic measurements and normal coordinate analyses on polyethylene have resulted in by far the most detailed and complete assignment of the normal vibrations of any polymer.[13,72,77-79] This assignment is summarized in Table I where the numbering of modes corresponds to the branch number given in Fig. 6, and where the descriptions of the methylene modes follow the designations in Fig. 4. Included in this table are assignments for the crystal, which will be considered in Section 3.1.5.2.2.

Less complete factor-group analyses have been carried out for many other regular polymer chains. Among these are isotactic polypropylene,[80,81] syndiotactic polypropylene,[82,83] polytetrafluoroethane,[84] poly(ethylene glycol),[85] polyoxymethylene,[86,87] poly(vinyl chloride),[88] trans-1,4-polybutadiene,[89] poly(vinyl fluoride),[90] polyglycine I and II,[91] and poly(L-alanine).[92]

The only polymer whose factor-group mode intensities have been calculated is polyethylene. A bond polarizability model has been used for the Raman intensities of polyethylene and perdeuteropolyethylene,[93] and a more elaborate electrooptical parameterized model has been used for both the Raman and infrared intensities of these same polymers.[94]

3.1.4.3.2. DETERMINATION OF DISPERSION CURVES. As has been emphasized, the infrared and Raman spectra of the infinitely long regular chain provide no data on dispersion curves except at the center of the Brillouin zone. Consequently other approaches must be used to determine the shapes of dispersion curves.

[78] R. G. Snyder, S. L. Hsu, and S. Krimm, *Spectrochim. Acta Part A* **34**, 395 (1978).

[79] P. J. Hendra, H. P. Jobic, E. P. Marsden, and D. Bloor, *Spectrochim. Acta, Part A* **33**, 445 (1977).

[80] R. G. Snyder and J. H. Schachtschneider, *Spectrochim. Acta* **20**, 853 (1964).

[81] T. Miyazawa, *J. Polym. Sci., Part C* **7**, 59 (1964).

[82] J. H. Schachtschneider and R. G. Snyder, *Spectrochim. Acta* **21**, 1527 (1965).

[83] T. Miyazawa, *J. Chem. Phys.* **43**, 4030 (1965).

[84] F. J. Boerio and J. L. Koenig, *J. Chem. Phys.* **52**, 4826 (1970).

[85] H. Matsuura and T. Miyazawa, *Bull. Chem. Soc. Jpn.* **41**, 1798 (1968).

[86] H. Sugeta and T. Miyazawa, *Rep. Prog. Polym. Phys. Jpn.* **9**, 177 (1966).

[87] H. Tadokoro, M. Kobayashi, Y. Kawaguchi, A. Kobayashi, and S. Murahashi, *J. Chem. Phys.* **38**, 703 (1963).

[88] W. H. Moore and S. Krimm, *Makromol. Chem., Suppl.* **1**, 491 (1975).

[89] S. L. Hsu, W. H. Moore, and S. Krimm, *J. Appl. Phys.* **46**, 4185 (1975).

[90] M. Kobayashi, K. Tashiro, and H. Tadokoro, *Macromolecules* **8**, 158 (1975).

[91] Y. Abe and S. Krimm, *Biopolymers* **11**, 1817 and 1841 (1972).

[92] K. Itoh and T. Shimanouchi, *Biopolymers* **9**, 383 (1970).

[93] R. G. Snyder, *J. Mol. Spectrosc.* **36**, 222 (1970).

[94] S. Abbate, M. Gussoni, G. Masetti, and G. Zerbi, *J. Chem. Phys.* **67**, 1519 (1977).

TABLE I. Assignment of the Fundamentals of the Polyethylene Chain and Crystal

Description[a]	Isolated chain			Crystal	
	Branch[b]	Species[c]	ν^d (cm^{-1})	Species[e]	ν(obs) (cm^{-1})
Raman					
Antisymmetric CH stretch	$\nu_6(0)$	$b_{1g}(xy)$	2880	A_g $B_{3g}(XY)$ }	2880
Symmetric CH stretch	$\nu_1(0)$	a_g	2845	A_g $B_{3g}(XY)$ }	~2845
CH$_2$ bend (scissors)	$\nu_2(0)$	a_g	~1440	A_g	1416
				$B_{3g}(XY)$	{ 1466 1442
CH$_2$ wag	$\nu_3(\pi)$	$b_{2g}(zx)$	1370	$B_{1g}(YZ)$ $B_{2g}(ZX)$	1368 1373
CH$_2$ twist	$\nu_7(\pi)$	$b_{3g}(xy)$	1295	$B_{1g}(YZ)$ $B_{2g}(ZX)$	1296 1294
CH$_2$ rock	$\nu_7(0)$	$b_{1g}(xy)$	1174	A_g $B_{3g}(XY)$ }	1171
Symmetric CC stretch	$\nu_4(0)$	a_g	1132	A_g $B_{3g}(XY)$ }	1132
Antisymmetric CC stretch	$\nu_4(\pi)$	$b_{2g}(zx)$	1064	$B_{1g}(YZ)$ $B_{2g}(ZX)$	1063 1065
Rotatory	$\nu_9(0)$	$b_{1g}(xy)$	0	A_g $B_{3g}(XY)$	136 108
Infrared					
Antisymmetric CH stretch	$\nu_6(\pi)$	$b_{2u}(y)$	2920	$B_{1u}(X)$ $B_{2u}(Y)$ }	2920
Symmetric CH stretch	$\nu_1(\pi)$	$b_{3u}(z)$	2850	$B_{1u}(X)$ $B_{2u}(Y)$	2847 2853
CH$_2$ bend (scissors)	$\nu_2(\pi)$	$b_{3u}(z)$	1470	$B_{1u}(X)$ $B_{2u}(Y)$	1473 1462
CH$_2$ wag	$\nu_3(0)$	$b_{1u}(x)$	1176	A_u $B_{3u}(Z)$	— 1176
CH$_2$ twist	$\nu_8(0)$	a_u	1050	A_u $B_{3u}(Z)$	— 1050
CH$_2$ rock	$\nu_8(\pi)$	$b_{2u}(y)$	725	$B_{1u}(X)$ $B_{2u}(Y)$	731 720
Translatory	$\nu_9(\pi)$	$b_{2u}(y)$	0 }	$B_{1u}(X)$ $B_{2u}(Y)$	75 106
Translatory	$\nu_5(\pi)$	$b_{3u}(z)$	0 }	$B_{1u}(X)$ $B_{2u}(Y)$	0 0
Translatory	$\nu_5(0)$	$b_{1u}(x)$	0	A_u $B_{3u}(Z)$	(54 calc)[f] 0

[a] Methylene vibrations are shown in Fig. 4.

[b] Branches are numbered as in Fig. 6.

[c] The x and z axes are in the plane defined by the carbon skeleton; x is parallel to the chain direction.

[d] Values are estimated from the crystal frequencies.

[e] The X and Y axes are parallel respectively to the a and b axes of the crystal.

[f] Calculated value from J. Barnes and B. Fanconi, J. Phys. Chem. Ref. Data 7, 1309 (1978).

These curves may be calculated from the structural parameters and force constants of a regular chain, and this in fact has been done in a number of cases where the necessary data were available. For example, it has been possible to determine accurately most branches of the polymethylene chain because reliable force constants had been evaluated from detailed normal coordinate analyses of the vibrations of a series of extended n-paraffins. In lieu of a computational analysis, the shapes of the dispersion curves can be estimated directly through the analysis of the vibrational spectra of a homologous series of molecules whose longest member is the polymer chain itself. Finally, there exist other kinds of spectra that are less constrained by selection rules and consequently provide more information. Examples are two-phonon infrared and Raman spectra, and neutron inelastic scattering spectra.

3.1.4.3.2.1. Calculation. A general method for the calculation of the normal vibrations of helical polymer chains was originally outlined by Higgs,[11] and has since been reformulated in ways more suitable for numerical calculation.[95,96] The method will be briefly outlined.

All regular polymers may be thought of as helices: each repeat unit contains one or more identical subunits, related by a screw axis. The phase angle difference between the vibrations of adjacent subunits is designated as ϕ. In passing we note that if the repeat unit contains p subunits in q turns of the helix, the condition defining a factor-group mode ($K = 0$) is

$$\phi = \pi i/q, \qquad i = 0, 1, 2, \cdots, p - 1. \qquad (3.1.31)$$

The theory of normal vibrations of the polymer diverges from that of a finite molecule with the definition of a set of phonon coordinates:

$$\mathbf{S}(\phi) = \eta \sum_{n=-\infty}^{\infty} \mathbf{r}^n e^{-in\phi}, \qquad (3.1.32)$$

where n designates a unit cell, $\mathbf{r}^n$ designates a vector of internal coordinates in that cell, and η is a normalization factor. The F and G matrices, used earlier to express the potential and kinetic energy of finite molecules in terms of internal coordinates [Eqs. (3.1.9) and (3.1.10)], are now written in terms of the phonon coordinates:

$$\mathbf{F}(\phi) = \mathbf{F}° + \sum_{s=1}^{\infty} (\mathbf{F}^s e^{is\phi} + \mathbf{F}'^s e^{-is\phi}), \qquad (3.1.33)$$

$$\mathbf{G}(\phi) = \mathbf{G}° + \sum_{s=1}^{\infty} (\mathbf{G}^s e^{is\phi} + \mathbf{G}'^s e^{-is\phi}), \qquad (3.1.34)$$

[95] H. Tadokoro, *J. Chem. Phys.* **33**, 1558 (1960).
[96] T. Miyazawa, *J. Chem. Phys.* **35**, 693 (1961).

where the primes indicate transposed matrices. In this equation $\mathbf{F}^s$ and $\mathbf{G}^s$ refer to potential and kinetic coupling between subunits m and n, where $s = |m - n|$. This simplication is possible because the interaction between subunits depends solely on their relative separation, indicated by s.

The dispersion curve can then be calculated from the equation

$$\mathbf{G}(\phi)\mathbf{F}(\phi)\mathbf{L}(\phi) = \mathbf{L}(\phi)\mathbf{\Lambda}(\phi), \qquad (3.1.35)$$

which is analogous to Eq. (3.1.12) for the finite molecule.

For computational purposes there are some advantages to expressing the phonon coordinates in Cartesian displacement coordinates,[97] although otherwise it makes little difference as to which set is used as a basis, since internal and Cartesian coordinates are linearly related.

Polymers whose dispersion curves have been calculated in the isolated chain approximation include polyethylene,[98] isotactic polypropylene,[99] orthorhombic and hexagonal polyoxymethylene,[98] poly(ethylene glycol),[100] poly(vinyl chloride),[101] trans-1,4-polychloroprene,[102] poly(vinylidene fluordie),[90] polytetrafluorethylene,[84,103] polyglycine I[104] and II,[105,106] and α-helical poly(L-alanine).[92,107,108]

3.1.4.3.2.2. Finite Chains. If a regular chain is finite in length, its normal vibrations will resemble those of the infinite chain, and if the chain is sufficiently long so that its standing waves are not significantly perturbed by end effects, the correspondence will be very close. Series of such standing waves can be observed in infrared and Raman spectra of regular chain molecules since the activity of the normal vibrations for the molecule is, of course, determined by its point-group symmetry and not by the $K = 0$ condition. Thus, vibrational frequencies can be measured across the Brillouin zone.

There is a phase angle difference associated with the standing wave. For a chain molecule consisting of one-dimensional coupled oscillators

[97] L. Piseri and G. Zerbi, *J. Mol. Spectrosc.* **26**, 254 (1968).
[98] L. Piseri and G. Zerbi, *J. Chem. Phys.* **48**, 3561 (1968).
[99] G. Zerbi and L. Piseri, *J. Chem. Phys.* **49**, 3840 (1968).
[100] H. Matsuura and T. Miyazawa, *Bull. Chem. Soc. Jpn.* **42**, 372 (1969).
[101] A. Rubčić and G. Zerbi, *Macromolecules* **7**, 754 (1974).
[102] D. L. Tabb and J. L. Koenig, *J. Polym. Sci., Polym. Phys. Ed.* **13**, 1159 (1975).
[103] L. Piseri, B. M. Powell, and G. Dolling, *J. Chem. Phys.* **57**, 158 (1973).
[104] V. D. Gupta, A. K. Gupta, and M. V. Krishnan, *Chem. Phys. Lett.* **6**, 317 (1970).
[105] E. W. Small, B. Fanconi, and W. L. Peticolas, *J. Chem. Phys.* **52**, 4369 (1970).
[106] R. D. Singh and V. D. Gupta, *Spectrochim. Acta, Part A* **27**, 385 (1971).
[107] B. Fanconi, E. W. Small, and W. L. Peticolas, *Biopolymers* **10**, 1277 (1971).
[108] M. V. Krishnan and V. D. Gupta, *Chem Phys. Lett.* **6**, 231 (1970).

the relation between phase angle and frequency has a particularly simple form. If the number of oscillators is n, it can be shown[109] that

$$\lambda_k = A_0 + 2 \sum_{l=1}^{\infty} A_l \cos l\phi_k, \tag{3.1.36}$$

where

$$\phi_k = k\pi/(n + 1), \qquad k = 1, 2, 3, \ldots, n, \tag{3.1.37}$$

and where λ_k is the frequency parameter of the kth mode. The A_l are functions of F and G elements. If the oscillator has more than one degree of freedom, in which case the A_l may be complex functions of the force constants and structural parameters, Eq. (3.1.36) is still applicable. And in the limit of an infinitely long chain, this equation becomes an expression for a dispersion curve.

Chain molecules terminate with a group that is more or less different from the groups that make up the chain. The presence of terminal groups results in a vibrational perturbation not accounted for by Eq. (3.1.36). This perturbation can be quite large for shorter chains, and this has motivated efforts to modify the equation to take into account end effects.[110]

Series of bands associated with Eq. (3.1.36) can be observed in the infrared spectra of molecules containing polymethylene chains. Band progressions have been observed, for example, in the spectra of the crystalline fatty acids.[111] In the case of the crystalline n-paraffins,[112] these progressions have been identified with the dispersion curves of polyethylene.[109,113]

If a progression is associated with an optic branch, band spacing tends to converge as the limiting (zone center) frequency is approached, as may be inferred from Eq. (3.1.30). At the same time, band intensities tend to increase or decrease monotonically in proceeding away from the head band, i.e., the band most nearly corresponding to the factor-group mode of the infinite chain. As the chains get longer there are more bands in the progression, but relative to an active head band, they become less intense; in the limit, only the head band remains.

An example of an infrared active optic progression is shown in Fig. 8 for crystalline $n-C_{24}H_{50}$. The bands belong to a series of methylene rocking–twisting modes, every other member of which is infrared active. This progression is associated with dispersion curve ν_8 of the extended

[109] R. G. Snyder and J. H. Schachtschneider, *Spectrochim. Acta* **19**, 85 (1963).

[110] H. Matsuda, K. Okada, T. Takase, and T. Yamamoto, *J. Chem. Phys.* **41**, 1527 (1964).

[111] R. N. Jones, A. F. McKay, and R. G. Sinclair, *J. Am. Chem. Soc.* **74**, 2575 (1952).

[112] J. K. Brown, N. Sheppard, and D. M. Simpson, *Philos. Trans. R. Soc. London, Ser. A* **247**, 35 (1954).

[113] R. G. Snyder, *J. Mol. Spectrosc.* **4**, 411 (1960).

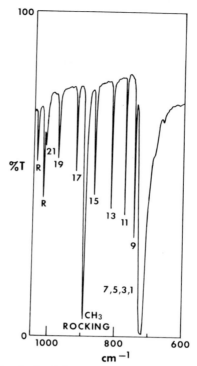

FIG. 8. Optic progression in the infrared spectrum of crystalline n-$C_{34}H_{50}$: rocking–twisting modes associated with branch ν_8; bands not belonging to this progression are indicated: CH_3 rocking and C–C stretching (R).

polyethylene chain (Fig. 6) and has been observed in the infrared spectrum of each of the n-paraffins, propane through n-$C_{30}H_{62}$.[109] A display of these band progressions is shown in Fig. 9. After the bands are assigned (by simply numbering them), the ϕ_k can be calculated [Eq. (3.1.37)] and plotted against the observed frequency. The dispersion curve derived from these data is shown in Fig. 10. Frequency perturbations due to end effects (in this case arising primarily from interactions with the rocking mode of the methyl group) are in evidence especially for the shorter chains. Data from the longer chains, however, give a good definition to the ν_8 branch. We note that for polyethylene only the head band at 720 cm^{-1} persists. Similar analyses of other n-paraffin progressions have helped establish the dispersion curves of polyethylene either directly, as illustrated here for ν_8, or indirectly, by providing assignments used in normal coordinate analyses, which in turn provide force constants used to calculate these curves.[109,113]

There appears in the low-frequency region of the Raman spectra of the

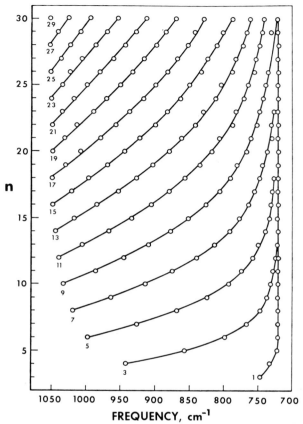

FIG. 9. Observed array of infrared active rocking-twisting modes of the crystalline n-paraffins n-C_3H_8 through n-$C_{30}H_{62}$ [R. G. Snyder and J. H. Schachtschneider, *Spectrochim. Acta* **23**, 85 (1963)].

n-paraffins a progression of bands associated with the longitudinal acoustic branch ν_5.[114,115] In contrast to the optic case, these bands do not converge in approaching the head band, but rather tend to be regularly spaced as indicated by Eq. (3.1.28). This behavior can be seen in Fig. 11 where the low-frequency Raman spectrum of n-$C_{29}H_{60}$ is shown.

3:1.4.3.2.3. Two-phonon Spectra. Although transitions between adjacent vibrational energy levels are responsible for most of the dominant features of infrared and Raman spectra, the effects of higher-order transi-

[114] S. Mizushima and T. Simanouti, *J. Am. Chem. Soc.* **71**, 1320 (1949).
[115] R. F. Schaufele and T. Shimanouchi, *J. Chem. Phys.* **47**, 3605 (1967).

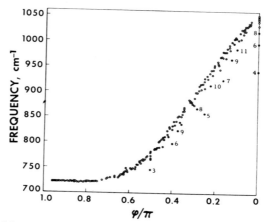

FIG. 10. Plot of frequency vs. phase difference for the infrared active rocking–twisting modes of the crystalline *n*-paraffins (Fig. 9); numbered points indicate specific short chains [R. G. Snyder and J. H. Schachtschneider, *Spectrochim. Acta* **23**, 85 (1963)].

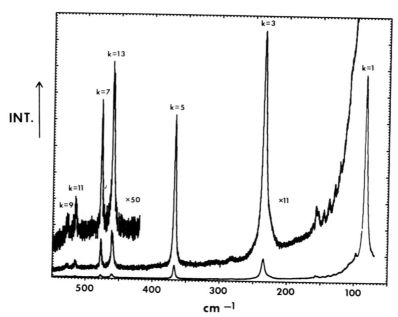

FIG. 11. Acousting progression in the Raman spectrum of crystalline n-$C_{29}H_{60}$ (figure courtesy of J. R. Scherer).

tions are often in evidence, and their study can be informative. Most of the weak bands appearing above the main fundamental region (usually above 1600 cm^{-1}) are associated with transitions involving a change of two quanta of vibrational energy. These binary combinations are weak because the origin of their intensity lies in the higher order terms in the expansion of the energy and of the dipole moment of polarizability. For a finite molecule, the frequency $\nu_{kk'}$ of the binary combination band is normally observed near the sum of the frequencies of the two fundamentals that are involved:

$$\nu_{kk'} \cong \nu_k + \nu_{k'}. \tag{3.1.38}$$

The same phenomenon occurs for the infinite chain, where the observed binary combination bands are sometimes called two-phonon bands. The potential complexity that might be anticipated as a result of combining two continua of phonon states, is alleviated to a large degree by a restriction on which phonons can or cannot combine. The condition is that the wavevectors of the two phonons must be equal and of opposite sign. Since $\nu(K) = \nu(-K)$, an expression analogous to Eq. (3.1.38) may be written for the infinite chain:

$$\nu_{kk'}(0) \cong \nu_k(K) + \nu_{k'}(K), \tag{3.1.39}$$

where k and k' refer, respectively, to two branches, and where $\nu_{kk'}(0)$ signifies a zone center mode. This mode may have infrared and Raman activity, since its wavevector is zero.

Since for polymers, two branches are involved in the double transition, a continuum of binary states results, and this gives the bands their characteristic breadth and asymmetry. Maxima occur at critical points in the density of vibrational states. In the case of polyethylene some infrared active two-phonon bands have been interpreted in these terms.[116,117]

There exists for finite chains a phase-difference condition, analogous to the wavevector condition for the infinite chain, which determines those binary combinations that will appear with appreciable intensity in the spectra of these molecules. The effect is to greatly reduce the number of binary combinations that can be observed, as may be seen in Fig. 12 for that region of the infrared spectrum of crystalline n–$C_{24}H_{50}$ where there appear combinations associated with the ν_7 and ν_8 branches. In this particular situation the phase condition is that ϕ_k must be the same for each fundamental. In Fig. 12 is indicated the value of k associated with the fundamentals contributing to each binary. For very long chains, such as

[116] J. R. Nielsen and R. F. Holland, *J. Mol. Spectrosc.* **6**, 394 (1961).
[117] R. G. Snyder, *J. Chem. Phys.* **68**, 4156 (1978).

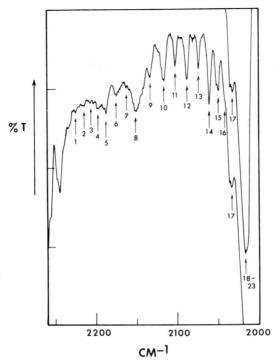

FIG. 12. Binary combinations ($\nu_7 + \nu_8$) in the infrared spectrum of n-$C_{24}H_{50}$ [R. G. Snyder, *J. Chem. Phys.* **68**, 4156 (1978)].

occur in the polymer, bands such as these crowd together to form continua that have maxima at the critical points. In another region of the infrared spectra of the crystalline n-paraffins are found binary combinations involving the ν_3 and ν_5 branches: since the ν_3 branch is known, Eq. (3.1.39) may be used to determine ν_5 in a region of the Brillouin zone not directly accessible.[117]

Similarly detailed analyses of the Raman spectra of finite-chain molecules have not been reported because of the difficulty in detecting two-phonon Raman scattering. In the case of polyethylene itself, some two-phonon Raman bands have been characterized,[118] and a theory of intensities applicable to two-phonon scattering from ν_5 has been reported.[119]

3.1.4.3.2.4. Neutron Inelastic Scattering. In this technique monoenergetic thermal neutrons are inelastically scattered by a polymer sample.

[118] R. G. Snyder, S. J. Krause, and J. R. Scherer, *Bull. Am. Phys. Soc.* [2] **22**, 257 (1977).
[119] J. W. Halley, P. Kleban, and J. Onffroy, *J. Polym. Sci., Polym. Phys. Ed.* **16**, 189 (1978).

As in the Raman effect, the incident particle loses (or gains) an amount of energy corresponding to that of a phonon. Unlike the Raman effect, all phonons are more or less eligible for interaction so that the spectrum of inelastically scattered neutrons can be transformed into a density of vibrational states. In addition to energy transfer, a momentum transfer occurs between the neutron and the phonon. As a consequence, scattering from an oriented sample is coherent, and data from inelastic coherent neutron scattering experiments can be used to determine a specific dispersion curve. Neutron scattering methods are most effective in the frequency region below 700 cm^{-1}.

This technique has been applied to crystalline polyethylene to study its ν_5 and ν_9 branches. The density of vibrational states of this polymer has been determined by several groups,[120–122] and critical point frequencies have been found that are in general agreement with calculated values. A coherent neutron scattering study on oriented perdeuteropolyethylene[123] revealed the shape of its longitudinal acoustic branch ν_5 across the entire Brillouin zone. The longitudinal acoustic branches of polytetrafluoroethylene and deuteropolyoxymethylene have been measured in the region near the zone center.[124] Neutron inelastic scattering is dealt with in detail in Chapter 3.4.

3.1.5. Ordered Polymer Systems

3.1.5.1. Introduction. Polymer molecules in the crystalline state are highly ordered. Although most of the prinicpal features of their vibrational spectra can be explained in terms of an isolated periodic chain, there always remain some bands or band complexities that cannot be accounted for. Their presence is ordinarily a result of interchain interaction or of departures from chain periodicity. The effects of interchain interaction in the crystal are easier to fathom and will be discussed first. The relation between structural irregularity and its effect on vibrational spectra is a complex, much less well developed subject, but an important one. Defects greatly affect polymer properties and diffraction methods are not well suited to their study.

3.1.5.2. Vibrations of Perfect Polymer Crystals. 3.1.5.2.1. CRYSTAL EFFECTS ON THE VIBRATIONS OF THE ISOLATED CHAIN. When

[120] W. R. Myers, G. C. Summerfield, and J. S. King, *J. Chem. Phys.* **44**, 184 (1966).

[121] S. Trevino and H. Boutin, *J. Macromol. Sci., Chem.* **1**, 723 (1967).

[122] W. R. Myers and P. D. Randolph, *J. Chem. Phys.* **49**, 1043 (1968).

[123] L. A. Feldkamp, G. Venkataraman, and J. S. King, *in* "Inelastic Scattering of Neutrons," Vol. 2, p. 159. IAEA, Vienna, 1968.

[124] J. W. White, *in* "Structural Studies of Macromolecules by Spectroscopic Methods" (K. J. Ivin, ed.), Chapter 4. Wiley, New York, 1976.

isolated polymer chains form a crystal, their spectra are affected in two ways. There are static effects associated with the crystalline environment, and there are dynamic effects associated with interchain coupling of vibrational modes.

3.1.5.2.1.1. Static Effects. Static effects tend to be small for optic modes since intermolecular forces are much weaker than intramolecular forces. Band positions always undergo frequency shifts when the environment of a molecule is changed. For example, small displacements of infrared and Raman bands are observed when a substance is transformed from the gaseous state to the solid state or, in dissolution, upon a change of solvent. Normally the frequency shifts are less than 2%, although the effect on intensities can be considerably greater. There are cases involving specific interaction where frequency shifts and intensity changes may be very large, the most important case being the effect of hydrogen bonding on the $O-H$ stretching vibration.[125]

Another effect of the crystal field is to modify selection rules. This occurs when the symmetry of the chain environment is lower than that of the chain itself. As a result, modes that were previously degenerate in frequency because of symmetry may become separated, and modes that were previously inactive may become active.[126,127] An example of the latter occurs for polyethylene. The isolated chain has symmetry D_{2h}, but in the crystal its environmental (or site) symmetry is C_{2h}, a subsymmetry of D_{2h}. The twisting mode at 1065 cm^{-1}, which is inactive for the isolated chain, is now allowed activity and does appear, though weakly, in the infrared spectrum of the crystalline polymer. For most systems, however, these effects are subtle, and tend to go unnoticed in the complexity of the spectra.

The crystal environment has a large effect on modes associated with certain acoustic branches. Chains can no longer rotate freely because of the presence of neighboring chains; unconstrained translation can occur only if it is a cooperative motion involving a translation of the whole crystal, i.e., a motion where all chains translate in phase. Thus the acoustic branches of isolated chains tend to be displaced as a result of the crystal environment, and one of the zone center modes, the rotatory mode, no longer has a zero frequency. To describe further the effects of the crystal on the acoustic branches, dynamic coupling must be considered.

3.1.5.2.1.2. Dynamic Effects. The vibrations of chains in a crystal can

[125] G. C. Pimentel and A. L. McClellan, "The Hydrogen Bond." Freeman, San Francisco, California, 1960.

[126] D. F. Hornig, *J. Chem. Phys.* **16**, 1063 (1948).

[127] H. Winston and R. S. Halford, *J. Chem. Phys.* **17**, 607 (1949).

couple as a result of intermolecular forces, and this coupling is analogous to the (much stronger) coupling between the repeat units of the isolated chain. In the intrachain case the interaction is parallel to the chain axis, while in the interchain case, it is perpendicular.

The dynamics of this system correspond to those of a three-dimensional crystal.[97] The repeat unit of the isolated chain now becomes a unit cell that may contain more than a single chain; the one-dimensional wavevector is replaced by a three-dimensional **K**. Dispersion curves must be defined in accordance with the direction of **K**. They are usually calculated for **K** parallel to the chain axis. In the case of polyethylene, there are two chains per unit cell so that each branch of the isolated chain has two components in the crystal. For optic branches, the components are only slightly separated, and they follow closely the branch associated with the isolated chain. In contrast, the acoustic branches, as shown in Fig. 13, are significantly displaced near the zone center (cf. Fig. 6) where vibrations formerly having zero frequencies become lattice modes that are appropriately described in terms of hindered translations and rotations of the chains.

Relatively few calculations of dispersion curves in three dimensions have been made for polymer crystals. Such calculations are not only difficult, but there are limited experimental data available for comparison. For polyethylene the separation of the two components of ν_8 across the zone may be inferred from the spectra of the orthorhombic crystalline n-paraffins.[128] Data related to the acoustic branches have also been reported for these molecules.[115,129,130] The extensive spectroscopic, neutron inelastic scattering, and heat capacity measurements on polyethylene and perdeuteropolyethylene have provided incentives for studying the lattice dynamics of this polymer.[131,132] Dispersion curves and the density of vibrational states of polyglycine-I have been calculated for a structure consisting of a hydrogen-bonded two-dimensional lattice.[133] A comparison of the heat capacity calculated from the density of states, with experimental values in the range 1–20 K, showed excellent agreement and accounted for an apparent anomaly observed at 8 K.[134] The dispersion curves of the polytetrafluoroethylene lattice have been considered in relation to the coherent inelastic neutron scattering spectrum of this polymer.[103]

[128] R. G. Snyder, *J. Mol. Spectrosc.* **7**, 116 (1961).
[129] H. G. Olf and B. Fanconi, *J. Chem. Phys.* **59**, 534 (1973).
[130] B. Fanconi, *J. Appl. Phys.* **46**, 4124 (1975).
[131] T. Kitagawa and T. Miyazawa, *Adv. Polym. Sci.* **9**, 335 (1972).
[132] D. I. Marsh and D. H. Martin, *J. Phys. C., Suppl.* **5**, 2309 (1972).
[133] B. Fanconi, *J. Chem. Phys.* **57**, 2109 (1972).
[134] B. Fanconi and L. Finegold, *Science* **190**, 458 (1975).

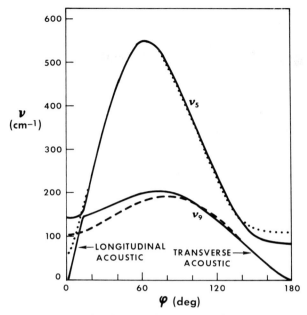

FIG. 13. Dispersion curves for the acoustic branches of crystalline polyethylene.

3.1.5.2.2. FACTOR-GROUP ANALYSIS. As in the case of the isolated chain, the only vibrations that can appear in the infrared or Raman spectra of the crystal are the factor-group modes, i.e., those modes in which all unit cells vibrate in phase. The symmetries of these modes are related to the symmetry of the unit cell, which is expressed in terms of the factor-group. Procedures for classifying factor-group modes into symmetry species are essentially identical to those described for molecular crystals.[14,75] If the unit cell of the crystal contains more than one chain, each factor-group fundamental of the isolated chain will have a multiplicity equal to the number of chains traversing the unit cell. In the common case where the unit cell contains two chains, the fundamental modes that appear in the spectrum may have two components, usually only slightly separated in frequency. This phenomenon is called factor-group or correlation splitting. More often than not, factor-group splitting is not observed, even though the unit cell of the polymer contains more than one chain. This happens because only one of the components is active in the spectrum or, more often, because the two components are not resolved.

Crystalline polyethylene displays factor-group splitting in both its infrared and Raman spectra. Its unit cell is orthorhombic and contains two chains and four methylene groups.[8] The 12 atoms in this cell give rise to

33 factor-group modes, which are classified and assigned to observed bands in Table I. It is fortuitous (and sometimes confusing) that for this polymer the isomorphous point group D_{2h} is the same for both the isolated chain and the crystal.

The factor-group modes of the optic branches of the isolated chain are doubled for the crystal; one component is associated with in-phase motion of the two chains in the unit cell, and the other component with out-of-phase motion. The separation between the components is less than 15 cm^{-1} for all fundamentals except $\nu_2(0)$. Relative to the frequencies of the fundamentals these separations are small and are of the magnitude expected for intermolecular forces. The splitting of the intense methylene bending (1465 cm^{-1}) and rocking (725 cm^{-1}) bands is clearly evident in the room temperature infrared spectrum (Fig. 7). Except for the bending fundamental near 1440 cm^{-1}, the separation of the components in the Raman spectrum is too small to be resolved at room temperature. Correlation splitting has, however, been observed for the sample at low temperature.[79,135]

The intermolecular forces responsible for the splitting in polyethylene have their origin primarily in repulsion between nonbonded hydrogens. A simple three-parameter model based on hydrogen atom–atom repulsion between three pairs of atoms is found to account semiquantitatively for most of the observed separations as well as for the frequencies of the lattice modes.[136] This intermolecular force field is satisfying in that the largest terms occur between nearest hydrogens. More elaborate models have since been used to achieve better agreement, especially for the lattice modes.[137]

Data on intermolecular effects for other crystalline polymers are much less complete. Some of the fundamentals in the Raman spectrum of polytetrafluoroethylene at low temperature ($-180°C$) have been reported to be split[138] although the origin of the splitting is not clear.[139] Lattice modes have also been observed for this polymer.[140–142] Correlation splitting and lattice modes have been observed and analyzed for the orthorhombic

[135] F. J. Boerio and J. L. Koenig, *J. Chem. Phys.* **52**, 3425 (1970).

[136] M. Tasumi and T. Shimanouchi, *J. Chem. Phys.* **43**, 1245 (1965).

[137] J. Barnes and B. Fanconi, *J. Phys. Chem. Ref. Data* **7**, 1309 (1978).

[138] F. J. Boerio and J. L. Koenig, *J. Chem. Phys.* **54**, 3667 (1971).

[139] G. Masetti, F. Cabassi, G. Morelli, and G. Zerbi, *Macromolecules* **6**, 700 (1973).

[140] G. W. Chantry, J. W. Fleming, E. A. Nicol, H. A. Willis, and M. E. A. Cudby, *Chem. Phys. Lett.* **16**, 141 (1972).

[141] R. G. Jones, E. A. Nicol, J. R. Birch, G. W. Chantry, J. W. Fleming, H. A. Willis, and M. E. A. Cudby, *Polymer* **17**, 153 (1976).

[142] G. W. Chantry, E. A. Nicol, R. G. Jones, H. A. Willis, and M. E. A. Cudby, *Polymer* **18**, 37 (1977).

form of polyoxymethylene[143,144] and for crystalline trans-1,4-polybutadiene.[89]

Specific interaction between the vibrations of the amide groups in polypeptides results in a significant frequency dependence of the characteristic amide I (mostly C=O stretching) and amide II (mostly C–N stretching and N–H bending) bands on conformation. This dependence enables the identification of the principal conformational structures (α-helix, β-form, disordered state, etc.) of the polypeptide components in proteins, from their infrared and Raman spectra.[15,17] The coupling is intramolecular in the case of the α-helix and intermolecular for the β-form. Because the amide I and II modes are highly localized, the effects of interaction can be calculated using a perturbation theory based on a model in which specific intrachain and interchain coupling terms can be defined and their values determined from known structures.[145] The magnitude of the interaction force constants can be accounted for in terms of transition dipole coupling.[146]

3.1.5.3. Studies on the Structure of Polymer Crystals. 3.1.5.3.1. INTRODUCTION. It is well known that crystalline polymers are not perfectly ordered. When grown from solution, and to a lesser extent when crystallized from the melt, many synthetic polymers form platelike crystals.[147] These lamellae have thicknesses in the approximate range 100–1000 Å, with lateral dimensions of the order of microns. The faces of the crystals are perpendicular to the chain direction, or nearly so. Since the thickness of a lamella is much less than the length of the chain, the chains either fold back at the lamellar surface and reenter the crystal, or wander away to enter either another crystal or an amorphous region. The part of the polymer molecule represented by chain segments within the crystal tends to be highly ordered, but at the surface of the crystal there is a discontinuity in the conformational regularity of each segment. In addition there may be other more or less ordered phases apart from the lamellae. Biopolymer crystals have their own characteristic complexities. Many proteins, for example, crystallize into well-defined structures in which the chain has an α-helical form. The helix, however, does not maintain its direction over long distances but undergoes abrupt turns at irregular intervals to produce a compact structure in which there is the possibility of cross links between chain segments. In short, the structure of crystalline polymers is complex and imperfectly understood. Vibra-

[143] V. Zamboni and G. Zerbi, *J. Polym. Sci., Part C* **7**, 153 (1964).
[144] G. Zerbi and P. Hendra, *J. Mol. Spectrosc.* **27**, 17 (1968).
[145] T. Miyazawa, *J. Chem. Phys.* **32**, 1647 (1960).
[146] W. H. Moore and S. Krimm, *Proc. Natl. Acad. Sci. U.S.A.* **72**, 4933 (1975).
[147] P. H. Geil, "Polymer Single Crystals." Krieger, New York, 1973.

tional spectroscopy is one of the techniques useful for the study of the secondary and tertiary structures of such systems.

3.1.5.3.2. RAMAN LAM MODES: CRYSTAL PROPERTIES AND MOR- PHOLOGY. In an ideal lamellar crystal, a straight chain segment has a well-defined length: its ends are determined by the abrupt onset of confor- mational disorder near the crystal surface. From the standpoint of cer- tain of its normal vibrations, this segment may act as a finite molecule, and its spectrum may exhibit band progressions. The bands of optic pro- gressions are not observed because there is a rapid convergence in spacing as the zone center is approached; furthermore only the modes very near the zone center have appreciable intensity. The situation for an acoustic progression is different. Here the bands tend to be evenly spaced, and their intensities are more evenly distributed.[115,148] One such progression has been observed at low frequencies in the Raman spectra of crystalline polymers.

The bands of this progression are referred to as longitudinal acoustic modes (LAM modes) because they are associated with the longitudinal acoustic dispersion curve of the infinite chain. They can also be described as internal modes since they are associated with finite chain segments. The forms of the LAM-k modes are very simple. The head band, LAM-1, is often referred to as an "accordion" mode since it is an extension–contraction of the chain segment. The form of this mode together with some of the higher-frequency members of this progression are depicted in Fig. 14 in terms of longitudinal displacement from equilib- rium. The integer k is then seen to be equal to the number of nodes; the nodes indicate the points of maximum or minimum density. In the case of the n-alkanes the odd-numbered modes are Raman active, and the even- numbered modes are infrared active. Raman active LAM modes for crystalline $n-C_{29}H_{60}$ may be seen in Fig. 11.

The frequencies of these modes are related to the length L of the straight-chain segment and to the elastic modulus E of the crystal. For a regular isolated chain with free ends, the relation is

$$\nu(\text{LAM}-k) = (k/2cL) (E/\rho)^{1/2}, \qquad k = 1,3,5, \ldots , \qquad (3.1.40)$$

where ρ is the density. This expression applies only when the wave- length of the LAM mode is much greater than the length of the repeat unit, and is equivalent to the expression given in Eq. (3.1.28), which ap- plies to acoustic branches for small values of ϕ.

Although polyethylene was the first polymer for which LAM-1 was ob-

[148] R. G. Snyder, S. Krause, and J. R. Scherer, *J. Polym. Sci., Polym. Phys. Ed.* **16**, 1593 (1978).

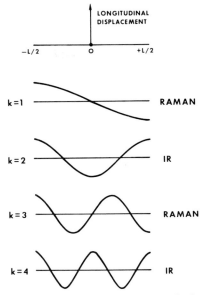

FIG. 14. Form of LAM vibrations in terms of longitudinal displacement from equilibrium.

served,[149] these modes had been observed previously in the Raman spectra of the crystalline n-paraffins, where the linearity between ν(LAM-1) and $1/L$ indicated in Eq. (3.1.40) was verified.[115] In addition to the n-paraffins, a linear relationship has since been found for a series of perfluoro n-alkanes[150] and for a number of crystalline polymers for which L was determined from small-angle X-ray scattering.[151–153] A series of data points may be obtained for a given polymer by measuring samples whose lamellar thicknesses differ as a result of the method of crystal preparation or as a result of annealing. Within experimental error, there is a linear relation in all cases so far reported: polyethylene,[149,151] isotatic polypropylene,[152,153] and polyoxymethylene.[153]

In Fig. 15 is shown the low-frequency Raman spectrum of polyoxy-

[149] W. L. Peticolas, G. W. Hibler, J. L. Lippert, A. Peterlin, and H. Olf, *Appl. Phys. Lett.* **18**, 87 (1971).

[150] J. F. Rabolt and B. Fanconi, *Polymer* **18**, 1258 (1977).

[151] H. G. Olf, A. Peterlin, and W. L. Peticolas, *J. Polym. Sci., Polym. Phys. Ed.* **12**, 359 (1974).

[152] S. L. Hsu, S. Krimm, S. Krause, and G. S. Y. Yeh, *J. Polym. Sci., Polym. Lett. Ed.* **14**, 195 (1976).

[153] J. F. Rabolt and B. Fanconi, *J. Polym. Sci., Polym. Lett. Ed.* **15**, 121 (1977).

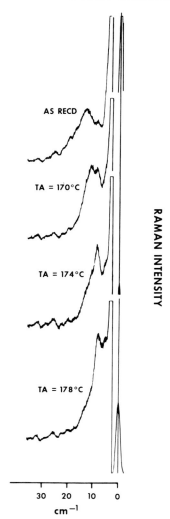

FIG. 15. LAM-1 mode of polyoxymethylene after the sample was annealed at successively higher temperatures (TA) [J. F. Rabolt and B. Fanconi, *J. Polym. Sci., Polym. Lett. Ed.* **15,** 121 (1977)].

methylene. Here one sees the LAM-1 band decrease in frequency as the fold period is increased by annealing.

Only the lowest-frequency member of the series is normally observed for polymers. This is largely the consequence of two factors: first, the low value of the frequency (< 30 cm^{-1}) of LAM-1 causes the observed intensity of this band to be high relative to other LAM modes as a result of

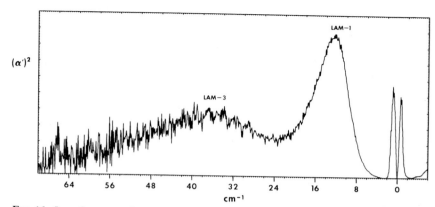

FIG. 16. Low-frequency Raman spectrum of an extruded sample of polyethylene displayed in terms of the scattering activity [R. G. Snyder, S. J. Krause, and J. R. Scherer, *J. Polym. Sci., Polym. Phys. Ed.* **16**, 1593 (1978)].

the $1/\nu$ and the Boltzmann factors appearing in the intensity expression in Eqs. (3.1.16) and (3.1.17); second, for any polymer there is a more or less broad distribution of chain segment lengths, and this causes the breadth of the LAM-k bands to increase as k increases. If the Raman spectra are plotted in terms of the scattering activity instead of observed intensity, higher-order LAM-k modes may become more apparent. In this way, LAM-3 and even LAM-5 have been observed for polyethylene.[148] Figure 16 shows a Raman spectrum of an extruded sample of polyethylene, plotted in terms of the scattering activity [Eq. (3.1.15)] in order to bring out LAM-3.

At a qualitative level of application, the presence of a LAM-1 band indicates some lamellar or lamellarlike structure. For example, the observation of the LAM-1 mode for bulk crystallized and cold-drawn polyethylene, both linear and branched, demonstrates the presence of long period folding in these samples.[154] In other studies, changes in the character of lamellar structure have been monitored during the processes of crystallization,[155,156] deformation,[157] and annealing.[157,158]

In addition to the use of this technique as a semiquantitative method to study and monitor crystal morphology, Eq. (3.1.40) indicates the possibil-

[154] A. Peterlin, H. G. Olf, W. L. Peticolas, G. W. Hibler, and J. L. Lippert, *J. Polym. Sci., Part B* **9**, 583 (1971).

[155] P. J. Hendra, E. P. Marsden, M. E. A. Cudby, and H. A. Willis, *Makromol. Chem.* **176**, 2443 (1975).

[156] P. J. Hendra, H. P. Jobic, and K. Holland-Moritz, *J. Polym. Sci., Polym. Lett. Ed.* **13**, 365 (1975).

[157] G. V. Fraser, P. J. Hendra, M. E. A. Cudby, and H. A. Willis, *J. Mater. Sci.* **9**, 1270 (1974).

[158] J. L. Koenig and D. L. Tabb, *J. Macromol. Sci., Phys.* **9**, 141 (1974).

ity of determining the longitudinal elastic modulus. Values of E derived from LAM measurements on polyethylene,[115] isotactic polypropylene,[152,153] poly(ethylene oxide),[159] and copolymers of polytetrafluoroethylene[150] are all significantly larger than the values obtained by an X-ray method whereby changes in the lattice constants are measured as a function of stress.[160] The reasons for the differences are not clear. The X-ray values may be too low because the assumption of uniform stress is a poor approximation. On the other hand, Eq. (3.1.40) is valid only for an isolated chain with free ends, and while there is little evidence that lateral chain interaction is important, end effects may be. There have been calculations on generalized models that attempt to take end effects into account,[151,161] and for the n-paraffins, part of the discrepancy between the values of the elastic moduli derived from the LAM and X-ray methods can be accounted for in these terms.[162]

It has been observed[151] that the shape of the LAM-1 band varies between specimens depending on their processing history. The shape of this band can be related quantitatively to the distribution of straight chain segments. The (unnormalized) distribution can be obtained from the relation

$$f_L \propto [1 - \exp(-hc\nu/kT)]\nu^2 I_\nu, \qquad (3.1.41)$$

where f_L is proportional to the number of chains of length L, ν is the corresponding LAM-1 frequency, T the sample temperature, and I_ν the observed Raman intensity at the frequency ν. Techniques for measuring I_ν and computing the distribution of linear chain segment lengths have been reported for samples of "solid-state" extruded and bulk crystallized polyethylene.[148] The distribution derived from the spectrum in Fig. 16 is shown in Fig. 17. This curve, which is for an extruded sample of polyethylene, is notable for its tail on the long segment length side, suggesting the presence of extended-tie chains.

Observation of LAM bands has so far been restricted to synthetic polymers. Analogous modes are predicted to occur in the straight chain segments of certain biopolymer systems such as protein crystals. For example, calculations indicate that a segment of α-helical poly(L-alanine) composed of 50 residues should have a LAM-1 mode near 10 cm^{-1}.[163]

3.1.5.3.3. MIXED CRYSTAL STUDIES ON CHAIN ORGANIZATION. There are unsolved structural problems concerning chain reentry. In the

[159] A. Hartley, Y. K. Leung, C. Booth, and I. W. Shepherd, *Polymer* **17**, 355 (1976).
[160] I. Sakurada, T. Ito, and K. Nakamae, *J. Polym. Sci., Part C* **15**, 75 (1966).
[161] S. L. Hsu and S. Krimm, *J. Appl. Phys.* **47**, 4265 (1976).
[162] G. R. Strobl and R. Eckel, *J. Polym. Sci., Polym. Phys. Ed.* **14**, 913 (1976).
[163] T. Shimanouchi, Y. Koyama, and K. Itoh, *Prog. Polym. Sci., Jpn.* **7**, 273 (1974).

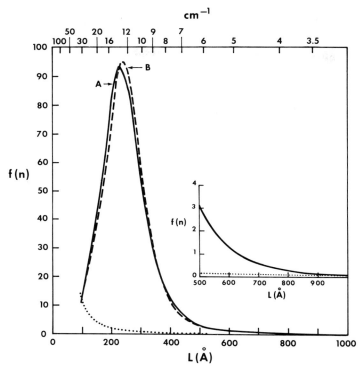

FIG. 17. Distribution of straight chain length segments in extruded polyethylene as derived from the spectrum shown in Fig. 16; the dashed curve (B) has been corrected for the effect of the finite width of LAM-1 for a given segment length; the dotted curve represents the estimated error in the distribution [R. G. Snyder, S. J. Krause, and J. R. Scherer, *J. Polym. Sci., Polym. Phys. Ed.* **68**, 4156 (1978)].

case of crystals grown from solution, a chain, after emerging from the core of a crystal, usually reenters the same crystal. One important question is where the reentry occurs.†

Two idealized models have been proposed. In the random reentry model the chain reenters the crystal at a place determined statistically. In this model it is unlikely that two straight chain segments immediately adjacent would share a common fold. On the other hand, the adjacent reentry model assumes this will always be the case. This model assumes that not only is reentry adjacent, but that the pattern of reentry is regular. The spectra of mixtures of polyethylene (PEH) and perdeuteropolyethylene (PED) manifest the effects of intermolecular coupling in different

† See also Chapter 5.4 (this volume).

ways for these two models, and this forms a basis for distinguishing them[164,165]

The methylene bending and rocking factor-group modes appear in the infrared spectrum of crystalline PEH at room temperature as intense doublets at 1473 and 1463 cm^{-1}, and at 731 and 720 cm^{-1}, respectively (Fig. 7); for PED the methylene bending mode components appear at 1091 and 1084 cm^{-1}. The same frequencies are observed for the orthorhombic or monoclinic n-paraffins both of which have essentially the same structure as crystalline polyethylene.   The splitting is due to interchain coupling (Table I), and as has been pointed out, the observed frequency separations can be calculated in the sense that physically reasonable values of intermolecular force constant parameters result when these parameters are adjusted to reproduce the observed separations.[136,166]

If all the molecules save one are deuterated in an orthorhombic n-paraffin crystal, the undeuterated molecule no longer undergoes a significant vibrational coupling with its neighbors,[167] and consequently its methylene bending and rocking modes will appear as single bands. This situation can be approximated by diluting an n-paraffin with its deuterated analog. For a solid solution consisting of 5% $n-C_{36}H_{74}$ in $n-C_{36}D_{74}$, i.e., 1H/20D, single bands are observed for the bending mode (1466 cm^{-1}) and rocking mode (723 cm^{-1}); similarly, for 1D/20H solution, the bending mode of the deuterated molecule also appears as a single band (1086 cm^{-1}).[165]

An intermediate situation arises when a molecule in dilute solution has one adjacent neighboring molecule of its own isotopic kind. In this case coupling, and hence band splitting, will occur, but the frequency separation between the components will be less than that for the isotopically pure crystal, and will be affected by how the two molecules are oriented with respect to one another. Using interchain potential constants derived from the isotopically pure crystals, the magnitude of the splitting can be estimated for different coupling situations that correspond to different kinds of chain-folding organization. A distinction can be made between the random and adjacent reentry models for polyethylene from the observed spectrum of an isotopically diluted chain. Splitting in the spectrum of the dilute component will not occur if chain reentry is random since the probability is low that a neighboring chain will be of the same isotope. Splitting will occur if reentry is adjacent since in this case there will always be neighboring chains of the same isotope and thus vibrational

[164] M. Tasumi and S. Krimm, J. Polym. Sci., Part A-2 6, 995 (1968).
[165] M. I. Bank and S. Krimm, J. Polym. Sci., Part A-2 7, 1785 (1969).
[166] M. Tasumi and S. Krimm, J. Chem. Phys. 46, 755 (1967).
[167] G. L. Hiebert and D. F. Hornig, J. Chem. Phys. 20, 918 (1952).

coupling will occur; furthermore, the magnitude of the splitting will be different depending on whether the folding is along (110) or (200) lattice planes.[166]

Such studies indicate[165] that for crystals grown in dilute solution, the folding is (110) and random reentry does not occur to an appreciable extent. These conclusions have been questioned on the grounds that aggregation of the minor isotopic constituent may occur due to the difference between the melting points of PEH and PED.[168] Further studies addressed to this question appear to show that the effects of aggregation can be distinguished from the effects of folding, and that the above conclusions are not affected.[169] This same mixed crystal technique has been used to study "cilia" (loose ends of a polymer chain not incorporated into the crystal)[170] and chain folding in the bulk crystallized polymer.[171]

3.1.5.3.4. STUDIES ON CHAIN CONFORMATION IN THE FOLDS. Proposed models of the conformation of the chain in the fold region vary from the well-defined, such as might occur in the case of "tight" folding in crystals grown from dilute solution, to the poorly defined, such as might be associated with "loose" folding in bulk-crystallized samples.

There are bands in the spectra of polymers that are characteristic of aperiodic chains. Although these are properly the subject of Section 3.1.6, it is appropriate to mention here that bands of this type have been used to characterize fold structure. In the infrared spectrum of polyethylene, bands associated with the presence of gauche bonds appear in the region $1400-1300$ cm^{-1}. Their relative intensities are sensitive to conformation, and there are significant differences in this region of the infrared spectrum between samples crystallized differently. In one study[172] the absorptivities of bands at 1350 and 1304 cm^{-1} were measured for two samples, one of which was prepared in a way that minimized folding, while the second was prepared in such a way that chain disorder could be assumed to be confined mostly to the folds. The values of the absorptivities thus established were used to compare samples as to their degree of folding. The spectra of a slow-cooled bulk-crystallized polyethylene and solution-crystallized polyethylene were compared, and on this basis, it was concluded that the folding is essentially the same in the two cases. More precise studies of this kind have been made using the infrared Fourier transform spectrometer, and heretofore unrecognized

[168] F. C. Stehling, E. Ergos, and L. Mandelkern, *Macromolecules* **4**, 672 (1971).
[169] S. Krimm and J. H. C. Ching, *Macromolecules* **5**, 209 (1972).
[170] M. I. Bank and S. Krimm, *J. Appl. Phys.* **40**, 4248 (1969).
[171] J. H. C. Ching and S. Krimm, *J. Appl. Phys.* **46**, 4181 (1975).
[172] J. L. Koenig and D. E. Witenhafer, *Makromol. Chem.* **99**, 193 (1966).

spectral features characteristic of the fold conformation have been revealed.[173] The spectrum of previously undried single crystals of polyethylene suspended in a liquid has been compared with the spectrum of
dried mattes.[174] From this study it was concluded that drying causes a
conformational change in the amorphous and fold structure, and results
in some deterioration of lateral order in the crystal core.

The infrared spectrum of polyethylene containing a low concentration
of randomly distributed $-CD_2-$ or $-CD_2CD_2-$ groups can be used to
identify the conformational states of CC bonds.[175] This technique,
described in more detail in Section 3.1.6.4.1, has been used to investigate
conformational structure in the crystalline state. The spectrum of
(CD_2CD_2)-doped polyethylene displays a number of bands each of which
can be associated with some specific conformational state of a sequence of
three adjoining CC bonds. A number of such bands are identified in Fig.
18, which shows the spectrum of a bulk crystallized sample. Comparison
of the spectra of bulk- and solution-crystallized polyethylene reveals significant differences in the degree and the nature of the conformational disorder. A structure involving a trans–gauche–trans (TGT) sequence, for
example, is associated more with the solution-crystallized polymer than
with the bulk-crystallized polymer.

3.1.5.3.5. STUDIES ON CHAIN "DEFECTS." The dynamics of a predominantly regular polymer chain that contains a low concentration of
"defects" may be viewed in terms of a perturbation of the vibration of the
ideally periodic chain. A number of different kinds of defects may be envisioned. They may be associated with mass, as when a small amount of
an isotope is incorporated randomly in a regular polymer; or they may be
chemical in nature, as when a foreign entity is incorporated in the chain.
In addition, there are two kinds of structural defects: configurational, as
head-to-tail inversion, and conformational, as exemplified by the various
defect structures postulated to account for peaks in mechanical relaxation
spectra.

For such an approach to be applicable, the perturbation must be localized. Thus, a suitable perturbation would not include the fold structure
associated with single crystals but would include a low concentration of
CD_2 groups in crystalline polyethylene, or limited conformational and
configurational disorder in, say, poly(vinyl chloride). Some insight into

---

[173] P. C. Painter, J. Havens, W. W. Hart, and J. L. Koenig, *J. Polym. Sci., Polym. Phys.
Ed.* **15**, 1223 (1977).
[174] P. C. Painter, J. Runt, M. M. Coleman, and I. R. Harrison, *J. Polym. Sci., Polym.
Phys. Ed.* **15**, 1647 (1977).
[175] R. G. Snyder and M. W. Poore, *Macromolecules* **6**, 708 (1973).

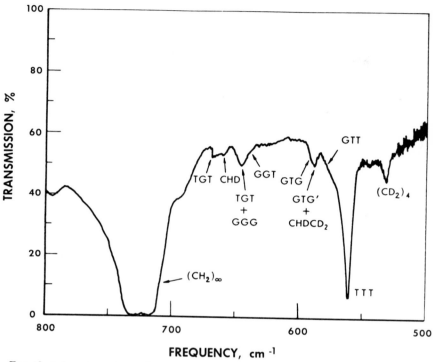

FIG. 18. Infrared spectrum of a pressed film of CD_2CD_2-doped polyethylene at 77 K in the rocking mode region. Reprinted with permission from R. G. Snyder and M. W. Poore, *Macromolecules* **6**, 708 (1973). Copyright by the American Chemical Society.

how these irregularities affect spectra may be derived from a simple model.[176]

Such a model consists of a system of one-dimensional coupled oscillators, all identical save for one oscillator, "the defect." This oscillator has a "natural" frequency (the frequency it would have if it were uncoupled) that is different from the rest of the oscillators. How the unperturbed system is affected depends on several factors. These include the relation between the natural frequency of the defect and the frequency region covered by the dispersion curve of the unperturbed system, as well as the strength of the coupling between the defect and the chain.

If the natural frequency of the defect is well removed from the dispersion curve, as is often the case for chemical impurities (e.g., carbon–carbon double-bond stretching from residual unsaturation in polyethyl-

[176] C. G. Opaskar and S. Krimm, *J. Polym. Sci., Part A-2* **7**, 57 (1969).

ene), the vibration associated with the defect remains essentially unaffected and does not interact appreciably with the rest of the system. If the frequency is near the dispersion curve, vibrations other than factor-group modes are allowed to appear in a spectrum. As a result, some infrared and Raman bands may be modified in shape, or new bands may appear. In either case, these spectral perturbations are related to the density of vibrational states of the regular chain. If the natural frequency of the defect falls within the region of dispersion, the situation is complex and the effects are difficult to predict. A likely result is the appearance of new bands whose frequencies may be inside or outside the frequency region of the dispersion curve.

At present, a quantitative analysis of these effects for real polymers is difficult. First it must be assumed that a sufficiently accurate normal coordinate calculation for the unperturbed chain is available. Then, if data on the intramolecular forces associated with the defect are also available, the frequencies and normal coordinates of modes associated with, or affected by, the defect may be calculated. In order to make a quantitative comparison with the observed spectrum it is necessary to weigh the calculated modes by intensity factors. In spite of these formidable difficulties, this approach has provided insight into the complex effects of conformational and configuration disorder on the C–Cl stretching vibrations of poly(vinyl chloride),[177] and has been used to explore the effects of structural disorder on the spectrum of polytetrafluoroethylene.[139,178]

A related structural problem concerns whether noncrystallizable counits are, or are not, included in the crystalline domains of a copolymer.[179,180] This question has been approached experimentally in a study on trans-1,4-polychloroprene.[181] The spectrum of the crystalline fraction of the polymer was obtained using the accurate spectral subtractive capabilities afforded by the Fourier transform infrared spectrometer. This spectrum, shown in Fig. 19, exemplifies the power of the technique. Many bands of the crystalline fraction were found to undergo small frequency shifts (<6 cm^{-1}), which depended on the concentration of structural defects in the chain. These shifts were interpreted as perturbations resulting from the incorporation of structural irregularities into chains that have predominantly the trans-1,4 structure.

3.1.5.3.6. DETERMINATION OF CRYSTALLINITY. The vibrational spec-

[177] A. Rubčić and G. Zerbi, *Macromolecules* **7**, 759 (1974).
[178] G. Zerbi and M. Sacchi, *Macromolecules* **6**, 692 (1973).
[179] P. J. Flory, *Trans. Faraday Soc.* **51**, 848 (1955).
[180] I. C. Sanchez and R. K. Eby, *J. Res. Natl. Bur. Stand., Sect. A* **77**, 353 (1973).
[181] D. L. Tabb, J. L. Koenig, and M. M. Coleman, *J. Polym. Sci., Polym. Phys. Ed.* **13**, 1145 (1975).

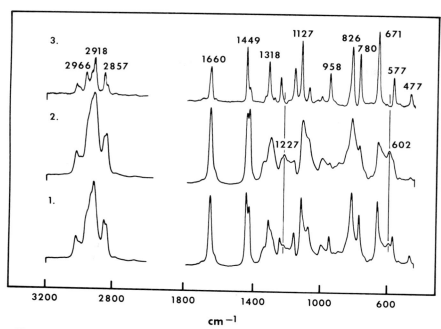

FIG. 19. Infrared absorbance spectra of trans 1,4-polychloroprene: (1) at room temperature; (2) at 80°C; (3) difference between spectra 1 and 2, showing the spectrum of crystalline component [M. M. Coleman, P. C. Painter, D. L. Tabb, and J. L. Koenig, *J. Polym. Sci., Polym. Lett. Ed.* **12**, 577 (1974)].

tra of the crystalline and amorphous fractions of a polymer are different, and this difference can be utilized to estimate the degree of crystallinity. Measurements are usually made in the infrared, where the procedure is simple if bands can be found that are characteristic of the crystalline and amorphous components, and if reference samples consisting of the predominantly crystalline and amorphous forms are available. If quantitative results are needed, the infrared measurements are calibrated using another technique, usually X ray. A tabulation of the absorption frequencies commonly used for crystallinity measurements has been assembled.[182]

It should be emphasized that the spectroscopic technique does not actually measure crystallinity. The differences between the spectra of a crystalline and amorphous polymer are the result of the chain being conformationally regular in the one case and disordered in the other. The

[182] B. Wunderlich, "Macromolecular Physics," Vol. 1, p. 418. Academic Press, New York, 1973.

spectrum of the regular chains in the amorphous fraction will be very similar to, if not indistinguishable from, the spectrum of the crystalline fraction. It may happen also that there are variations in the type and degree of conformational disorder associated with the amorphous fraction. This will lead, of course, to variations in the spectra of the amorphous component. An example has already been cited [172] where the spectra of certain amorphous fractions of polyethylene were found to differ according to how samples were processed.

3.1.6. Disordered Polymer Systems

3.1.6.1. Introduction. Indirect methods must be used for investigating the conformational structure of highly disordered chain molecules such as occur in solution, in a melt, or in the amorphous fraction of a solid. Diffraction methods, supreme in their ability to render structures displaying spatial periodicity, are relatively ineffective when applied to disordered systems. Spectroscopic methods fare better, but their approach is indirect since energy levels, not structural parameters, are measured. The dependence of the energy levels on conformational structure is complex, and the effectiveness with which spectroscopic data can be used to determine structure rests on how well this dependence is understood.

It is apparent that a complete vibrational analysis of a long aperiodic chain molecule is impractical because the number of vibrational modes of a chain in a given conformation is enormous, as is the number of conformationally different chains. How rapidly the difficulty of interpreting spectra increases as the chain length increases is vividly illustrated by the n-alkanes in the liquid state.[56] Both n-butane and n-pentane exist in more than one rotameric form, but the number of forms is small. The spectra of these n-alkanes can be interpreted completely in that virtually every observed band in their infrared and Raman spectra can be identified with one or more of the rotamers. One reason assignments for these small molecules in the liquid state are easier than for the larger molecules is that their spectra can be measured at temperatures lower than are possible for the longer chains and, thus, bands are better resolved, as seen in Fig. 20. The next member of the series, n-hexane, is the last for which most of the observed bands can be associated with most of the possible rotamers: the prominent features of this spectrum can be confidently assigned to the more abundant rotamers although in the case of some of the less intense bands the assignments are not without ambiguities, and a few of the rotamers that are predicted to be present cannot be identified in the spectra. For longer chains our ability to identify rotamers deteriorates rapidly while at the same time the observed spectra appear to became simpler. This apparent simplicity is evident in the spectrum of n-C$_{17}$H$_{36}$ shown in

Fig. 20. In the case of molten polyethylene, also shown in this figure, nearly all band structure has disappeared. As a result, the observed spectrum of the polymer is relatively uninformative. It consists of a plethora of overlapping bands whose individual identities are lost along with most of the conformational information that could otherwise be derived from them. This situation is a representative one: the infrared and Raman spectra of polymer chains in highly disordered states tend to show diffuse, broad bands.

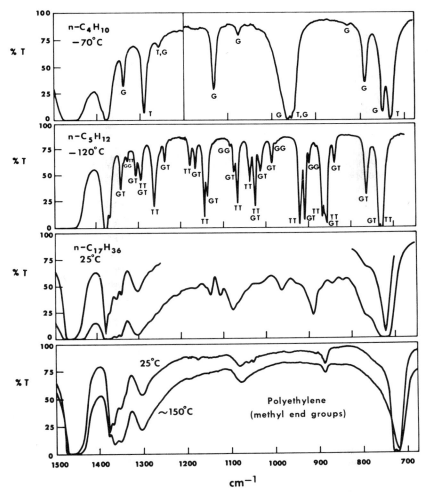

FIG. 20. Infrared spectra of polymethylene chains of different lengths in the liquid state and crystalline polyethylene at 25°C [R. G. Snyder, *J. Chem. Phys.* **47**, 1316 (1967)].

3.1.6.2. Classification of Bands of a Disordered Chain. 3.1.6.2.1.
BASIS OF CLASSIFICATION. Changes in the spectrum of an ordered polymer are observed to occur when, as a result of melting or dissolution, the
polymer chain becomes conformationally disordered. The apparent behavior of bands through this transition provides a basis for classification.
There are two general classes of bands.

When a chain becomes disordered, some factor-group modes observed
in the spectrum of the regular chain persist, albeit with some modification
that may include the loss of fine structure (e.g., multiplicity), small frequency shifts in the peak positions, and some changes in the breadth and
shape of the band. These bands, characterized by insensitivity to conformational change, constitute one class.

There is a second class of bands that are sensitive to conformational
change. Included are bands suffering large decreases in intensity (even
to the point of appearing to vanish) when the chain becomes disordered,
and new bands not present in the spectrum of the regular chain.

Although this classification must necessarily be imprecise, it serves as a
framework for qualitative understanding of why changes occur in going to
a disordered system. Before discussing this classification in more detail,
some general remarks relating to the dynamics of a disordered system are
appropriate.

Conformational disorder leads to a loss of periodicity in the vibrations
of the chain. The periodicity is lost because the coupling between certain
internal coordinates is dependent on conformation. Although coupling
between coordinates involves both potential and kinetic interaction
terms, it is the kinetic energy terms that play the more important role
since they alone are spatially sensitive, i.e., they are sensitive to the
changes in the relative spatial orientation of interacting coordinates that
result from changes in bond conformation.

Unlike the periodic chain, whose vibrational modes are propagated
along the whole chain, regardless of its length, the vibrations of the
aperiodic chain are localized within finite segments of the chain, the
length of which depends on the type of vibration involved. Thus, the
spectrum of a disordered chain displays bands due to modes that are more
or less localized, and an analysis of such a spectrum focuses on the
strength of the coupling between vibrating units and how this is affected
by conformational change.

The observed spectrum cannot be interpreted without considering intensities. Related to the loss of symmetry is the attendant loss of formal
selection rules that constrain infrared or Raman activity. However, even
if all modes appeared in the spectrum with equal intensity (which they do
not), the observed spectrum would still show maxima reflecting an uneven
distribution of vibrational states.

It is observed that the infrared spectrum and the Raman spectrum of a disordered chain differ from one another, and that the difference is no less marked than that observed between the infrared and Raman spectra of the ordered chain. This can be largely understood on the basis of local symmetry. The observed intensities are determined by the distribution of vibrational states, each state being weighted by its intrinsic intensity. A specific vibrational mode will have an intensity in the infrared different from that in the Raman because infrared intensity depends on dipole derivatives and Raman intensity on polarizability derivatives. The magnitude of these derivatives is determined in large part by local symmetry—by the presence of a local inversion center, for example. A case in point is the stretching of a $C–C$ bond in the polymethylene chain. The center of inversion associated with this bond ensures that any mode involving primarily $C–C$ stretching will be strong in the Raman and weak in the infrared since by analogy with a covalent homopolar diatomic molecule the stretching mode is allowed and is normally intense in the Raman, but is not allowed in the infrared. This will tend to be the case independent of conformation since to a first approximation the inversion center associated with this bond is present for both the trans and gauche states. In short, even though there are no selection rules for the disordered chain, local symmetry plays the primary role in determining vibrational intensities, and arguments based on local symmetry can account qualitatively for intensity changes.

3.1.6.2.2. BANDS INSENSITIVE TO CONFORMATION. *3.1.6.2.2.1. Localized Modes.* Highly localized vibrations are normally little affected by conformational change. These modes are weakly coupled so that their dispersion curves are characteristically flat. Often these vibrations are of the kind closely associated with certain chemical groups. For example, if the backbone of a polymer has pendant groups, the characteristic vibrations of these groups will tend to be retained. Such is the case for isotactic polystyrene. The repeat unit of the helical chain of this polymer contains three subunits $[–CH_2–CH(C_6H_5)–]$ per turn. The selection rules indicate that any vibration characteristic of the phenyl group (i.e., a mode localized within it) will have two components in the spectrum of the polymer: one is totally symmetric (species A) and consists of all phenyls vibrating in phase; the other is degenerate (species E) and consists of adjacent phenyls vibrating with a phase difference $\phi = 120°$. Analysis of the spectrum of the helical chain[183] shows that in some cases the result of coupling between the characteristic phenyl vibrations is to separate their A and E components by only a few wavenumbers, so that it is not

[183] P. C. Painter and J. L. Koenig, *J. Polym. Sci., Polym. Phys. Ed.* **15,** 1885 (1977).

surprising that these bands have essentially the same frequencies in the spectrum of the crystalline, dissolved, and molten states of the polymer.

Characteristic frequencies are of special interest in the analysis of the structure of biopolymers.[15,184] Some nine vibrations more or less characteristic of the secondary amide group have been identified, and the structural factors affecting their frequencies have been studied extensively. In special cases it is possible to approximate the observed spectrum of a protein by adding together the spectra of its amide constituents.[185] Similar approaches are used to interpret the spectra of the nucleic acids, though these systems are more complex, being made up of bases, sugars, and phosphates.[24,186]

A band found to be insensitive to conformation may be associated with a characteristic frequency that has not been previously established. A simple system where this occurs is polypropylene. The infrared spectra of crystalline isotactic polypropylene[187] and the two crystalline forms of syndiotactic polypropylene[188] differ from one another in many ways. These spectra have in common, however, two intense bands, near 1155 and 975 cm^{-1}. After the polymers are melted, these bands remain and are now the most prominent features in the spectra below 1300 cm^{-1}, all other bands having disappeared or undergone a large reduction in intensity. It has been shown through normal coordinate calculations[81] that the 1155 and 975 cm^{-1} bands are localized vibrations and that they are associated with methyl rocking. This is consistent with the fact that methyl groups in saturated hydrocarbons are known to have rocking mode frequencies in this general region. In this way the existence of a conformationally insensitive mode of a $-CH_2CH(CH_3)-$group in polypropylene chain was established.

3.1.6.2.2. Nonlocalized Modes. There is a situation, perhaps not an uncommon one, in which a band in the spectrum of the disordered chain can appear at the same frequency as in the regular chain even though the mode involves strong intrachain coupling. This can occur if the dispersion curve associated with the mode is nearly flat in the region nearest the zone center, and if at the same time the disordered chain has an appreciable concentration of segments, even short segments, that have the conformation of the regular chain. Under these circumstances the mode in question will occur in the ordered segments at frequencies near the factor-group frequency of the ordered chain. An example is the infrared

[184] R. C. Lord, *Appl. Spectrosc.* **31**, 187 (1977).
[185] R. C. Lord and N. Yu, *J. Mol. Biol.* **50**, 509 (1970); **51**, 203 (1970).
[186] M. Tsuboi, *Appl. Spectrosc. Rev.* **3**, 45 (1969).
[187] J. P. Luongo, *J. Appl. Polym. Sci.* **3**, 302 (1960).
[188] M. Peraldo and M. Cambini, *Spectrochim. Acta* **21**, 1509 (1965).

active methylene rocking band, which occurs near 720 cm^{-1} in the spectrum of the regular polyethylene chain (Fig. 7) and persists in the liquid (Fig. 20) even though the internal coordinates associated with the mode are known to be highly coupled. Although in the liquid state the population of gauche bonds may approach 50%, a significant fraction of segments have three or more trans bonds in sequence. These sequences have an "in-phase" rocking mode for which, according to Eq. (3.1.37), $\phi > 135°$, so that the frequency of this mode is in the region 730–720 cm^{-1} as indicated by the dispersion curve for ν_8 shown in Fig. 6. Thus a band appears to persist at nearly the same frequency as for the regular chain. The same situation occurs in the Raman for the intense 1300 cm^{-1} methylene twisting band. This band appears in the spectra of both the ordered and disordered polymer.

3.1.6.2.3. BANDS SENSITIVE TO CONFORMATION. *3.1.6.2.3.1. Localized Modes.* A chain vibration is usually conformationally sensitive only when the coupling between the neighboring internal coordinates associated with the vibration changes significantly with conformation.

There is one exceptional case, however, where bands associated with localized, weakly coupled modes can be conformationally dependent. This comes about in an indirect way due to the interaction of the localized mode with a binary combination (two-phonon state) that itself is conformationally dependent. This type of interaction, called Fermi resonance interaction, is well known for small molecules and takes place between a fundamental vibration and a binary combination, provided that the frequencies of the fundamental and the combination are sufficiently close, and that certain symmetry requirements are met.[3] The interaction mixes the wavefunctions of the two states and has two effects: there is a frequency displacement of the two states that increases their separation; the binary state increases its intensity at the expense of the fundamental. The same phenomenon also occurs for polymers. In the case of a periodic chain, the interaction is between a fundamental and not just one but a continuum of binary states, namely, those given by Eq. (3.1.39). If the interaction is strong, the result can be a gross alteration of the band shape of the fundamental. The degree of interaction depends on the frequency separations between the unperturbed fundamental and the binary states, as well as on the symmetry and the density of the binary states. These factors are in turn dependent on the crystal structure of the regular chain and are affected by changes in conformation. Hence the band shape of the fundamental is sensitive to changes in crystal structure and conformation.

Such effects have been observed for the polymethylene chain. For the crystal, the shape of the intense Raman band attributed to the symmetric

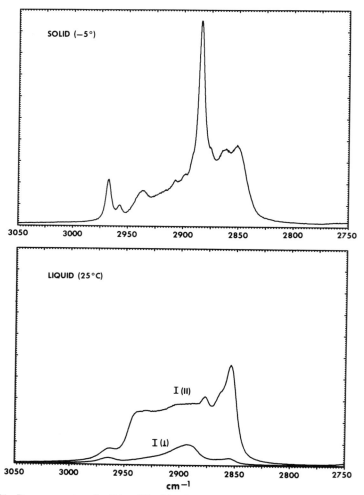

FIG. 21. Raman spectra of solid and liquid n-$C_{16}H_{34}$ in the C–H stretching region (figure by courtesy of J. R. Scherer).

C–H stretching fundamental of the methylene group is found to be sensitive to chain packing. This results from interaction between the factor-group fundamental and the binary combinations involving methylene scissoring modes since the shapes of the dispersion curves of these latter modes are sensitive to interchain coupling.[78] In addition, the effect of melting is to drastically change the Fermi resonance interaction and in consequence to change the band structure in this region as may be seen in Fig. 21. This sensitivity to packing and conformation makes Raman

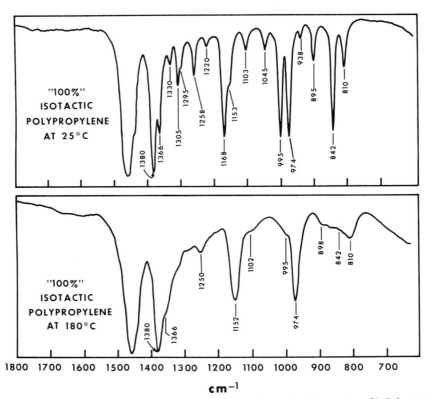

FIG. 22. Infrared spectra of crystalline and molten isotactic polypropylene [J. P.Luongo, *J. Appl. Polym. Sci.* **3,** 301 (1960)].

spectroscopy useful for investigating and monitoring the structure of hydrocarbon chains in phospholipid membrane systems.[189-191]

3.1.6.2.3.2. Nonlocalized Modes. With few exceptions, bands due to highly coupled nonlocalized factor-group modes "disappear" when conformational disorder is introduced. These modes remain only if there is present an appreciable concentration of segments that have the conformational structure of the regular chain, and that are sufficiently long to give ϕa a value so small that the $k = 1$ mode of Eq. (3.1.36) approximates a factor-group mode. An appreciable concentration of such segments is not likely to exist in a complex, highly disordered system. An example is isotactic polypropylene, whose infrared spectra are shown in Fig. 22.

[189] B. J. Bulkin and K. Krishnamachari, *J. Am. Chem. Soc.* **94,** 1109 (1972).
[190] K. Larsson and R. P. Rand, *Biochim. Biophys. Acta* **326,** 245 (1973).
[191] B. P. Gaber and W. L. Peticolas, *Biochim. Biophys. Acta* **465,** 260 (1977).

Upon melting the crystalline polymer, its spectrum is "simplified," in that the majority of its bands have disappeared or lost most of their intensity. Even those bands seeming to persist are probably associated with the presence of a low concentration of helical segments.

In the transition from order to disorder, the spectra of simpler polymers usually undergo greater changes than the spectra of more complex systems since for simpler polymers a greater proportion of their vibrations are sensitive to conformation. For example, polypropylene consists of only two simple groups (CH_2 and $CHCH_3$), whose modes are more or less highly coupled through a bond whose conformation is determined statistically when the chains is in a disordered state.

3.1.6.2.3.3. New Bands. In the spectrum of a disordered chain, there appear bands that have no counterpart in the spectrum of the regular chain. The positions and intensities of these bands are dependent on conformation, but a detailed interpretation is usually difficult and made the more so by their weakness and diffuseness.

These bands are associated with segments of the disordered polymer, within which are localized vibrations characteristic of the conformation of the segment.

A relatively simple system that has been analyzed in some detail is the complex of wagging modes of the polymethylene chain.[56] The wagging coordinates of adjacent methylenes are highly coupled when the CC bond connecting them is in the trans conformation. The coupling is greatly reduced, however, when this bond assumes a gauche conformation. Thus, in terms of methylene wagging coordinates, the disordered chain consists of trans segments that are highly coupled within but separated and uncoupled from other trans segments by one or more gauche bonds. These segments may be designated as $G^+T_mG^\pm$, where $m = 0.1, . . .$, where G^+ and G^- refer to right and left gauche states. Corresponding to each m there is at least one wagging mode characteristic of the segment. A number of these conformations have been identified with bands in the infrared spectra (Fig. 20) of the liquid *n*-paraffins and polyethylene in the region $1400-1250$ cm^{-1}, e.g., GG at 1350 cm^{-1}, GTG$^-$ at 1367 cm^{-1}. These bands are also present in the spectrum of crystalline polyethylene and, as mentioned earlier, are useful for studying conformational structure of the folds.

Some types of polymer structure are conducive to vibrational modes that have an especially simple dependence on conformation. An example of such a structure is $(-XCH_2CH_2-)_n$, where X is part of the skeleton but is not a CH_2 group. Here the methylenes are coupled in pairs, and the frequency of certain vibrational modes is determined by the conformational state of the CC bond joining them. A system of this type which has been studied is poly(ethylene glycol).[192]

The C–C stretching vibrations of the polymethylene chain are less localized than methylene wagging since coupling between the C–C stretching coordinates is nearly the same regardless of the conformation of the CC bonds. As a result these modes extend over long segments of the disordered chain. Although the vibrational dynamics of these modes are complex and incompletely analyzed, the intense Raman bands attributable to them are commonly used in conformational studies. The Raman spectrum of the extended chain shows two very intense factor-group bands at 1130 and 1065 cm^{-1}. Conformational disorder reduces their intensity and at the same time leads to a new band at 1080 cm^{-1}. The changes in the intensities of these bands have been used to estimate chain disorder. For example, changes in the conformational structure of the hydrocarbon component of phospholipid membranes can be followed by monitoring these bands.[191]

3.1.6.3. Analysis of Spectra by Computational Methods. There have been a few attempts to correlate the observed vibrational spectra of the disordered chains of simple polymers with spectra calculated from model systems.

In one approximation it is assumed that each segment of the disordered chain can be associated with an ordered polymer that has essentially the same conformation as the segment. A frequency comparison is then made between bands in the observed spectrum and the calculated factor-group modes of the ordered polymer. In this way bands in the Raman and infrared spectra of liquid poly(ethylene glycol) have been identified with one, or more than one, of the six conformationally lowest energy forms of the regular chain.[192]

Another approach involves analyzing the rotameric forms of oligomers of the polymer. Certain bands associated with the disordered conformation of a polymer chain appear also in the spectrum of a disordered chain of a much shorter oligomer. Since the vibrations responsible for these bands of the oligomer are necessarily localized within the molecule, it follows that the corresponding vibrations in the polymer are localized within a segment essentially no longer than the oligomer chain. The relation between the frequencies of these vibrations and the various conformations of the oligomer can be analyzed through normal coordinate calculations. This approach has been used to study the rotameric forms of the simpler liquid n-paraffins, and many of the conformationally dependent

[192] H. Matsuura and T. Miyazawa, *J. Polym. Sci., Part A-2* **7,** 1735 (1969).

bands of disordered polyethylene have been thus identified.[56] This approach is less effective for vibrations, such as C–C stretching modes, which extend over long segments, since it is necessary to analyze correspondingly long oligomers.

A more comprehensive approach entails calculating the normal modes of a segment of a chain that has a statistically random conformation and that is long enough to approximate a real polymer. The advantage here is that the results are not significantly affected by end effects, and that the less localized vibrations are included. A calculation of this kind has been performed for a random polymethylene chain consisting of 200 CH_2 groups.[193] The calculated frequencies are expressed in terms of a density of vibrational states taken at intervals of 10 cm^{-1} or less. These results are useful in explaining some aspects of the observed spectrum, and there is some correlation between the density of states and the infrared spectrum of molten polyethylene. For the purpose of comparison, however, the calculated density of states must be weighted by intensity factors. The importance of these is brought out when the infrared and Raman spectrum of this polymer in the molten state are compared. Although the same density of states underlies each, the two observed spectra do not resemble one another.

The above methods for interpreting the spectra of disordered polymer chains are based on techniques for calculating normal coordinates. These lie in the domain of the specialist and furthermore rely on the availability of accurate force constants. Experimental methods will now be considered. Here the emphasis is less on the interpretation of spectra, and more on the determination of chain conformation from spectra.

3.1.6.4. Determination of Conformation. 3.1.6.4.1. ANALYSIS BY DILUTE ISOTOPIC SUBSTITUTION. It has been emphasized that much of the spectroscopic data informative of disordered chain conformation is lost because the multitude of bands from these large molecules overlap to form continua. In some situations this complexity can be bypassed by using deuterium substitution to uncouple and isolate a single vibration whose frequency exhibits a conformational dependence. In this way some aspects of the conformational structure of polyethylene are revealed in the infrared spectra of samples containing a low concentration (5%) of CD_2 groups.[175] In this case the vibration of interest is a methylene rocking mode (Fig. 4), which for a CH_2 group is essentially a motion of the hydrogens. The mass effect resulting from deuterium substitution thus reduces the frequency of this mode by a factor of approximately $1/\sqrt{2}$ so that the rocking mode of an isolated CD_2 group in an extended chain

---

[193] G. Zerbi, L. Piseri, and F. Cabassi, *Mol. Phys.* **22**, 241 (1971).

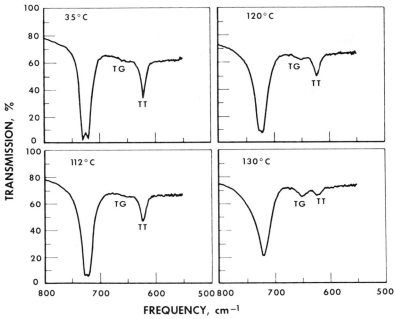

FIG. 23. Infrared spectra of CD_2-doped polyethylene at temperatures below and above the melting point [R. G. Snyder and M. W. Poore, *Macromolecules* **6**, 708 (1973)].

occurs at a frequency (~ 620 cm^{-1}) well below ν_8, the dispersion curve for methylene rocking in the undeuterated polymer chain. The CD_2 is almost but not quite uncoupled from the chain. Because of mild coupling with its two immediate CH_2 neighbors, the frequency of the CD_2 rocking mode is significantly dependent upon the conformation of the bonds connecting these neighbors to the CD_2 group. If the adjoining bonds are both trans, the frequency is 622 cm^{-1}, whereas if one is trans and the other gauche, it is 652 cm^{-1}. Spectra of CD_2-doped polyethylene at temperatures below and above its melting point are shown in Fig. 23. As the temperature is increased there is observed a significant increase in the concentration of TG pairs even before the melting point of the polymer is reached. In the liquid, the ratio of the concentration of TT and TG pairs can be determined, and is found to be somewhat higher than that calculated using the rotational isomeric state model.[53] Thus the method can be used to identify bond conformers and to measure their concentration.

As already mentioned in Section 3.1.5.3.4, the same technique has been employed to study conformation in crystalline polyethylene. For this purpose the spectra of samples doped with CD_2CD_2 entities provide additional information in that the rocking mode frequencies of this isotopic impurity are dependent on the conformation of three adjoining bonds.[175]

3.1.6.4.2. DETECTION OF CONFORMATIONAL ORDER. Since the spectra of ordered and disordered chain molecules are different, the degree of order in a partially disordered system can often be estimated. The analysis is, of course, simplest when the system consists of a perfectly ordered phase, and a perfectly disordered phase such as is sometimes nearly the case in the solid state.

The existence of conformational order in polymer chains in the liquid state is of great interest. In the case of molten isotactic polypropylene, it has been shown on the basis of infrared evidence supported by normal coordinate analysis, that segments of the three-fold helix, extending over at least five monomer units, are present in low concentration.[194] The evidence is the persistence of the band at 973 cm^{-1} in the infrared spectrum (Fig. 22) of the molten polymer. This band is identified with a conformationally sensitive factor-group mode of the helical chain, and its appearance in the melt signifies the presence of segments of the helical chain. Additional evidence for helical order is afforded by the Raman spectrum, which displays many of the same bands observed in the spectrum of the crystalline polymer, though with considerably reduced intensity.[195]

Similar approaches can be used to study the conformation of globular proteins in aqueous solution. The Raman spectrum of a crystalline protein whose structure is known from X-ray diffraction, may be compared with the spectrum of the same sample in solution. Changes in the backbone structure may be revealed by differences in the spectra.[196] The structure of transfer ribonucleic acid has been studied in this way.[197]

If present in the liquid, straight chain segments may be expected to manifest LAM-1 bands in the Raman. The low-frequency region where these bands are observed for crystalline polymers is largely obscured in liquids by depolarized scattering from intermolecular interaction. On the other hand if the segments are short enough, their LAM modes will occur in a higher, more accessible region. Such bands are observed in the melt phase of the shorter perfluoro n-alkanes.[150] These bands indicate that the chains exist as rigid rods in the liquid state, in contrast to the situation for the n-alkanes. For the latter, Raman spectra of chains longer than n-C_9H_{20} show a broad band that is characteristic of the disordered

[194] G. Zerbi, M. Gussoni, and F. Ciampelli, *Spectrochim. Acta, Part A* **23**, 301 (1967).

[195] G. V. Fraser, P. J. Hendra, D. S. Watson, M. J. Gall, H. A. Willis, and M. E. A. Cudby, *Spectrochim. Acta, Part A* **29**, 1525 (1973).

[196] N.-T. Yu and B. H. Jo, *J. Am. Chem. Soc.* **95**, 5033 (1973).

[197] M. C. Chen, R. Giege, R. C. Lord, and A. Rich, *Biochemistry* **14**, 4385 (1975).

chain.[198] This band appears in the region 250–200 cm¹, the lower limit being identified with the longest chains.

Infrared dichroism has been observed in aqueous solutions of certain synthetic polypeptides. Orientation was achieved by subjecting such a solution to a large rate of shear. In this way the existence of the α-helix form in solution was confirmed by observing the polarization of the carbonyl stretching band.[199]

Other studies have been designed to detect and monitor the conformational changes accompanying changes in such variables as temperature and concentration, or accompanying changes that may occur in going from one solvent to another. For example, when the temperature of a dilute solution of isotactic polypropylene or isotactic polystyrene is lowered, its infrared spectrum is altered to resemble a spectrum that is characteristic of helical conformation of the polymer.[200] The influence of a solvent on chain conformation is apparent in comparing the Raman spectra of poly(ethylene glycol) dissolved in water and in chloroform: the spectrum of the aqueous solution is more nearly like the crystalline polymer, while the spectrum of the chloroform solution more closely resembles the molten polymer.[16]

3.1.6.4.3. CONFORMATIONAL CHANGE FROM RAMAN DEPOLARIZATION RATIOS. Changes in conformation can be measured from variations in band intensities. This technique is especially suitable for complex systems where it is very difficult to relate specific bands to specific conformations. Intensity changes are usually small, however, and absolute intensity measurements are usually difficult. In the Raman, these changes can be measured in a relative way using the depolarization ratio ρ, which is defined in Eqs. (3.1.3) and 3.1.22) and which can be measured in the simple manner described in Section 3.1.3.4.2. The value of ρ for highly polarized bands (in which case $\overline{\alpha}'^2 \neq 0$ and ρ is small) is sensitive to environmental factors, reflecting the differences in the sensitivities of $\overline{\alpha}'$ and $\overline{\beta}'$ to these same factors. Depolarization ratios have been measured for poly(dimethyl siloxane)[201,202] as a function of strain and temperature, and for dilute solutions of poly(dimethyl siloxane) and polystyrene as a function of temperature.[203] In both cases the observed changes were related to chain extension and to changes in conformational structure, and

[198] R. F. Schaufele, *J. Chem. Phys.* **49**, 4168 (1968).
[199] G. R. Bird and E. R. Blout, *J. Am. Chem. Soc.* **81**, 2499 (1959).
[200] M. Kobayashi, K. Tsumura, and H. Tadokoro, *J. Polym. Sci., Part A-2* **6**, 1493 (1968).
[201] J. Maxfield and I. W. Shepherd, *Chem. Phys. Lett.* **19**, 541 (1973).
[202] A. J. Hartley and I. W. Shepherd, *J. Polym. Sci., Polym. Phys. Ed.* **14**, 643 (1976).
[203] R. Speak and I. W. Shepherd, *J. Polym. Sci., Polym. Phys. Ed.* **13**, 997 (1975).

the results were compared with the prediction of the rotational isomeric state model. In a similar way the helix-coil transition and changes in the coil conformation of ionized poly(L-lysine) dissolved in aqueous methanol were studied; the stabilization energy of the helix and energy differences between the rotational isomeric states of the coil were estimated.[204]

[204] I. W. Shepherd, *Biochem. J.* **155,** 543 (1976).

3.2. Inelastic Electron Tunneling Spectroscopy

By H. W. White and T. Wolfram

3.2.1. Introduction

3.2.1.1. General Remarks. Inelastic electron tunneling spectroscopy (IETS) is a new technique that provides a versatile and sensitive method for measuring the vibrational spectrum of a molecular species adsorbed on the surface of a metal oxide. The phenomenon of IETS was discovered by Jaklevic and Lambe in 1966.[1,2] It involves the measurement of the electric current I associated with electrons that tunnel through an oxide film having an adsorbed molecular layer on its surface. The electrons, which inelastically tunnel through the oxide, excite the characteristic vibrations of the adsorbed molecules. Graphs of the bias voltage V vs. the second derivative d^2I/dV^2 display large peaks centered at the voltage $V_m = h\nu_m/e$, where h is Planck's constant, e the electronic charge, and ν_m a molecular vibrational frequency. These graphs show peaks that can be associated with the presence of both infrared and Raman-like modes. The spectra can be used to identify molecular species on the oxide layer, estimate molecular orientation, and detect chemisorption bonding to the oxide layer.

The molecular species to be studied is placed in contact with an oxidized metal film. Another metallic film (usually lead) is evaporated over the "doped" oxide to form a metal–insulator–metal tunnel junction. The junction is then cooled to liquid helium temperatures and the spectra measured. One of the primary advantages of IETS over other spectroscopic techniques is its sensitivity. Coverages of a monolayer or less over an area less than 1 mm^2 can easily be studied. Its main disadvantages are that the preparation of good samples free of unwanted contamination is not easy and that liquid helium temperatures are required for good resolution.

A number of large molecules have been studied by IETS but few attempts have been made to look at polymers. The goal of this section is to present the basic experimental and theoretical features of IETS as a new

[1] R. C. Jaklevic and J. Lambe, *Phys. Rev. Lett.* **17**, 1139 (1968).
[2] J. Lambe and R. C. Jaklevic, *Phys. Rev.* **165**, 821 (1968).

149

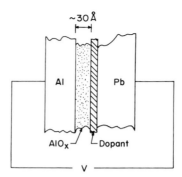

FIG. 1. Cross section of a tunnel junction.

spectroscopic tool. The extent to which it can be exploited to study the vibrational spectra of polymers is not fully known at the time of writing.

The topics to be discussed in the introduction include those on the general features of IETS and the spectra obtained. Subsequent sections cover experimental methods, experimental results, theory, and applications.

More detailed information on theoretical models for elastic and inelastic tunneling, film preparation, and experimental procedures can be obtained from several books.[3-6] An excellent review that includes several applications has been given recently by Hansma.[7] These and other sources have been used liberally in the preparation of this Chapter.

3.2.1.2. General Features. Figure 1 shows a cross section of the type of tunnel junction most commonly used in IETS. Aluminum is deposited on a glass or ceramic substrate (not shown) for structural support. The oxide thickness ranges from 10 to 100 Å and forms a barrier to current flow. The dopant is the molecular species to be studied and is usually about one monolayer in thickness. The lead film acts as a protective coating for the dopant and serves as the second electrode during measurement of the spectra.

If a variable bias is applied across the junction terminals, only a small current of the order of milliamperes will flow across the junction since the oxide layer is an excellent insulator. This current is due to electrons that

[3] C. B. Duke, *Solid State Phys., Suppl.* **10**, (1969).

[4] E. Burstein and S. Lundquist, eds., "Tunneling Phenomena in Solids." Plenum, New York, 1969.

[5] R. V. Coleman, ed., "Methods of Experimental Physics," Vol. 11, Chapters 4 and 12. Academic Press, New York, 1974.

[6] T. Wolfram, ed., "Inelastic Electron Tunneling Spectroscopy," Springer Series in Solid-State Sciences, Vol. 4. Springer-Verlag, Berlin and New York, 1978.

[7] P. K. Hansma, *Phys. Rep.* **30**, 145 (1977).

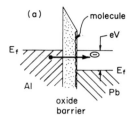

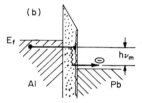

FIG. 2. Illustration of the (a) elastic tunneling process and (b) inelastic tunneling process.

tunnel (in the quantum-mechanical sense) through the insulating oxide barrier. This tunneling current has contributions from both elastic and inelastic processes. Figure 2a illustrates the elastic tunneling process in which an electron tunnels through the oxide barrier from an occupied state below the Fermi level in the aluminum to an unoccupied electron state above the Fermi level of the lead film. Electrons that tunnel without loss of energy contribute to the elastic tunneling current. Figure 2b illustrates an inelastic tunneling process. In this case the tunneling electron interacts with a molecule adsorbed on the oxide and excites one of the characteristic vibrational modes of the molecule. If the molecular vibrational frequency is ν_m then the energy lost by the tunneling electron is $h\nu_m$. It is evident from the diagram that the threshold bias for such an inelastic tunneling event is $eV_{\text{threshold}} = h\nu_m$. There is a different threshold voltage for each vibrational mode of the molecules on the oxide.

3.2.1.3. Peak Location. At a threshold voltage there is an abrupt change in the slope of the current–voltage characteristic of the tunnel junction since a new channel for the inelastic tunneling process is then available. This change is illustrated in Fig. 3a. Electronic modulation methods are used to record the junction conductance $G = dI/dV$ and the second derivative d^2I/dV^2 in order to enhance the detection of this effect. The opening up of a new channel changes the conductance by 1% or so, as illustrated in Fig. 3b. The effect in d^2I/dV^2 is very large, as shown in Fig. 3c, since the only sharp structure in the conductance is due to inelastic tunneling. Experiments are almost always performed at liquid helium

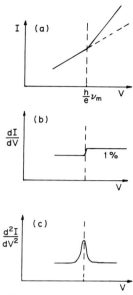

FIG. 3. Changes in the I, dI/dV, and d^2I/dV^2 vs. V curves associated with the opening up of a new tunneling channel.

temperatures in order to eliminate thermal smearing of tunneling structure.

The appearance of a d^2I/dV^2 vs. V curve is very similar to an infrared absorption spectrum in that peaks in the IETS and dips in the infrared correspond to the location of vibrational modes. There are significant differences, however. All the vibrational modes appear in an electron tunneling spectrum but only those vibration modes that have a dipole moment will be observed in an infrared spectrum. For example, some symmetric vibrational modes are observed in tunneling but not in infrared absorption. The reasons for differences in selection rules between infrared and electron tunneling are as follows: (1) the de Broglie wavelength of the tunneling electron is of the order of the molecular dimensions so that the dipole approximation is invalid, (2) the electron is a strongly interacting particle and polarizes the molecules, and (3) the electron interacts through both the molecular dipole and the polarizability tensor.

The useful range of energy that can be scanned without encountering difficulties is about 50–500 meV. This range corresponds roughly to 400–4000 cm^{-1} (1 meV $\simeq$ 8.06 cm^{-1}). Scans below 50 meV show large peaks due to lattice phonons and geometrical resonances. In Section 3.2.2.3 reference is made to a bridge that can be used to eliminate some of these problems. Both Raman and infrared spectra are useful for assignment of peaks in IETS spectra and for identification of the molecular species present.

During the past few years the potential of IETS as a diagnostic tool has begun to be realized. Since the technique is sensitive to the vibrations of molecules in close proximity to an oxide surface it provides critically needed information to a large class of problems associated with molecule–molecule interactions on metallic oxide surfaces and on molecule–surface interactions.

Reported applications include studies of molecular adsorption,[8,9] identification of trace substances[10] and biological molecules,[11,12] radiation damage of biological molecules,[13] molecular kinetics,[14] studies of catalytic particles,[15] and adhesives.[16] Studies on polymer thin films have been recently reported.[17] Applications of IETS to other areas are currently being explored. Section 3.2.5 discusses applications to polymers and some large molecules.

3.4.1.4. Peak Resolution. Thermal and modulation voltage broadening contribute to the linewidth above that expected from the natural width. Lambe and Jaklevic[2] showed that the full-width at half-maximum voltage associated with thermal broadening is approximately $5.4kT/e$ for normal electrodes (k is Boltzmann's constant). At 4.2 K this contribution would be about 0.35 meV, or 2.5 cm^{-1}. Having at least one electrode superconducting further reduces the thermal broadening, but only by a small amount.

Klein and co-workers[9] found the effect of modulation voltage broadening to be $0.707V_m$, where V_m is the rms value of the modulation voltage used in obtaining spectra. Values for V_m are usually 1 mV or larger in order to facilitate obtaining the spectra in a resonable time period. When using a 1 mV modulation signal for a junction at 4.2 K the broadening due

[8] I. W. N. McMorris, N. M. D. Brown, and D. G. Walmsley, *J. Chem. Phys.* **66**, 3952 (1977).

[9] J. Klein, A. Leger, M. Berlin, D. DeFourneau, and M. J. L. Sangster, *Phys. Rev. B* **7**, 2336 (1973).

[10] Y. Skavlatos, R. C. Barker, G. L. Haller, and A. Yelon, *Surf. Sci.* **43**, 353 (1974).

[11] M. G. Simonsen and R. V. Coleman, *Phys. Rev. B* **8**, 5875 (1973).

[12] P. K. Hansma and R. V. Coleman, *Science* **184**, 1369 (1974).

[13] P. K. Hansma and M. Parikh, *Science* **188**, 1304 (1975).

[14] B. F. Lewis, M. Mosesman, and W. H. Weinberg, *Surf. Sci.* **41**, 142 (1974).

[15] P. K. Hansma, W. Kaska, and R. Laine, *J. Am. Chem. Soc.* **98**, 6064 (1976).

[16] H. W. White, L. M. Godwin, and T. Wolfram, *in* "Inelastic Electron Tunneling Spectroscopy" (T. Wolfram, ed.), Springer Series in Solid-State Sciences, Vol. 4, p. 70. Springer-Verlag, Berlin and New York, 1978.

[17] R. Magno and J. G. Adler, *Thin Solid Films* **42**, 237 (1977).

to the modulation voltage is about twice that due to thermal smearing. If these two effects are minimized by using the lowest practical values for V_m and T, then the combined contribution is of the order of 1 meV, or 10 cm^{-1}.

3.2.1.5. Peak Intensity. There is a strong orientational dependence of the intensity of vibrational modes that are observed by IETS. It has been theorized[7] that when the oscillating molecular dipole is oriented parallel to the oxide surface that the corresponding IETS intensity will be weak and primarily due to Raman-like vibrational modes. The intensity is expected to be large when the molecular dipole is perpendicular to the oxide surface because of strong excitation of infrared-like modes. These orientational effects have been verified in a number of IETS experiments including studies of benzoic acid[7] and aromatics. Korman and Coleman[18] have used intensity data to deduce information about the orientation of a variety of adsorbed nitro-substituted single aromatic ring compounds. They also were able to derive information about the chemisorption bonds of the aromatic series.

The information contained in the IETS intensities is usually not used except in a qualitative manner. The primary reason for this situation is the lack of a theory whereby the tunneling current can be calculated. No one has yet calculated even the relative intensities for all of the peaks in a measured tunneling spectrum. Information on the theoretical intensities would be of enormous help in interpreting the spectra obtained.

3.2.2. Experimental Methods

3.2.2.1. General Remarks. The basic equipment requirements for IETS are vacuum facilities for preparing the junctions, a liquid helium cryostat, and a sensitive spectrometer for making the necessary $I-V$ measurements. The techniques required for the preparation of high quality, reproducible samples are those associated with the preparation of thin films, with special attention being given to cleanliness, growth of the oxide film, and the manner of doping. A good, clean vacuum system that is free of any organics that might contaminate the oxide layer is a necessity. Most of the spectrometers and bridges use commercially available lock-in detectors to obtain dV/dI and d^2V/dI^2 vs. V.

The next two sections discuss methods of sample preparation and measurement of spectra.

3.2.2.2. Sample Preparation. The first step in the preparation of a tunnel junction is to evaporate a thin metallic film onto a glass or ceramic substrate. It is very important that the substrate material be smooth and

[18] C. S. Korman and R. V. Coleman, *Phys. Rev. B* **15**, 1877 (1977).

clean. A mask is used to form a strip about 0.2–2 mm wide and several millimeters long. Aluminum is used most often since it forms a suitable oxide layer more easily than other metals. Magnesium and lead have also been used with success. The evaporation should be done in a clean vacuum system with a pressure of 10^{-6} Torr, or less. Both cryogenic/ion pumping systems, which are relatively oil-free, and oil diffusion/mechanical pumping systems have been used to obtain good results. The latter type requires that cryogenic traps be used to keep oil from contaminating the vacuum chamber. Optimum film thickness is 1000–2000 Å. A deposition rate monitor is very useful in controlling film thickness.

The film is then oxidized by one of several methods to form an oxide layer 20–30 Å thick. A glow discharge method is commonly used since it gives oxide barriers having good integrity. In this method 0.02–0.2 Torr of pure oxygen gas is admitted to the sealed vacuum chamber. A glow discharge is then formed by passing 10–50 mA of current at 300–500 V for a few minutes to promote oxide growth. The values for the current, voltage, and the rate of oxide growth are very system-dependent and must be established by trial and error. Other methods of oxide growth include exposure to air or a stream of oxygen.[19] One advantage of the glow discharge method over some others is that it can be done without exposure to room air. A glow discharge for several hours duration followed by pumping can also be used to help clean a vacuum chamber contaminated with hydrocarbons.

The next step in the fabrication process is to "dope" the oxide layer with the molecular species of interest. This doping can be done in several ways. The two most commonly used techniques are chemical vapor deposition and liquid doping. The vapor doping method[11] has been used extensively for materials that can be evaporated without decomposing. The dopant material is volatilized by a resistance heater inside the vacuum chamber and the molecular stream impinges on the oxide barrier. With this method the oxide does not have to be exposed to the atmosphere before doping and evaporation of the second electrode.

The liquid doping process can be used when the molecular species to be studied can be dissolved in a suitable solvent. A drop of the solution is placed on the oxide layer and the excess is removed by spinning.[19] Its main advantages are that it usually gives spectra with high resolution and is sometimes more convenient to use than vapor doping. The concentration of the dopant in the solvent must be determined by trial and error for each species but is usually in the range 0.1–1 mg/liter. Low molecular weight solvents such as water, ethanol, and chloroform that leave little

[19] M. G. Simonsen, R. V. Coleman, and P. K. Hansma, *J. Chem. Phys.* **61**, 3789 (1974).

residual contamination are preferred over those whose spectra would interfere with the spectra of the species being studied.

After the dopant molecule has been placed on the oxide, the second electrode is evaporated to form a metal/oxide/metal junction as shown in Fig. 1. Lead is most often used as the second electrode because it is superconducting at 4.2 K, thereby decreasing the thermal broadening, and because it causes the smallest vibrational frequency shifts from those values obtained by infrared and Raman methods.

Recently a new technique of doping has been presented by Jaklevic and Gaerttner.[20] It is called "external doping" because the junction is doped after the metal/oxide/metal junction has been fabricated. The junctions are doped by proper exposure of the fabricated junctions to the vapor of the dopant/water mixture. Spectra were obtained for a number of organic compounds. By using D_2O as a tag molecule they showed that water can diffuse into and out of junctions. The OH groups may act as carriers for the organics, perhaps by diffusing through microcracks in the lead electrode. Development of this technique could enhance the usefulness of IETS as an analytical tool.

To protect junctions from contamination and deterioration they should be stored in a clean, dry atmosphere.

3.2.2.3. Measurement of Spectra. Each junction should be screened using a low-power ohmmeter to check its room temperature resistance. Resistance values should be in the range 10–5000 Ω, depending on cross-sectional area, with doped junctions having values approximately 10 times higher than undoped ones. Higher resistance junctions are usually noisy. Since the resistance is proportional to exp (t/t_0), where t_0 is approximately 1 Å, it is apparent why oxide thickness t and uniformity are important. Thin spots in the oxide cause the current to pass through only a small region of the surface. Spectra from these junctions show little detail.

Junctions with acceptable room temperature resistances should be checked near 4.2 K to see if a tunneling current exists. This can most easily be done if one of the electrodes is superconducting, in which case the first derivative will show peaks due to the presence of a superconducting energy gap centered about the Fermi energy. Good quality junctions should show a small increase in resistance when taken from 300 to 4.2 K. A decrease indicates the presence of metallic shorts. These junctions should not be used.

The cryostat used to cool the junctions can be a rather simple He⁴ Dewar system with some means available to lower the junctions into the

[20] R. C. Jaklevic and M. R. Gaerttner, *Appl. Phys. Lett.* **30**, 646 (1977).

liquid helium bath. The junctions can be fastened to a stainless steel tube and leads attached by use of indium solder or mechanical contacts. Care should be taken not to heat the junctions if solder is used. Measurements at 4.2 K are usually adequate for most spectra.

Several spectrometer designs have been used to obtain good quality IETS spectra. Most utilize commercially available lock-in amplifiers, which have first and second harmonic detection capabilities. The essential features of a spectrometer system for measuring dV/dI and d^2V/dI^2 vs. V are illustrated in the block diagram in Fig. 4. A dc bias voltage is applied across two leads of the junction, using a sweep generator with time constants selectable from 10 to 10,000 sec. Faster sweep speeds are usually employed when using time-averaging techniques for data collection. Superimposed on the dc bias is an ac modulation signal $V_m \cos 2\pi ft$, where the frequency f is usually in the range 500–5000 Hz. The modulation source must be very stable and have very low distortion since harmonic detection is employed. Filters are often used to block undesirable first or second harmonic signals at a number of points in the circuit. Attenuators, not shown, are used to adjust the ac and dc levels so that approximately 0.5–5 mV rms can be applied to the junction while the dc bias is varied from -500 to $+500$ mV. The signal developed across the junction is supplied to ac and dc preamplifiers. The dc voltage can then be presented to the x axis of an x-y recorder. The ac signal is routed to a lock-in amplifier. If the lock-in amplifier is operated in the first harmonic mode then the output is proportional to dV/dI. In the second harmonic mode the output is proportional to d^2V/dI^2. One of these outputs is presented to the y axis of the recorder. Reasonable spectra can be ob-

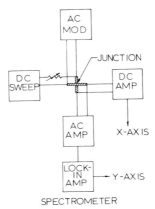

FIG. 4. A block diagram showing the essential features of a spectrometer system for measuring dV/dI and d^2V/dI^2 vs. V.

tained in a manner of minutes. High-resolution spectra may require several hours. Signal-averaging techniques are very useful for improving resolution by helping to eliminate the effects of low-frequency noise. It is very important to follow grounding procedures appropriate for low signal levels since the signal voltages developed across the junction are 0.01 mV, or less.

Using commercial lock-in amplifiers it is easier to measure the quantities dV/dI and d^2V/dI^2 vs. V than it is to measure dI/dV and d^2I/dV^2 vs. V. However, it is the latter two quantities that are of most interest when making comparisions with theoretical calculations. For a junction with constant resistance over the measured voltage range, the quantities d^2V/dI^2 and d^2I/dV^2 are related by the expression $d^2I/dV^2 = -(dI/dV)^3$ d^2V/dI^2. Since there are no reported theoretical calculations of even the relative intensities in an IETS spectrum, most of the data are reported in arbitrary units of d^2V/dI^2. This information is sufficient to give vibrational frequencies for peak locations. Also, dI/dV is a fairly smoothly varying function over much of the energy range in IETS so that d^2V/dI^2 is a fairly good representation of the shape of d^2I/dV^2. It is possible to obtain the desired quantities, however. Adler, Chen, and Straus[21,22] have developed a bridge for obtaining calibrated spectra of dI/dV and d^2I/dV^2 vs. V.

Tunneling spectra do contain undesirable background structure due to metal electrodes and the oxide layer. Colley and Hansma[23] have developed a bridge that minimizes the peaks due to these other vibrations. It measures the normalized difference in d^2V/dI^2 for two junctions as a function of voltage. Usually one is doped and the other is an undoped control. It allows measurements of molecular vibrations in the important range 20–50 meV (160–400 cm^{-1}). Further, it removes structure due to aluminum oxide phonons in the region from 60 to 125 meV, and decreases the background slope over the entire spectrum.

3.2.3. Experimental Results

3.2.3.1. Spectral Features. IETS spectra shown as either d^2V/dI^2 or d^2I/dV^2 vs. the bias voltage V are analogs to the infrared and Raman spectra in that vibrational frequencies are associated with peak structure. A tunneling spectrum of a junction liquid-doped with L-phenylalanine is shown in Fig. 5. The resolution is typical of that often found in IETS spectra. A large number of the vibrational modes have been identified by

[21] J. G. Adler, T. T. Chen, and J. Straus, *Rev. Sci. Instrum.* **42**, 362 (1971).

[22] J. G. Adler and J. Straus, *Rev. Sci. Instrum.* **46**, 158 (1975).

[23] S. Colley and P. K. Hansma, *Rev. Sci. Instrum.* **48**, 1192 (1977).

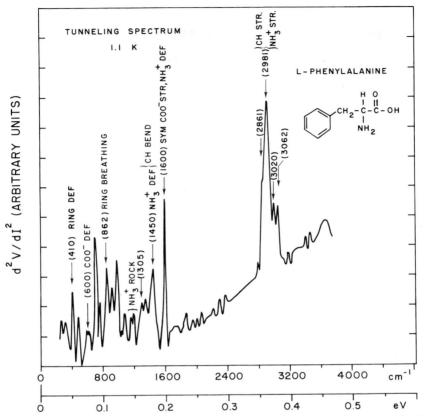

FIG. 5. Tunneling spectrum of a junction liquid-doped with L-phenylalanine from distilled water.[19]

comparison with infrared data and are shown in the figure. The energy range 50–500 meV (400–4000 cm⁻¹) covers most of the stretch and bend modes of many organic compounds. One prominent feature in any IETS spectrum for a molecule having C–H stretch modes is a large peak near 360 meV. Smaller peaks at 200 meV and below are associated with C–H bend and torsional modes. A number of OH stretch modes are found near 450 meV since H_2O is present in the vacuum chambers used for junction preparation.

   Figure 6 shows the tunneling spectrum of a junction vapor-doped with L-phenylalanine. The infrared spectrum is also shown below the tunneling curves.[19] By comparison of the liquid and vapor doped curves it can be seen that the main peaks occur at the same wavenumber. The resolu-

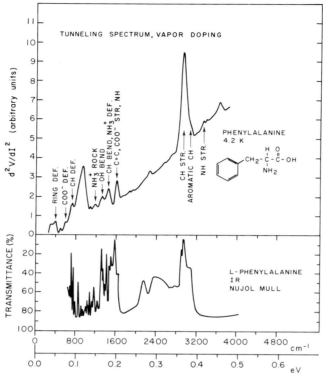

FIG. 6. Tunneling spectrum of a junction vapor doped with L-phenylalanine. The infra-red spectrum is shown below the tunneling curve.[19]

tion of the peaks in the liquid doped curve are superior, however. The infrared spectrum has higher resolution than the IETS curves in the region below 200 meV.

When comparing infrared and IETS spectra as shown here it is impor-tant to note both similarities and differences in the two curves. Peak broadening, and shifting with respect to the infrared values, can and do occur in IETS spectra because the molecule can interact with the oxide surface and the top metal electrode, thereby affecting resolution. Simi-larly, in infrared measurements the mulling oil can contribute to the ad-sorption. In both types of measurements these effects must be recog-nized. The information obtained from IETS spectra should be viewed as being complementary to that obtained from infrared, Raman, and other spectra.

3.2.3.2. Peak Shifts. The locations for the vibration frequencies ob-tained from observed IETS spectra can differ from those obtained from infrared or Raman data. The presence of the top metal electrode causes the measured values to be shifted downward in energy. The size of the correction for several modes of benzoic acid has been measured for a number of top metal electrodes by Kirtley and Hansma.[24] Pb was found to cause the least shift and this fact accounts, in part, for its widespread use. They found that vibrational modes that involve little or no deformation of the C–H bonds have mode energy shifts too small to be measured (less than 0.1 meV). Vibrational modes involving C–H bond deformation had shifts of about 0.5 eV. Vibrational modes involving O–H ions had relatively large shifts of approximately 14 meV. These results were discussed in terms of a simple image dipole model in which the oscillating charge distribution in a molecule induces an image of itself in the metal surface. The dipole interacts with the image to modify the resonant frequency.

Kirtley and Hansma[24] also reported careful measurements on the shift to higher energies of the observed IETS peaks due to superconducting electrode effects. With one electrode superconducting, this shift is nominally equal to Δ_1, one-half the energy gap. If both electrodes are superconducting it is approximately one-half the sum $\Delta_1 + \Delta_2$. They found the shifts to higher energies to be somewhat less than the above values and dependent on the magnitude of the modulation signals. As an example, with Al and Pb at 1.1 K, an upward shift of 1.45 meV would be expected. For a 1 mV modulation signal the shift was only 1.05 meV. Thermal smearing is reduced, however, when at least one electrode is superconducting so that most IETS data are taken with at least one electrode superconducting.

3.2.3.3. Sensitivity. The primary advantage of IETS over infrared and Raman techniques is its sensitivity. Coverages of a monolayer or less of a molecular species can give repeatable, identifible peaks. A very significant study of IETS peak heights for benzoic acid coordinated with surface concentration studies has been reported by Langan and Hansma.[25] The surface concentration was determined using radioactively labeled benzoic acid. The results showed that coverages down to about 1/30 of a monolayer could be detected. The data also suggested that peak height decreased more rapidly than surface concentration below saturation coverage. This latter result was thought to be due to an increase in the ratio of current flowing through "undoped" regions compared with that through "doped" areas when the surface concentration falls below a monolayer. This increase in the current ratio will be faster than the change in surface

[24] J. Kirtley and P. K. Hansma, *Phys. Rev. B* **13**, 2910 (1976).
[25] J. D. Langan and P. K. Hansma, *Surf. Sci.* **52**, 211 (1975).

coverage since the regions without coverages have a thinner barrier; hence, a larger current density. This argument suggests that the degree of nonlinearity is greatest for large molecules. This conjecture is consistent with the fact that a number of researchers have found that successfully doping with large molecules is considerably more difficult than with smaller species.

3.2.4. Theory

3.2.4.1. Theory of Intensities. Compared with the experimental situation the theory of IETS has received relatively little attention. At the present time there does not exist a detailed model for calculating or predicting the intensities of IETS or for interpreting the IETS data of typical molecules (e.g., benzene ring compounds).

The first theory was given by Scalapino and Marcus[26] shortly after the discovery of IETS by Jaklevic and Lambe.[1,2] Scalapino and Marcus provided an explanation of the order of magnitude of the IETS peak intensities based on the interaction between the tunneling electron and an oscillating dipole. Lambe and Jaklevic[2] subsequently considered the excitation of molecular vibrations by a polarizability mechanism similar to that of Raman scattering. Their results indicated that the Raman-like mechanism could produce observable peaks in IETS spectra. Additional theoretical treatments of several aspects of tunneling have been given by a number of authors.[27-30]

To calculate $I(V)$ or any of its derivatives, first-order time-dependent perturbation theory is used to write the rate of transitions between initial and final states. The result is known as Fermi's golden rule. In the notation of Kirtley, Scalapino, and Hansma[31] the expression can be written as,

$$W_{KK'} = (2\pi/\hbar)|M_{KK'}|^2\delta(\epsilon_K - \epsilon_{K'} - \hbar\omega), \qquad (3.2.1)$$

where the matrix elements $M_{KK'}$ are defined as

$$M_{KK'} = \int_0^l d^3x \Psi_{K'}^* V_1(\mathbf{r}, \omega t)\Psi_K. \qquad (3.2.2)$$

The wavefunctions $\Psi_{K'}$ and Ψ_K are for the initial and final states and have standard WKB waveforms within the tunneling region. The perturbation V_1 that causes the inelastic tunneling processes in Eq. (1) is due to the

[26] D. J. Scalapino and S. M. Marcus, *Phys. Rev. Lett.* **18**, 459 (1967).
[27] A. J. Bennett, C. B. Duke, and S. D. Silverstein, *Phys. Rev.* **176**, 969 (1968).
[28] J. W. Gadzuk, *J. Appl. Phys.* **41**, 286 (1970).
[29] T. E. Feuchtwang, *Phys. Rev. B* **10**, 4121 (1974).
[30] C. Caroli, R. Combescot, P. Noziers, and C. Saint-James, *J. Phys. C* **5**, 21 (1972).
[31] J. Kirtley, D. J. Scalapino, and P. K. Hansma, *Phys. Rev. B* **14**, 3177 (1976).

modulation of the charge density of the molecule as it vibrates. The current density j_m associated with a given mode of vibration m is obtained by summing Eq. (1) over all K,K' states. The difficulty in the method lies in obtaining a realistic expression for the electron–molecule interaction potential V_I.

Kirtley, Scalapino, and Hansma employed a transfer Hamiltonian method and characterized the electron–molecule interaction in terms of the Coulomb interactions of the tunneling electron with a set of fictitious oscillating "partial charges" located on each of the atoms of the molecule. The magnitudes of these partial charges and the effective amplitudes of oscillation were treated as empirical parameters to be determined from experimental data.

Although empirical in nature, their results are extremely important because they show (1) that the correct order of magnitude of IETS intensities is obtained from a weak-interaction model and first-order perturbation theory and (2) that the IETS intensities should depend strongly on the orientation of the oscillating dipoles relative to the oxide barrier surface.

In another recent calculation Rath and Wolfram[32] reported results from a model that is similar to that above except that the construction of V_I is different. In their model the electron–molecule potential is calculated from molecular wavefunctions and charge densities expressed as functions of the atomic displacements associated with each mode. A significant result from this model is that the potentials calculated for modes of the ethylene molecule only roughly resemble dipoles. The angular and radial dependences of the potentials are not dipolar.

3.2.4.2. Selection Rules. In both models the calculation of d^2I/dV^2 involves an integration over all initial and final states for which the orientation of the molecule with respect to the oxide must be specified. The resulting intensities depend strongly on the molecular orientation. The theory of Scalapino and Marcus predicted that the maximum intensity would occur for vibration modes in which the oscillating dipole moment was perpendicular to the oxide surface. The intensity would be zero for moments parallel to the surface. Kirtley, Scalapino, and Hansma's results showed that this orientational selection rule is weakened slightly by the effects of off-axis scattering. There is experimental evidence for such IETS orientational selection rules.[18,19,24]

The theory also indicates that the Raman mode intensities should be nearly the same size as the infrared mode intensities. Further it predicts that modes forbidden in both Raman and infrared spectra should be

[32] J. Rath and T. Wolfram, in "Inelastic Electron Tunneling Spectroscopy" (T. Wolfram, ed.), Springer Series in Solid-State Sciences, Vol. 4, p. 92. Springer-Verlag, Berlin and New York, 1978.

observable with IETS. Tunneling spectra on anthracene by Simonsen, Coleman and Hansma[19] indicate that both infrared and Raman modes are observed with comparable intensities. To our knowledge, however, there are no observations of modes in IETS which are forbidden to both infrared and Raman spectroscopy.

In summary, it is easily appreciated that a truly rigorous treatment of the electron–molecule interaction problem is well beyond the state of the art, not to mention the difficulties inherent in the theory of tunneling itself. One must necessarily adopt a model with numerous approximations in order to make progress. From a utilitarian point of view, a theory capable of predicting at least the relative intensities and the dependence of the intensities on the orientation of the adsorbed molecule would be of great value for interpreting IETS spectra. The need for a quantitative and predictive theory of IETS is quite obvious.

3.2.5. Applications

3.2.5.1. General Remarks. A number of applications of IETS to the study of smaller molecules have been done, some of which have been mentioned earlier. There have been fewer applications to the study of very large molecules since it is more difficult to obtain junctions with good signal-to-noise ratios. The number of individual modes is large and the spectra can appear as an envelope rather than well-resolved peaks. Nevertheless, some very interesting work has been done on large molecules, including polymers.

Applications of IETS to the study of biochemical, adhesive, and polymer molecules are discussed in the following three subsections.

3.2.5.2. Biological Molecules. The high sensitivity of IETS and its orientational selection rules have made it possible to use it as a tool for identification of biochemical species. Clark and Coleman[33] have shown that very similar nucleic acid derivatives can be differentiated with tunneling spectroscopy. Coleman, Clark, and Korman[34] have reviewed the application of IETS to studies on amino acids, purine and pyridine bases, nucleosides, nucleotides, DNA, RNA, and proteins. The development of the technique involves consideration of the effect of adsorption of the molecule on the oxide substrate. Chemical reactions of side groups and orientation of groups with respect to the oxide affect the intensities and resolution that can be obtained for a particular molecular species.

One application is illustrated by the upper curve in Fig. 7, from Clark

[33] J. M. Clark and R. V. Coleman, *Proc. Natl. Acad. Sci. U.S.A.* **73**, 1598 (1976).

[34] R. V. Coleman, J. M. Clark, and C. S. Korman, *in* "Inelastic Electron Tunneling Spectroscopy" (T. Wolfram, ed.), Springer Series in Solid-State Sciences, Vol. 4, p. 34. Springer-Verlag, Berlin and New York, 1978.

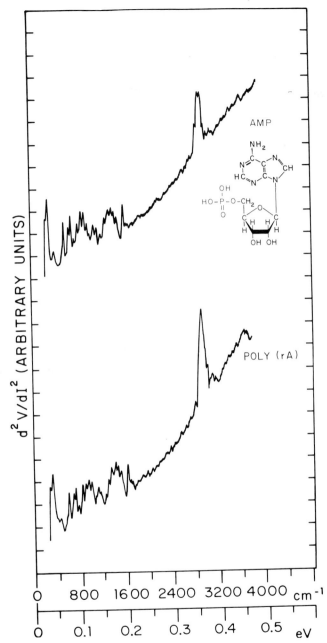

FIG. 7. Tunneling spectrum of adenosine-5′-monophosphate (upper curve) and poly(adenylic acid) (lower curve). Junction resistances were 1030 Ω for AMP and 993 Ω for Poly(rA). Both were doped from H_2O solution. IETS resolution is approximately the same for both the mononucleotide and the polynucleotide (J. M. Clark and R. V. Coleman, unpublished).

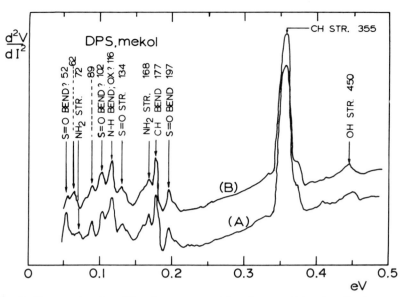

FIG. 8. Two spectra, A and B, taken on a tunnel junction containing the adhesive curative diamino diphenyl sulfone (DPS). Spectrum B was taken nine days after that of A.[16]

and Coleman,[35] which is the tunneling spectrum of adenosine-5'-monophosphate (AMP). The inset shows the molecular structure of AMP with the adenine and monophosphate groups attached to the ribose ring. Most of the strong modes between 500 and 1600 cm[-1] are from the adenine ring. Vibrational modes involving the phosphate group are expected near 980 and 1090 cm[-1]. The modes in the range below 600 cm[-1] provide information about small variations in the molecular structure such as the attachment site of the phosphate on the sugar, the type of sugar, and the presence or absence of the phosphate groups. The lower curve in Fig. 7 will be discussed in Section 3.2.5.4.

3.2.5.3. Adhesive Molecules. Application of IETS to the study of adhesive molecules has been made by White, Godwin, and Wolfram.[16,36] They obtained tunneling spectra on the components and mixture of a high-performance commercial epoxy having two molecular components. The component molecules were diamino diphenyl sulfone (DPS) and tetraglycidyl 4,4'-diamino diphenyl methane (DPM). Figure 8 shows two tunneling spectra, labeled A and B, taken on a junction containing the curative component DPS doped from a solution of methyl ethyl ketone.

[35] J. M. Clark and R. V. Coleman, unpublished.
[36] H. W. White, L. M. Godwin, and T. Wolfram, *J. Adhes.* **9**, 237 (1978).

The peaks were assigned by comparison with IR, other IETS data, and normal mode frequencies and eigenvectors calculated using literature values for the molecular force constants. Spectrum B was measured nine days after A was taken. The two spectra are the same except in the low-energy range from 55 to 75 meV. The change in this region is associated with an aging effect and is speculated to be due to NH_2 groups on the ends of the DPS molecules having chemisorbed to the oxide layer.

The adhesive formed by the reaction of the epoxy and curative molecules is polymeric in nature. One important feature of using IETS to study adhesives is that it allows *in situ* studies of the molecules in contact with the oxide layer.

3.2.5.4. Polymers. The applications of IETS to the study of polymers are very limited and have been done only recently. Tunneling spectra have been obtained for polymerized ethylene and benzene films formed by glow discharge and for poly(adenylic acid).

Magno and Adler[17] have used IETS to study the molecular structure of barriers formed on aluminum and magnesium in a glow discharge of ethylene or benzene. Figures 9a and b show the tunneling spectra for junctions of Al–ethylene–Pb and Al–benzene–Pb, respectively. Figure 9c is a tunneling curve for an Al–oxygen discharge–Pb junction, which is shown to illustrate the absence of hydrocarbon bonds that might arise from hydrocarbon contamination in the vacuum system. It should not be taken as a background curve for the ethylene and benzene curves. The data show that the barriers contain many molecular groups that may be associated with substituted forms of ethylene and benzene and other groups that must come from the breaking of these molecules. Magno and Adler discuss a number of peaks in these spectra in detail. The results indicate that both the ethylene and benzene molecules were broken up and recombined to form aliphatic $-CH_3$ and $-CH_2$ groups. The ethylene junctions also had a peak at 1635 cm^{-1}, which indicated the presence of olefinic C=C groups. The molecular structure of the barriers formed on magnesium base films was similar to those formed on aluminum films. Magno and Adler have also used IETS to systematically study the techniques used to grow these barriers.[37,38] Their results indicated that IETS is a useful tool for investigating materials produced by glow discharge polymerization. It can be used to study the way in which the initial polymerization depends on the various types of surfaces and should be useful in studying polymerization reactions initiated by ultraviolet light or electron bombardment.

[37] R. Magno and J. G. Adler, *Phys. Rev. B* **13**, 2262 (1975).
[38] R. Magno and J. G. Adler, *Phys. Rev. B* **15**, 1744 (1977).

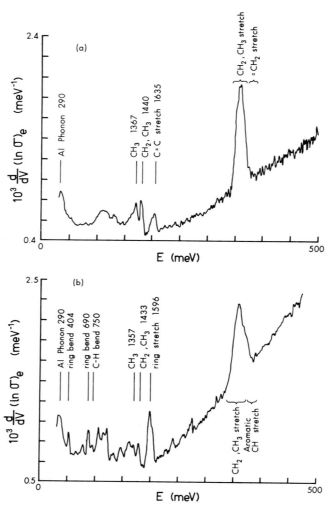

FIG. 9. Tunneling spectra showing (d/dV) (ln σ)$_0$ for (a) Al–ethylene–Pb, (b) Al–benzene–Pb, and (c) Al–oxygen–Pb. The numbers with the identified peaks are the energies in units of cm^{-1}.[17]

More recently, the tunneling spectrum for poly(adenylic acid) (Poly(rA)) has been obtained by Clark and Coleman[35] and is shown as the lower curve in Fig. 7. Poly(rA) is a polynucleotide formed from units of adenosine-5'-monophosphate (AMP). The structure of AMP is illustrated in the inset of Fig. 7. The bonding necessary for formation of the polymer occurs between an OH site on the monophosphate group to one

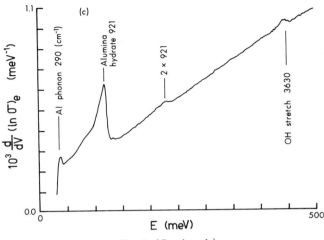

Fig. 9. (Continued.)

of the OH groups on the ribose ring. A careful comparison of the upper and lower spectra shows that the IETS resolution for the nomonucleotide and the polynucleotide is approximately the same. Both junctions were doped from H_2O solutions using the liquid doping technique.

Many of the modes occurring in the AMP also occur in the Poly(rA), as might be expected from the earlier results on adenine derivates reported by Clark and Coleman.[33] There are some obvious differences between the two spectra. To obtain more specific information from the Poly(rA) spectra would require further analysis using detailed wave-numbers and expanded spectra.

3.2.6. Summary

The basic experimental and theoretical features of IETS as a spectroscopic tool for the study of molecular vibrations have been presented. Several applications to the study of large molecules and polymers have been given. The fact that quality IETS spectra have been obtained on ethylene and benzene polymerized films and on the polymer polyadenylic acid should provide incentive for additional IETS studies on these and other polymers. Further development will be required to fully assess the potential of this unique spectroscopic tool for the study of polymers.

Acknowledgments

The authors wish to thank R. V. Coleman, R. Magno, J. G. Adler, P. K. Hansma, R. C. Jaklevic, and J. Lambe for valuable discussions. The authors are grateful to J. M. Clark and R. V. Coleman for providing spectra prior to publication.

3.3. Rayleigh–Brillouin Scattering in Polymers

By G. D. Patterson

3.3.1. Introduction

Rayleigh–Brillouin scattering in dense media is due to fluctuations in the local dielectric tensor ϵ.[1,2] The total intensity is proportional to

$$I \propto \mathrm{Tr}\langle|\Delta\epsilon|^2\rangle, \tag{3.3.1}$$

where Tr denotes the trace of the square of the magnitude of the fluctuation of the local dielectric tensor $\Delta\epsilon$ and the brackets signify an ensemble average.

In 1922 Brillouin[3] predicted that thermal acoustic phonons would lead to such fluctuations and hence to light scattering. In addition, the scattered light should be shifted in frequency because the phonons are moving. The frequency shift $\Delta\omega$ is given by[3]

$$\pm\Delta\omega/\omega_0 = 2n(V/C)\sin\theta/2, \tag{3.3.2}$$

where ω_0 is the incident frequency, n the refractive index, V the phonon velocity, C the speed of light in a vacuum, and θ the scattering angle in the scattering plane.

Acoustic phonons in dense media are attenuated by a variety of mechanisms. The calculation of the spectrum due to scattering by damped phonons was first carried out by Leontovich.[4] The Brillouin peaks were predicted to have a half-width at half-height Γ given by

$$\Gamma = \alpha V/2\pi, \tag{3.3.3}$$

where α is the phonon attenuation coefficient and Γ is measured in hertz.

In addition to the shifted peaks due to scattering by thermal acoustic phonons there are peaks centered at the incident frequency with finite widths. The central peaks are called Rayleigh scattering. A typical Rayleigh–Brillouin spectrum is shown in Fig. 1.

The basic experimental arrangement is shown in Fig. 2. Light of fre-

[1] A. Einstein, *Ann. Phys.* (*Leipzig*) [4] **33**, 1275 (1910).

[2] M. Smoluchowski, *Ann. Phys.* (*Leipzig*) [4] **25**, 205 (1908).

[3] L. Brillouin, *Ann. Phys.* (*Paris*) [9] **17**, 88 (1922).

[4] M. A. Leontovich, *Z. Phys.* **72**, 247 (1931).

METHODS OF EXPERIMENTAL PHYSICS, VOL. 16A

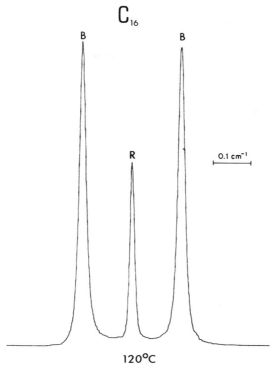

FIG. 1. Rayleigh–Brillouin spectrum of *n*-hexadecane at 120°C. Reprinted with permission from G. D. Patterson and J. P. Latham, *Macromolecules* **10,** 736 (1977). Copyright by the American Chemical Society.

quency ω_0 and wavevector $\mathbf{q}_i$ is incident on the sample. The light interacts with the medium and is scattered through an angle θ. The scattered light has a frequency distribution $I(\mathbf{q}, \omega)$ and a wavevector $\mathbf{q}_s$, where $\mathbf{q} = \mathbf{q}_s - \mathbf{q}_i$. For Rayleigh–Brillouin scattering, the frequency shifts are small in comparison to ω_0 so that the magnitude of $\mathbf{q}$ is accurately given by

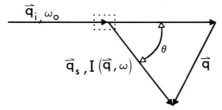

FIG. 2. Scattering diagram for Rayleigh–Brillouin spectroscopy. Light with frequency ω_0 and wavevector $\mathbf{q}_i$ is incident upon the sample. Light is scattered by fluctuations with wavevector $\mathbf{q}$ and the scattered light has a frequency distribution $I(\mathbf{q}, \omega)$ and a wavevector $\mathbf{q}_s$.

$$|\mathbf{q}| = (4\pi n/\lambda) \sin \theta/2, \tag{3.3.4}$$

where λ is the wavelength of the incident light in a vacuum. The incident and scattered wavevectors define the scattering plane. It is customary to polarize the incident light either perpendicular (vertical) or parallel (horizontal) with respect to the scattering plane. The scattered light is then analyzed with a polarizer set in the vertical or horizontal direction.

The spectrum of the scattered light is given by the fluctuations in the local dielectric tensor

$$I(\mathbf{q}, \omega) \propto \mathrm{Tr}\langle \Delta\boldsymbol{\epsilon}(\mathbf{q}, \omega) \, \Delta\boldsymbol{\epsilon}^*(0)\rangle, \tag{3.3.5}$$

where $\Delta\boldsymbol{\epsilon}(\mathbf{q}, \omega)$ is the amplitude of the dielectric tensor fluctuation with wavevector $\mathbf{q}$ and frequency ω. If we let the incident light propagate along the x axis and define the scattering plane as the xy plane, then there will be three types of scattered spectra:

$$
\begin{aligned}
I_{VV}(\mathbf{q}, \omega) &\propto \langle \Delta\epsilon_{zz}(\mathbf{q}, \omega) \, \Delta\epsilon_{zz}^*(0), \\
I_{HV}(\mathbf{q}, \omega) &\propto \langle \Delta\epsilon_{yz}(\mathbf{q}, \omega) \, \Delta\epsilon_{yz}^*(0)\rangle, \\
I_{HH}(\mathbf{q}, \omega) &\propto \langle [\Delta\epsilon_{yx}(\mathbf{q}, \omega) \sin \theta \\
&\quad - \Delta\epsilon_{yy}(\mathbf{q}, \omega) \cos \theta][\Delta\epsilon_{yx}^*(0) \sin \theta \\
&\quad - \Delta\epsilon_{yy}^*(0) \cos \theta]\rangle,
\end{aligned}
\tag{3.3.6}
$$

where $\Delta\epsilon_{ij}$ is the component of the local dielectric tensor fluctuation in the laboratory frame.

The theory of Rayleigh–Brillouin scattering attempts to calculate $\Delta\boldsymbol{\epsilon}(\mathbf{q}, \omega)$ in terms of the mechanical and thermal variables of the medium. The general relaxation theory of Rayleigh–Brillouin scattering developed by Rytov[5] is presented in Section 3.3.2. The experimental apparatus necessary to measure Rayleigh–Brillouin spectra is discussed in Section 3.3.3. Applications of Rayleigh–Brillouin spectroscopy to polymers is the subject of Section 3.3.4. Only bulk polymers will be considered in the present work.

3.3.2. Theory

3.3.2.1. General Theory. A general theory of Rayleigh–Brillouin scattering by a viscoelastic medium has been developed by Rytov.[5] The average dielectric tensor $\langle \boldsymbol{\epsilon}\rangle$ is a scalar for an isotropic amorphous medium. For a single-component fluid the value of $\langle \boldsymbol{\epsilon}\rangle$ depends only on the equilibrium density $\langle \rho\rangle$ and temperature $\langle T\rangle$. However, the fluctuation $\Delta\boldsymbol{\epsilon}$ may depend on a whole set of scalar and tensor variables. The

---

[5] S. M. Rytov, *Sov. Phys. —JETP (Engl. Transl.)* **31**, 1163 (1970).

two fluctuating variables that correspond to ρ and T are the displacement gradient tensor $\nabla \mathbf{S}$ and the temperature fluctuation ΔT. In addition, there will be a set of scalar variables, which we denote by $\{\xi^{(j)}\}$ and a set of tensor variables $\{\boldsymbol{\zeta}^{(k)}\}$. The dielectric tensor fluctuation is then given by

$$\Delta \boldsymbol{\epsilon} = \left(\frac{\partial \boldsymbol{\epsilon}}{\partial \nabla \mathbf{S}}\right) \cdot \nabla \mathbf{S} + \left(\frac{\partial \boldsymbol{\epsilon}}{\partial T}\right) \Delta T + \sum_j \left(\frac{\partial \boldsymbol{\epsilon}}{\partial \xi^{(j)}}\right) \xi^{(j)} + \sum_k \left(\frac{\partial \boldsymbol{\epsilon}}{\partial \boldsymbol{\zeta}^{(k)}}\right) \boldsymbol{\zeta}^k. \quad (3.3.7)$$

The tensor $\Delta \boldsymbol{\epsilon}$ can be decomposed as

$$\Delta \boldsymbol{\epsilon} = \Delta \bar{\epsilon} \, \mathbf{E}_3 + \boldsymbol{\epsilon}^{\mathrm{A}} + \hat{\boldsymbol{\epsilon}}, \quad (3.3.8)$$

where $\Delta \bar{\epsilon} = (\tfrac{1}{3}) \operatorname{Tr} \Delta \boldsymbol{\epsilon}$ is the mean scalar part, $\epsilon_{ij}^{\mathrm{A}} = (\Delta \epsilon_{ij} - \Delta \epsilon_{ji})/2$ the antisymmetric part, and $\hat{\epsilon}_{ij} = (\Delta \epsilon_{ij} + \Delta \epsilon_{ji})/2 - \Delta \bar{\epsilon} \, \delta_{ij}$ the symmetric traceless part. The quantity $\hat{\boldsymbol{\epsilon}}$ is also called the dielectric anisotropy tensor. The most general theory should include the possibility that $\Delta \boldsymbol{\epsilon}$ is antisymmetric, but for most bulk polymer materials $\boldsymbol{\epsilon}^{\mathrm{A}}$ is negligible and we shall ignore the antisymmetric part.

The displacement gradient tensor $\nabla \mathbf{S}$ can be represented as

$$\nabla \mathbf{S} = \frac{\nabla \cdot \mathbf{S}}{3} \mathbf{E}_3 + \frac{\nabla \mathbf{S} - (\nabla \mathbf{S})^{\mathrm{T}}}{2} + \left[\frac{\nabla \mathbf{S} + (\nabla \mathbf{S})^{\mathrm{T}}}{2} - \frac{\nabla \cdot \mathbf{S}}{3} \mathbf{E}_3\right] \quad (3.3.9)$$

$$= u\mathbf{E}_3 + \boldsymbol{\phi} + \hat{\mathbf{u}},$$

where u is the compression, $\boldsymbol{\phi}$ the antisymmetric pure rotation tensor, and $\hat{\mathbf{u}}$ the symmetric traceless pure shear tensor. A completely general theory[6] of Brillouin scattering would allow the displacement gradient tensor to be antisymmetric, but for isotropic amorphous media we shall neglect $\boldsymbol{\phi}$. We shall also require all the $\{\boldsymbol{\zeta}^{(k)}\}$ to be symmetric and shall separate them into mean scalar parts $\bar{\zeta}^k$ and pure anisotropy tensors $\hat{\boldsymbol{\zeta}}^{(k)}$. The $\bar{\zeta}^{(k)}$ will then be included in the set of scalar fluctuation variables.

In order to calculate the fluctuations we need to express the excess free energy Ψ associated with the fluctuations in terms of the mechanical and thermal variables. In the theory of Rytov a linear coupling approximation is made and Ψ is expressed as a sum of pair products of the scalar variables and bilinear invariants of the tensor variables:

$$\Psi = \frac{1}{2} \frac{\partial^2 \Psi}{\partial u^2} u^2 + \frac{1}{2} \operatorname{Tr} \frac{\partial^2 \Psi}{\partial \hat{\mathbf{u}}^2} \hat{\mathbf{u}}^2$$

$$+ \frac{1}{2} \frac{\partial^2 \Psi}{\partial \Delta T^2} \Delta T^2 + \frac{1}{2} \sum_j \frac{\partial^2 \Psi}{\partial \xi^{(j)2}} \xi^{(j)2}$$

[6] D. F. Nelson and M. Lax, *Phys. Rev. B* **3**, 2778 (1971).

$$+ \frac{1}{2} \operatorname{Tr} \sum_k \frac{\partial^2 \Psi}{\partial \hat{\zeta}^{(k)2}} \hat{\zeta}^{(k)2} + \frac{\partial^2 \Psi}{\partial u \, \partial \Delta T} u \, \Delta t$$

$$+ \sum_j \frac{\partial^2 \Psi}{\partial u \, \partial \xi^{(j)}} u \xi^{(j)} + \sum_j \frac{\partial^2 \Psi}{\partial \Delta T \, \partial \xi^{(j)}} \Delta T \xi^{(j)}$$

$$+ \operatorname{Tr} \sum_k \frac{\partial^2 \Psi}{\partial \hat{u} \, \partial \hat{\zeta}^{(k)}} \hat{u} \hat{\zeta}^{(k)}. \qquad (3.3.10)$$

The second derivatives of the free energy are the appropriate high-frequency limiting moduli for the system. In general, quantities such as $(\partial^2 \Psi / \partial \hat{u}^2)$ are fourth-rank tensors, but for an isotropic medium we shall take all the moduli to be scalars. The derivatives are given by

$$\frac{\partial^2 \Psi}{\partial u^2} = K_\infty, \qquad \frac{\partial^2 \Psi}{\partial \hat{u}^2} = 2\mu_\infty, \qquad \frac{\partial^2 \Psi}{\partial \Delta T^2} = -\frac{\rho C_{V\infty}}{T},$$

$$\frac{\partial^2 \Psi}{\partial u \, \partial \Delta T} = -K_\infty \alpha_\infty, \qquad \frac{\partial^2 \Psi}{\partial u \, \partial \xi^{(j)}} = L_j,$$

$$\frac{\partial^2 \Psi}{\partial \Delta T \, \partial \xi^{(j)}} = M_j, \qquad \frac{\partial^2 \Psi}{\partial \hat{u} \, \partial \hat{\zeta}^{(k)}} = N_k, \qquad (3.3.11)$$

where K is the modulus of compression, μ is the shear modulus, C_V is the specific heat at constant volume, α is the thermal expansion coefficient, and the subscript ∞ denotes the high-frequency limit. The quantities $\{L_j\}$, $\{M_j\}$, and $\{N_k\}$ are coupling amplitudes that are related to the relaxing parts of K, μ, C_V, and α. If we introduce a set of relaxation times $\{\tau_j\}$ and $\{\tau_k\}$ associated with each variable, then

$$K = K_\infty - \sum_j \frac{L_j^2}{1 + i\omega\tau_j}, \qquad \mu = \mu_\infty - \frac{1}{2} \sum_k \frac{N_k^2}{1 + i\omega\tau_k},$$

$$K\alpha = K_\infty \alpha_\infty + \sum_j \frac{L_j M_j}{1 + i\omega\tau_j}, \qquad C_V = C_{V\infty} + \frac{T}{\rho} \sum_j \frac{M_j^2}{1 + i\omega\tau_j}. \qquad (3.3.12)$$

Thus the mechanical and thermal moduli are explicit functions of frequency.

There will also be a corresponding set of dielectric moduli as given in Eq. (3.3.7). Again, quantities such as $(\partial \epsilon / \partial \hat{u})$ are fourth-rank tensors, but we shall treat them as scalars. Also, the high-frequency limiting value is taken:

$$\left(\frac{\partial \epsilon}{\partial \hat{u}}\right)_\infty = x, \qquad \left(\frac{\partial \epsilon}{\partial u}\right)_\infty = y, \qquad \left(\frac{\partial \epsilon}{\partial \Delta T}\right)_\infty = z,$$

$$\left(\frac{\partial \epsilon}{\partial \xi^{(j)}}\right)_\infty = m_j, \qquad \left(\frac{\partial \epsilon}{\partial \hat{\zeta}^{(k)}}\right)_\infty = n_k. \qquad (3.3.13)$$

From the above quantities we can define a set of frequency-dependent dielectric moduli

$$X = x - \sum_k \frac{n_k N_k}{1 + i\omega\tau_k}, \qquad Y = y - \sum_j \frac{m_j L_j}{1 + i\omega\tau_j},$$

$$Z = z - \sum_j \frac{m_j M_j}{1 + i\omega\tau_j}. \tag{3.3.14}$$

The scattered spectra can then be expressed entirely in terms of the moduli defined above. The results are

$$
\begin{aligned}
I_{VV}(\mathbf{q}, \omega) = \frac{-k_B T}{2\pi i\omega} \Bigg[&\frac{1}{\Delta} \left(\frac{X^2 C q^2}{9} - \frac{2XYCq^2}{3} + \frac{2XZK\alpha q^2}{3} + Y^2 C q^2 \right. \\
&\left. - 2YZK\alpha q^2 - Z^2(A + Bq^2) \right) + \sum_j \frac{m_j^2}{1 + i\omega\tau_j} \\
&+ \frac{2}{3} \sum_k \frac{n_k^2}{1 + i\omega\tau_K} - \text{c.c.} \Bigg],
\end{aligned}
$$

$$\tag{3.3.15}$$

$$I_{HV} = I_{VH} = \frac{-k_B T}{2\pi i\omega} \left[\frac{X^2 q_2^2}{4A} + \frac{1}{2} \sum_k \frac{n_k^2}{1 + i\omega\tau_k} - \text{c.c.} \right],$$

$$
\begin{aligned}
I_{HH} = \frac{-k_B T}{2\pi i\omega} \Bigg[&\frac{q_2^2}{\Delta} \left(\frac{X^2 C}{2} \left(1 - \frac{\cos\theta}{3} \right) + 2X(YC - ZK\alpha)\cos\theta \right) \\
&+ \frac{1}{2} \sum_k \frac{n_k^2}{1 + i\omega\tau_k} \sin^2\theta - \text{c.c.} \Bigg] + I_{VV} \cos^2\theta,
\end{aligned}
$$

where k_B is Boltzman's constant,

$$q_2 = -\frac{2\pi n}{\lambda} \sin\theta, \qquad A = \mu q^2 - \rho\omega^2, \qquad B = K + \frac{\mu}{3},$$

$$C = \frac{\rho C_V}{T} + \frac{\kappa q^2}{Ti\omega}, \qquad \Delta = (A + Bq^2)C + K^2\alpha^2 q^2,$$

κ is the thermal conductivity, and c.c. denotes the complex conjugate. The general expressions can be reduced to a particular model calculation by choosing the form of the moduli in a straightforward manner.

The linear fluctuation theory of Rytov[5] correctly describes the general features of the Rayleigh–Brillouin spectrum of isotropic amorphous media. However, scattering due to multiple fluctuations has been neglected. A full nonlinear theory is not presently available, but a qualitatively correct approach is presented here.

In the linear theory, the spectrum is calculated based on the equilibrium average values of the moduli. In a real experiment the observed spectrum is obtained by summing the spectra from a large number of small volumes. If we require the linear dimensions of the small volume to be comparable to q^{-1} and require the scattering volume to contain many such small volumes, then we can represent the observed spectrum as

$$I(\mathbf{q}, \omega) = \frac{1}{N} \sum_{i=1}^{N} I_i(\mathbf{q}, \omega, \{\Xi\}), \qquad (3.3.16)$$

where the set $\{\Xi\}_i$ represents the average values of the moduli in volume i at the time of the scattering event. The average values of the moduli will depend on the values of the wavevectors $\mathbf{q}_f$ associated with the fluctuations. The total number of thermal fluctuations is proportional to $k_B T$. However, only those fluctuations with wavevector comparable to or less than $\mathbf{q}$ will appreciably affect the average modulus in the small volume. The fraction of fluctuations with wavevector equal to or less than $\mathbf{q}$ will be a decreasing function of temperature. Thus, there will be some intermediate range of temperature where the effect of multiple fluctuations must be taken into account. The effect of spatial inhomogeneity on the Brillouin linewidths near the glass transition will be calculated in Section 3.3.4.3.

3.3.2.2. Model Theories. Light scattering can be separated into an isotropic component whose magnitude depends only on the scalar fluctuations and an anisotropic component that depends on the tensor fluctuations. We shall consider first the isotropic Rayleigh–Brillouin scattering.

The isotropic intensity depends on the values of Y, Z, and $\{m_j\}$. The dependence of ϵ on temperature at constant density is usually small for amorphous isotropic media and thus we shall neglect terms that depend on Z. We shall also assume that $\{m_j\} = 0$. The isotropic spectrum is then given by

$$I_{\mathrm{iso}}(\mathbf{q}, \omega) = \frac{-k_B T}{2\pi i \omega} \left[\frac{Y^2 C q^2}{\Delta} - \text{c.c.} \right]. \qquad (3.3.17)$$

The calculation of I_{iso} now depends only on the model chosen for the moduli. The simplest assumption is the purely elastic case with $\alpha = 0$, $K = K_\infty$, and $\mu = \mu_\infty$. The resulting spectrum contains infinitely narrow lines at $\omega = \pm q[(K + \frac{4}{3}\mu)/\rho]^{1/2}$. The simplest model that gives realistic results is for a viscous fluid where $K = K_0$, $\mu = i\omega\eta$, $\alpha = 0$, the subscript 0 denotes the low-frequency or thermodynamic limit, and η is the shear viscosity. The spectrum is calculated to be

$$I_{\mathrm{iso}}(\mathbf{q}, \omega) = \frac{k_B T}{\pi} Y^2 q^2 \frac{\frac{4}{3}\eta q^2}{(\frac{4}{3}\eta \omega q^2)^2 + (K_0 q^2 - \rho \omega^2)^2}. \qquad (3.3.18)$$

This spectrum has two peaks with splitting

$$\Delta\omega_l = q\left[\frac{K_0}{\rho} - \left(\frac{2}{3}\frac{\eta q}{\rho}\right)^2\right]^{1/2}$$

and width

$$\Gamma_l = \frac{2}{3}\frac{\eta q^2}{\rho} \qquad (3.3.19)$$

where the subscript l denotes the spectrum due to longitudinal acoustic phonons.

Most fluids have a finite value for the thermal expansion coefficient. The spectrum then becomes much more complicated; the exact result is

$$I_{iso}(q, \omega) = \frac{k_B T}{\pi} Y^2 q^2 \frac{Kq2E/\omega - \rho C_V D}{D^2 + E^2}, \qquad (3.3.20)$$

where

$$D = -\omega^2(\tfrac{4}{3}\rho C_V \eta q^2 + \rho\kappa q^2) + \kappa K_0 q^4$$

and

$$E = -\omega^3 \rho^2 C_V + \omega(\rho C_V K_0 q^2 + \tfrac{4}{3}\eta\kappa q^4 + TK_0^2\alpha^2 q^2).$$

The exact result can be understood by considering the roots of the equation $\Delta = 0$. For the case discussed above, there will be three roots given by

$$\omega_1 = i\Gamma_C, \qquad \omega_{2,3} = i\Gamma_l \pm qV_l, \qquad (3.3.21)$$

where

$$\Gamma_C \approx \frac{Kq^2}{\rho C_V}\left(1 + \frac{K_0\alpha^2 T}{\rho C_V}\right)^{-1},$$

$$\Gamma_l \approx \frac{\tfrac{4}{3}q^2\eta}{2\rho} + \frac{Kq^2}{2\rho C_p}(\gamma - 1),$$

where $\gamma = C_p/C_V$, C_p is the specific heat at constant pressure, and

$$V_l \approx \left[\frac{K_0}{\rho}\left(1 + \frac{K_0\alpha^2 T}{\rho C_V}\right) - \frac{\Gamma_l^2}{q^2}\right]^{1/2}.$$

The first root corresponds to a central peak with width Γ_C and the other two roots give the Brillouin peaks with splitting $\pm\Delta\omega_l = qV_l$ and width Γ_l.

For a low-viscosity fluid, the width of the central peak is described quite well by Γ_C and the Brillouin splitting is accurately given by qV_l.

However, the Brillouin lines are always wider than predicted above. This fact can be taken into account by including a volume viscosity so that $K = K_0 + i\omega\eta_V$. The Brillouin linewidth is then given by

$$\Gamma_l = \frac{q^2(\eta_V + \frac{4}{3}\eta_S)}{2\rho} + \frac{Kq^2}{2\rho C_p}(\gamma - 1). \tag{3.3.22}$$

The integrated intensity of the central peak is proportional to $1 - \gamma^{-1}$ and the sum of the two Brillouin peaks to γ^{-1}. The ratio of the central peak intensity to the Brillouin intensity is called the Landau–Placzek ratio[7]

$$R_{LP} = I_C/2I_B = \gamma - 1. \tag{3.3.23}$$

The pure viscous theory outlined above is sufficient for simple liquids,[8] but for polymers the theory breaks down very badly. The width of the Brillouin peaks is predicted to increase with increasing viscosity. For polymers near the glass–rubber relaxation this would lead to such wide peaks that they would not be observable. In fact, the Brillouin peaks are quite sharp near T_g. The value of $\gamma - 1$ for polymers is quite small, but the ratio of the central peak intensity to the sum of the Brillouin intensities usually exceeds unity for polymer fluids. A relaxation theory is required to account for the observed behavior.

The longitudinal modulus is defined as $M = K + \frac{4}{3}\mu$. The simplest theory that incorporates relaxation represents the longitudinal modulus as

$$M = M_\infty - M_R/(1 + i\omega\tau), \tag{3.3.24}$$

where $M_R = M_\infty - K_0$ is the relaxing part of the modulus. There will still be a central peak with width equal to Γ_C, but in order to simplify the analysis we shall ignore the effect of thermal expansion on the Brillouin peaks. The resulting spectrum is

$$I_{iso}(\mathbf{q}, \omega) = \frac{kT}{\pi} Y^2 q^2 \frac{\dfrac{M_R q^2 \tau}{1 + \omega^2\tau^2}}{\left(\dfrac{M_R q^2 \omega\tau}{1 + \omega^2\tau^2}\right)^2 + \left(M_\infty q^2 - \dfrac{M_R}{1 + \omega^2\tau^2} - \rho\omega^2\right)^2}. \tag{3.3.25}$$

Again, the spectral features are determined by the roots of the equation $\Delta = 0$. There will be three roots:

$$\omega_1 = i\Gamma_\tau, \qquad \omega_{2,3} = i\Gamma_l \pm qV_l, \tag{3.3.26}$$

---

[7] L. Landau and G. Placzek, *Phys. Z. Sowjetunion* **5**, 172 (1934).
[8] R. D. Mountain, *Rev. Mod. Phys.* **38**, 205 (1966).

where

$$\Gamma_\tau \equiv \frac{1}{\tau} - 2\Gamma_l,$$

$$V_l \equiv \left(\frac{M_\infty}{\rho} + \frac{3\Gamma_l^2}{q^2} - \frac{2\Gamma_l}{\tau q^2}\right)^{1/2},$$

$$\Gamma_l \approx \frac{M_R q^2/2\rho\tau}{(1/\tau^2) + (M_\infty q^2/\rho)},$$

where for Γ_l we have assumed that $\Gamma_l \ll \Delta\omega_l$. Thus as the relaxation time becomes long and the viscosity becomes very high, the Brillouin peaks become narrow.

It is still a good approximation to let the intensity of the central peak with width Γ_C be proportional to $1 - \gamma^{-1}$. The central peak with width Γ_τ has an intensity proportional to $\gamma^{-1}[(V_l^2 - V_l^2(0))/V_l^2]$ and the sum of the two Brillouin peaks is proportional to $\gamma^{-1}[V_l^2(0)/V_l^2]$, where $V_l(0) = (K_0/\rho)^{1/2}$. The apparent Landau–Placzek ratio is given by

$$R'_{LP} = \frac{(1 - \gamma^{-1}) + \gamma^{-1}[(V_l^2 - V_l^2(0))/V_l^2]}{\gamma^{-1}[V_l^2(0)/V_l^2]}. \qquad (3.3.27)$$

Thus, even though $\gamma - 1$ is very small, the apparent Landau–Placzek ratio can be quite large.

The simple relaxation theory can be extended to include different relaxation times for compression and shear, and the effect of thermal expansion on the Brillouin peaks can be taken into account. However, the number of parameters rapidly exceeds the information actually obtained from the spectrum. In practice, the isotropic spectrum is fit to the form of Eq. (3.3.25) plus a central peak with width Γ_C.

The isotropic scattering contributes only to the I_{VV} spectrum. The anisotropic scattering gives rise to a spectrum for all the polarizations. The major source of anisotropic scattering for most polymer fluids is the inherent optical anisotropy of the molecular subunits. The anisotropic scattering is due to fluctuations in orientation of the bonds in the molecules. There will be a central peak with width $\Gamma_{or} = (2\pi\tau_{or})^{-1}$ for Γ measured in gigahertz for all the spectra, where τ_{or} is the relaxation time for the collective orientation variable of the medium.

If the longitudinal and transverse phonons couple to molecular reorientation, then the parameter X will have a finite value when $\omega > 0$. The anisotropic contribution to the spectrum due to longitudinal phonons is usually small relative to the isotropic contribution in amorphous polymers and hence we shall neglect it. The spectrum due to transverse phonons is entirely anisotropic in character and occurs only in the $I_{HV} = I_{VH}$ spectrum.

When the relaxation time for the shear modulus $\tau \ll 10^{-10}$ sec, the transverse phonon velocity is imaginary and no transverse Brillouin peaks are observed. However, shear fluctuations do occur and they couple with molecular reorientation. The I_{HV} spectrum is given by[9]

$$I_{HV}(\mathbf{q}, \omega) \propto \frac{\Gamma_{or}}{\Gamma_{or}^2 + \omega^2} \sin^2 \frac{\theta}{2} + \Gamma_{or} \cos^2 \frac{\theta}{2}$$

$$\frac{(q^4 \eta_S^2/\rho^2)(1 - R) + \omega^2}{\left(\Gamma_{or} \dfrac{q^2 \eta_S}{\rho} - \omega^2\right)^2 + \omega^2 \left(\Gamma_{or} + \dfrac{q^2 \eta_S}{\rho}(1 - R)\right)^2}, \quad (3.3.28)$$

where η_S is the shear viscosity and R is a parameter that is equal to the fraction of the shear viscosity that is due to coupling to molecular reorientation. The depolarized (I_{HV}) Rayleigh–Brillouin spectrum of n-hexadecane at 65°C is shown in Fig. 3. The coupling leads to a central dip with width $q^2 \eta_S / \rho$.

When the relaxation time becomes longer, the simple result given in Eq. (3.3.28) is no longer valid and the full expression given in Eq. (3.3.15) must be used. The spectrum will then depend on the roots of the equation $Gq^2 - \rho\omega = 0$. If we consider only a single shear relaxation time τ_S, then there will be three roots

$$\omega_1 = 0, \qquad \omega_2 = i\Gamma_t \pm qV_t, \qquad (3.3.29)$$

where

$$\Gamma_t = 1/2\tau_S, \qquad V_t = \left(\frac{G_\infty}{\rho} - \frac{(1/2\tau_S)^2}{q^2}\right)^{1/2}.$$

The full spectrum is given by

$$I_{HV}(\mathbf{q}, \omega) \propto \frac{\Gamma_{or}}{\Gamma_{or}^2 + \omega^2} + \frac{1}{2} \frac{G_{or}\omega^2 \tau_{or}^2}{1 + \omega^2 \tau_{or}^2} q_2^2$$

$$\times \frac{(F/\omega) + \rho\tau_S\omega^2}{F^2 + \rho^2\omega^4}, \qquad (3.3.30)$$

where $F = G_\infty \tau_S q^2 \omega - \rho\tau_S\omega^3$ and G_{or} is the fraction of the high-frequency shear modulus due to coupling to molecular reorientation. If τ_S is long enough, the transverse phonons will lead to well-defined Brillouin peaks with splitting $\pm\Delta\omega_t = qV_t$ and width Γ_t. The Rayleigh–Brillouin spec-

[9] G. R. Alms, D. R. Bauer, J. I. Brauman, and R. Pecora, *J. Chem. Phys.* **59**, 5304 (1973).

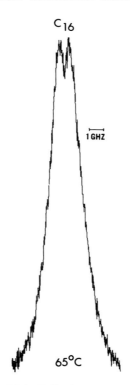

FIG. 3. Depolarized (I_{HV}) Rayleigh–Brillouin spectrum of n-hexadecane at 65°C.

trum ($I_{VV} + I_{HV}$) of bisphenol-A polycarbonate showing both longitudinal and transverse peaks is shown in Fig. 4.

The form of the Rayleigh–Brillouin spectrum in viscoelastic media is now well understood. However, the number of variables that affect the spectrum in a particular case still requires detailed analysis. A number of examples will be given in Section 3.3.4.

3.3.3. Experimental

The scattering diagram in Fig. 2 pictured the incident beam as monochromatic and unidirectional. No real source has these properties, but a laser can provide an intense source of well-collimated light with a narrow frequency distribution. Typical Brillouin splittings are in the range 10^8–10^{10} Hz. In order to measure $\Delta\omega_l$, the incident frequency width should be less than 10^8 Hz. Rayleigh linewidths can range from a few

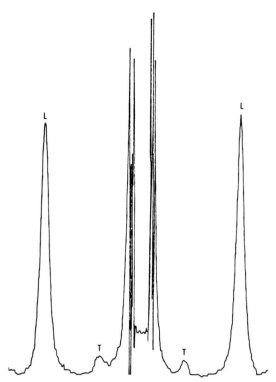

FIG. 4. Rayleigh–Brillouin ($I_{VV} + I_{HV}$) spectrum of bisphenol-A polycarbonate showing both longitudinal (L) and transverse (T) Brillouin peaks.[42]

hertz to 10^{11} Hz. The gain curve of an argon ion laser at 5145 Å has a half-width at half-height of approximately 3 GHz. The actual output consists of a sequence of much narrower lines separated by $C/2l$, where l is the laser cavity length. For a typical 1 m cavity this gives a separation of 150 HMz. A single-cavity mode can be selected by introducing an etalon into the laser cavity with a free spectral range of 10 GHz. The linewidth of the single mode is typically 10 MHz and is due primarily to jitter in the overall cavity length. The linewidth can be reduced to 1 MHz if the cavity length is feedback controlled or the acoustic jitter is minimized. Spectral linewidths greater than 10 MHz can be determined directly. Rayleigh linewidths in the $1{-}10^6$ Hz range must be determined indirectly using photon correction spectroscopy.[10]

[10] H. Z. Cummins and E. R. Pike, eds., "Photon Correlation and Light Beating Spectroscopy." Plenum, New York, 1974.

Laser emission is usually linearly polarized due to the use of Brewster angle windows on the plasma tube. The polarization of the incident light can be adjusted to be vertical or horizontal with respect to the scattering plane with a double Fresnel rhomb polarization rotator. For observation of the I_{VV} spectrum, the laser light is sufficiently polarized to give acceptable spectra, but for the study of the I_{HV} spectrum it is often necessary to include an additional polarizer in the incident beam to eliminate leakage of the I_{VV} spectrum. The scattered light can be analyzed with a prism polarizer.

Except near a critical point, Rayleigh–Brillouin scattering is a weak effect. In order to increase the power density in the scattering volume, the incident light is usually focused to a diameter in the 100–200 μm range. The optimal beam diameter can be chosen to equal the image size of the pinhole resolving element in the Fabry–Perot interferometer used to analyze the spectrum.

A variety of sample configurations is possible. For most amorphous samples above the glass transition, a square fluorimeter cell is convenient for observation at $90°$. If the sample is to be cooled below T_g a square cell is often not strong enough to withstand the strain created by the polymer and a round test tube has been found to be preferable. If the sample is to be studied only near or below the glass transition no container is necessary or desirable. However, great care must be taken in the preparation of glassy polymer samples for Rayleigh–Brillouin scattering studies. The above point will be discussed in more detail in Section 3.3.4.

Most polymer samples cannot be prepared as clear amorphous blocks suitable for Rayleigh scattering studies. However, many polymers can be prepared as films and the Brillouin spectrum can be obtained.[11] The main problem with films is the definition of the true value of $\mathbf{q}$. There are two experimentally convenient geometries for examining films. In the first case the incident beam is normal to the film surface and the scattered light is observed at some angle less than $90°$. The true scattering angle is then given by $\theta = \arcsin(\sin \theta'/n)$, where θ' is the nominal scattering angle and n is the refractive index of the film. Another geometry is to illuminate the rear surface of the film placed at $45°$ to the incident beam and to observe the light scattered at a nominal angle of $90°$. The true scattering angle is then $\theta = 2 \arcsin(0.707/n)$. Thus a knowledge of the refractive index of the film is necessary to relate measured values of $\Delta\omega$ to the phonon velocity V.

Successful spectra have also been obtained from tensile bars, extruded rods, and other common mechanical specimens. The main requirement is that the true scattering geometry be well defined.

[11] G. D. Patterson, *J. Polym. Sci., Polym. Phys. Ed.* **14**, 143 (1976).

Since the Rayleigh–Brillouin spectra are explicit functions of q it is very important to define the actual distribution of scattering angles observed in a real experiment. If the angular distribution at 90° is ± 1°, then there will be approximately a 1% spread in q. If the Brillouin splitting is 10^{10} Hz, then there will be a 10^8 Hz contribution to the linewidth due to the range of scattering angles actually observed. At lower scattering angles the relative effect of finite scattering solid angle is even more severe. Great care must be taken to restrict the scattering solid angle so that only a negligible broadening of the observed spectrum occurs.

If the characteristic linewidths of the scattered spectrum are less than 10^6 Hz, then the technique used to resolve the spectrum is photon correlation spectroscopy.[10] Reference 10 contains an extensive discussion of this approach and no further discussion will be given here. The only application of photon correlation spectroscopy to a bulk polymer was reported by Jackson *et al.*[12] for poly(methyl methacrylate). They observed the very narrow central line with width Γ_r due to structural relaxation near the glass–rubber relaxation. The problems associated with the measurement of central lines in bulk polymers will be discussed in Section 3.3.4.

The instrument most commonly used to resolve the Rayleigh–Brillouin spectrum is the Fabry–Perot interferometer. This device consists of a pair of highly reflective, optically polished mirrors. For most routine work, the mirrors are plane parallel glass plates with dielectric coatings. If higher resolving power is desired, the mirrors may be spherical or confocal in geometry. A brief description of a plane-parallel Fabry–Perot is given below.

The transmission function for a plane-parallel Fabry–Perot interferometer is

$$\frac{I(\omega)}{I_0} = \frac{(T^2/(1 - R)^2)}{1 + (4F^2/\pi^2)\sin^2(\pi n d/C)}, \tag{3.3.31}$$

where T is the transmission of the mirrors, R is the reflectivity, F is the finesse, n is the refractive index of the medium between the plates, d is the distance between the plates, and C is the speed of light in a vacuum. When $\omega = (\pi C N/nd)$, where N is an integer, the interferometer will have its maximum transmission

$$\frac{I_{max}}{I_0} = \frac{T^2}{1 - R} = \left(1 - \frac{A}{1 - R}\right)^2, \tag{3.3.32}$$

where A is the absorption of the mirrors. With modern dielectric coatings, the maximum transmission can be quite high, even for 98% reflectivity.

[12] D. A. Jackson, E. R. Pike, J. G. Powles, and J. M. Vaughan, *J. Phys. C* **6**, L55 (1973).

The spacing between maxima is equal to $\pi C/nd$ and is called the free spectral range (FSR). If we assume $n = 1$ for air-spaced plates and express the FSR in hertz instead of radians/second the result is FSR $= C/2d$. The finesse is equal to the FSR divided by the full-width at half-height of the transmission function. The finesse is determined by many factors, but the two most important variables are the reflectivity and the flatness figure of the mirrors. The total finesse is given by

$$F_{tot} = \left[\sum_i (1/F_i^2) \right]^{-1/2}, \tag{3.3.33}$$

where the F_i are the contributions from the various sources that determine the finesse. The reflectivity finesse is given by

$$F_R = \pi\sqrt{R}/(1 - R) \tag{3.3.34}$$

and the flatness finesse by

$$F_F = M/2,$$

where M is from the surface figure expressed as λ/M. Total finesses of the order of 50 can routinely be obtained for nominally $\lambda/200$ plates with $R \approx 98\%$. Higher reflectivities only lower the transmission without appreciably increasing the finesse because other factors become dominant.

The other important characteristic of a Fabry–Perot interferometer is the contrast. At the minimum in the transmission function the intensity relative to the maximum transmission is $(1 + 4F^2/\pi^2)^{-1}$. Thus the contrast is accurately given by $4F^2/\pi^2$ and a contrast near 10^3 is routinely obtained. However, for many bulk polymer samples particulate impurities, partial crystallinity, or anisotropic light scattering may lead to Rayleigh peak intensities that are 10^4–10^7 times greater than the Brillouin intensities. The Brillouin peaks will then be hidden under the intense wings of the central peak. The problem of contrast can be overcome by the use of the multipass interferometer.[13]

When the light is passed through the interferometer more than one time, the composite transmission function is the product of the single-pass transmission functions

$$I_{tot}(\omega) = I_1(\omega)I_2(\omega) \; \cdots \; I_n(\omega). \tag{3.3.35}$$

A typical system employs three passes, but two- and five-pass arrangements have also been used. If the transmission functions of each pass are identical then total contrasts greater than 10^8 can be obtained from a three-pass interferometer. However, if the peak transmission of each

[13] J. R. Sandercock, *Opt. Commun.* **2**, 73 (1970); *RCA Rev.* **36**, 89 (1975).

pass is slightly different, then the total maximum transmission may easily fall to 10^{-3} of the single-pass maximum transmission and no good spectra will be obtained. In practice it is preferable to use lower reflectivity plates to optimize the maximum transmission. The total finesse remains high because the multipass finesse is

$$F_n = F_1/(2^{1/n} - 1). \tag{3.3.36}$$

For a three-pass system with $R \approx 90\%$ and plates flat to $\lambda/200$, the total finesse was observed to be 60 and the contrast approximately 10^7.

For a three-pass system, a pair of high-quality corner cube retrore-flectors is employed. The light enters through a hole in the front cube and after three traverses exits through a hole in the rear cube. This is the most convenient arrangement since the interferometer need not be moved from its normal single-pass position. Alignment of a multipass interfer-ometer is very critical and it is useful to align the system for maximum single-pass finesse before attempting to operate in the multipass mode. While a small misalignment in single pass will not appreciably affect the maximum transmission, the same error in a multipass system may elimi-nate the signal altogether. However, the rewards for using the high-contrast interferometer for studying polymers far outweigh the aggrava-tions.

The interferometer must be scanned in order to be used as a spectrum analyzer. This is accomplished by changing the optical path length nd between the plates. The earliest studies[14,15] using a laser source accom-plished this by changing the pressure of the gas between the plates. This technique can give very high quality results, but it is very tedious and multiple rapid scans are not possible. Most modern Fabry–Perot inter-ferometers use piezoelectric transducers to mechanically change the dis-tance between the plates. Commercial versions of such devices are now in a high state of perfection.

The resolving element in the spectrometer is a pinhole, which is imaged on the scattering volume. It must be small enough to maintain a high fi-nesse but should be as large as the scattering volume if possible. A 200-μm pinhole is useful unless the FSR is less than about 5 GHz. For higher resolutions the Fabry–Perot rings are closer together and the pin-hole should be smaller than the ring spacing divided by the intrinsic finesse.

Because the Fabry–Perot interferometer is a periodic filter, an addi-tional narrow bandpass interference filter is necessary to isolate the

[14] R. V. Chiao and B. P. Stoicheff, *J. Opt. Soc. Am.* **54**, 1286 (1964).
[15] D. H. Rank, E. M. Kiess, V. Fink, and T. A. Wiggins, *J. Opt. Soc. Am.* **55**, 925 (1965).

Rayleigh–Brillouin spectrum and reject the Raman scattering, background fluorescence, or other stray light.

The light that passes through the pinhole and interference filter is imaged on a photomultiplier tube. Because the beam diameter is very small, a minimal size photocathode can be used to reduce the dark current without sacrificing signal. Because the signal levels associated with Brillouin scattering are very low, a photon counter is essential to analyze the output of the photomultiplier. It is very convenient to record the spectrum with a multichannel analyzer. The digital spectrum can then be displayed on an oscilloscope or analyzed with a computer. A variety of hard copy outputs are also available. A block diagram of a typical Fabry–Perot spectrometer is shown in Fig. 5.

If the sample is free from impurities and only intrinsic Rayleigh–Brillouin scattering is obtained, the spectrum can be fit with a computer and all the characteristic parameters obtained. Because the Fabry–Perot interferometer is a periodic filter, the spectrum should be fit to a sum of many orders. A typical spectrum will scan over two free spectral ranges,

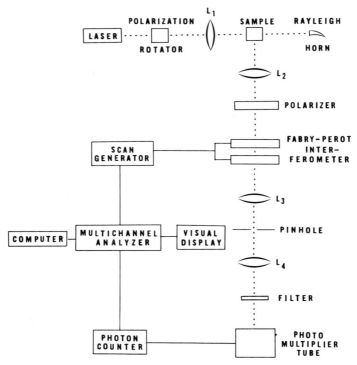

FIG. 5. Block diagram of a Fabry–Perot spectrometer.

but the intensity at each point is the sum over all orders passed by the interference filter. The I_{VV} spectrum of n-hexadecane at 60°C is shown in Fig. 6. The spectrum contains 1024 points in approximately two orders. The spectrum was fit to a central peak with width Γ_C, a central peak with width Γ_{or}, and two shifted Brillouin peaks with splitting $\pm \Delta \omega_l$ and width Γ_l. The value of $I(\omega)$ at each point was obtained by summing over six Fabry–Perot orders. The calculated spectrum is shown by the solid line and every fifth data point is indicated by a plus. The agreement is excellent.

In order to obtain the true values of Γ_C, Γ_{or}, and Γ_l, it is necessary to deconvolute the effect of Fabry–Perot resolution, finite frequency distribution in the incident light, and finite-scattering solid angle. The first two effects can be taken into account by recording the spectrum of an elastic scatterer and convoluting it with the intrinsic spectral form. The resulting spectra can then be fit with the same programs used to fit the experimental spectra and the two results compared. This leads to very accurate deconvolutions. The effect of finite-scattering solid angle is best taken into account experimentally and eliminated by using a sufficiently small aperture in the scattered beam.

Most polymer samples give rise to a large central peak due to elastic scattering by impurities. However, the Brillouin spectrum can still be obtained. The Brillouin splitting can be determined directly from the digital spectrum as displayed on an oscilloscope using the multichannel analyzer. The linewidth can also be obtained with moderate accuracy ($\pm 10\%$) by counting points. If the Brillouin peaks are well separated from the central line, they can also be fit with a computer and a more accurate linewidth obtained.

3.3.4. Applications

3.3.4.1. Low-Viscosity Fluids. When the structural relaxation times are shorter than 10^{-10} sec, the I_{VV} Rayleigh–Brillouin spectrum can be described as shown in Fig. 6. The isotropic spectrum contains a central line with width Γ_C and two shifted Brillouin peaks with width Γ_l. The anisotropic contribution to I_{VV} is a single Lorentzian line with width Γ_{or}. If the value of Γ_{or} is comparable to or greater than the free spectral range, then the anisotropic peak will appear only as a broad background as illustrated in Fig. 1.

The ratio of the central isotropic peak intensity to the Brillouin intensities is equal to the Landau–Placzek[7] ratio $\gamma - 1$. For the n-hexadecane spectrum shown in Fig. 6 the Landau–Placzek ratio is determined to be

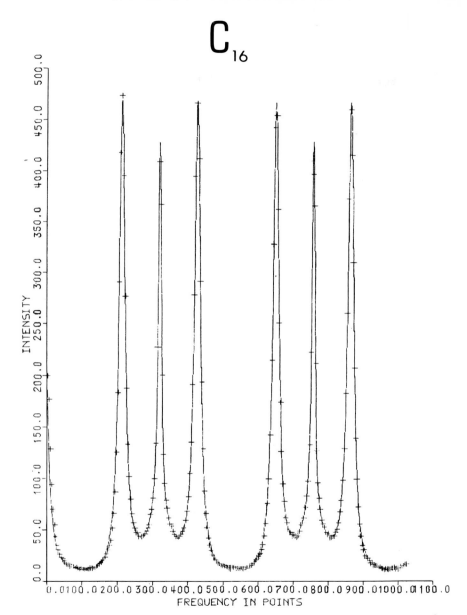

FIG. 6. Rayleigh–Brillouin I_{VV} spectrum of n-hexadecane at 60°C showing computer fit to the experimental data. The calculated spectrum is shown by the solid line and every fifth data point by +.

0.219. Huang and Wang[16] measured $R_{LP} = 0.25$ for a very clean sample of low molecular weight (425 gm/mole) poly(d,l-propylene glycol). If the intrinsic Rayleigh–Brillouin spectrum can be obtained and the isotropic component extracted, then the determination of R_{LP} is a good method for obtaining the ratio of specific heats γ. If Γ_{or} is too narrow to be resolved and Γ_C is also not resolved, it is still possible to obtain the anisotropic intensity from the I_{HV} spectrum and to subtract $\frac{4}{3}I_{HV}$ from the I_{VV} central peak intensity. For most polymer samples elastic scattering precludes meaningful measurement of R_{LP}.

The Brillouin splitting in the low viscosity regime is given by $\pm \Delta\omega_l \approx q(K_0\gamma/\rho)^{1/2}$, where $K_0 = \beta_T^{-1}$, the reciprocal of the isothermal compressibility. If the values of γ and ρ are known, measurements of $\Delta\omega_l$ can be used to determine β_T. For the n-hexadecane spectrum shown in Fig. 1, the isothermal compressibility is calculated[17] to be $\beta_T = 1.6 \pm 0.1 \times 10^{-10}$ cm^2/dyne, in good agreement with the directly measured value.[18]

The linewidth Γ_C of the central peak in the isotropic spectrum is equal to $\kappa q^2/\rho C p$. The thermal conductivity of organic fluids is typically 3×10^{-4} cal/sec cm^2 (°C/cm). At 90° scattering angle the value of Γ_C is calculated to be only a few megahertz. Thus measurement of Γ_C is difficult but possible. No values of Γ_C for a polymer have yet been reported. The contribution of thermal conductivity to the Brillouin linewidths is ($\gamma - 1)\Gamma_C/2$ and hence we shall ignore its effect in all further discussion.

The Brillouin linewidth Γ_l is then given by $q^2(\eta_V + \frac{4}{3}\eta_S)/2\rho$. If the shear viscosity is known, measurements of Γ_l can be used to determine the volume viscosity η_V. Such an analysis has been carried out by Champion and Jackson[19] for the lower n-alkanes. They find that the volume viscosity determined in the above manner is essentially constant with temperature.

In Section 3.3.2.2, the volume viscosity was introduced in an ad hoc manner, but the general theory presented in Section 3.3.2.1 included the volume viscosity explicitly. A general expression for Γ_V is

$$\Gamma_V = \frac{q^2}{2\rho} \sum_j \frac{L_j^2\tau_j}{1 + \omega^2\tau_j^2}, \tag{3.3.37}$$

where $\omega = \Delta\omega_l$. When a longitudinal sound wave propagates in a fluid, both the local displacement gradient and temperature are changed. The volume viscosity reflects the finite rate of equilibration for those variables

[16] Y. Y. Huang and C. H. Wang, *J. Chem. Phys.* **62**, 120 (1975).
[17] G. D. Patterson and J. P. Latham, *Macromolecules* **10**, 736 (1977).
[18] R. A. Orwoll and P. J. Flory, *J. Am. Chem. Soc.* **89**, 6814 (1967).
[19] J. V. Champion and D. A. Jackson, *Mol. Phys.* **31**, 1169 (1976).

that depend on the compression and temperature. Fluids with a low shear viscosity are expected to have a correspondingly low contribution to the volume viscosity due to structural changes upon compression. Also, the structural volume viscosity should have a temperature dependence that is similar to the shear viscosity. The ratio η_S/η_V is often observed to be constant with temperature.[20]

If the temperature is changed, the energy gain or loss must be distributed into all the modes of the system. The translational degrees of freedom should come into equilibrium very fast in a low-viscosity liquid. There will also be a set of kinetic orientational degrees of freedom. For a chain-molecule fluid, there is a strong coupling between translational and orientational motions so that all the kinetic modes should equilibrate rapidly. There will also be a set of fully quantized vibrational modes. In chain-molecule fluids, the relaxation times for the vibrational modes should be very fast. Finally, there will be a set of rotational isomeric states (RIS). There will be a whole spectrum of relaxation times for changes from one equilibrium distribution of rotational isomeric states to another. Rotations about the initial or terminal bond in a polymethylene chain do not change the RIS. Rotations about the internal bonds in n-butane and n-pentane have been studied using ultrasonic attenuation by Piercy and Rao.[21] They were able to describe the relaxation time for this process by

$$\tau = \tau_0 \exp(E_a/RT), \tag{3.3.38}$$

where $\tau_0 \approx 10^{-13}$ sec and E_a is approximately 3.5 kcal/mole. Thus at 300 K the relaxation time would be approximately 10^{-10} sec. The appropriate frequency is in the range $\omega = 3 \times 10^{10}$ rad/sec. This means that $\omega\tau > 1$ and the dynamic volume viscosity due to conformational changes of the second and $(n-1)$th bonds will actually fall with decreasing temperature. The relaxation time for other internal bonds will be even slower and hence they should make a negligible contribution to Γ_V at hypersonic frequencies.

Champion and Jackson[19] also found that the magnitude of the volume viscosity was inversely proportional to the chain length of the n-alkane. This is fully consistent with the analysis presented above since only the penultimate bonds contribute significantly to the hypersonic volume viscosity, and the lower n-alkane fluids will contain more of these bonds per unit volume.

The depolarized (I_{HV}) Rayleigh–Brillouin spectrum of low viscosity

[20] K. F. Herzfeld and T. A. Litovitz, "Absorption and Dispersion of Ultrasonic Waves." Academic Press, New York, 1959.
[21] J. E. Piercy and M. G. S. Rao, *J. Chem. Phys.* **46**, 3951 (1967).

fluids depends on the reorientational motions of the chains and on those other modes of motion that couple to reorientation such as shear. The spectral form is given by Eq. (3.3.28). An analysis of the spectrum shown in Fig. 3 gives $\Gamma_{or} = 1.30 \pm 0.05$ GHz, $q^2\eta_S/\rho = 196 \pm 20$ MHz, and $R = 0.33 \pm 0.02$. The overall width Γ_{or} corresponds to a relaxation time for molecular reorientation of 1.2×10^{-10} seconds. The shear viscosity calculated from the I_{HV} spectrum is in good agreement with the directly measured value.[22]

The general expression for the shear viscosity is

$$\eta_S = \sum_k \frac{N_k^2 \tau_k}{1 + \omega^2 \tau_k^2}. \tag{3.3.39}$$

Since $0 < R < 1$, there must be at least two mechanisms that relax the shear fluctuations in n-alkane fluids. Molecular translation is one obvious shear relaxation process. Overall molecular reorientation will relax the shear. And conformational state changes can relieve the stress. In the depolarized Rayleigh spectrum of low-viscosity fluids the transverse velocity is imaginary and the appropriate frequency in Eq. (3.3.39) is $\omega = 0$. Thus the measured viscosity is equal to the zero shear viscosity. However, the above analysis suggests that the dynamic shear viscosity at the longitudinal phonon frequency will be lower since overall molecular rotation or rotational isomeric state changes will not be fast enough to equilibrate with the hypersonic waves. This means that estimates of Γ_S based on the zero shear viscosity $\eta_S(0)$ will be too high and analyses such as that carried out by Champion and Jackson[19] are subject to some reservation. The full description of the Rayleigh–Brillouin spectrum in terms of the molecular dynamics of chain-molecule fluids is indeed very complicated, but the general features are now well understood.

3.3.4.2. Viscoelastic Fluids. When the structural relaxation times of the system are comparable to 10^{-10} seconds, the fluid exhibits significant viscoelasticity in the hypersonic frequency range. The isotropic spectrum can be described by Eq. (3.3.25) plus a central peak with width Γ_C. The anisotropic contribution to the I_{VV} spectrum is a narrow central peak with width Γ_{or}. A comparison of the Rayleigh–Brillouin spectrum of a liquid in its low-viscosity and viscoelastic regimes is shown in Fig. 7. In the viscoelastic case the Brillouin peaks become very broad and an additional central peak with width Γ_τ is apparent. The Landau–Placzek ratio at 100°C is 0.22, while the sum of the two central peak intensities divided by the Brillouin intensities is 0.49 at 23°C. The behavior of the apparent Landau–Placzek ratio R'_{LP} is given by Eq. (3.3.27).

[22] F. D. Rossini, "Selected Values of Physical and Thermodynamic Properties of Hydrocarbons and Related Compounds," API Res. Proj. 44. Carnegie Press, Pittsburgh, Pennsylvania, 1953.

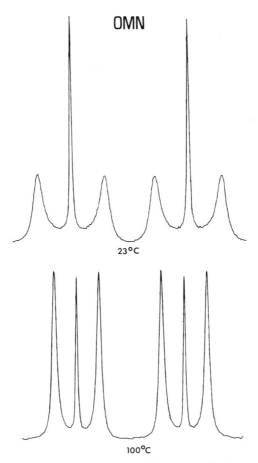

FIG. 7. Rayleigh–Brillouin spectra of 2,2,4,4,6,6,8,8-octamethyl nonane at 23 and 100°C.

As the fluid is cooled, the structural relaxation times become longer and the values of K_0, K_R, and G_∞ increase. As a result the phonon velocity increases with decreasing temperature. Because the experiments are usually carried out at essentially fixed q, the frequency of the phonons also changes with temperature and no dramatic rise in $\Delta\omega_l$ is observed in the viscoelastic regime. However, a plot of $\Delta\omega_l$ vs. temperature does change slope in this region. Eventually the slope of the line is determined by the change of $(M_\infty/\rho)^{1/2}$ with temperature, while at high temperatures the line follows $(K_0\gamma/\rho)^{1/2}$. The limiting high-frequency modulus rises faster with decreasing temperature than K_0.

If we adopt the results for the dynamic shear and volume viscosities

presented in Section 3.3.4.1, the longitudinal Brillouin linewidth will be
given by

$$\Gamma_l = \frac{q^2}{2\rho} \left[\sum_j \frac{L_j^2 \tau_j}{1 + \omega^2 \tau_j^2} + \frac{4}{3} \sum_k \frac{N_k^2 \tau_k}{1 + \omega^2 \tau_k^2} \right]. \qquad (3.3.40)$$

The exact dependence of Γ_l on temperature could be very complicated,
but the linewidth rises to a region of maxima and then falls at lower temperatures.

Hypersonic relaxation data are usually presented in terms of the loss
tangent, where

$$\tan \delta = 2\Gamma_l \, \Delta\omega_l / (\Delta\omega_l^2 - \Gamma_l^2). \qquad (3.3.41)$$

Measurements of $\Delta\omega_l$ and tan δ for polyisobutylene[23] (PIB) are presented
in Fig. 8. The Brillouin frequency falls monotonically with temperature
and tan δ goes through a maximum at approximately 200°C and a frequency of 4.95 GHz. Hypersonic relaxation results for poly(d,l-

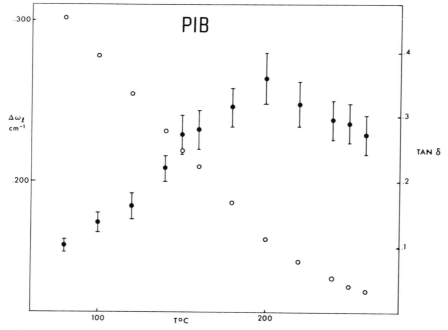

FIG. 8. Brillouin frequency $\Delta\omega_l$ (○) and loss tangent (●) vs. temperature for polyisobutylene (PIB).[24]

[23] G. D. Patterson, *J. Polym. Sci., Polym. Phys. Ed.* **15**, 455 (1977).

propylene glycol) (PPG) are presented in Fig. 9.   In this case two maxima are observed.   The lower temperature maximum occurs at 50°C and 5.43 GHz and the higher one at 100°C and 4.40 GHz.   Temperatures of maximum loss have been determined for many polymers and the results are presented in Table I.[23-30]   For the two crystallizable polymers (poly-carbonate and polyethylene oxide) the temperatures of maximum loss are comparable to or greater than the melting point for these materials.   For other polymers such as polyethylene, the temperature of maximum loss must be below T_m since PE behaves as a low hypersonic viscosity fluid in the melt.   Any general correlation between the melting point and the temperature of maximum loss observed by Brillouin scattering is unwarranted.

Dynamic loss data are often presented as a transition map where the log of the frequency is plotted vs. the reciprocal of the temperature of maximum loss.   A large number of transition maps for polymers have been presented by McCrum, Reed, and Williams[31] and by McCall.[32]   Although the data were obtained by mechanical, dielectric, and nuclear magnetic relaxation, the results group very well into discrete transition lines.   For

TABLE I.   Temperatures of Maximum Hypersonic Loss

Polymer	T_{max}(°C)	$\Delta\omega_l$ (GHz)	References
Poly(methyl acrylate)	170	4.3	23
Polyisobutylene	200	4.95	24
Polypropylene	180	4.11	24
Poly(dimethyl siloxane)	40	3.8	24
Poly(vinyl acetate)	200	4.5	24
Polystyrene	240	5.50	25
Polycarbonate	280	5.43	26
Poly(propylene glycol)	50	5.43	27–29
	100	4.40	27
Poly(ethylene oxide)	60	6.06	30

[24] Y. Y. Huang, E. A. Friedman, R. D. Andrews, and T. A. Hart, in "Light Scattering in Solids" (M. Balkanski, ed.), p. 488.   Flammarion, Paris, 1971.

[25] G. D. Patterson, J. Polym. Sci., Polym. Phys. Ed. 15, 579 (1977).

[26] G. D. Patterson, J. Macromol. Sci. Physics B 13, 647 (1977).

[27] G. D. Patterson, D. C. Douglass, and J. P. Latham, Macromolecules 11, 263 (1978).

[28] C. H. Wang and Y. Y. Huang, J. Chem. Phys. 64, 4847 (1976).

[29] S. M. Lindsay, A. J. Hartley, and I. W. Shepherd, Polymer 17, 501 (1976).

[30] G. D. Patterson and J. P. Latham, Macromolecules 10, 1414 (1977).

[31] N. G. McCrum, B. E. Read, and G. Williams, "Anelastic and Dielectric Effects in Polymeric Solids."   Wiley, New York, 1967.

[32] D. W. McCall, Natl. Bur. Stand. (U.S.) Spec. Publ. 301, 475–537 (1969).

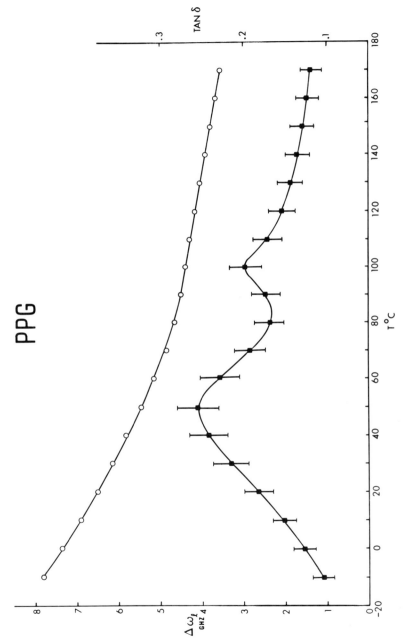

Fig. 9. Brillouin frequency $\Delta\omega_l$ (○) and loss tangent (●) vs. temperature for poly(d,l-propylene glycol) (PPG).

linear amorphous polymers, there are usually two such lines. One line corresponds to the primary glass–rubber relaxation and can be described by a relation[33]

$$f = f_0 \exp[-E/R(T - T_0)]. \qquad (3.3.42)$$

The other line represents the secondary main chain glass–rubber relaxation. These data follow a simple Arrhenius behavior. A transition map for polyisobutylene is shown in Fig. 10. The hypersonic result falls on an extrapolation of the secondary relaxation line above the region where the primary and secondary transitions become unresolvable.

At high temperatures, the primary and secondary transitions merge and it is usually not possible to resolve them at hypersonic frequencies. However, as shown in Fig. 9, two loss maxima are observed for PPG. A transition map for PPG is shown in Fig. 11. The lower-temperature loss falls on an extrapolation of the secondary relaxation line. The higher temperature maximum agrees very well with the high-frequency dielectric

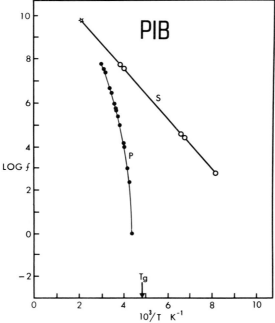

FIG. 10. Transition map for polyisobutylene. The Brillouin result is indicated by the open star at the highest frequency.

[33] M. L. Williams, R. F. Landel, and J. D. Ferry, *J. Amer. Chem. Soc.* **77**, 3701 (1955).

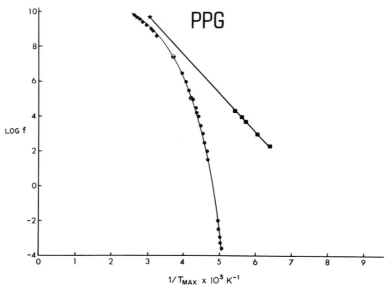

FIG. 11. Transition map for poly(*d,l*-propylene glycol). The Brillouin results are indicated by stars.

relaxation study of Yano *et al.*,[34] which falls on the primary relaxation line. Thus Brillouin scattering is an important complementary method for studying relaxation at high frequencies.

The hypersonic viscosities have been shown to depend on many variables for chain-molecule fluids. In order to characterize the intrinsic properties of a polymer fluid it is desirable to use a high enough molecular weight so that overall molecular reorientation is not contributing significantly to the hypersonic loss. The effect of chain length on the temperature of maximum loss has been studied by Wang and Huang[28] in PPG and by Patterson[25] in polystyrene. A plot of $10^3/T_{max}$ vs. $10^6/M^2$, where M is the molecular weight, is shown for polystyrene in Fig. 12. In this case the value of T_{max} is near its asymptotic value for a degree of polymerization near 20. This means that moderate length oligomers can be used to characterize the limiting hypersonic behavior of polymers. This fact has been used to advantage in the case of polyethylene oxide[30] where the melting point was depressed below the temperature of maximum hypersonic loss by using a 1000 molecular weight oligomer. Similar attempts in the case of polyethylene were unsuccessful.

[34] S. Yano, R. R. Rahalkar, S. P. Hunter, C. H. Wang, and R. H. Boyd, *J. Polym. Sci., Polym. Phys. Ed.* **14**, 1877 (1976).

The depolarized (I_{HV}) Rayleigh–Brillouin spectrum in the viscoelastic regime is given by Eq. (3.3.30). If the average shear relaxation time τ_S is less than approximately 3×10^{-11} seconds, then the transverse phonon velocity observed at 90° scattering angle for a typical polymer fluid is imaginary and no Brillouin peaks will be observed. The spectrum will consist of a narrow central peak due to overall molecular reorientation and a broader central peak due to overdamped shear fluctuations. The width of the transverse Brillouin peaks will remain greater than the splitting until the relaxation time is greater than approximately 4×10^{-11} seconds. The linewidth should fall to 100 MHz when $\tau \approx 8 \times 10^{-10}$ seconds. It is worth noting that the transverse linewidth is not a function of q.

As noted above, the hypersonic viscosity in the viscoelastic regime is dominated by the fluid structural relaxation. There will be contributions of both an intermolecular and intramolecular character. It is also apparent that there must be at least two structural relaxation times corresponding to the primary and secondary glass–rubber relaxations. In order for the fluid to adopt a new equilibrium volume and a state of zero shear, the chain segments must change their positions and free volume must be transported. Free volume transport is a cooperative process as is overall shear relaxation. The rates of the cooperative processes will be limited by the kinetics of the fundamental local structural rearrangements. The rate of the fundamental step will depend on the shape and local flexibility of the chains and on the details of the liquid structure. Since these processes must take place essentially without changes of rotational isomeric state, only the distribution of rotational isomers and the flexibility within each rotational isomeric state are important for the hypersonic

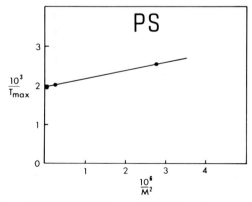

FIG. 12. Plot of $10^3/T_{max}$ vs. $10^6/M^2$ for polystyrene.[25]

relaxation in the viscoelastic regime.   Only by studying the high-frequency relaxation where the nature of the dynamical process can more easily be identified will the true nature of the glass–rubber relaxation be elucidated.   Brillouin scattering is a vital tool in this area of research.

3.3.4.3. The Glass Transition. When the structural relaxation time $\tau \gg 10^{-10}$ seconds, the phonon velocities reach their limiting values $V_{l\infty}$ and $V_{t\infty}$.   These quantities depend only on the moduli K_{∞} and G_{∞} and the density.   The temperature dependence of $V_{l\infty}$ has been analyzed by Brody et al.[35] in terms of the Gruneisen parameter γ_l:

$$- \frac{1}{V_{l\infty}} \left(\frac{\partial V_{l\infty}}{\partial T} \right)_P = \alpha \gamma_l . \tag{3.3.43}$$

It is observed that the Brillouin splitting above the glass transition is a linear function of the density, $V_{l\infty} = a\rho + b$.

As long as the fluid remains in equilibrium the observed Brillouin splitting should change smoothly with temperature because the density is changing smoothly.   Measurements[36] of $\Delta\omega_l$ for poly(methyl methacrylate) (PMMA) and for two polystyrenes are plotted vs. temperature near their glass transitions in Fig. 13.   The samples were allowed to reach

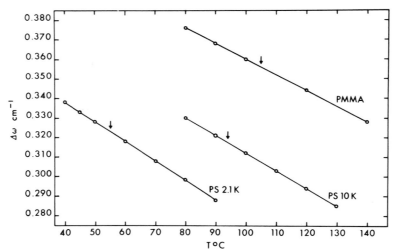

FIG. 13. Brillouin splittings $\Delta\omega_l$ vs. temperature near the nominal glass transition of PMMA, 10,000 molecular weight polystyrene and 2100 molecular weight polystyrene.   The arrows indicate the value of T_g determined with a differential scanning calorimeter at 10°C/min.[36]

[35] E. M. Brody, C. J. Lubell, and C. L. Beatty, J. Polym. Sci., Polym. Phys. Ed. **13**, 295 (1975).

[36] G. D. Patterson, J. Polym. Sci., Polym. Lett. Ed. **13**, 415 (1975).

equilibrium at each point and thus there is no change in slope at the nominal glass transition temperature T_g determined by differential scanning calorimetry at 10°C/minute.

The apparent Landau–Placzek ratio R'_{LP} should also reach a limiting value defined by Eq. (3.3.27) with $V_l = V_{l\infty}$. The value of R'_{LP} should continue to increase smoothly with decreasing temperature because $V_{l\infty}$ is rising more rapidly than V_{l0}.

The Brillouin splittings $\Delta\omega$ and the apparent Landau–Placzek ratio R'_{LP} are given accurately in the region above the glass transition by the linear fluctuation theory outlined in Section 3.3.2. However, the observed linewidths[23,37] near T_g are much larger than those predicted by Eq. (3.3.26). The longitudinal Brillouin linewidths for PMMA are plotted versus temperature near Tg in Fig. 14.

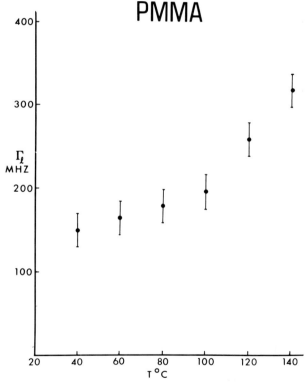

FIG. 14. Longitudinal Brillouin linewidth Γ_l vs. temperature for PMMA near T_g.

[37] D. A. Jackson, H. T. A. Pentecost, and J. G. Powles, *Mol. Phys.* **23**, 425 (1972).

When the relaxation time exceeds the time scale over which measurements are carried out, the fluid may not reach equilibrium when it is cooled. This leads to the formation of a glass with a density lower than that for an equilibrium fluid at the same temperature and pressure. As a result the observed Brillouin splitting will be lower and a plot of $\Delta\omega$ vs. temperature will change slope in this region. The density in the glassy state will again change smoothly with temperature and hence the Brillouin splittings will follow. The longitudinal and transverse Brillouin splittings[38] for an amorphous film of poly(ethylene terephthalate) (PET) are plotted vs. temperature in Fig. 15. There is a change in slope at a temperature of 70°C, which corresponds to the glass transition.

The behavior of the apparent Landau–Placzek ratio R'_{LP} in the glass transition region has been the subject of considerable controversy.[39] Below T_g, the true value of R'_{LP} is accurately given by Eq. (3.3.27) with $V_l = V_{l\infty}$ and V_{l0} taken at T_g. This is illustrated by Friedman et al.[40] for

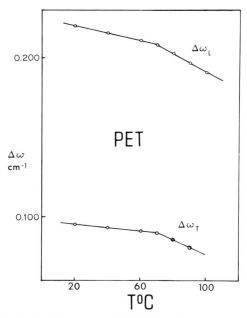

FIG. 15. Longitudinal and transverse Brillouin splittings vs. temperature for an amorphous film of poly(ethylene terephthalate) (PET). The glass transition is observed at 70°C.[38]

[38] G. D. Patterson, J. Polym. Sci., Polym. Phys. Ed. 14, 1909 (1976).
[39] See G. D. Patterson and J. P. Latham, Macromol. Rev. to be published.
[40] E. A. Friedman, A. J. Ritger, and R. D. Andrews, J. Appl. Phys. 40, 4243 (1969).

PMMA and Coakley et al.[41] for polystyrene. Because $V_{l\infty}$ continues to rise in the glassy state, R'_{LP} also rises and there may be only a slight change in the slope of a plot of R'_{LP} vs. temperature at T_g.

Experimentally it is very difficult to study R'_{LP} in polymers near the glass transition. In order to eliminate particulate impurities the sample must be filtered. Filtration of high molecular weight bulk polymers is not feasible for most samples. The approach used in Friedman et al.[40] and Coakley et al.[41] was to polymerize filtered monomer. The careful treatment of the final polymer samples gave satisfactory results. They were annealed well above T_g until the fluid reached a true equilibrium state and all residual strains caused by chain entanglements during polymerization were relieved.

The presence of finite strain in the samples can lead to a large enhancement of the central peak intensity. Low molecular weight polymer samples can be filtered, but strain induced by the cuvette as the polymer is cooled below T_g usually invalidates the measurements of R'_{LP}. Measurements of the apparent Landau–Placzek ratio near T_g should be viewed with considerable caution.

In the glassy state well below T_g, Brillouin scattering should measure the same mechanical moduli as would be obtained by a quasi-static technique. If the transverse Brillouin peaks are visible, then Poisson's ratio can be obtained according to

$$\sigma = \frac{\frac{1}{2}(\Delta\omega_l/\Delta\omega_t)^2 - 1}{(\Delta\omega_l/\Delta\omega_t)^2 - 1}. \qquad (3.3.44)$$

The results obtained by Brillouin scattering for bisphenol-A polycarbonate[42] (0.38) and PET[38] (0.38) are in good agreement with the directly measured values of σ.

The inhomogeneous broadening of the Brillouin lines can be used to study polymer–polymer compatibility. PMMA and poly(vinylidene fluoride) (PVF_2) have been shown[43] to be compatible above the melting point of PVF_2 (170°C). Clear glassy films can be prepared by rapid quenching from the melt. The Brillouin spectra of these films display sharp peaks, which confirms that they are a homogeneous amorphous glass.

[41] R. W. Coakley, R. S. Mitchell, J. R. Stevens, and J. L. Hunt, J. Appl. Phys. 47, 4271 (1976).

[42] G. D. Patterson, J. Polym. Sci., Polym. Phys. Ed. 14, 741 (1976).

[43] G. D. Patterson, T. Nishi, and T. T. Wang, Macromolecules 9, 603 (1976).

Measurements of the Rayleigh–Brillouin spectrum of bulk amorphous polymers can be used to determine a large number of thermodynamic and kinetic quantities. Even morphological studies can be carried out. The future use of Rayleigh–Brillouin scattering as a tool in polymer science is very promising.

3.4. Inelastic Neutron Scattering Spectroscopy†

By C. V. Berney and Sidney Yip

3.4.1. Introduction

Neutron scattering is an experimental tool in both diffraction,[1] where it is complementary to X-ray scattering, and molecular spectroscopy,[2-5] where it is complementary to infrared absorption and Raman scattering. In comparing neutron scattering with these techniques, important differences should be noted along with the similarities. The essential properties that make neutron scattering a unique probe are its sensitivity to hydrogen, its ability to measure structure and dynamics on the atomic scale of wavelengths and frequencies, and its ability to probe magnetic properties in addition to nuclear positions and motions. Because of these advantages, neutron scattering can provide information that is not available by other means. But at the same time, to exploit these advantages effectively requires sophisticated experimental facilities and extensive computational efforts.

It is the purpose of this chapter to describe those aspects of the basic principles and experimental techniques of inelastic neutron scattering that are useful for the investigation of polymers. While our discussions are also intended to serve as a brief introduction to inelastic neutron scattering, a number of important topics in the field could not be included.

[1] G. E. Bacon, "Neutron Diffraction," 3rd ed. Oxford Univ. Press, London and New York, 1975.

[2] H. Boutin and S. Yip, "Molecular Spectroscopy with Neutrons." MIT Press, Cambridge, Massachusetts, 1968.

[3] G. E. Bacon, "Neutron Scattering in Chemistry." Butterworth, London, 1977.

[4] C. V. Berney, in "Spectroscopy in Biology and Chemistry" (S.-H. Chen and S. Yip, eds.), p. 297. Academic Press, New York, 1974.

[5] J. W. White, in "Chemical Applications of Thermal Neutron Scattering" (B. T. M. Willis, ed.), p. 49. Oxford Univ. Press, London and New York, 1973; in "Polymer Science" (A. D. Jenkins, ed.), Vol. 2, p. 1843. North-Holland Publ., Amsterdam, 1972. see also G. Allen and C. J. Wright, *Phys. Chem., Ser. Two,* **8,** 233 (1975).

† See also Vol. 3B (Molecular Physics) in this series, in particular Chapter 7.3.

METHODS OF EXPERIMENTAL PHYSICS, VOL. 16A

One of these is magnetic scattering,[6,7] a field in which the neutron is uniquely useful because of its spin of $\frac{1}{2}\hbar$. Although organic and biological polymers are nonmagnetic, polymeric metal phosphinates have recently been prepared[8] and their magnetic properties measured.[9] Preliminary theoretical analysis of the magnetic scattering from such substances has shown that scattering properties dependent on the momentum transfer can yield useful information about the polymer conformation statistics.[10]

A general comparison of neutron scattering with other related techniques can be made in terms of the frequency and wavevector regions where each technique can operate. Such a diagram[11] is shown in Fig. 1, where frequency is given in energy units of reciprocal centimeters (cm^{-1}) and wavevector is given in units of reciprocal angstroms ($Å^{-1}$). The diagram also represents energy and momentum transfers between the probe and the sample system. In terms of neutron energy E and wavevector $\mathbf{k}$, the energy transfer $\hbar\omega$ and momentum transfer $\hbar Q$ are

$$\hbar\omega = E_i - E_f, \tag{3.4.1}$$

$$\mathbf{Q} = \mathbf{k}_i - \mathbf{k}_f, \tag{3.4.2}$$

where subscripts i and f indicate pre- and postcollision variables. One sees that neutron scattering can probe the frequency region important in molecular spectroscopy and it can also probe the wavevector region important in crystallography and molecular structure studies. In Fig. 1 no distinction is made between the underlying interactions in neutron and photon measurements. Although one can list eight different types of neutron interactions with atoms,[12] the process relevant to the present discussion is nuclear scattering, the direct interaction between the neutron and the nucleus. Thus in neutron-scattering measurements the probe is coupled directly to the nucleus, whereas in optical measurements one observes the nuclear motions through the coupling of photons with the elec-

[6] Yu. A. Izyumov and R. P. Ozerov, "Magnetic Neutron Diffraction." Plenum, New York, 1970.

[7] W. Marshall and S. W. Lovesey, "Theory of Thermal Neutron Scattering." Oxford Univ. Press, London and New York, 1971.

[8] P. Nannelli, B. P. Block, J. P. King, A. J. Saraceno, O. S. Sprout, N. D. Peschko, and G. H. Dahl, *J. Polym. Sci., Polym. Chem. Ed.* **11**, 2691 (1973).

[9] J. C. Scott, A. F. Garito, A. J. Heeger, P. Nannelli, and H. D. Gillman, *Phys. Rev. B* **12**, 356 (1975).

[10] M. F. Thorpe, *Phys. Rev. B* **13**, 2186 (1976).

[11] G. Dolling, *in* "Dynamical Properties of Solids" (G. K. Horton and A. A. Maradudin, eds.), Vol. 1, p. 541. North-Holland Publ., Amsterdam, 1974.

[12] C. G. Shull, *Trans. Am. Crystallogr. Assoc.* **3**, 1 (1967).

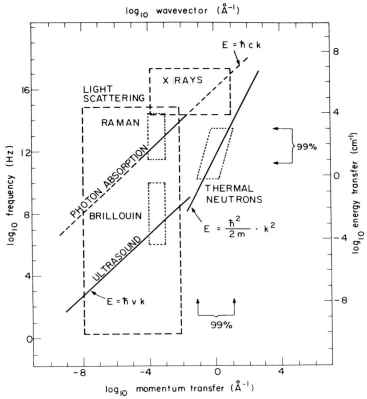

FIG. 1. Energy- and momentum-transfer diagram with regions accessible to various experimental techniques (legends at the top and left refer to phonon properties; those at the bottom and right refer to the variables in a scattering experiment). In this volume, Raman and infrared spectroscopy are discussed in Chapter 3.1, Rayleigh–Brillouin spectroscopy in Chapter 3.3, X-ray diffraction in Chapter 6.1 (this series, Part B), and ultrasound techniques in Chapter 12.1 (this series, Part C). The solid segment of the photon absorption line denotes the infrared region. The brackets labeled 99% indicate the frequency and wavevector ranges within which most phonons in typical crystals are found.[11]

trons. This direct coupling greatly simplifies the interpretation of intensities in neutron experiments.

The intensity of inelastic neutron scattering that one observes in an actual experiment is determined by many factors, some involving only the properties of the neutron and others involving only the properties of the sample system.[7] An elementary factor, but one that is of central importance in determining what kind of information may be extracted from the experiment, is the total scattering cross section of each type of nucleus present in the sample. This cross section is defined as $\sigma = 4\pi b^2$, where the scattering length b is a purely nuclear property of the constituent atoms of the sample and is independent of its physical state. The cross section is the sum of two contributions, coherent and incoherent, which correspond to scattering processes in which the scattering lengths of the nuclei in the sample either remain constant or change in value from one nucleus to another. The scattering length can be viewed as a fluctuating property when the nucleus has a nonzero spin. This is because the neutron–nucleus interaction is spin dependent, and therefore the interaction will change depending on whether the relative orientation of the neutron and nuclear spins is parallel or antiparallel. Fluctuations also occur when the sample contains two or more isotopes since different isotopes can have very different scattering lengths. Incoherent scattering, which arises from the fluctuations in the scattering length, may be called single-particle scattering in the sense that interference effects, the essence of diffraction phenomena, are absent. By contrast, coherent scattering occurs when scattering lengths do not fluctuate. Briefly stated, coherent scattering occurs when the nature of the sample is such that the wavelike properties of the neutron are predominant, while in incoherent scattering, the neutron acts as a particle.

The coherent and incoherent scattering cross sections of several nuclei of interest in polymer studies are listed in Table I.[1] Hydrogen is seen to have a large incoherent cross section, while carbon and oxygen are coherent scatterers with cross sections typical of many elements. It follows that incoherent neutron scattering is particularly sensitive to hydrogen motions. The large difference between the hydrogen and deuterium cross sections means that vibrations involving significant hydrogen motion can be readily identified by repeating the measurements using a selectively deuterated sample.[13,14] Since deuterium has the larger coherent cross section, hydrogen positions can be more easily determined in diffraction measurements if a deuterated sample is used.

In addition to being coherent or incoherent, neutron scattering can be described as elastic ($\omega = 0$) or inelastic. These two modes of classification can be used to divide the field of neutron scattering into four areas as indicated in Table II. In general, when a beam of neutrons is directed onto a sample, all four types of scattering take place. Elastic events are much more frequent than inelastic, so that in diffraction studies, total scattering may be measured and treated as if it were entirely elastic. Quasi-elastic scattering is distinguished from true elastic scattering in that small energy transfers (brought about by Doppler shifts in the scattering

[13] J. S. Higgins, G. Allen, and P. N. Brier, *Polymer* **13**, 157 (1972).
[14] C. V. Berney and J. W. White, *J. Am. Chem. Soc.* **99**, 6878 (1977).

TABLE I. Bound-Atom Thermal Neutron Scattering Cross Sections
of Selected Nuclei

Nucleus	σ_{coh}	σ_{inc}	σ_{tot}
H	1.8	79.7	81.5
D	5.8	2.0	7.8
C	5.5	0	5.5
N	10.5	0.9	11.4
O	4.2	0	4.2
F	4.0	0	4.0
Cl	11.5	3.5	15.0

event) do in fact take place. We shall briefly discuss each of the four divisions indicated in Table II. Our main concern in this chapter, however, will be incoherent inelastic neutron scattering (INS).

3.4.1.1. Bragg Scattering. The classic experimental application of Bragg scattering is the determination of atomic positions in crystalline materials. In this context, neutron diffraction is exactly analogous to x-ray diffraction (see Chapter 6.1, in this volume, Part B). The main advantage of using neutrons is that the scattering lengths for hydrogen and deuterium are comparable with those of the heavier elements, so that their positions can be accurately determined. For this reason, neutron diffraction results are of great interest in the study of hydrogen-bonded systems.[15,16] This technique has not been widely applied in the study of

TABLE II. Principal Areas of Experimental Neutron Scattering and Types
of Information Obtained

	Coherent	Incoherent
Elastic	*Bragg scattering* structural information (single-crystal diffraction studies, powder diffraction studies, and small-angle neutron scattering)	*Quasi-elastic scattering* diffusion constants mean thermal displacements residence times relaxation times
Inelastic	*Phonon scattering* lattice modes phonon dispersion curves dynamics of phase transformations	*Incoherent inelastic neutron scattering (INS)* amplitude-weighted vibrational density of states $G(\omega)$

[15] J. C. Speakman, in "Chemical Applications of Thermal Neutron Scattering" (B. T. M. Willis, ed.), p. 201. Oxford Univ. Press, London and New York, 1973.
[16] T. F. Koetzle, in "Spectroscopy in Biology and Chemistry" (S.-H. Chen and S. Yip, eds.), p. 177. Academic Press, New York, 1974.

polymer structures, probably because larger single crystals are required than for a comparable X-ray study, and large, perfect single crystals of most polymeric materials are rare. On the other hand, the technique of small-angle neutron scattering, which explores structural domains of 30–3000 Å, is uniquely suited to the study of conformational and disorder effects in polymers, and is consequently enjoying a period of explosive growth. This area is reviewed by King in Chapter 5.4 in this volume.

3.4.1.2. Quasi-Elastic Scattering. If a beam of monoenergetic neutrons is directed onto a sample and the scattered neutrons are energy-analyzed, the resulting plot will have a strong central peak, analogous to the Rayleigh peak in Raman spectroscopy. The neutrons contributing to this peak scatter without interacting with any phonons in the sample. They may, however, undergo small energy shifts due to diffusive or reorientational motions in the sample; for this reason the scattering is described as *quasi-elastic*. Information about mean amplitudes of motion, diffusion constants, and residence times may be obtained by determining how the intensity and width of the quasi-elastic peak changes with Q. This area has recently been reviewed by Springer,[17,18] and specific applications to polymers have been discussed by Allen and Higgins.[19] We discuss a particular example (segmental diffusion in rubbers) in Section 3.4.4.3.

3.4.1.3. Phonon Scattering. Phonons are vibrational excitations in condensed phases, characterized by a wavevector $\mathbf{q}$ and by a frequency $\omega(\mathbf{q})$, which in general is a function of the wavevector. They may also be described in terms of the vibrational displacements of the atoms in the unit cell (the eigenvectors of the secular equation) and the phase difference δ between adjacent unit cells (δ, however, is not an independent parameter, but is proportional to $|\mathbf{q}|$).

If a cold neutron impinges on a solid, a quantum of energy equal to $\hbar\omega$ may be transferred, resulting in annihilation of the phonon. If the neutron comes in with kinetic energy $\geq \hbar\omega$, it may either create or annihilate a phonon of frequency ω. Since phonon creation reduces the energy of the neutron, this process is referred to as *downscattering;* conversely, phonon annihilation results in *upscattering* (both processes, of course, are inelastic). A spectrum obtained in the downscattering mode is analogous to the Stokes branch of a Raman spectrum (see Section 3.1.2) and an upscattering spectrum corresponds to the anti-Stokes branch.

[17] T. Springer, "Quasielastic Neutron Scattering for the Investigation of Diffusive Motions in Solids and Liquids." Springer-Verlag, Berlin, and New York, 1972.

[18] T. Springer, *in* "Dynamics of Solids and Liquids by Neutron Scattering" (S. W. Lovesey and T. Springer, eds.), p. 255. Springer-Verlag, Berlin, and New York, 1977.

[19] G. Allen and J. S. Higgins, *Rep. Prog. Phys.* **36**, 1073 (1973).

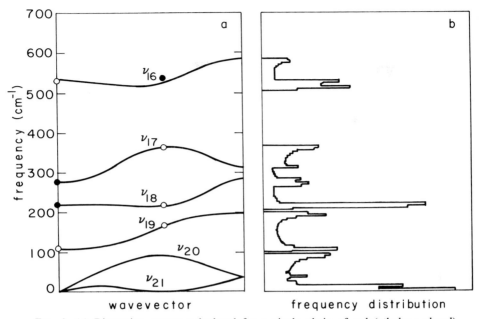

FIG. 2. (a) Dispersion curves calculated for a single chain of poly(ethylene glycol) $(-CH_2CH_2O-)_n$. Observed Raman frequencies are indicated by filled circles, infrared frequencies by open circles. (b) Frequency distribution obtained from the dispersion curves.[21,22]

If the scattering is coherent, momentum conservation conditions lead to the restriction

$$Q = 2\pi\tau \pm q,$$

where τ is a reciprocal lattice vector[20] (see Section 3.4.2). This condition, together with the energy conservation condition, makes coherent inelastic scattering a very restrictive process. For a given scattering angle, only phonons of a particular q and $\omega(q)$ can be observed. By incrementally changing spectrometer settings for k_i and k_f, Q (and thus q) may be varied and dispersion curves $\omega_j(q)$ (where j is the phonon branch index) may be mapped out. The procedure is described in more detail in Section 3.4.3.3. A set of calculated dispersion curves for poly(ethylene glycol)[21,22] is shown in Fig. 2a.

[20] W. Marshall and S. W. Lovesey, "Theory of Thermal Neutron Scattering," p. 83. Oxford Univ. Press, London and New York, 1971.
[21] H. Matsuura and T. Miyazawa, *Bull. Chem. Soc. Jpn.* **42**, 372 (1969).
[22] H. Matsuura and T. Miyazawa, *J. Chem. Phys.* **50**, 915 (1969).

3.4.1.4. Incoherent Inelastic Scattering. If the scattering is incoherent, the momentum conservation conditions no longer apply to the scattering as a whole, though of course momentum is conserved in each individual scattering event. The effect is that of relaxation of a selection rule, and contributions to the spectrum (scattered intensity vs. energy transfer) are made from each phonon branch throughout the Brillouin zone. The result (Fig. 2b) is a spectrum showing the vibrational density of states (the frequency distribution) weighted by the vibrational amplitude squared times the incoherent scattering cross section of each nucleus (see Section 3.4.2 for a more detailed description).

Note that the peaks in the frequency distribution (Fig. 2b) correspond to regions where the slope of a particular branch in the dispersion curve approaches zero. (These regions are sometimes termed "critical frequencies," "cutoff frequencies," or "van Hove singularities.") The same is in general true of the weighted frequency distribution observed by incoherent inelastic neutron scattering. Thus peak positions in neutron-scattering spectra are in principle different from those observed for the same vibrations in infrared and Raman spectroscopy, although for flat branches the frequencies may be nearly the same.

3.4.2. Basic Principles

The scattering of thermal neutrons by an isolated nucleus is an example of the well-known problem of potential scattering involving only s-waves. In order to describe neutron scattering in a macroscopic sample, it is necessary to treat the atoms as spatially and dynamically correlated.[7] The analysis is now more involved, but it is possible to formulate a general theory by assuming that the neutron–sample interaction can be expressed as a sum of neutron–nucleus interactions, each characterized by the appropriate scattering length and localized at the position of the nucleus. This assumption, the so-called Fermi pseudopotential approximation, seems quite reasonable on physical grounds since nuclear forces are very short range (of order 10^{-12} cm) on the scale of interatomic distances and thermal neutron wavelengths. However, it has not been examined in detail except in a few special cases.[23,24] For example, it has been shown that in elastic and inelastic scattering by a harmonically bound proton, the Fermi approximation will introduce an error of about 0.5% or less. The assumption of the Fermi pseudopotential is generally not considered to pose a limitation in a practical sense, since less accurate approximations concerning the dynamics of the sample system generally have to be introduced.

[23] P. Plummer and G. C. Summerfield, *Phys. Rev.* **131**, 1153 (1963).
[24] G. C. Summerfield, *Ann. Phys.* (*N.Y.*) **26**, 72 (1964).

The most useful aspect of the theory of neutron scattering in solids is that which relates the observed scattering intensity to the normal modes of vibration of the solid. Assuming the solid is a harmonic crystal, one can describe the normal modes in terms of phonon frequencies $\omega_j(\mathbf{q})$, where $\mathbf{q}$ is the phonon wavevector and j the branch index, and associated polarization vectors $\mathbf{e}(\mu|\mathbf{q}j)$, where μ is an atom label.[25] Although it is beyond the scope of the present discussion to go into the theory of lattice dynamics, it is pertinent to recall here the connection between the phonon properties $\omega_j(\mathbf{q})$ and $\mathbf{e}(\mu|\mathbf{q}j)$ and the interatomic forces. The connection lies in the normal mode problem[25]

$$m_\mu \omega_j(\mathbf{q}) e_\alpha(\mu|\mathbf{q}j) = \sum_{\mu'\beta} D_{\alpha\beta}^{\mu\mu'}(\mathbf{q}) e_\beta(\mu'|\mathbf{q}j) \qquad (3.4.3)$$

where the dynamical matrix is defined as

$$D_{\alpha\beta}^{\mu\mu'}(\mathbf{q}) = \sum_l \left[\frac{\partial^2 U}{\partial u_\alpha(l\mu)\, \partial u_\beta(l'\mu')} \right]_0 \exp[i\mathbf{q} \cdot (\mathbf{R}_{l'} - \mathbf{R}_l)], \qquad (3.4.4)$$

and the polarization vectors have the properties

$$\sum_\mu \mathbf{e}^*(\mu|\mathbf{q}j) \cdot \mathbf{e}(\mu|\mathbf{q}j') = \delta_{jj'} \qquad (3.4.5)$$

$$\sum_j e_\alpha^*(\mu|\mathbf{q}j) e_\beta(\mu'|\mathbf{q}j) = \delta_{\alpha\beta}\, \delta_{\mu\mu'}. \qquad (3.4.6)$$

In Eq. (3.4.4), U is the potential energy of the crystal, $u_\alpha(l\mu)$ the αth component of displacement of atom μ in the lth unit cell, $\mathbf{R}$ the unit cell position, and subscript zero indicates that the enclosed quantity is to be evaluated at the equilibrium configuration of the crystal.

Given the normal modes, one can then calculate the double-differential scattering cross section $d^2\sigma/d\Omega\, dE_f$, the quantity that governs the intensity of scattering into an element of solid angle $d\Omega$ of neutrons with incident wavevector $\mathbf{k}_i$ and outgoing wavevector $\mathbf{k}_f$. In particular one is interested in that part of the cross section that corresponds to the excitation and deexcitation of a normal mode of vibration, the so-called one-phonon scattering process. The one-phonon incoherent and coherent scattering cross sections are of the form[7,25]

$$\left[\frac{d^2\sigma}{d\Omega\, dE_f}\right]_{\text{inc}}^{(1)} = \frac{k_f}{k_i} \sum_{\mathbf{q}j} \sum_\mu |f_\mu(\mathbf{Q}\mathbf{q}j)|^2 \{ n_j \delta[\omega + \omega_j(\mathbf{q})]$$

$$+ (n_j + 1)\, \delta[\omega - \omega_j(\mathbf{q})]\}, \qquad (3.4.7)$$

[25] G. Venkataraman, L. A. Feldkamp, and V. C. Sahni, "Dynamics of Perfect Crystals." MIT Press, Cambridge, Massachusetts, 1975.

$$\left[\frac{d^2\sigma}{d\Omega\, dE_f}\right]_{\text{coh}}^{(1)} = \frac{k_f}{k_i}\frac{(2\pi)^3}{v_a}\sum_{qj}\sum_{\tau}|F(\mathbf{Q}qj)|^2$$

$$\{n_j\delta[\omega + \omega_j(\mathbf{q})]\delta(\mathbf{Q} + \mathbf{q} - 2\pi\boldsymbol{\tau})$$
$$+ (n_j + 1)\delta[\omega - \omega_j(\mathbf{q})]\,\delta(\mathbf{Q} - \mathbf{q} - 2\pi\boldsymbol{\tau})\}. \quad (3.4.8)$$

Here v_a is the unit cell volume and the summation over $\boldsymbol{\tau}$ extends to all vectors in the reciprocal lattice. The two terms in each cross section correspond to neutron energy gain, or upscattering, and energy loss, or downscattering, respectively. The intensity ratio of these two processes is $n_j/(n_j + 1)$, where $n_j = \{\exp[\hbar\omega_j(\mathbf{q})/2k_BT] - 1\}^{-1}$ is the average population of phonon $(\mathbf{q}j)$ at temperature T.

Aside from the kinematic factor k_f/k_i, each cross section is proportional to the modulus squared of an intensity factor. For incoherent scattering the factor is

$$f_\mu(\mathbf{Q}qj) = \left[\frac{\hbar}{2m_\mu\omega_j(\mathbf{q})}\right]^{1/2} b_\mu^{\text{inc}}\,[\mathbf{Q}\cdot\mathbf{e}(\mu|\mathbf{q}j)]\exp[- W_\mu(Q)], \quad (3.4.9)$$

where m_μ is the mass of atom μ, b_μ^{inc} the incoherent scattering length of the nucleus, and $\exp(-2W\mu)$ is the Debye–Waller factor,

$$2W_\mu(Q) = \sum_{qj}\frac{\hbar}{2m_\mu N\omega_j(\mathbf{q})}|\mathbf{Q}\cdot\mathbf{e}(\mu|\mathbf{q}j)|^2\coth\left[\frac{\hbar\omega_j(\mathbf{q})}{2k_BT}\right]. \quad (3.4.10)$$

For coherent scattering, the corresponding factor is

$$F(\mathbf{Q}qj) = \sum_\mu f_\mu(\mathbf{Q}qj)\left(\frac{b_\mu^{\text{coh}}}{b_\mu^{\text{inc}}}\right)\exp(i\mathbf{Q}\cdot\mathbf{x}_\mu), \quad (3.4.11)$$

where b_μ^{coh} is the coherent scattering length and $\mathbf{x}_\mu$ the equilibrium position of atom μ measured within the unit cell. For a crystal consisting of N unit cells, each containing s atoms, there are N allowed values of $\mathbf{q}$ in the first Brillouin zone and there are $3s$ values of j.

The essential difference between incoherent and coherent scattering is clearly illustrated by Eqs. (3.4.7) and (3.4.8). Energy conservation conditions for up- and downscattering appear in both cross sections, but for coherent scattering an additional wavevector conservation condition must be satisfied as well. In order to observe one-phonon excitations in coherent scattering, the neutron frequency and wavevector transfers must match the frequency and wavevector of the phonon. Thus by determining separately $\omega_j(\mathbf{q})$ and $\mathbf{q}$, one obtains the phonon dispersion relation for branch j. From Eq. (3.4.7) one sees that the incoherent scattering intensity is given by the sum of $|f_\mu|^2$, the individual contribution from each atom. On the other hand, Eq. (3.4.8) shows that coherent scattering intensity depends on $|F|^2$, and so the contributions are first summed over atoms with appropriate phase factor $\exp(i\mathbf{Q}\cdot\mathbf{x}_\mu)$ and then squared. This distinction means that position or structure correlations can be directly measured by coherent scattering but not by incoherent scattering.

The most useful information that one can obtain from incoherent neutron-scattering measurements is contained in the one-phonon cross section, Eq. (3.4.7). The scattering intensity contribution from each nucleus is proportional to the cross section σ_μ^{inc} and inversely proportional to the mass and the vibrational frequency. Inelastic neutron scattering is therefore most sensitive to low-frequency vibrations that involve light atoms. In the case of hydrogenous samples, the observed scattering spectrum will invariably be dominated by hydrogen motions, and it is often a good approximation to analyze the data considering only the hydrogen contributions to Eq. (3.4.7).

The observed scattering spectrum of solid samples can generally be separated into an elastic-scattering portion and an inelastic region. The elastic-scattering part of the spectrum will have a line shape that is the convolution of the incident neutron spectrum and the spectrometer resolution function. No broadening due to the scattering will be observed unless there is significant amount of diffusive motion in the sample system.[17] Since the intensity of elastic scattering is modulated by the Debye–Waller factor, one can deduce the factor $\exp(-2W)$ in Eqs. (3.4.7) and Eq. (3.4.8) by measuring the variation of integrated intensity with wavevector transfer Q. The inelastic region is composed of one-phonon scattering as well as multiphonon contributions.[26] Multiphonon effects tend to give rise to a spectrum that is relatively smooth compared to one-phonon scattering; experimentally, they can be minimized by working at low temperatures and small Q values. However, as will be seen in Section 3.4.4, multiphonon contributions in typical experiments are not negligibly small, and they can account for some of the structure in the measured scattering spectrum.

The preceding discussion serves to emphasize that while incoherent inelastic scattering allows one to observe atomic vibrational frequencies, the measured spectrum is in general not the vibrational density of states one associates with a crystalline solid. In terms of the phonon frequencies, the latter can be defined as[25]

$$g(\omega) = \frac{1}{3sN} \sum_{\mathbf{q}j} \delta[\omega - \omega_j(\mathbf{q})]. \tag{3.4.12}$$

For the special case of the cubic Bravais lattice, one can use the symmetry property[27]

$$\sum S(\mathbf{q}j) \, |\mathbf{Q} \cdot \mathbf{e}(\mathbf{q}j) = \frac{Q^2}{3} \sum_{\mathbf{q}j} S(\mathbf{q}j), \tag{3.4.13}$$

[26] A. Sjölander, Ark. Fys. 14, 315 (1958).
[27] G. Placzek and L. Van Hove, Phys. Rev. 93, 1207 (1954).

for arbitrary $S(\mathbf{q}j)$ and $\mathbf{Q}$, to show that Eq. (3.4.7) is proportional to $g(\omega)$. Equation (3.4.13) does not hold for an arbitrary crystal, and so one cannot directly relate the one-phonon incoherent cross section to the density of states.

When special directions are present in the sample, one can exploit such properties by rewriting Eq. (3.4.7), which is valid for a single crystal. A case of particular interest is that of oriented chain polymers, in which different chains are effectively aligned along the chain-axis direction. By performing the scattering measurement with $\mathbf{Q}$ either parallel or perpendicular to the direction of orientation, one can observe longitudinal vibrations (where atomic displacements are along the chain axis), separately from the transverse vibrations.[28] It is then useful to identify in Eq. (3.4.7) the tensor

$$G_\mu^{\alpha\beta}(\omega) = \frac{1}{3sN} \sum_{\mathbf{q}j} e_\alpha^*(\mu|\mathbf{q}j)e_\beta(\mu|\mathbf{q}j)\delta[\omega - \omega_j(\mathbf{q})], \qquad (3.4.14)$$

where α and β denote Cartesian components. We thus define the directional longitudinal and transverse frequency spectra

$$G_\mu^{L}(\omega) = G_\mu^{\hat{c}\hat{c}}(\omega), \qquad (3.4.15)$$

$$G_\mu^{T}(\omega) = \tfrac{1}{2}[G_\mu^{\hat{a}\hat{a}}(\omega) + G_\mu^{\hat{b}\hat{b}}(\omega)], \qquad (3.4.16)$$

where $\hat{c}$ denotes the chain-axis direction and $\hat{a}$ and $\hat{b}$ the directions perpendicular to the chain.

The reduction of Eq. (3.4.7) for polycrystalline samples is also of considerable interest. The measured cross section in this case is an average of Eq. (3.4.7) over all crystal directions, or equivalently, the orientation averaging can be carried out as an average over all directions of $\mathbf{Q}$. If one assumes that the averaging operation can be approximated as

$$\langle |\mathbf{Q} \cdot \mathbf{e}|^2 e^{-2W} \rangle \cong \langle |\mathbf{Q} \cdot \mathbf{e}|^2 \rangle e^{-2\langle W \rangle} = \frac{Q^2}{3} |\mathbf{e}|^2 \, e^{-2\langle W \rangle}, \qquad (3.4.17)$$

then Eq. (3.4.7) becomes

$$\left[\frac{d^2\sigma}{d\Omega \, dE_f}\right]_{\text{inc}}^{(1)} = \frac{k_f}{k_i} \frac{sQ^2}{4\pi|\omega|} \sum_\mu \frac{\sigma_{\text{inc}}^\mu \exp(-2\langle W_\mu \rangle)}{m_\mu}$$

$$\cdot [\bar{n}G_\mu^{+}(\omega) + (\bar{n} + 1)\bar{G}_\mu(\omega)], \qquad (3.4.18)$$

where

$$G_\mu^{\pm}(\omega) = \frac{1}{3sN} \sum_{\mathbf{q}j} |e(\mu|\mathbf{q}j)|^2 \delta[\omega \pm \omega_j(\mathbf{q})], \qquad (3.4.19)$$

[28] G. C. Summerfield, J. Chem. Phys. 43, 1079 (1965).

$$2\langle W_\mu \rangle = \frac{\hbar Q^2}{2m_\mu} \int_0^\infty d\omega \, \frac{G_\mu(\omega)}{\omega} \coth\left(\frac{\hbar\omega}{2k_B T}\right), \tag{3.4.20}$$

and $\bar{n} = [\exp(\hbar|\omega|/2k_B T) - 1]^{-1}$. Equation (3.4.18) is a rather transparent result that shows that all the information about the dynamics of atom μ is contained in the amplitude-weighted frequency spectrum $G_\mu(\omega)$. Notice that while sharp features in $g(\omega)$ are expected to show up in $G_\mu(\omega)$ as well, the relative intensity of different vibrational peaks in the spectrum can be significantly altered by variations in $|e(\mu|\mathbf{q}j)|^2$.

In the analysis of incoherent inelastic-scattering experiments, Eq. (3.4.7) is probably the most useful result. One sees that any quantitative interpretation will require a knowledge of the phonon frequencies and the associated polarization vectors. This means that normal-mode calculations based on an appropriate force-field or potential-function model have to be carried out and the results used to evaluate the one-phonon cross section and the multiphonon contributions. If these results are unavailable, one can attempt to extract the amplitude-weighted frequency spectrum directly from the measured scattering intensity. Two problems are then encountered. Since the Debye–Waller factor also depends on $G(\omega)$, one would have to use an assumed $G(\omega)$ to evaluate $\exp(-2W)$ in order to deduce an approximate $G(\omega)$ from the data, and iterate this process until convergence is achieved. The second problem is the multiphonon contribution, which one may be forced to either ignore completely or estimate using an assumed $G(\omega)$. Because the n-phonon process is proportional to $(2W)^n$, one can expect multiphonon effects to be important when $2W$ is not small compared to unity.[26] For this reason, measurements should be made at low sample temperatures and small scattering angles.

We conclude this section with a comment on anharmonic effects.[29] In principle, the width of the one-phonon peaks observed in coherent scattering on single crystals provides a measure of anharmonic effects. One can extract from such widths the phonon lifetime, although in practice this is difficult because of the presence of multiphonon scattering.[30] In incoherent scattering, anharmonic effects are generally not considered explicitly. Since the force constants used to calculate $G(\omega)$ usually have been adjusted to give frequencies in agreement with optical data (or even the neutron data) at the sample temperature, one can say that whatever renormalization effects due to anharmonic interactions are present, they have been taken into account implicitly through the force constants.[31]

[29] See, for example, G. K. Horton and A. A. Maradudin, ed., "Dynamical Properties of Solids," Vol. 1. North-Holland Publ., Amsterdam, 1974.

[30] J. Skalyo, V. J. Minkiewicz, G. Shirane, and W. B. Daniels, *Phys. Rev. B* **6**, 4766 (1972).

[31] D. C. Wallace, "Thermodynamics of Crystals." Wiley, New York, 1972.

3.4.3. Experimental Techniques

3.4.3.1. Spectrometer Components. The principles governing the design of neutron spectrometers are basically the same as those of optical spectrometers (Section 3.1.2), with certain modifications due to the differing properties of the neutron and photon.

A large number of different types of neutron spectrometers have been developed over the last 20 years. Concerns common to all of them are a source of neutrons, monochromation (or energy identification), collimation, and detection. Space limitations in the present review preclude more than a cursory discussion of the most common types of spectrometers. More extensive material may be found elsewhere.[11,32-34]

3.4.3.1.1. SOURCE. The source of neutrons for most instruments is a fission reactor. Neutrons are released by the fission of ^{235}U nuclei at energies around 2 MeV, and are then slowed down by repeated collisions with the moderating material around the reactor core. Most reactors emit a beam of thermal neutrons whose fractional energy distribution is fairly well described by a Maxwellian equation

$$n(E)\ dE = \left[\frac{E}{\pi(k_B T)^3}\right]^{1/2}\ \exp(-E/k_B T)\ dE, \qquad (3.4.21)$$

with a characteristic temperature T in the neighborhood of 300 K. The source is thus comparable to the "globar" in a commercial infrared spectrometer, except that the effective temperature is lower (and, of course, the source is much larger and more expensive). The thermal neutron flux delivered to the spectrometers varies from $\sim 10^{13}$ neutrons $cm^{-2}sec^{-1}$ for the smaller research reactors to $\sim 10^{15}$ for the larger reactors. The practical limit for steady-state fission reactors seems to be $\sim 10^{16}$. Pulsed sources using spallation reactions in a heavy metal (uranium or tungsten) initiated by high-energy protons or electrons are currently in the construction or planning stage at several laboratories.[35] The pulsing reduces the heat-transfer problems that set the upper limit to fluxes at steady-state reactors. Time-of-flight (tof) spectrometers (Section 3.4.3.2) can make particularly efficient use of the radiation from pulsed sources.

3.4.3.1.2. MONOCHROMATION. Three strategies for preparing a beam of approximately monoenergetic neutrons are filtration, velocity selection, and diffraction. These techniques are discussed below. Energy

[32] H. Maier-Leibnitz, in "Inelastic Scattering of Neutrons in Solids and Liquids," p. 681. IAEA, Vienna, 1972.

[33] B. N. Brockhouse, in "Inelastic Scattering of Neutrons in Solids and Liquids," p. 113. IAEA, Vienna, 1961.

[34] "Instrumentation for Neutron Inelastic Scattering." IAEA, Vienna, 1972.

[35] J. M. Carpenter, Nucl. Instrum. & Methods 145, 91 (1977).

identification is an alternative technique used in tof spectrometers. Particularly when used with pulsed neutron sources, it can be more efficient than brute-force monochromation.

Polycrystalline beryllium is probably the most commonly used neutron filter, although quartz, MgO, BeO, and silicon have been used as well. All these materials have a characteristic maximum lattice spacing d_{max}. Neutrons of wavelength $\lambda \leqslant 2d_{max}$ are scattered, and over a sufficiently long pathlength, effectively eliminated from the beam. This process establishes a lower bound on the wavelength of neutrons transmitted and thus an upper bound on their energy, giving a bandpass of $0 - E_{cutoff}$, where $E_{cutoff} = (C/2d_{max})^2$ ($C = 25.6865$ for E in cm^{-1} and d_{max} in Å).

Velocity selection may be carried out with a series of phased choppers, arranged so that of the polychromatic neutrons passed by the first chopper, only those of a given velocity arrive in phase with the openings of subsequent choppers. A rotating tube with helical vanes uses the same principle to provide an essentially continuous beam of roughly monochromatic neutrons; such devices are frequently used with small-angle scattering spectrometers.

Diffraction (Bragg reflection) from a suitable single crystal is a very selective means of monochromating neutrons. The process is governed by the familiar equation

$$n\lambda = 2d \sin \theta_{Bragg}, \tag{3.4.22}$$

where λ is the wavelength of the reflected neutron, $n = 1, 2, 3, \ldots$ the order of the reflection, and d the spacing of the planes doing the scattering. The crystal functions exactly as the grating in an infrared or Raman spectrometer, and a spectrum can be scanned by changing θ_{Bragg}. As in the infrared spectrometer, order contamination can be a problem, and filters are frequently used to eliminate unwanted orders.

3.4.3.1.3. COLLIMATION. Since collimation is essentially the act of throwing away all neutrons outside a given angular divergence, it is always carried out at the expense of signal intensity. Thus the degree of collimation should always be kept as loose as is consistent with acceptable resolution. The collimator usually used in neutron spectrometers is the Soller slit, a set of vanes coated with a neutron-absorbing material (cadmium or gadolinium) fixed in a holder.

3.4.3.1.4. DETECTION. Neutrons are usually detected using tubes filled with gaseous mixtures including BF_3 or ^{3}He. The tubes are wired so that the charged particles produced by the neutron-capture reactions, ^{10}B(n, α)^{7}Li and ^{3}He(n, p)^{3}H, give rise to pulses, which are then amplified and counted. The tubes range from 1 to 10 cm in diameter and from 10 to 50 cm in length. The ^{3}He tubes can be made somewhat smaller, and so are desirable where spatial resolution is a significant factor. Detection efficiencies can be made as high as 100%. Detectors with low efficiencies (10^{-3}–10^{-4}) are used as monitors, to measure the neutron flux at a given point without substantially reducing it. The active element in these devices is usually a thin film of uranium, which provides charged particles by undergoing a fission reaction. Linear[36] and two-dimensional[37] position-sensitive detectors locate the neutron-absorption event along an axis or in a plane by comparing pulse rise time or resistivities. By gathering data simultaneously over a range of angles they can improve instrumental efficiency, though thus far they have had a greater impact on the design of small-angle scattering spectrometers than on INS instruments.

3.4.3.2. Time-of-Flight (tof) Spectrometers. In contrast to the case of the photon, the velocity of a neutron is not a universal constant, a fact ascribable to its nonzero rest mass, and one that is exploited in the construction of tof spectrometers. In the basic tof experiment, a beam of neutrons is periodically interrupted (chopped), usually by a rotating mechanical device (a chopper). As the neutron burst leaves the chopper, an electronic pulse is sent to a multichannel analyzer, which sequentially records in successive channels n the number of neutrons detected during the time interval between $(n - 1)\,\Delta t$ and $n\,\Delta t$, where Δt is the channel width. The average velocity of the neutrons tallied in channel n is thus $L/(n - 0.5)\,\Delta t$, where L is the length of the chopper-to-detector flight path, and the average final energy (in cm^{-1}) is

$$E_f(n) = 8065.5 \left(\frac{0.723\,L}{n\,\Delta t + d} \right)^2 . \tag{3.4.23}$$

Here the factor 0.723 incorporates a number of constants, including the mass of the neutron and conversion from ergs to election volts. The quantity d is -0.5 plus any electronic delay built into the counting circuit. In this expression, L is in centimeters and Δt and d are in microseconds.

Since the object of the inelastic scattering experiment is to determine the transfer of energy and momentum, the spectrometer must incorporate some monochromating device to supplement the tof measurement. One of the earliest spectrometers[38] used a refrigerated beryllium filter to provide approximate monochromation. Polycrystalline beryllium has a wavelength cutoff of 3.95 Å, giving an energy bandpass of 0–42 cm^{-1}. In the Brookhaven instrument,[38] the beryllium filter was placed just before the sample, so the beam striking the sample consisted of the low-energy

[36] C. J. Borkowski and M. K. Kopp, *Rev. Sci. Instrum.* **39**, 1515 (1968).

[37] R. W. Hendricks, *Trans. Am. Crystallogr. Assoc.* **12**, 103 (1976).

[38] C. M. Eisenhauer, I. Pelah, D. J. Hughes, and H. Palevksy, *Phys. Rev.* **109**, 1056 (1958).

tail of the Maxwellian [Eq. (3.4.21)]. The beam of neutrons scattered at 90° was chopped and then intercepted by a set of detectors 5 m away. This device was used in one of the earliest neutron-scattering investigations of polyethylene.[39] A similar spectrometer, constructed at the Army Materials and Mechanics Research Center, Watertown, Massachusetts, was used for a number of other INS investigations of polymers, including polyethylene, polyoxymethylene, polyacrylonitrile, and poly(ethylene glycol).[40]

Several characteristics of the experimental data obtained from a beryllium filter tof spectrometer are evident in Fig. 3. The upper curve (dashed) represents the INS spectrum of crystalline polyethylene at room temperature, while the lower curve (full line with experimental points) is the same sample at 100 K. Since the spectrometer irradiates the sample with cold neutrons, the only inelastic scattering process possible is upscattering, in which energy is transferred to the neutron by excitations in the solid. Transition probabilities are weighted by the detailed-balance factor, $\exp(\hbar\omega/2k_BT)$, and so the apparent intensity of peaks in an upscattering spectrum (for which $\omega < 0$) falls off rapidly as $|\omega|$ increases, the more so as the temperature is lowered. This makes it difficult to get adequate statistics on small or weakly scattering samples at low tempera-

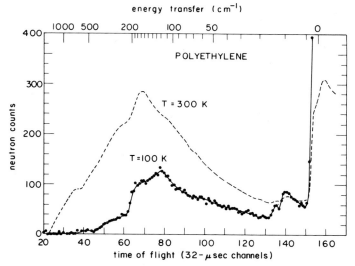

FIG. 3. INS spectra of polyethylene (Marlex 6050) taken on a beryllium-filter tof instrument (scattering angle $\phi = 90°$).[39]

[39] H. R. Danner, G. J. Safford, H. Boutin, and M. Berger, *J. Chem. Phys.* **40**, 1417 (1964).
[40] S. Trevino and H. Boutin, *J. Macromol. Sci., Chem.* **1**, 723 (1967).

tures. Another feature of interest in Fig. 3 is the appearance of the
quasi-elastic peak ($\omega \sim 0$) to the right of the beryllium cutoff around
channel 150. It increases in sharpness and relative intensity as the temperature is lowered. Note also that the energy scale (at the top of the figure) compresses toward the left, since the energy width of channel n
varies approximately as n^{-2} [Eq. (3.4.23)]. Resolution in tof spectrometers can be partially described as the energy width of the channel con-
volved with factors pertaining to the monochromating device. At high
energy transfer (low channel number) the former effect becomes dominant. Thus, tof spectrometers are intrinsically best adapted to looking at
low energy-transfer regions (though introduction of a delay in the
counting circuit gives some flexibility). For polymer studies, where
acoustic and interchain modes are important, this may be an advantage.

Much more sophisticated tof instruments have been built more recently. One of them, the TNTOF (thermal neutron time-of-flight) spectrometer at Argonne National Laboratory,[41] is shown in Fig. 4. Monochromation is achieved by Bragg reflection from a pair of matched
crystals inside the shielding. Fast-neutron and gamma-ray contamination
of the beam is reduced by the double reflection, and by the diversion of
the diffracted beam from the axis of the incident beam. Three pairs of
crystals are available for insertion into the momochromator: Cu(220),
giving a beam of neutrons with $\lambda = 1.5$ Å, $E_i = 293$ cm^{-1}; Cu(111), $\lambda =$
2.5 Å, $E_i = 106$ cm^{-1}; and pyrolytic graphite (002), $\lambda = 4.1$ Å, $E_i =$
39 cm^{-1}. The monochromated beam is chopped just before it strikes the
sample, so that neutrons scattered over a wide range of angles ($\sim 130°$)
may be counted by the bank of detectors. The detectors are ganged into
a number of groups so that data for up to 32 scattering angles may be acquired simultaneously. The instrument thus collects data very efficiently
and very rapidly, and a dedicated PDP-8 computer is used to assist in their
handling. The efficiency of the instrument can be further increased by
correlation chopping, a technique in which the "open time" of the
chopper is increased to $\sim 50\%$ by using a chopping disk with a pseudorandom array of open and closed elements.[42] Experimental spectra
must be decorrelated before they are interpretable, but this can easily be
done with the dedicated computer.

A number of other ingenious variations on the theme of tof spectrometry have been realized. In one,[43] Bragg reflection from a rotating
crystal simultaneously monochromates and chops the beam. In an-

[41] R. Kleb, G. E. Ostrowski, D. L. Price, and J. M. Rowe, *Nucl. Instrum. & Methods* **106,**
221 (1973).
[42] D. L. Price and K. Sköld, *Nucl. Instrum. & Methods* **82,** 208 (1970).
[43] O. K. Harling, *Rev. Sci. Instrum.* **37,** 687 (1966).

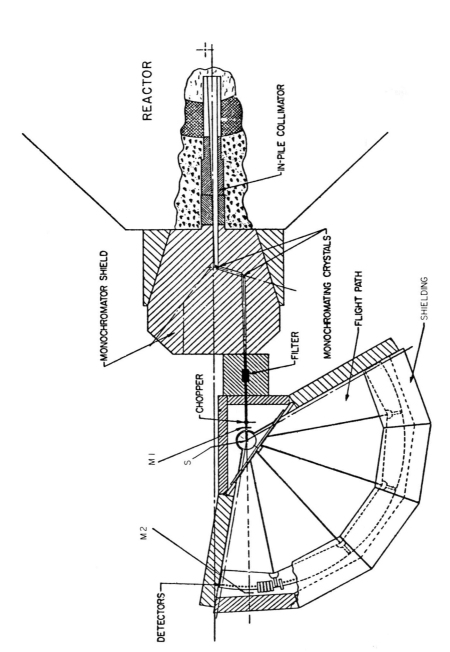

REACTOR

IN-PILE COLLIMATOR

MONOCHROMATOR SHIELD

MONOCHROMATING CRYSTALS

FILTER

FLIGHT PATH

CHOPPER

SHIELDING

M1

S

M2

DETECTORS

Fig. 4. Diagram of an advanced tof spectrometer (TNTOFS, Argonne National Laboratory). M1 and M2 are beam monitors, and the sample is positioned at S.[41]

other,[44] a chopped beam of white (unmonochromated) neutrons falls on the sample, and only those downscattered into a "window" of fixed final energy (~ 25 cm^{-1}) are diffracted into the detectors by an array of graphite crystals. To varying degrees, all of these later instruments can be used for down-scattering experiments, giving increased freedom with respect to sample temperatures. Further details about tof systems in general may be found in reviews by Brugger[45] and Dolling.[11]

3.4.3.3. Triple-Axis Spectrometers. The most versatile tool available to the neutron scatterer is the triple-axis spectrometer, developed around 1960 by Brockhouse.[33] Figure 5 is a schematic diagram of a typical instrument. A beam of thermal neutrons from the reactor falls on the first crystal, X1, where it is monochromated by Bragg reflection. The monochromated beam falls on a sample S, usually a single crystal with one of its axes fixed at an angle Ψ with respect to the incident beam. Neutrons scattered through an angle ϕ by the sample are analyzed with respect to wavelength by Bragg reflection over a number of angles from a second crystal, X2, into the detector. The instrument derives its name from the fact that the Bragg angles of the monochromator and analyzer and the scattering angle ϕ are all independently programmable (in practice, the

[44] S.-H. Chen, J. D. Jorgensen, and C. V. Berney, *J. Chem. Phys.* **68,** 209 (1978).

[45] R. M. Brugger, *in* "Thermal Neutron Scattering" (P. A. Egelstaff, ed.), p. 53. Academic Press, New York, 1965.

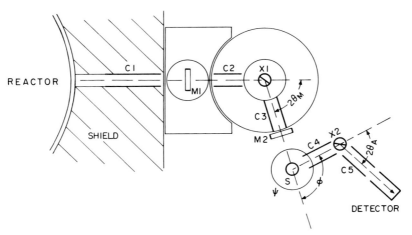

FIG. 5. Diagram of a representative triple-axis spectrometer. C1–C5 are collimators, M1 and M2 are beam monitors, X1 is the monochromator crystal, and X2 is the analyzer crystal. The sample is mounted at S with a given crystallographic axis at an angle ψ with respect to the beam from the monochromator.[11]

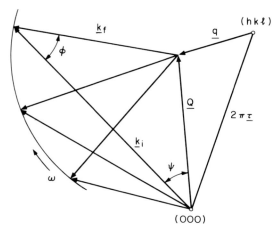

FIG. 6. Vector diagram for coherent inelastic scattering (constant-Q method). Energy transfer ω is scanned by programmed changes in ϕ, ψ, $\mathbf{k}_i$ and $\mathbf{k}_f$; three representative values of these variables are shown.[30]

sample angle Ψ can also be varied). The experimenter thus has control of the energy and momentum of neutrons falling on the sample and of the energy and momentum of scattered neutrons reaching the detector. As a result, he can select the length and orientation of the momentum transfer vector $\mathbf{Q}$ he wishes to use.

The experiment for which the triple-axis spectrometer is uniquely suited is the tracing of a phonon branch through the Brillouin zone, as described in Section 3.4.1. The method usually used is the constant-Q method, first described by Brockhouse.[33,46] The basis of the method is diagramed in Fig. 6. The Bragg angle of the monochromating crystal (θ_m), the sample angle Ψ, and the scattering angle ϕ are all changed in a predetermined way so that a number of frequency values ω are observed. During the scan, $\mathbf{k}_f$ (the wavevector of neutrons scattered into the detector) changes direction but is constant in magnitude, while $\mathbf{Q}$ remains entirely constant. The scan thus represents a vertical cut at some q through a dispersion-curve diagram such as Fig. 2a. When the cut intersects a phonon branch, the conditions of momentum conservation and energy conservation are simultaneously fulfilled, and a neutron group (a peak in the count rate) is observed. Several branches may be intercepted at one value of $\mathbf{Q}$. In practice, calculations and results from preliminary runs are used to predict branch positions so that the time spent searching for neutron groups is minimized.

[46] B. N. Brockhouse, in "Phonons and Phonon Interactions" (T. A. Bak, ed.), p. 221. Benjamin, New York, 1964.

While dispersion curves have been calculated for a number of polymers,[47] experimental observations have been reported only for deuterated† polyethylene[48-50] (see Section 3.4.4), for polytetrafluoroethylene,[51-53] and for polyoxymethylene.[54]

3.4.3.4. Modified Triple-Axis Instruments (INS Spectrometers). A triple-axis spectrometer can be programmed to vary the incident neutron energy E_i while holding E_f fixed. The inelastic scattering cross section of a given sample can then be recorded as a function of energy transfer. If the scattering is primarily incoherent, an INS spectrum is obtained. This technique was used by Myers *et al.*[55] to investigate a sample of unoriented polyethylene. A variation of the technique that trades broadened resolution for increased signal is to replace the analyzer crystal with a beryllium filter, so that E_f is fixed within a range of $0-42$ cm^{-1}.

A triple-axis instrument at the PLUTO reactor at Harwell has been modified in this way. Figure 7 shows an INS spectrum of formic acid obtained there.[56] The data were originally presented in terms of parameters directly related to the experiment, i.e., an abscissa scale showing the angle of the monochromating crystal (X1 in Fig. 5) and an ordinate scale showing the number of counts accumulated over a period of 400 sec at each monochromator setting. This plot is shown in Fig. 7a. The data are distorted in two ways: (1) the abscissa is not linear in energy transfer, and (2) peaks at higher energy transfer are reduced in apparent intensity because the incident flux from the reactor falls off. Figure 7b shows the same data plotted on a linear energy-transfer scale, with intensities calculated for a given number of monitor counts rather than for a fixed counting period. The $1/v$ efficiency of the monitor cancels the k_f/k_i factor in the double-differential cross section [Eq. (3.4.18)], so the plotted intensity is

[47] T. Kitagawa and T. Miyazawa, *Adv. Polym. Sci.* **9**, 335 (1972).

[48] L. A. Feldkamp, G. C. Venkataraman, and J. S. King, *in* "Inelastic Scattering of Neutrons," p. 159. IAEA, Vienna, 1968.

[49] J. F. Twisleton and J. W. White, *in* "Neutron Inelastic Scattering," p. 301. IAEA, Vienna, 1972.

[50] G. Pépy and H. Grimm, *in* "Neutron Inelastic Scattering 1977." Vol. 1, p. 605. IAEA, Vienna, 1978.

[51] V. LaGarde, H. Prask, and S. Trevino, *Discuss. Faraday Soc.* **48**, 15 (1969).

[52] J. F. Twisleton and J. W. White, *Polymer* **13**, 40 (1972).

[53] L. Piseri, B. M. Powell, and G. Dolling, *J. Chem. Phys.* **58**, 158 (1973).

[54] J. W. White, *in* "Dynamics of Solids and Liquids by Neutron Scattering" (S. W. Lovesey and T. Springer, eds.), p. 223. Springer-Verlag, Berlin and New York, 1977.

[55] W. Myers, J. L. Donovan, and J. S. King, *J. Chem. Phys.* **42**, 4299 (1965).

[56] M. F. Collins and B. C. Haywood, *J. Chem. Phys.* **52**, 5740 (1970).

† Deuterated rather than hydrogenous polyethylene is used for these measurements because the larger coherent scattering cross section of deuterium increases the intensity of the neutron groups, while its much lower incoherent scattering cross section reduces the isotropic background of incoherently scattered neutrons.

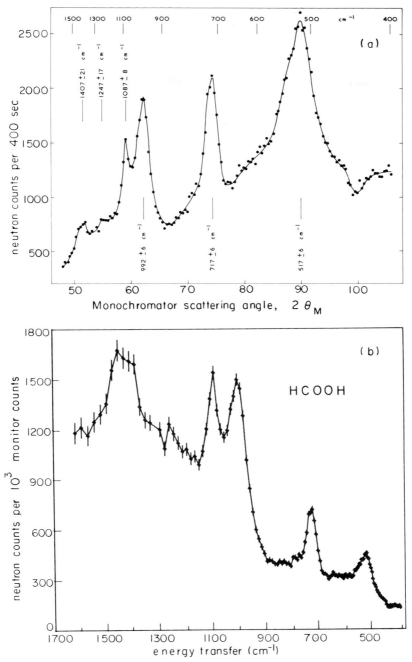

FIG. 7. INS spectrum of formic acid, HCOOH, taken on a beryllium filter spectrometer. The scattering angle ϕ is 81°, the temperature ~77 K. (a) Data as originally presented. (b) Same data, replotted so that the abscissa is linear in energy transfer and the ordinate is proportional to the scattering law $S(\omega)$.[56]

proportional to $S(\omega)$ rather than to $d^2\sigma/d\Omega \, d\omega$. The peaks in Fig. 7b are readily assigned to intramolecular vibrations,[14] and their relative intensities can be used as constraints in force-field calculations.[57] Note particularly that the peak around 500 cm^{-1}, which appears to dominate the spectrum in Fig. 7a, is reduced to rather modest proportions in Fig. 7b, in line with its assignment[57] as a two-phonon transition. Note also that in Fig. 7b the peaks appear superimposed on a background that increases rapidly with ω; this is apparently due to the multiphonon contribution discussed in Section 3.4.2. The resolution (FWHM) of the instrument used in this study varies from 5 cm^{-1} at $\omega = 70$ cm^{-1} to 80 cm^{-1} at 1500 cm^{-1}.

Reynolds *et al.*[58] have published INS data taken with a spectrometer of the MARX (multiangle reflecting crystal) design. This instrument uses an array of analyzer crystals to fix E_f and a linear position-sensitive detector (Section 3.4.3.1) to gather data over a range of $\mathbf{Q}$.

3.4.3.5. Backscattering Spectrometers. An ingenious device that combines features of triple-axis and tof neutron spectrometers with those of a Mössbauer spectrometer is the backscattering (π) spectrometer.[59,60] The neutron beam from the reactor is backscattered (reflected at an angle of 180° or π radians) from a monochromating crystal and directed onto the sample. The advantage of backscattering is that it eliminates the angle-dependent part of the energy resolution of a crystal monochromator, which is usually the dominant contribution, and makes $\Delta E/E$ entirely dependent on the crystal mosaic spread. Neutrons scattered from the sample are backscattered onto the detector by the analyzer, a mosaic of small crystals with a spherical figure, having the sample at the focus and the detector immediately behind. The analyzer is constructed of crystals identical to the monochromating crystal, so that with the monochromator stationary, only elastically scattered neutrons will be counted. In operation, however, the monochromator crystal is moved back and forth along the beam axis by a mechanical drive, so that the wavelength of the reflected neutrons is modulated by the Doppler effect. Thus the energy of the neutrons incident on the sample varies sinusoidally about the fixed final energy required for reflection from the (stationary) analyzer into the detector, and (except at the extremes of the drive cycle) a neutron must be upscattered or downscattered by the sample to be counted. The signal from the detector is fed into a multichannel analyzer, the channel being controlled by a signal from the monochromator

[57] D. H. Johnson, C. V. Berney, S. Yip, and S.-H. Chen, *J. Chem. Phys.* **71**, 292 (1979).
[58] P. A. Reynolds, J. K. Kjems, and J. W. White, *J. Chem. Phys.* **56**, 2928 (1972).
[59] M. Birr, A. Heidemann, and B. Alefeld, *Nucl. Instrum. & Methods* **95**, 435 (1971).
[60] B. Alefeld, *Kerntechnik* **14**, 15 (1972).

drive in such a way that a plot of counts vs. channel number gives the profile of the quasi-elastic peak.

Spectrometers of this type have been built at Munich, Jülich and Grenoble. They are capable of energy resolution on the order of 0.1 μeV (~ 0.001 cm^{-1}) over a range of 8 μeV (~ 0.06 cm^{-1}), making them ideal for studies of quasi-elastic scattering. This high resolution allows them to bridge the gap in frequency space between Mössbauer and NMR spectrometers on the one hand, and conventional neutron-scattering spectrometers on the other. The application of this type of instrument to the study of proton jump motions in polyethylene samples of varying crystallinities is discussed in Section 3.4.4.3.

3.4.4. Selected Applications

Polymeric materials in various physical states present a variety of dynamical problems suitable for inelastic neutron-scattering investigations.[5,19,47,54,61,62] Among the dynamical properties that have been studied are vibrational frequency spectra and phonon dispersion curves in crystalline chain-polymers, side-group motions in amorphous polymers, and segmental diffusion in polymers in the rubber state. Although we shall discuss applications in each of these cases, we shall be mostly concerned with the study of crystalline polyethylene, and our discussions are intended mainly to illustrate the strengths and limitations of the inelastic neutron-scattering technique and the type of information that can be obtained.

3.4.4.1. Low-Frequency Vibrations in Crystalline Polyethylene. Polyethylene is the polymer that has been most thoroughly studied by neutron-scattering techniques.[47,54,61,62] Incoherent scattering measurements have been carried out on unoriented and oriented samples, and amplitude-weighted frequency spectra have been extracted from the data. Coherent scattering from oriented deuterated samples also has been performed to obtain phonon dispersion curves. In addition, extensive normal-mode and cross-section calculations have been made to analyze the data in terms of the quantities discussed in Section 3.4.2.

The crystal structure of polyethylene at room temperature is orthorhombic.[63,64] The primitive unit cell consists of two planar zigzag chains with axes parallel to the $\hat{c}$ axis. With two methylene groups per chain there are 12 atoms in the unit cell and therefore 36 vibrational modes.

[61] J. S. King, in "Spectroscopy in Biology and Chemistry, Neutron X-Ray Laser" (S.-H. Chen and S. Yip, eds.), Chapter 7. Academic Press, New York, 1974.

[62] J. S. King, J. Macromol. Sci., Phys. 12, 13 (1976).

[63] C. W. Bunn, Trans. Faraday Soc. 35, 482 (1929).

[64] P. R. Swan, J. Polym. Sci. 56, 403 (1962).

Symmetry analyses and normal-mode calculations have been carried out in detail, and the optical data have been used to generate force-field models of interchain as well as intrachain interactions.[47] We shall be interested only in the two lattice modes denoted as ν_5 and ν_9. The former may be visualized as a purely longitudinal stretch-bend motion of the C–C–C skeleton and the latter as an out-of-plane torsional motion of methylene groups about the C–C bond.

The first inelastic scattering measurement on normal polyethylene was carried out on an unoriented sample using a tof spectrometer and a beryllium-filtered incident beam[39] (Section 3.4.3). The spectrum, observed in the upscattering mode, revealed a distinct peak at 190 cm^{-1} and less prominent peaks at around 140 and 330 cm^{-1}. Peaks were also observed at higher energy transfers, but the data were questionable because of low counting statistics and decreased instrumental resolution, which becomes worse as ω increases. The same sample was studied again on a triple-axis spectrometer in a downscattering measurement.[55] The two results were found to agree in the frequency region below 250 cm^{-1}, but there were substantial discrepancies at higher energy transfers. These could not be resolved since neither experiment had the necessary resolution for a definitive measurement in this region.

The interpretation of the neutron data was facilitated by the availability of normal-mode analyses both of the isolated chain and the three-dimensional structure. The peak at 190 cm^{-1} was assigned as the critical frequency of ν_9. On the same basis one expected a peak around 570 cm^{-1} corresponding to the ν_5 vibrations. Since ν_5 is purely longitudinal and ν_9, in the isolated-chain model, is transverse, measurements were then made on an oriented sample of polyethylene to measure separately the directional frequency spectra G^L and G^T discussed in Section 3.4.2. With the momentum-transfer vector $\mathbf{Q}$ placed parallel and perpendicular to the chain axis one could directly verify the assignment of ν_5 and ν_9 in the incoherent neutron-scattering data.

Polyethylene samples used in neutron scattering have been obtained from the melt; these samples (spherulitic PEH) have crystallinities of 80–96%.[61] The melt form could be plastically stretched up to 18 times its original length. Beyond approximately fivefold deformation the polycrystals are oriented so that the chain axes $\hat{c}$ of all the molecules in each polycrystal are aligned to within 3–9° of the stretch direction. The $\hat{a}$ and $\hat{b}$ axes meanwhile remain randomly oriented. A triple-axis spectrometer operating in the constant-$\mathbf{Q}$ mode was used in making the downscattering measurements on an oriented polyethylene sample.[65] The resulting directional frequency spectra are shown in Fig. 8.

[65] W. Myers, G. C. Summerfield, and J. S. King, *J. Chem. Phys.* **44**, 184 (1966).

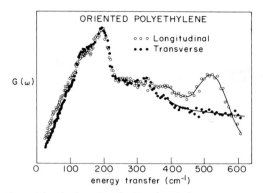

FIG. 8. Amplitude-weighted directional vibrational frequency spectra of a stretch-oriented sample of polyethylene at 100 K, as measured with a triple-axis spectrometer. Open and closed circles represent the longitudinal and transverse components $G^L(\omega)$ and $G^T(\omega)$ of the amplitude-weighted frequency spectrum.[65]

The main features of Fig. 8 are peaks at 190 and 525 cm^{-1} when $\mathbf{Q}$ was placed along $\hat{c}$, and a peak at 190 cm^{-1} when $\mathbf{Q}$ was perpendicular to $\hat{c}$. In addition, a shoulder at 150 cm^{-1} was observed in both spectra. It should be noted that the intensities of the two spectra have been normalized at 190 cm^{-1}; otherwise the 190 cm^{-1} peak in G^T would be about twice as intense as in G^L. The results of Fig. 8 have since been confirmed by two independent tof measurements[66,67] where, because the data were taken at fixed scattering angles, $\mathbf{Q}$ would change with energy transfer and therefore was not held constant. However, in these cases the scattering geometry was arranged so that deviations from purely longitudinal or transverse placement were no more than 10% in the relative magnitude of the unwanted component.

In analyzing the data on oriented polyethylene, the available force-field models in the literature were used to calculate both the one-phonon scattering and multiphonon effects.[47,68] A comparison of the calculated and measured directional frequency spectra is shown in Fig. 9. The theoretical results represent both the one- and two-phonon contributions to the scattering intensity.[68] On the basis of such comparisons, one can conclude that the 525 cm^{-1} peak indeed corresponds to the longitudinal ν_5 vibrations, while the 190 cm^{-1} peak corresponds to ν_9, which shows up strongly in the transverse spectrum but is also observable in the longitu-

[66] S. F. Trevino, *J. Chem. Phys.* **45**, 757 (1966).

[67] W. R. Myers and P. D. Randolph, *J. Chem. Phys.* **49**, 1043 (1968).

[68] J. E. Lynch, G. C. Summerfield, L. A. Feldkamp, and J. S. King, *J. Chem. Phys.* **48**, 912 (1968).

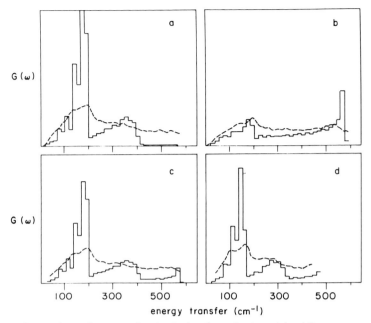

FIG. 9. Comparison of measured and calculated amplitude-weighted frequency spectra of polyethylene. (a) Transverse directional frequency spectrum $G^T(\omega)$ of oriented PEH at 100 K. (b) Longitudinal directional frequency spectrum $G^L(\omega)$ of oriented PEH at 100 K. (d) Frequency spectrum $G(\omega)$ of unoriented PED at 78 K.[68]

dinal spectrum due to mixing caused by the crystalline field. That the calculations are able to give correctly the relative intensities means that the force-field model used is quite realistic.

In Figs. 9c and d the calculated frequency spectra $G(\omega)$ for unoriented PEH and PED are compared with the experimental results.[68] One sees that deuteration causes a downward shift in peak frequencies but no change in relative intensities in $G(\omega)$. Recall from our discussion in Section 3.4.2 that $G(\omega)$ in general is not the same as the frequency distribution $g(\omega)$ because of the weighting by the square of the polarization vectors. The calculated $g(\omega)$ in the frequency region below 600 cm^{-1} for the orthorhombic polyethylene crystal is shown in Fig. 10.[69] Comparing this spectrum with the calculations in Fig. 9c, one finds that the peak intensities below 200 cm^{-1} are indeed affected by the polarization vector weighting. Moreover, the absence of any structure in Fig. 10 in the region between 200 and 500 cm^{-1} indicates that the calculated intensity

[69] T. Kitagawa and T. Miyazawa, *J. Polym. Sci., Part B* **6**, 83 (1968).

around 350 cm^{-1} in Figs. 9a, c, and d must be due to two-phonon transitions.

Although the two-phonon and higher order processes are of no interest in the context of molecular dynamics studies, it is nevertheless important to demonstrate that their effects are well understood and can be readily calculated. In Fig. 11 the individual one-, two-, and three-phonon contributions to the differential cross section for the experimental conditions corresponding to Fig. 8 are shown.[47] The one-phonon cross section for transverse vibrations is large in the frequency region below 200 cm^{-1}. Because of the sharp peaks in $G^T(\omega)$, the two-phonon contribution is seen to be appreciable over a range extending out to 400 cm^{-1}. By contrast, the one-phonon cross section for longitudinal vibrations is relatively smooth and less intense, and so the corresponding two-phonon contribution is much smaller and completely featureless. This comparison is a good example for showing the smoothing characteristics of multiphonon transitions in general. Notice that the three-phonon transverse cross section is nonnegligible in the region around 400 cm^{-1}, whereas the longitudinal cross section is too small to be shown in the figure. From these results one can conclude that in the region 200–400 cm^{-1} multiphonon effects in polyethylene samples are important corrections in measuring transverse vibrations, but they are relatively unimportant in measuring longitudinal vibrations.

Another quantity that should be considered in evaluating neutron scattering data is the Debye–Waller factor. The magnitude of $2W$ provides a good indication of the importance of multiphonon contributions in the

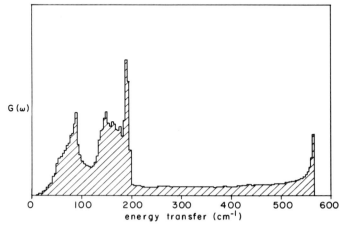

FIG. 10. Calculated vibrational frequency spectrum in the region 0–600 cm^{-1} of orthorhombic polyethylene.[47]

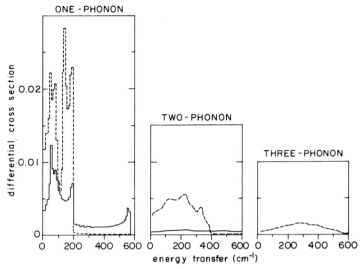

FIG. 11. Calculated contributions to the double-differential scattering cross section of an oriented polyethylene sample at 100 K from one-, two-, and three-phonon scattering processes. Final neutron energy E_f was fixed at 240 cm^{-1} and the scattering angle ϕ at 90°. Solid and dashed lines denote longitudinal and transverse orientations, respectively.[47]

sense that the cross section for n-phonon transitions is proportional to $(2W)^n \exp(-2W)$.[26] In tof measurements, the scattered spectrum is usually observed at fixed scattering angles, so that over a range of energy transfer of several hundred wavenumbers the momentum transfer Q, and therefore the Debye–Waller factor, can vary significantly. Under such conditions the relative intensities of peaks in the observed scattering spectrum can be quite different from those in $G(\omega)$. The variations with energy transfer of the calculated anisotropic Debye–Waller factors, $\exp(-2W^L)$ and $\exp(-2W^T)$, are shown in Fig. 12 for two temperatures.[47] Over the range of energy transfers that we have been discussing, one sees that $\exp(-2W)$ can vary by a factor of 2 to 3. Also, the temperature dependence in $\exp(-2W)$ is considerable. The results of Fig. 12 clearly indicate that it is important to keep the sample temperature as low as possible. Debye–Waller factors in polyethylene have been extracted from the observed elastic-scattering cross section. The values determined in this way agree with the calculated results to within 15% or better.[65,68]

Thus far we have considered only incoherent inelastic-scattering studies of low-frequency lattice vibrations. More detailed investigations of the ν_5 and ν_9 modes have been carried out through phonon dispersion curve measurements by coherent inelastic scattering.[48,49] Results of the

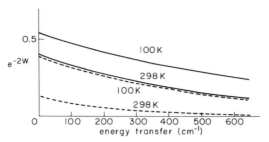

FIG. 12. Debye–Waller factors $\exp(-2W)$ of an oriented polyethylene sample at two temperatures and the same scattering conditions as in Fig. 11. Solid and dashed lines denote longitudinal and transverse orientations, respectively.[47]

first measurement[48] on an oriented 98% perdeuterated sample of polyethylene are shown in Fig. 13. The sample had been stretched 500% with a 9° mosaic. A triple-axis spectrometer was used with $\mathbf{Q}$ placed along the stretch direction. The neutron results are in fair agreement with force-field model calculations; however, the measured dispersion curve apparently did not flatten out as $\zeta \to 1.0$, as predicted theoretically. Further measurements have shown that the last four data points actually belong to the ν_9 branch, and that indeed a gap exists between the ν_5 and ν_9 branches.[50] On the other hand, the observed temperature dependence of the ν_5 branch in this region still remains to be explained.

3.4.4.2. Torsional Motions of Side Groups in Amorphous Polymers. In the glassy state of polymers, the backbone of each chain is frozen in a random configuration while the side chains are still free to undergo local motions. Neutron scattering is particularly effective for

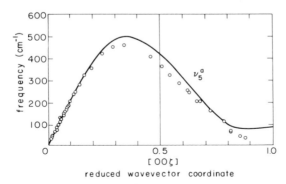

FIG. 13. Longitudinal phonon dispersion branch along the $\hat{c}$ axis. The solid curve denotes calculated results and the points denote experimental data obtained using a triple-axis spectrometer and an oriented deuterated polyethylene sample at 298 K.[48]

studying the torsional dynamics of $-CH_3$ side groups in glasses.[13] Because the dispersion of the torsional mode is weak and the amplitude of proton motions is large, one can expect the corresponding peak in the scattering spectrum to be sharp and relatively intense. Although the torsional mode is also optically active in infrared and Raman spectra, the bands are difficult to measure because they tend to be quite weak.

Time-of-flight measurements of methyl-group torsions have been carried out on a number of amorphous polymers.[13] Figure 14 shows a frequency spectrum $p(\omega)$ obtained in the case of poly(propylene oxide) and a deuterated sample. The spectrum $p(\omega)$ was obtained by the extrapolation procedure.

$$p(\omega) = \omega \lim_{Q \to 0} \frac{S(Q, \omega)}{Q^2} \exp\left(\frac{\hbar\omega}{2k_B T}\right), \qquad (3.4.24)$$

where the scattering law $S(Q, \omega)$ is related to the measured tof differential cross section $d^2\sigma/d\Omega\, d\tau$ through the relation

$$S(Q, \omega) = \frac{\hbar}{b^2} \frac{k_i}{k_f} \frac{\tau^3}{m} \frac{d^2\sigma}{d\Omega\, d\tau}, \qquad (3.4.25)$$

where m is the neutron mass and τ is the time of flight in microseconds per meter. Aside from the Debye–Waller factor, $p(\omega)$ is essentially proportional to the amplitude-weighted frequency spectrum $G^+(\omega)$.

It can be seen in Fig. 14 that deuterium substitution has a very striking effect. The dominant peak at 228 cm^{-1} in the spectrum is completely eliminated, thus showing clearly that this peak corresponds to the CH_3 torsions. From the peak frequency a potential barrier for internal rotation can be deduced by assuming the CH_3 group is an independent rotor attached to a chain of infinite mass. One obtains a barrier of 13.8 kJ/mol in this case, a value considered typical for methyl groups attached to aliphatic main chains. Neutron-scattering results of this type also have been obtained for acrylic polymers.[13]

3.4.4.3. Segmental Diffusion in Rubbers. At high temperatures, polymers that do not decompose on melting exist in the rubber state, where the chains undergo rapid and continuous random changes in conformations. The motions of the chain segments associated with these rapid fluctuations in chain conformations may be viewed as a type of diffusion with characteristic frequencies considerably lower than the vibrations and torsions discussed above. Such motions can be observed by neutron scattering through the broadening of the quasielastic peak. On the basis of a simple diffusion model and calculation of the incoherent scattering cross section, it can be easily seen that the width of the quasielastic peak is given by $2DQ^2$, where D is an effective diffusion coefficient.[17]

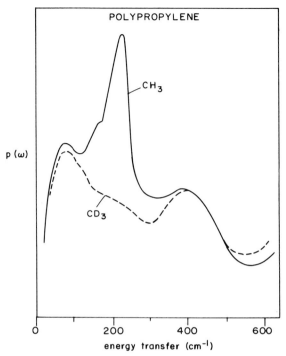

FIG. 14. Incoherent inelastic neutron scattering spectra of poly(propylene oxide) (solid curve) and poly(propylene oxide-d_3) (dashed curve) as obtained from measurements using a tof spectrometer.[13]

Incoherent quasi-elastic scattering measurements have been performed on several high molecular weight polymers in the raw rubber states.[70] These polymers have low glass transition temperatures and small bulk viscosities. A tof spectrometer was used to measure the double-differential scattering cross section. The data were first converted to scattering laws $S(Q, \omega)$, which were then symmetrized by multiplying by the factor $\exp(\hbar\omega/2k_BT)$ as in Eq. (3.4.24). A typical quasi-elastic peak obtained in this fashion for polypropylene oxide$-CD_3$ is shown in Fig. 15a, along with the instrumental resolution function.[70] The variation of the half-width with Q^2 at several temperatures is shown in Fig. 15b. The broadening is seen to vary linearly with Q^2, a behavior taken to imply that one is observing simple diffusion motions.

The temperature dependence of the results in Fig. 15b, as well as that of

[70] G. Allen, J. S. Higgins, and C. J. Wright, *J. Chem. Soc., Faraday Trans. II* **70**, 348 (1974).

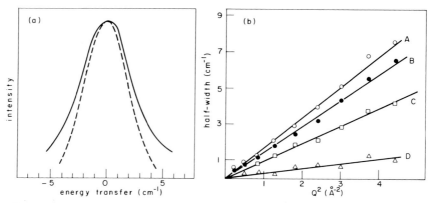

FIG. 15. Incoherent quasi-elastic neutron scattering in poly(propylene oxide-d_3). (a) Typical line shape of quasi-elastic scattering (solid curve) and instrumental resolution function (dashed curve). (b) Experimental line widths (symbols) at four temperatures: A, 392; B, 377; C, 344; D, 295 K.[70]

the other polymers studied in the same way, was found to follow a simple Arrhenius behavior

$$D(T) = D_0 \exp(-\Delta E/RT). \qquad (3.4.26)$$

Within the experimental error all the data could be represented by Eq. (3.4.26) with $D_0 = 7.2 \times 10^{-3}$ cm²/second and $\Delta E/R = 19$ kJ/mol.[70] The activation energies derived from inelastic neutron scattering are smaller than those obtained from dielectric and NMR relaxation measurements, which in turn are smaller than those extracted from viscosity data. It has been suggested that these differences have to do with the different frequency regions probed by these techniques.[70] At one extreme, neutron scattering operates on very short time scales (cf. Fig. 1) so that the measurements are most sensitive to local segmental motions. At the opposite extreme of long time scales, one has viscosity measurements in which significant effects due to motions of the center of mass of the entire molecule can be expected.

In a more recent application of incoherent quasi-elastic scattering, proton jump motions in linear polyethylene samples of high (80%) and low (48%) crystallinity were studied by combining the measurements using a high-resolution backscattering spectrometer with the data obtained with a tof spectrometer.[71] Energy resolutions ranged from 2–8 × 10^{-3} cm⁻¹ in the backscattering spectrometer results to about 3.4 cm⁻¹ in the tof data.

[71] T. Peterlin-Neumaier and T. Springer, *J. Polym. Sci.* **14**, 1351 (1976).

By analyzing the temperature dependence of the quasi-elastic line broadening and the Q dependence of the integrated intensity of the purèly elastic contribution, it was possible to show that the relaxation times for CH_2-group reorientations are consistent with NMR measurements of γ relaxations and that these motions take place predominantly in the noncrystalline regions of the polymer.

3.4.5. Prospects and Perspectives

As we have seen in the previous sections of this chapter, inelastic neutron scattering at present contributes to the study of polymers through three basic types of experiments: (1) the mapping of phonon dispersion curves by coherent inelastic scattering, (2) the acquisition of amplitude-weighted density-of-states spectra by incoherent inelastic scattering, and (3) the acquisition of information about diffusional and reorientational motions by (incoherent) quasi-elastic scattering. Only a relatively small number of polymer samples have been studied by each of these methods, and the future of the field will undoubtedly include extension to other samples. Improvements in the techniques of polymer sample preparation (improved crystallinity, reduction of orientation mosaics, and the like) will enable more quantitative interpretation of the experimental data.

Further advances may come from the improvement of neutron sources. For example, a pulsed neutron source such as that proposed for construction at Argonne[35,72] would provide increased fluxes at the high-energy end of the thermal range, allowing extension of the current limits of maximum energy and momentum transfers from 1500 cm^{-1} and 10 Å^{-1} to about 8000 cm^{-1} and 50 Å^{-1}. One also would like to reach lower wavevector transfers than the current limit of ~ 0.1 Å^{-1} in order to measure long-range spatial correlations. With a small-angle scattering facility it is possible to reach as low as 10^{-4} Å^{-1}. Plans are currently underway to establish a national center for small-angle neutron scattering at Oak Ridge.[73] Construction of new source facilities undoubtedly will be accompanied by advances in instrumentation, which will further enhance the capability of the classic experimental techniques described above (e.g., by extending the frequency range over which phonon branches may be mapped or INS spectra acquired, or by improving resolution).

With increases in neutron flux, wavelength and frequency regions of observation, and precision of measurements, one can expect significant contributions to polymer research from neutron diffraction and

[72] "Uses of Advanced Pulsed Neutron Sources," Rep. ANL-76-10, Vol. 2. Argonne Natl. Lab., Lemont, Illinois, 1976.
[73] A. L. Robinson, *Science* **199**, 673 (1978).

inelastic-scattering studies.[72] Examples are the measurements of radii of gyration for crystalline polymers, orientation effects in drawn bulk polymers, and ordering in solutions of copolymers, the extension of phonon dispersion curve measurements to higher energy excitations, and the study of phase transitions in crystalline polymers. Besides bulk properties one can also study the conformation of absorbed polymers and response properties such as the relaxation of polymeric materials under temperature and pressure shocks and swelling effects.

Due to the considerable cost of high-flux neutron sources and the associated spectrometers, the future of neutron scattering as a research technique lies in the establishment and the use of national or international centers. At present, the premier example of a successful large neutron research establishment is the Institut Laue-Langevin in Grenoble. The ILL is funded by the governments of France, Germany, and England, and its presence has greatly stimulated condensed matter research programs in these countries. The complexity of neutron-scattering techniques also means that the most successful experiments likely will be collaborative ones. In a typical arrangement, the scattering experiment could be handled by an instrumentalist on the staff of a research center, while the sample selection and preparation could be the responsibility of a materials scientist from industry or academia. Close collaboration of this type is essential to the planning and implementation of experiments of more than routine interest.

Acknowledgment

This work was partially supported by the National Science Foundation under Grant CHE-7616594.

4. HIGH-RESOLUTION NUCLEAR MAGNETIC RESONANCE SPECTROSCOPY

By J. R. Lyerla

4.1. Nuclear Magnetic Resonance Spectroscopy

4.1.1. Introduction

Since its concomitant discovery by Bloch[1] and Purcell[2] in 1946, nuclear magnetic resonance (NMR) spectroscopy has evolved as one of the more useful analytical techniques for the elucidation of the structural features and motional characteristics of molecules. The first NMR investigations of the properties of polymeric materials occurred soon (1947) after the initial observations of the phenomenon and involved a study[3] of the temperature dependence of the width of the proton resonance line arising from bulk samples of natural rubber and of carbon-loaded rubber. Such "wide-line" spectra provided information on molecular motion (through the linewidth) and formed the thrust of NMR applications to polymers until 1959, when a report on "high-resolution" proton magnetic resonance studies of polystyrene was published.[4] So-called high-resolution NMR (HR-NMR) had its origins in the observation that the characteristic resonant frequency of a nucleus depends to a small but measurable extent upon its chemical enivronment. This finding led directly to the rapid development of NMR as a tool for structural analysis.

In these chapters, the primary emphasis is placed on the utility of HR-NMR techniques in the investigation of the properties of macromolecular systems. By HR-NMR, we mean simply that measurement of spectra is made under conditions that allow magnetically nonequivalent nuclei in the same molecule to be distinguished—conditions generally attainable for polymer solutions, polymer melts, and some bulk polymers well above T_g. In Chapter 4.1, the NMR experiment and the origin of this more

[1] F. Bloch, W. W. Hansen, and M. Packard, *Phys. Rev.* **69,** 127 (1946).

[2] E. M. Purcell, H. C. Torrey, and R. V. Pound, *Phys. Rev.* **69,** 37 (1946).

[3] N. L. Alpert, *Phys. Rev.* **72,** 637 (1947).

[4] F. A. Bovey, "High Resolution NMR of Macromolecules." Academic Press, New York, New York, 1972.

common experimental parameters are reviewed and NMR instrumentation and experimental techniques discussed. Chapters 4.2 and 4.3 illustrate the applicability of HR-NMR to the determination of polymer structure and polymer chain dynamics, respectively. In these chapters, a majority of examples are drawn from ^{13}C NMR studies to reflect the significant increases in the number of investigations in which this nucleus has been employed as the probe of the molecular system. In addition, the examples are confined to synthetic polymers—the interested reader may consult Komoroski et al.[5] for a recent review of data on biopolymers. Chapter 4.4 is devoted to a discussion of techniques developed recently to obtain HR-NMR spectra of macromolecules in the solid state.

4.1.2. The NMR Experiment

In NMR spectroscopy, use is made of a nucleus possessing a magnetic moment μ and spin angular momentum $\mathbf{J}$ as a probe to investigate magnetic effects occurring at the molecular level. As these local magnetic fields are the resultant of the electronic structure and molecular motion of the system, NMR is amenable to the investigation of both molecular structure and molecular dynamics.

The magnetic moment and angular momentum behave as parallel vectors and thus are related by a scalar:

$$\mu = \gamma\mathbf{J} = \gamma\hbar\mathbf{I}. \tag{4.1.1}$$

The magnetogyric ratio γ (rad/G-sec) is characteristic for a nucleus. The replacement of $\mathbf{J}$ by $\hbar\mathbf{I}$ in (4.1.1), where $\mathbf{I}$ is the nuclear spin quantum number, defines explicitly the quantization condition on $\mathbf{J}$. Elementary quantum treatment of nuclear spin in which μ and $\mathbf{I}$ are treated as vector operators predicts the magnetic nucleus to have $2I + 1$ distinct "spin" states in which the component of angular momentum along any selected direction will have values from $+I$ to $-I$ in integral or half-integral multiples of $\hbar$. In the absence of an external applied field, the various states are degenerate. However, insertion of a nucleus with a magnetic moment into a uniform magnetic field $\mathbf{H}_0$, applied along the z direction of the laboratory reference frame, gives rise to interaction between μ and $\mathbf{H}_0$. The interaction removes the energy degeneracy and results in $2I + 1$ equally spaced nuclear spin energy levels with separation

$$\Delta E = \Delta(-\mu \cdot \mathbf{H}_0) = \gamma\hbar H_0 \tag{4.1.2}$$

[5] R. A. Komoroski, I. R. Peat, and G. C. Levy, in "Topics in Carbon-13 NMR Spectroscopy" (G. C. Levy, ed.), Vol. 2, Chapter 4. Wiley (Interscience), New York, 1976.

referred to as the nuclear Zeeman splitting. The basis of NMR experiments is to induce transitions between adjacent Zeeman levels by the absorption or emission of energy quanta. This is accomplished by applying an alternating magnetic field $\mathbf{H}_1(\omega t)$ perpendicular to the static field. From the Bohr relation, the frequency of radiation that induces transitions between adjacent levels is

$$\nu_0 = \Delta E/h = \gamma H_0/2\pi \quad \text{Hz} \tag{4.1.3}$$

or

$$\omega_0 = \gamma H_0 \quad \text{rad/sec.}$$

Expression (4.1.3) defines the fundamental condition for observing an NMR signal. For the typical range of laboratory magnetic fields 1–8.5 T (Tesla), ν_0 is in the region of radio frequencies (e.g., $\nu_0 = 42–360$ MHz for proton resonance in 1–8.5 T fields).

Because various aspects of NMR, in particular pulse techniques, are visualized more readily in classical, rather than quantum-mechanical terms, we now consider the classical motion of a magnetic dipole in a uniform static magnetic field. Briefly, the torque exerted on a magnetic dipole inclined at angle θ to H_0 causes the precession of $\boldsymbol{\mu}$ about $\mathbf{H}_0$ with angular velocity (the so-called Larmor precession frequency)

$$\boldsymbol{\omega}_0 = -\gamma \mathbf{H}_0. \tag{4.1.4}$$

To describe the motion of spins under the influence of $\mathbf{H}_0$ and a radiofrequency field $\mathbf{H}_1(\omega t)$, it is conventional in magnetic resonance to transform to a rotating frame representation. In the rotating frame, the spins are subject to an effective field $\mathbf{H}_{\text{eff}}$, given by

$$\mathbf{H}_{\text{eff}} = (H_0 - \omega/\gamma)\hat{z} + H_1\hat{x}, \tag{4.1.5}$$

where $\hat{x}$ and $\hat{z}$ are unit vectors along the x and z axes of the rotating frame (note that this z axis is coincident with the laboratory z direction) and ω is the frequency of H_1. Thus in the rotating-frame representation, there exist for $I = \frac{1}{2}$ two spin levels with Zeeman splitting $\omega_{\text{eff}} = \gamma H_{\text{eff}}$. Equivalently, $\boldsymbol{\mu}$ precesses in a cone about H_{eff} (see Fig. 1) at frequency and direction given by $\boldsymbol{\omega} = -\gamma \mathbf{H}_{\text{eff}}$. If the frequency of $\mathbf{H}_1$ is equal to the Larmor precession frequency, $\boldsymbol{\mu}$ then precesses about $\mathbf{H}_1$ in the rotating frame. The torque exerted causes large oscillations in the angle between $\boldsymbol{\mu}$ and $\mathbf{H}_0$ and thus large changes in the potential energy of $\boldsymbol{\mu}$ in the laboratory frame. During a precession cycle, there is then a periodic exchange of energy between the spin system and H_1. This condition, realized at $\omega = \omega_0 = \gamma H_0$, is the resonance phenomenon.

Because observations are made on macroscopic samples and not the iso-

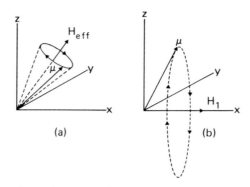

FIG. 1. Precession of a nuclear magnetic moment μ with $\gamma > 0$ about the effective magnetic field H_{eff} as observed from a coordinate system rotating at the frequency ω of the rotating magnetic field H_1: (a) $\omega < \omega_0$ (b) $\omega = \omega_0$ [N. Boden, *in* "Determination of Organic Structures by Physical Methods" (F. C. Nachod and J. J. Zuckerman, eds.), Chapter 2. Academic Press, New York, 1971].

late magnetic moments discussed above, NMR spectroscopy is concerned with the effect of $\mathbf{H}_0$ and H_1 on an ensemble of nuclear spins of a given element. If it is assumed that $I = \frac{1}{2}$ (e.g., ^{1}H, ^{13}C, ^{19}F, ^{31}P); that $\gamma > 0$; that the spin system is in thermal contact with its environment; and that nuclear spin–nuclear spin interactions are weak compared to the respective spin interaction with H_0, a Boltzmann distribution of spins between the two energy states (quantized along H_0) will be established at thermal equilibrium. That is, the ratio of occupation numbers for the two spin states (the lower state $(+\frac{1}{2})$ defined as that having parallel alignment of the nuclear moment with H_0 and the upper state $(-\frac{1}{2})$ that having antiparallel alignment) is given by

$$(n_{1/2})/(n_{-1/2}) = e^{2\mu H_0/kT} \sim 1 + (2\mu H_0/kT) \qquad (4.1.6)$$

when $kT > 2\mu H_0$, i.e., in the "high"-temperature approximation. At 25°C and $H_0 = 1$ T, the factor $2\mu H_0/kT$ is $<10^{-5}$, indicative of the low sensitivity of NMR. If we now define a macroscopic nuclear magnetization $\mathbf{M}$ as the magnetic moment per unit volume (given by the vector sum along $\mathbf{H}_0$ of the individual spin moments, $M_z = \Sigma_i \mu_i$, over unit volume), the thermal equilibrium magnetization M_0 is defined by[6]

$$M_0 = (N_I/3kT)\gamma^2\hbar^2I(I + 1)H_0 = \chi_0 H_0, \qquad (4.1.7)$$

where N_I is the number of spins per unit volume and χ_0 is the static magnetic susceptibility. The intensity of an NMR signal is proportional to M_0 and directly related to the magnitude of the applied field.

[6] A. Abragam, "The Principles of Nuclear Magnetism," Chapter I. Oxford Univ. Press, London and New York, 1961.

4.1.3. NMR Parameters

4.1.3.1. Spin–Lattice Relaxation Time.

Spin energy levels are degenerate in the absence of H_0; thus on a statistical basis, the two states corresponding to spin-$\frac{1}{2}$ nuclei must be equally occupied upon the initial insertion of the sample in H_0. As spontaneous emission is a negligible process in the region of NMR energies,[6] the redistribution of spin population required to establish a net magnetization must arise from interaction of the spins with the surroundings. Thus, a net number of transitions from the $(-\frac{1}{2})$ to $(+\frac{1}{2})$ spin state with concurrent release of energy is necessary to attain the distribution specified by (4.1.6). The process whereby the spin system exchanges energy and comes to thermal equilibrium with the surroundings is termed spin–lattice relaxation. Here the lattice refers to the translational, rotational, and other degrees of freedom of the molecular system (i.e., the surroundings). Except at very low temperature, the equilibrium temperature will be that of the lattice since the heat capacity of the spin system is negligible compared to that of the lattice.

The approach of the spin system toward equilibrium magnetization along H_0 is in many cases, particularly for neat liquids or solutions, governed by a single exponential function given by[7]

$$dM_z/dt = (M_0 - M_z)/T_1, \qquad (4.1.8)$$

where the time constant T_1 is defined as the spin–lattice relaxation time and is the time at which the difference between M_0 and M_z is reduced by a factor of e. Experimental values of T_1 range from $\sim 10^{-4}$ to 10^4 seconds and are usually of the order of several seconds for liquids free of paramagnetic impurities. This order of time constant clearly indicates that the spin system and lattice are only weakly coupled; nonetheless, it is only because of this coupling that spin relaxation occurs. The spin–lattice relaxation process and its characteristic time constant can be related to the motions of the molecular system; thus determination of T_1 values represents one method of learning about molecular dynamics in polymers. To illustrate this relation, we now present an expanded discussion of spin–lattice relaxation in liquids and solutions. (Relaxation in bulk polymers is considered in Chapter 4.4.)

The coupling of the nuclear spin system to the lattice is provided by the interactions between the spin moments and the fluctuating magnetic (and in some cases electrical) fields h_l produced by the lattice at the nuclear site. The time dependence of the local fields arises from the thermal motions of the molecules and therein lies the link between spin relaxation and molecular dynamics.[8] If the lattice fields have components at values

[7] F. Bloch, *Phys. Rev.* **70**, 460 (1946).

of the nuclear spin transition frequency and are in a direction perpendicular to H_0, energy can be transferred to the lattice and spin relaxation induced, i.e., the h_l play the role of the applied RF field H_1 mentioned earlier. In fluids, typical sources of local fields (referred to as relaxation mechanisms) are nuclear and electron dipoles, electric quadrupoles, anisotropy in the chemical shielding tensor, modulation of scalar coupling, and molecular rotation (spin–rotation).[9] The efficiency of any one of these sources in causing relaxation depends on the time scale of lattice motions and the magnitude of the fluctuations in the h_l.[10]

The mathematical description of the interaction of the nuclear moment with h_l is similar for each mechanism. As the majority of relaxation studies on polymers use the proton or ^{13}C nucleus as the probing nucleus, we consider only the nuclear dipole–nuclear dipole mechanism since this interaction is the primary source of spin relaxation for ^1H and ^{13}C in most polymer solutions and the melt state. (The explicit formulation and discussion of the other mechanisms, all of which have been found to be important in ^1H and ^{13}C relaxation in small molecules,[9] can be found in standard references.)[11,12] The instantaneous magnetic field produced at nucleus i by a magnetic nucleus j is given by[12]

$$h_l{}^i = \pm \mu_j \frac{3 \cos^2\theta_{ij} - 1}{r_{ij}^3}, \qquad (4.1.9)$$

where μ_j is the dipole moment of j, r_{ij} the nuclear separation, and θ_{ij} the angle of the vector r_{ij} relative to the external DC field. The $\pm$ sign expresses the fact h_l may add to or subtract from H_0. Thus the field experienced by i is a range of fields $H_0 \pm h_l{}^i$. In a solid in which there is a random arrangement of the interacting dipoles with respect to H_0, this range of local fields is realized in the spectrum yielding the broad resonance lines normally associated with NMR in solids (see Chapter 4.4). However, in solution the widths of resonance lines are reduced to only a few hertz because the nuclei due to molecular motion are no longer fixed in position with respect to H_0. For intramolecular dipole interactions (i.e., fixed nuclear separations), rotation imparts a time dependence to h_l

[8] A. Abragam, "The Principles of Nuclear Magnetism," Chapters II and VIII. Oxford Univ. Press, London and New York, 1961.

[9] J. R. Lyerla, Jr. and G. C. Levy, in "Topics in Carbon-13 NMR Spectroscopy" (G. C. Levy, ed.), Vol. 1, Chapter 3. Wiley, New York, 1974.

[10] H. S. Gutowsky, Tech. Org. Chem. 1, Part 4, 2663 (1960).

[11] A. Abragam, "The Principles of Nuclear Magnetism," Chapter VIII. Oxford Univ. Press, London and New York, 1961.

[12] A. Carrington and A. D. McLachlan, "Introduction to Magnetic Resonance," Chapter 11. Harper, New York, 1967.

through θ_{ij} variation; for intermolecular interactions, both θ_{ij} and r_{ij} are time dependent, owing to relative translation as well as rotational motion. For motion that is rapid on the time scale of the dipolar interaction ($\gtrsim 50$ kHz), h_l reduces to its average value of zero. Thus the interaction does not affect (to first order) the resonance linewidths in fluids; however, it still gives rise to relaxation effects. To explore further the dipolar field-induced relaxation process, we examine several features of h_l for liquids and solutions.

If it is assumed the molecules constituting the system can be described as Brownian particles, then the local dipolar field has the following properties.[13,14]

(1) $\langle h_l(t) \rangle = 0$, i.e., the mean value of h_l is zero since the field is random. The mean can be taken over an ensemble of spins or observation of a single spin over a sufficiently long period of time.

(2) $\langle h_l^2(t) \rangle \neq 0$, i.e., the mean of $h_l^2(t)$ is nonzero since the function is always positive. For an ensemble of spins the mean value of $h_l^2(t)$ is independent of time (when steady-state conditions prevail).

(3) $\langle h_l(t) \cdot h_l(t + \tau) \rangle = 0$ for τ sufficiently long, i.e., after a sufficient time interval τ, $h_l(t + \tau)$ does not depend on $h_l(t)$ or the two fields are not correlated. Of course, for short τ, the function does have nonzero value.

These characteristics of $h_l(t)$ are embodied in the autocorrelation function of h_l, $G(t, \tau)$. In a qualitative sense, such a function describes the persistence of a dynamical property of the molecular system before being averaged out by the molecular motion in the solution or fluid. It is particularly appropriate for describing weakly coupled physical systems such as the nuclear spin and its lattice.[13] $G(t, \tau)$ is given by the expression

$$G(t, \tau) = \langle h_l(t) \cdot h_l(t + \tau) \rangle_t, \qquad (4.1.10)$$

where the time dependence is that produced by the molecular motion and the average is over the nuclear spin ensemble at reference time t. This autocorrelation function is independent of the reference time and is both a real and even function of τ. Thus if the time origin is taken at $t = 0$, $G(t, \tau)$ becomes[11]

$$G(\tau) = \langle h_l(0) \cdot h_l(\tau) \rangle_0. \qquad (4.1.11)$$

The autocorrelation function then describes the time scale for decay of inherent motional order in the system. Information on the molecular

[13] R. J. Gordon, *Adv. Magn. Reson.* **3**, 1 (1968).
[14] W. T. Dixon, "Theory and Interpretation of NMR Spectra," Chapter 7. Plenum, New York, 1972.

system depends on $G(\tau) \neq 0$, and as $G(\tau)$ decays toward zero so does knowledge of the dynamical processes. In simplistic terms, $G(\tau)$ provides information on the "average" manner in which the collection of spins (or the collection of vectors between interacting nuclear dipoles) moves about.

The autocorrelation function forms a Fourier transform pair with the spectral density function $J(\omega)$.[11] Thus the Fourier inverse of $G(\tau)$ yields $J(\omega)$ and represents a transformation from the time to frequency domain, i.e.,

$$J(\omega) = \int_{-\infty}^{\infty} G(\tau)e^{-i\omega\tau}\, d\tau. \qquad (4.1.12)$$

$J(\omega)$ provides a means of characterizing the distribution of motional frequencies and their intensities in the autocorrelation function. Evaluation of (4.1.12) requires a description of the time decay of $G(\tau)$. For neat liquids and solutions, the decay process is assumed to be exponential. In particular, an exponential decay is consistent with the properties of $G(\tau)$ and h_l outlined in the above discussion. $G(\tau)$ then becomes

$$G(\tau) = \langle h_l^2(0)\rangle e^{-\tau/\tau_c}, \qquad (4.1.13)$$

where τ_C is the characteristic exponential time constant (correlation time) in which it is assumed motional order decays out of the system. The spectral density expression is now written as

$$J(\omega) = \int_{-\infty}^{\infty} \langle h_l(0) \cdot h_l(\tau)\rangle_0 e^{-i\omega\tau}\, d\tau. \qquad (4.1.14)$$

Since the local dipolar field h_L has x,y,z spatial components, $J(\omega)$ is

$$J_{ab}(\omega) = \int_{-\infty}^{\infty} \langle h_a(0) \cdot h_b(\tau)\rangle_0 e^{-i\omega\tau}\, d\tau. \qquad (4.1.15)$$

If h_L is truly random, directional components are not correlated (but are assumed to share the same correlation time) so that $J_{ab}(\omega)$ has components $J_{xx}(\omega)$, $J_{yy}(\omega)$, and $J_{zz}(\omega)$. If spatial isotropy is assumed $\langle h_x^2(0)\rangle = \langle h_y^2(0)\rangle = \langle h_z^2(0)\rangle = \langle h_l^2(0)\rangle = \frac{1}{3}\langle h_L^2(0)\rangle$, so that for all components

$$J(\omega) = \int_{-\infty}^{\infty} \langle h_l^2(0)\rangle e^{-\tau/\tau_c} e^{-i\omega\tau}\, d\tau \qquad (4.1.16)$$

when the exponential decay form of $G(\tau)$ is inserted. It should be recalled that only the x and y components of the local field (fluctuating at the appropriate frequency) induce spin–lattice relaxation.[15]

    [15] C. P. Slichter, "Principles of Magnetic Resonance," Chapter 5. Harper, New York, 1963.

Evaluation of the integral in (4.1.16) yields

$$J(\omega) = 2\langle h_l^2(0)\rangle \tau_C/(1 + \omega^2\tau_C^2). \qquad (4.1.17)$$

Clearly, $J(\omega)$ is a maximum at $\omega = 0$, is approximately constant over the range of ω such that $\omega^2\tau_C^2 \ll 1$, and begins to fall off as $\omega \to 1/\tau_C$. If we rewrite (4.1.17) and integrate the spectral density over ω, we see that

$$\frac{1}{2\langle h_l^2(0)\rangle} \int_0^\infty J(\omega)\, d\omega = \int_0^\infty \frac{\tau_C}{1 + \tau_C^2\omega^2}\, d\omega = \pi/2, \qquad (4.1.18)$$

i.e., the area under the $J(\omega)$ vs. ω curve is a constant and therefore independent of τ_C. Thus, variation in τ_C merely changes the distribution of spectral densities over the frequency spectrum. As an example, in Fig. 2 are plotted $J(\omega)$ vs. ω for several values of τ_C. $J(\omega)$ is also referred to as a power density and in this context represents the power available at ω to relax the spin system, i.e., the magnitude of local field at ω. In NMR, the frequencies of interest are the Larmor frequency, ω_0, and nearby frequencies. The spectral density is maximum at $\tau_C = 1/\omega_0$ and falls off at both long (slow motion) and short (fast motion) correlation times. Thus, the spin–lattice relaxation time (the inverse of the relaxation rate) will show a minimum when plotted against τ_C; the value of τ_C at which the minimum occurs is a function of the applied magnetic field. The flat portion of the spectral density curves (Fig. 2), i.e., where $\omega_0 \ll 1/\tau_C$, is termed the region of motional narrowing and (4.1.17) reduces to

$$J(\omega) \sim 2\langle h_l^2(0)\rangle \tau_C. \qquad (4.1.19)$$

in this range. Liquids or solutions in which τ_C is appropriate to the motional narrowing limit include most organic molecules at or above room

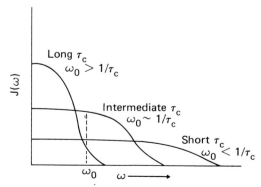

FIG. 2. Spectral density $J(\omega)$ as a function of frequency for various values of the correlation time τ_C (ω_0 corresponds to the nuclear Larmor frequency).

temperature as well as polypeptides and synthetic polymers in solution above 25°C and at magnetic fields $\sim 1-2$ T.[9] At superconducting field strengths, the values of τ_C for most polymers in solution are such that the motional narrowing condition is not valid.

In the above discussion, it has been assumed that $J(\omega_0) \propto 1/T_1$; we now turn to the explicit formulation of the relationship. Because the local magnetic fields that induce relaxation are such that $h_l \ll H_0$, the spin–lattice process can be treated via time-dependent perturbation theory[11] for a transition between spin states $\alpha \rightarrow \beta$. The transition probability $W_{\alpha\beta}$ is given by

$$W_{\alpha\beta} = \int_{-\infty}^{\infty} \langle \alpha| \mathcal{H}'(0)|\beta\rangle\langle \beta| \mathcal{H}'(\tau)|\alpha\rangle e^{-i\omega_{\alpha\beta}\tau}\, d\tau, \qquad (4.1.20)$$

where the average is over the statistical ensemble and $\mathcal{H}'(\tau)$ is a random operator on lattice variables. However, $\langle \alpha| \mathcal{H}'(0)|\beta\rangle\langle \beta| \mathcal{H}'(\tau)|\alpha\rangle$ is the correlation function $G_{\alpha\beta}(\tau)$ of $\langle \alpha| \mathcal{H}'(\tau)|\beta\rangle$, and thus

$$\langle W_{\alpha\beta}\rangle = \int_{-\infty}^{\infty} G_{\alpha\beta}(\tau)e^{-i\omega_{\alpha\beta}\tau}\, d\tau = J_{\alpha\beta}(\omega_{\alpha\beta}), \qquad (4.1.21)$$

where the latter identity follows by analogy from (4.1.12). This relationship between transition probability and spectral density is the link between T_1 and information on molecular motion. The explicit form of the perturbing Hamiltonian for the dipolar mechanism can be written as a product of a spin operator A with components (A_i) and second-order spherical harmonics $Y_{2,m}$. The latter are random functions of time and define the relative positions of the two interacting dipoles with respect to H_0. Thus, (4.1.21) becomes[16]

$$W_{\alpha\beta} = \sum_{m} J_m(\omega_{\alpha\beta})\langle \alpha|A_m|\beta\rangle^2, \qquad (4.1.22)$$

where

$$J_m(\omega_{\alpha\beta}) = \int_{-\infty}^{\infty} G_m(\tau)e^{-i\omega_{\alpha\beta}\tau}\, d\tau = \int_{-\infty}^{\infty} \frac{\langle Y_{2,m}(0)\, Y_{2,m}(\tau)\rangle}{r^6}\, e^{-i\omega_{\alpha\beta}\tau}\, d\tau, \qquad (4.1.23)$$

where r is the internuclear distance and $m = 0,1,2$ is the change in total spin quantum number for the transition.

The energy level diagram and associated transition probabilities for two unlike, spin-$\frac{1}{2}$ nuclei, I and S, are given in Fig. 3. As discussed, for spin–lattice relaxation, we are interested in the rate at which the magnetization approaches equilibrium along H_0. Solomon[17] has shown that the

[16] K. F. Kuhlmann, D. M. Grant, and R. K. Harris, *J. Chem. Phys.* **52,** 3439 (1970).

[17] I. Solomon, *Phys. Rev.* **99,** 559 (1955).

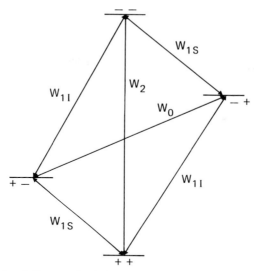

FIG. 3. Energy level diagram for I–S two-spin system. Transition probabilities are denoted by W_i, with i referring to the change in total spin quantum number upon transition. The first sign of a particular state denotes the I spin.

differential equation that characterizes the time dependence of M_z for the S species when relaxed by a dipolar process with species I is

$$\frac{d\langle M_z \rangle_S}{dt} = -\rho_S(\langle M_z \rangle_S - M_0^S) - \sigma(\langle M_z \rangle_I - M_0^I), \qquad (4.1.24)$$

where

$$\rho_S = W_0 + 2W_{1S} + W_2 = 1/T_1^S, \qquad (4.1.25)$$

$$\sigma = W_2 - W_0 = 1/T_1^{IS}, \qquad (4.1.26)$$

and M_0^I, M_0^S are the equilibrium values of magnetization. An analogous equation holds for $d\langle M_z \rangle_I/dt$. The second term in (4.1.24) expresses the fact that the approach to equilibrium of the S spins depends on the degree of polarization of the I spins. Hence, unlike in (4.1.8), the recovery process is governed by two relaxation times and is nonexponential. For the particular case of S = ^{13}C and I = ^{1}H, (4.1.24) simplifies since carbon resonance is usually observed under conditions of proton decoupling, in which case $\langle M_z \rangle_I = 0$. The recovery of $\langle M_z \rangle_S$ toward equilibrium is then exponential with time constant ρ_S. Hence the ^{13}C spin–lattice relaxation rate is given by

$$\frac{1}{T_1^C} = \frac{1}{20} \frac{\gamma_H^2 \gamma_C^2 \hbar^2}{r_{CH}^6} [J_0(\omega_H - \omega_C) + 3J_1(\omega_C) + 6J_2(\omega_H + \omega_C)] \qquad (4.1.27)$$

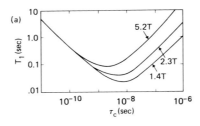

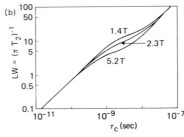

FIG. 4. (a) T_1 as a function of correlation time τ_C for various magnetic field strengths; (b) linewidth $(\pi T_2)^{-1}$ as a function of the correlation time τ_C for various magnetic field strengths.

when the transition probabilities in (4.1.25) are solved using (4.1.22) and (4.1.23). Figure 4a depicts a plot of T_1^C from this equation vs. τ_C at various values of H_0.

If motional narrowing holds, expression (4.1.27) reduces to

$$\frac{1}{T_1^C} = \frac{\gamma_H^2 \gamma_C^2 \hbar^2}{r_{CH}^6} \tau_C. \qquad (4.1.28)$$

To take into account interactions from more than one attached proton or protons on nearby atoms of the molecule, (4.1.28) can be written as the sum

$$\frac{1}{T_1^C} = \sum_{i=1}^{N} \frac{\gamma_H^2 \gamma_C^2 \hbar^2}{r_{CH_i}^6} \tau_C \qquad (4.1.29)$$

if τ_C is the appropriate correlation time for each interaction. Because the C–H bond length is short relative to nonbonded C–H distances, the inverse sixth-power dependence of the interaction allows contribution to T_1^C from nonbonded protons to be ignored for carbons with directly attached protons, so that (4.1.29) reduces to[9]

$$\frac{1}{T_1^C} = \frac{n_H \gamma_H^2 \gamma_C^2 \hbar^2}{r_{CH}^6} \tau_C, \qquad (4.1.30)$$

where n_H is the number of attached protons. These same features and the fact that carbon nuclei are almost never on the periphery of a molecule also result in intermolecular C–H dipolar interactions rarely contributing[18] to $T_1{}^C$. Finally, since the ^{13}C nucleus is only 1% abundant, the contribution of C–C dipolar interactions to relaxation can be ignored. (For very accurate work, contributions from nonbonded C–H interactions should be included when determining τ_C; however, the total contribution is usually smaller than the error in the T_1 measurement itself for carbons with directly bonded proton, and thus its neglect is usually justified.)

The expression for T_1 when $I = S$, i.e., homonuclear dipolar relaxation, is given by

$$\frac{1}{T_1} = \frac{3}{20} \frac{\gamma^4 \hbar^2}{r^6} [J_1(\omega) + 4J_2(2\omega)], \qquad (4.1.31)$$

which reduces in the case of motional narrowing and H–H interactions to

$$\frac{1}{T_1{}^H} = \frac{3}{2} \frac{\gamma_H{}^4 \hbar^2}{r_{HH}^6} \tau_C. \qquad (4.1.32)$$

Unlike the carbon case, interactions from other protons in the molecule contribute substantially to $T_1{}^H$. Because protons are at the periphery of the molecules, intermolecular H–H dipolar interactions can play a substantial role in the relaxation of a proton. Thus (4.1.32) must be modified to account for rotational and translational motion.[11]

Motional information is manifested in the molecular correlation times τ_C, and thus their determination from measured relaxation times is the final concern of relaxation experiments.[9,18] In a diffusional description of molecular reorientation, τ_C can be pictured to measure the period for molecular reorientation of a C–H vector, for example, through a given angular displacement. Taking into account a variety of factors, the average angular displacement in a time τ_C is ~ 1 rad. Factors governing the value of τ_C depend on whether the rotational reorientation, on the average, occurs via large or small angular displacements. For large displacements, angular momentum persists for relatively long time periods, owing to fewer intermolecular interactions ("collisions") that interrupt the molecular rotation. Under these conditions angular displacement comparable or larger than 1 rad may be realized and τ_C becomes bounded by the equipartition principle $(I_m/kT)^{1/2}$ for free rotation through 1 rad. For polymers, small angular displacements (random-walk processes) usually govern the C–H vector's reorientation.

4.1.3.2. Spin–Spin Relaxation Time. A spin system at thermal equilibrium with the lattice has no component of magnetization in the xy

[18] J. R. Lyerla, Jr. and D. M. Grant, *Phys. Chem.*, *Ser. One* **4**, 155 (1972).

plane because the spins precess in random phase about H_0. A perturbation that impresses a coherence on the motion of the spins, e.g., the observation of resonance, produces a magnetization M_{xy}. Return to equilibrium requires decay of M_x and M_y to zero. This process by which the spin system comes to internal thermal equilibrium is termed transverse or spin–spin relaxation and is governed by a rate constant $1/T_2$, i.e.,

$$\frac{dM_x}{dt} = \frac{-M_x}{T_2}, \qquad \frac{dM_y}{dt} = \frac{-M_y}{T_2}. \qquad (4.1.33)$$

The directional components of local dipolar fields that bring about T_2 relaxation are h_{xx} and h_{zz} for M_y and h_{yy} and h_{zz} for M_x. Viewed from the rotating frame,[19] the decay of M_x (M_y) must arise from "static" y (x) or z fields in this frame. Because the z axis of the rotating and laboratory frame coincide, the static laboratory component of h_{zz} causes transverse relaxation. For h_{xx} and h_{yy}, it is the laboratory component fluctuating at ω_0 that is important since these fields are static in the rotating frame.

The spin–spin relaxation time is related to the inverse of the resonance linewidth of an NMR signal. Because of spin–lattice processes, nuclear spin states have finite lifetimes that give rise to lifetime or nonsecular[20] broadening of resonance lines.[15] This contribution to linewidth is due to h_{xx} and h_{yy} and gives rise to a term $\frac{1}{2}T_1$ in the expression for the transverse relaxation rate. The contribution of h_{zz} is secular and arises from the spread in nuclear precession frequencies due to the fact that h_{zz} can add to or oppose H_0. From a random-walk model for the loss of phase coherence, it is straightforward[15] to show that the contribution to $1/T_2$ is $\gamma_i^2 h_{zz}^2 \tau$, where τ is time for a group of spins in phase at $t = 0$ to get out of phase by ~ 1 rad. The shorter τ or the faster the motion, the narrower the linewidth. This is the motional narrowing phenomenon discussed in the previous section. It results in line narrowing because it allows the spins to sample many fields h_{zz}, and hence dephasing to take place by random small steps each much less than 1 rad. The correlation time defined in our discussion of T_1 is the same as τ. Hence in the motional narrowing limit the contribution to transverse relaxation from h_{zz} is equivalent to that of h_{yy} or h_{xx}; therefore, $T_2 = T_1$. This is the case for fluids, and the resultant absorption line has a Lorentzian shape[21] with width at half-height equal to $1/\pi T_2$.

[19] R. T. Schumacher, "Introduction to Magnetic Resonance," Chapter 3. Benjamin, New York, 1970.
[20] Secular terms are those that commute with the Zeeman Hamiltonian; nonsecular terms are those that do not commute.
[21] A. Carrington and A. D. McLachlan, "Introduction to Magnetic Resonance," Chapter 1. Harper, New York, 1967.

If molecular motion becomes so slow that $T_2 \sim \tau$, a given spin will experience an essentially constant local field h_{zz}. A group of spins in phase at $t = 0$ rapidly gets out of phase since they are now insensitive to the fluctuations in h_{zz}. The transverse relaxation time becomes independent of τ_c. This results in the Gaussian-shaped temperature-independent rigid-lattice linewidth observed for crystalline solids.[15]

If we return to the case of fluids, we can use the spectral density function (4.1.16) and transition probabilities to write the expression for $1/T_2$ due to a dipolar process between ^{13}C and ^{1}H, i.e., the case of two unlike spins[17]:

$$\frac{1}{T_2{}^C} = \frac{\gamma_H{}^2 \gamma_C{}^2 \hbar^2}{20 r_{CH}^6} \left[2J(0) + \frac{J_0(\omega_H - \omega_C)}{2} + \frac{3J_1(\omega_C)}{2} \right.$$
$$\left. + 3J_1(\omega_H) + 3J_2(\omega_H + \omega_C) \right]. \quad (4.1.34)$$

This expression is plotted vs. τ_C for various magnetic field strengths in Fig. 4b. For two like spins, the expression for $1/T_2$ in the case of a ^{1}H–^{1}H dipolar process is given by

$$\frac{1}{T_2{}^H} = \frac{3}{40} \frac{\gamma^4 \hbar^2}{r_{HH}^6} [3J(0) + 10J_1(\omega_H) + 4J_2(2\omega_H)]. \quad (4.1.35)$$

In the region of motional narrowing, these expressions reduce to (4.1.28) and (4.1.32), respectively. However, the frequency independence of $J(0)$ results in the different behavior for T_2 as opposed to T_1 as a function of τ_C (see Fig. 4b) in the non-motionally narrowed region. At the value of τ_C that yields the rigid lattice T_2, (4.1.34) and (4.1.35) become invalid.

For fluids even under motional narrowing conditions, it is rarely the case that $T_1 = T_2$. This arises because the inhomogeneity of the applied magnetic field produces a spread in precession frequencies that "artificially" shorten T_2. Thus, the observed linewidth Δ is usually written as

$$\pi\Delta = 1/T_2 + 1/T_2' = 1/T_2^*, \quad (4.1.36)$$

where $1/T_2'$ is the contribution from magnet inhomogeneity. For the fluid state of small molecules and for dilute solutions, $1/T_2'$ usually dominates the linewidth.

4.1.3.3. Spin–Lattice Relaxation in the Rotating Frame.

As we have seen in previous discussion, T_1 is sensitive to molecular motions with correlation times $\tau_C \simeq (\omega_0)^{-1} \simeq 10^{-8}$ sec. In order to gain information on low-frequency (kilohertz) motions, as would be characteristic of solid polymers below T_g, it is necessary to observe spin–lattice relaxation in a magnetic field of a few gauss rather than kilogauss, so that energy states are separated by kilohertz frequencies. Of course, as is evident by examination of (4.1.7), the net magnetization in a DC field of a few gauss is quite small — making sensitivity problems overwhelming. To circumvent the sensitivity problems, Ailion and Slichter[22,23] devised a relaxation experiment in the rotating frame. First, equilibrium magnetization is attained in the usual large magnetic field, then a resonant H_1 field (as before, perpendicular to H_0) is applied in such a manner (see Section 4.1.4) that the magnetization established in H_0 is aligned and held or "spin-locked" along the much smaller H_1 field. Provided $T_2 \ll T_1$, the spin system will reach internal thermal equilibrium in H_1 long before relaxing back along H_0 by T_1 processes. Initially, the spin system is far from an equilibrium magnetization along H_1 since the magnetization is that established in H_0. The spin system relaxes toward equilibrium by absorption of energy from the lattice (i.e., spin–lattice interaction). Except at very low temperature, the amount of energy absorbed by the spin system from the lattice (in the range of microcalories) has no effect on the macroscopic temperature of the sample. Because the spin states in H_1 are separated by kilohertz, fluctuating local fields with components in this region induce the necessary transitions for relaxation. Or in equivalent terms, the characteristic time constant $T_{1\rho}$ governing the approach to equilibrium of the spin system and the lattice in H_1 (the so-called spin–lattice relaxation time in the rotating frame) depends on spectral density in the kilohertz region and thereby provides information on molecular motion in this frequency regime.[24] A schematic representation of a $T_{1\rho}$ vs. correlation time plot is given in Fig. 5 for the case of a dipole–dipole relaxation mechanism. Also depicted are the plots for T_1 and T_2, which allow direct comparison of the behavior of the various relaxation times.

4.1.3.4. Chemical Shift. The energy separation between magnetic spin states or resonance frequency is the primary source of information on molecular structure from HR-NMR experiments. The value of ν_0 in (4.1.3) depends to the extent of a few parts per million (ppm) for proton resonance to several hundred ppm for carbon resonance on the local electronic environment surrounding the nucleus. The origins of this variation in resonance frequency are electron currents induced in the sample by the presence of the external magnetic field. The induced orbital currents give rise to secondary magnetic fields at the nuclei of the molecule[25] that are

[22] C. P. Slichter and D. Ailion, *Phys. Rev. A* **135**, 1099 (1964).

[23] D. Ailion and C. P. Slichter, *Phys. Rev. A* **137**, 235 (1965).

[24] Viewed from the rotating frame, the relaxation is governed by T_1-like processes, even though in the laboratory frame it is a relaxation of transverse magnetization, a T_2-like process.

[25] J. A. Pople, W. G. Schneider, and H. J. Bernstein, "High Resolution Nuclear Magnetic Resonance," Chapter 1. McGraw-Hill, New York, 1959.

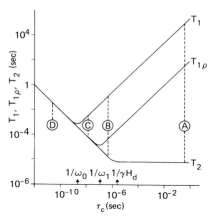

FIG. 5. Schematic dependence of relaxation times on molecular correlation time τ_c for relaxation determined by dipole–dipole interactions. Ordinate and abscissa values are only approximate. $\omega_0 = \gamma H_0$; $\omega_1 = \gamma H_1$; H_d = dipolar field. (A) Rigid lattice, $T_1 \gg T_{1\rho} \gg T_2$. (B) Nonrigid solid, $T_1 \gg T_{1\rho} \gg T_2$. (C) Viscous liquid, $T_1 > T_{1\rho} = T_2$. (D) Non-viscous liquid, $T_1 = T_{1\rho} = T_2$ (from T. C. Farrar and E. D. Becker, "Pulse and Fourier Transform NMR," Chapter 6. Academic Press, New York, 1971.)

proportional to the magnitude of H_0. Thus the net magnetic field H_N experienced by the nuclei can be written as

$$H_N = (1 - \sigma)H_0, \qquad (4.1.36)$$

where σ is a magnetic shielding constant that is independent of H_0 but dependent on the chemical (electronic) environment. As is evident from substituting the value of H_N for H_0 in (4.1.3), the effect of nuclear shielding is to decrease the separation between spin energy levels. At fixed field, resonance will thus occur at a lower frequency value than for a "bare" or unshielded nucleus.

The local screening field may not be isotropic, in which case the magnitude of the observed magnetic shielding is dependent on the orientation of the molecule with respect to H_0. Thus, the shielding is actually represented by a second-rank tensor and contributes a term[26]

$$\mathscr{H}_{CS} = \hbar \sum_i \gamma_i \mathbf{I}_i \cdot \hat{\sigma}_i \cdot \mathbf{H}_0 \qquad \text{(sum over } i \text{ nuclei)} \qquad (4.1.37)$$

to the Hamiltonian describing the NMR spectrum of a diamagnetic substance. Because the elements of the shielding tensor are small relative to unity (i.e., ppm magnitude), only the σ_{izz} term need be retained, where zz

[26] E. R. Andrew, *Prog. Nuc. Magn. Reson. Spectrosc.* **8,** 1 (1972).

defines the laboratory field direction. If the principal values of $\hat{\sigma}_i$ in the molecular reference frame are σ_{i1}, σ_{i2}, σ_{i3} and the direction cosines of its principal axes with respect to H_0 are λ_{i1}, λ_{i2}, λ_{i3}, then[26]

$$\sigma_{izz} = \lambda_{i1}^2 \sigma_{i1} + \lambda_{i2}^2 \sigma_{i2} + \lambda_{i3}^2 \sigma_{i3}. \qquad (4.1.38a)$$

As the isotropic average value of each λ_i^2 is $1/3$, the average value of σ_{izz} for a fluid where rapid molecular motions result in isotropic averaging is

$$\bar{\sigma}_{izz} = \tfrac{1}{3} \operatorname{tr} \hat{\sigma}_i = \sigma, \qquad (4.1.38b)$$

where σ is the scalar shielding constant defined in (4.1.36). This case, where $\hat{\sigma}_i$ reduces to a scalar, forms the basis of the majority of applications relating nuclear shielding to molecular structure.

The shielding constant can be decomposed into several contributions as follows[27]:

$$\sigma_i = \sigma_{ii}^d + \sigma_{ii}^p + \sum_{i \neq j} \sigma_{ij} + \sigma_i^{deloc}, \qquad (4.1.39)$$

where σ_{ii}^d is the contribution to the shielding of nucleus i by induced diamagnetic currents on atom i; σ_{ii}^p arises from induced paramagnetic currents on i as a consequence of mixing ground and excited electronic states under the influence of H_0; σ_{ij} is the screening due to either diamagnetic or paramagnetic local induced currents on atom j (or functional group j), i.e., a neighboring group effect; σ_i^{deloc} arises from induced currents involving delocalized electrons (e.g., π-electron systems). The magnitude of σ_{ii}^d is a function of the electron density about nucleus i and hence is an indirect function of the electronegativity of atoms or functional groups bonded to atom i; the σ_{ii}^p term occurs only for atoms having populated orbitals that possess intrinsic angular momentum (p, d oribtals, etc.) and hence is zero for protons but the dominant contributor to carbon shielding. This term gives rise to the greater dispersion for carbon shielding ($\sim$200 ppm) relative to proton ($\sim$ 10 ppm). The magnitude of σ_{ii}^p is strongly affected by charge polarization effects arising from groups bonded to atom i as well as remote electrostatic field effects and steric effects. The σ_{ij} term depends on the nature of j, the interatom or intergroup separation r_{ij}, and the spatial orientation of j with respect to i. In its simplest form (i.e., an axially symmetric case), σ_{ij} is given by the McConnell equation[28]

$$\sigma_{ij} = \frac{1}{3 N_0 r_{ij}^3} \Delta \chi_j (1 - 3 \cos^2 \theta), \qquad (4.1.40)$$

[27] D. J. Pasto and C. R. Johnson, "Organic Structure Determination," Chapter 5. Prentice-Hall, Englewood Cliffs, New Jersey, 1969.
[28] H. M. McConnell, *J. Chem. Phys.* **27**, 226 (1957).

where N_0 is Avogadro's number, $\Delta\chi_j$ the difference in parallel and perpendicular magnetic susceptibilities, and θ the angle between the symmetry axis of j and r_{ij}.

Clearly, through determination of the relative shielding of the magnetic nuclei in a molecule, one has an extremely sensitive probe of relative differences in local electronic environments. For macromolecules, this parameter can be related to such features as chain conformations, stereochemical configurations, and monomer sequences. There is a considerable literature on the calculation of chemical shielding constants using various molecular orbital schemes.[29] In general, proton shieldings are calculated with good agreement, while semiquantitative agreement is obtained for the case of carbon shielding constants.

Because the nuclear energy level spacing is proportional to the applied magnetic field [see Eq. (4.1.3)], there is no natural scale unit or natural reference in NMR spectroscopy. The change in resonance frequency (arising from differences in nuclear shielding) between nuclei of the same type is therefore expressed as a relative change of ν_0. These changes are referenced to the absorption frequency of an arbitrary standard (tetramethyl silane for ^{1}H and ^{13}C NMR studies) via the expression $(\nu_i - \nu_r)/\nu_r$. The displacements are in the ppm range and are termed chemical shifts. This method of referencing makes the chemical shift values independent of the applied field. The ^{13}C NMR spectrum of syndiotactic α-methyl styrene (Fig. 6) in solution demonstrates the spectral dispersion due to chemical shift effects.

4.1.3.5. Spin–Spin Coupling Constant. In addition to the direct nuclear dipole–nuclear dipole interaction discussed under spin relaxation, two magnetic nuclei in a molecule can interact through an indirect mechanism involving either a slight polarization of the spins or the orbital motions of the valence electrons of intervening bonds, or both.[30] This form of spin–spin coupling has a magnitude independent of the applied field and is not averaged to zero by rapid molecular tumbling. The interaction contributes a term

$$\mathcal{H}_J = \sum_{i<j} J_{ij}\mathbf{I}_i \cdot \mathbf{I}_j \qquad (4.1.41)$$

to the total nuclear spin Hamiltonian, where J_{ij} is the scalar coupling constant between nuclear spins i and j. In simple terms, the interaction may be conceived as a transmission of information on the spin state i to j and

[29] For example, see P. D. Ellis and R. Ditchfield, in "Topics in Carbon-13 NMR Spectroscopy" (G. C. Levy, ed.), Vol. II, Chapter 8. Wiley (Interscience), New York, 1976.

[30] F. A. Bovey, "Nuclear Magnetic Resonance Spectroscopy," Chapter IV. Academic Press, New York, 1969.

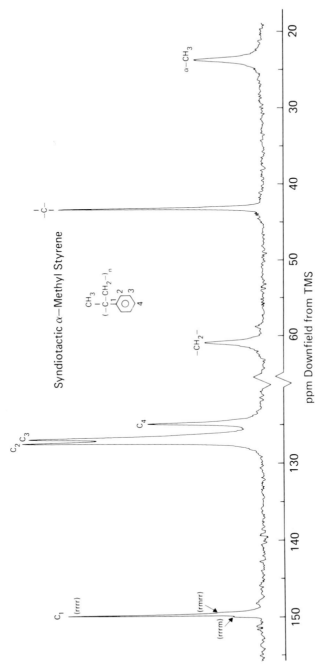

FIG. 6. The 20 MHz proton-decoupled ^{13}C NMR spectrum of syndiotactic poly(α-methyl styrene). Resonance line assignments are indicated.

vice versa via a scheme[31]: nucleus i interacting with electron i in turn interacting with electron j in turn interacting with nucleus j. Three types of electron–nuclear spin interactions can contribute to (4.1.41): (1) a nuclear dipole–electron dipole interaction; (2) an electronic current (due to electron orbital motion)—nuclear dipole interaction; (3) a Fermi contact interaction between electron and nuclear spins. The latter interaction is generally considered to be dominant ($\gtrsim 90\%$) for proton–proton and directly bonded proton–carbon scalar couplings. The contact term depends on a finite electron density at the nucleus (hence the appellative contact) and thus on the S character in the bonds between interacting nuclei as only S atomic states have finite electron density at the nucleus.

The spectral consequence of the indirect coupling interaction is the introduction of fine structure. For example, if two spin-$\frac{1}{2}$ nuclei are scalar coupled, each will split the other's resonance into a doublet whose separation is equal to J_{ij}. (Because the population differential between spin states is so small, the intensity of each doublet member is essentially equal.) In Fig. 7 are given the possible spin states seen by a nucleus A due to coupling to one, two, or three identical spin-$\frac{1}{2}$ nuclei (X) and one, two,

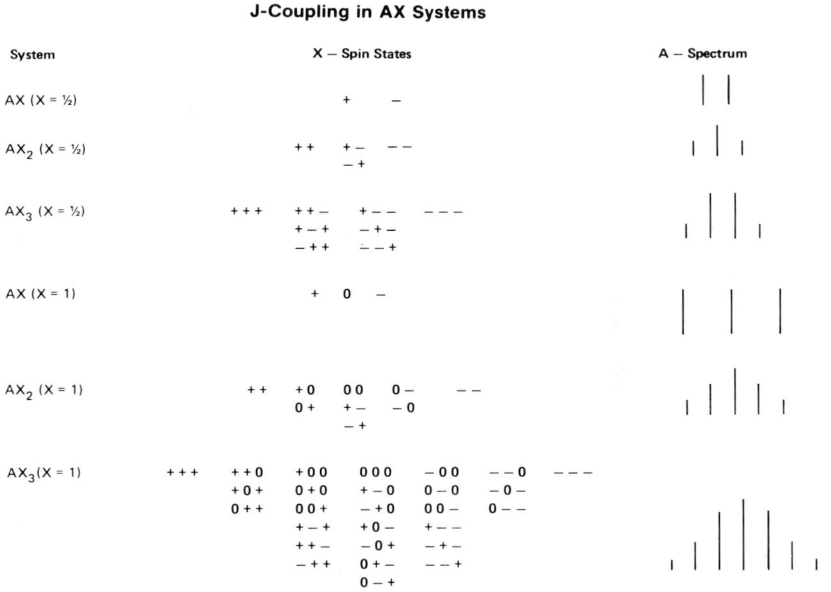

FIG. 7. Spectra of an A nucleus coupled to an X nucleus and spin states and statistical weights for the X nucleus for: (a) one spin-$\frac{1}{2}$ X nucleus; (b) two spin-$\frac{1}{2}$ X nuclei; (c) three spin-$\frac{1}{2}$ X nuclei; (d) one spin-1 X nucleus; (e) two spin-1 X nuclei; (f) three spin-1 X nuclei.

[31] J. Kowalewski, Prog. Nucl. Magn. Reson. Spectrosc. 11, 1 (1977).

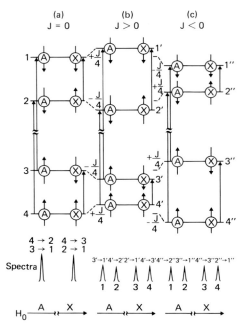

FIG. 8. Energy levels and schematic spectra of the AX system. The breaks in the arrows and schematic spectra are meant to indicate that the energy spacings between the 1,2 and the 3,4 levels are very large. The direction of H_0 is upward in the energy level representation (from F. A. Bovey, "Nuclear Magnetic Resonance Spectroscopy," Chapter IV. Academic Press, New York, 1969.)

or three identical spin-1 nuclei and the expected fine structure in the spectrum.

Returning to the case of two spin-$\frac{1}{2}$ nuclei and assuming the spins are not of the same type, for example, an ^{1}H–^{13}C interaction, it is possible to construct an energy level diagram like that in Fig. 8. The four spin states correspond to all possible arrangements of the moments. State 1, in which both moments are antiparallel to H_0, is of highest energy. In the absence of coupling, the two transitions of each nucleus are of the same energy and thus a one-line spectrum is produced for each nucleus. In the case of scalar interaction (for $J > 0$), states 1 and 4, in which the nuclear spins are parallel are raised in energy by $J/4$, and states 2 and 3, where the spins are paired, are lowered in energy by $J/4$. Thus for each nucleus, one transition has increased in energy by $J/2$ and one decreased by $J/2$ to yield spectra in which the resonance appears as a doublet separated by J (in hertz).

The spin system of Fig. 8 is an example of a so-called AX spin system in which $J/\Delta\nu$ (where $\Delta\nu$ is the frequency difference of the chemical shift between the nuclei) is very small and thus produces a "first-order" spectrum.[30] Spectra characterized by relatively large values of $J/\Delta\nu$ (clearly involving nuclei of the same type) are termed "strongly coupled." These spectra often appear as complex systems of lines haivng spacings with no obvious regularity and intensities that deviate widely from binomial. The analysis of many such spectra requires computer simulation schemes in which the energy levels and stationary-state wave functions for the system of coupled spins are determined and then perturbation methods and transition selection rules used to compute the transition probabilities between levels when a resonant H_1 field is applied.[32] The line positions are functions of the energy level separations and the intensities functions of the transition probabilities. The details of such calculations can be found in several texts and there exist various compilations of calculated spectra for typical spin systems.[30,32]

The usefulness of the J coupling in providing information on molecular structure arises because the magnitude of the coupling constant is a function of the number and nature of intervening bonds between interacting spins, the relative orientation of the two nuclei within the geometric framework of the molecule, and the type(s) of functional group(s) present in the molecule. An extensive literature exists on the characteristics and utilization of $^1H-^1H$ spin-coupling constants in molecular structure studies[30] and considerable data have also been obtained on $^{13}C-^1H$ couplings.[33] However, because J couplings have not been employed as extensively in the study of polymer structure as chemical shifts (due to resolution difficulties), we do not consider in further detail this parameter but only point out one useful relationship for deriving conformational information on vinyl polymers from vicinal couplings. Theoretical expressions have been developed for calculating the vicinal coupling constant $^3J_{HH}$ (the superscript designating the number of bonds between the coupled nuclei) as a function of the dihedral angle ϕ between the two C–H bonds (e.g., see Fig. 26a). The originai relationships were derived by Karplus[34] and are given (in hertz) by

$$J_{VIC} = \begin{array}{ll} 8.5 \cos^2\phi - 0.28 & (0 \le \phi \le 90°), \\ 9.5 \cos^2\phi - 0.28 & (90° \le \phi \le 180°), \end{array} \qquad (4.1.42)$$

and later refined to give[35]

$$J_{VIC} = 4.22 - 0.5 \cos\phi + 4.5 \cos^2\phi. \qquad (4.1.43)$$

[32] P. L. Corio, "Structure of High Resolution NMR Spectra." Academic Press, New York, 1966.

[33] J. B. Stothers, "Carbon-13 NMR Spectroscopy," Chapters 9 and 10. Academic Press, New York, 1972.

[34] M. Karplus, *J. Chem. Phys.* **30**, 11 (1959).

[35] M. Karplus, *J. Am. Chem. Soc.* **85**, 2870 (1963).

Experimental values of J from conformationally rigid systems with known bond angles are in good agreement with theory. For systems where rotation about the C–C bond is unhindered, the observed J values are time-averaged values of the couplings in the possible conformations, appropriately weighted.

4.1.3.6. The Nuclear Overhauser Enhancement. Despite the information available from the J coupling between two nuclei I and S, it is often expedient when attempting to observe the resonance of a low-sensitivity nucleus (e.g., S=^{13}C) to collapse the multiplet structure. This results in a gain in signal/noise by "concentrating" the total intensity of the S nucleus in one resonance line. The "decoupling" of spins I and S is readily accomplished by applying an RF field at the I resonance frequency while observing the resonance of S, i.e., a double-resonance experiment. If the magnitude of the field applied on I(H_2) is such that

$$\gamma H_2/2\pi \gg J_{IS}, \tag{4.1.44}$$

rapid transitions between I spin states are induced. This results in the lifetime of a particular state being short relative to $1/J_{IS}$ and thus a collapse of coupling. There are a number of variations on the double-resonance experiment, which include homonuclear decoupling and spin-tickling. The interested reader is referred to McFarlane[36] for a recent review of double resonance; in this chapter, we refer strictly to ^{13}C–{^{1}H} double resonance in the context of (4.1.44).

When a heteronuclear I–S dipolar mechanism contributes to the T_1 process for S, a direct consequence of double resonance is a signal enhancement of the S resonance above the sum of the multiplet intensity. If (4.1.24) is solved under steady-state conditions with $\langle M_z \rangle_I = 0$, the result is

$$\frac{\langle M_z \rangle_S}{M_0^S} = 1 + \frac{\sigma}{\rho_S} \frac{M_0^I}{M_0^S} = 1 + \frac{\sigma}{\rho_S} \frac{\gamma_I}{\gamma_S}, \tag{4.1.45}$$

where γ_I/γ_S gives the ratio of the Boltzmann populations. The term $(\sigma/\rho_S)\,\gamma_I/\gamma_S$ is defined as the nuclear Overhauser enhancement factor η and the expression $1 + \eta$ defined as the nuclear Overhauser enhancement (NOE). Solving for the transition probabilities in (4.1.25) and (4.1.26) gives for the ^{13}C–{^{1}H} case,[16]

$$\text{NOE} = 1 + \frac{\gamma_H}{\gamma_C}\left[\frac{6J_2(\omega_H + \omega_C) - J_0(\omega_H - \omega_C)}{J_0(\omega_H - \omega_C) + 3J_1(\omega_C) + 6J_2(\omega_H + \omega_C)}\right]. \tag{4.1.46}$$

In the general case, the NOE is frequency dependent like the relaxation rates (see Fig. 9) and becomes frequency invariant only in the

[36] W. McFarlane, *in* "Determination of Organic Structures by Physical Methods" (F. C. Nachod and J. J. Zuckerman, eds.), Vol. 4, Chapter 3. Academic Press, New York, 1971.

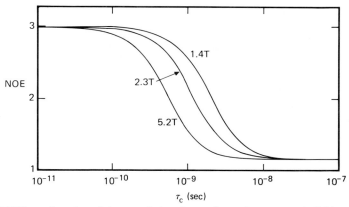

FIG. 9. NOE as a function of the correlation time τ_c for various magnetic field strengths.

motional-narrowing region. Unlike T_1, however, the NOE is independent of internuclear distance between the interacting dipoles. Equation (4.1.46) has a maximum value of 2.987 and a minimum of 1.15.[16] If relaxation mechanisms other than I–S dipolar contribute to the T_1 for S, the NOE is reduced from maximum even in the frequency-independent region and can be zero if dipolar relaxation is totally inefficient. This requires (4.1.46) to be reformulated to include the contributions to ρ_S from other mechanisms.[9,16] However, for polymer melts and solutions, usually the C–H dipolar mechanism dominates T_1 and (4.1.46) is valid.

4.1.4. Experimental Methods and Instrumentation

Two experimental methods are widely used to obtain nuclear magnetic resonance spectra: continuous wave (CW) or steady-state absorption techniques and pulse Fourier transform (FT) procedures. CW techniques have long been the observation method employed in HR-NMR, while pulse techniques have been the standard experimental method used in broad-line NMR. In CW NMR spectroscopy, a weak RF field (~0.1 mG) is applied to the sample and the frequency of the field varied. The energy absorbed by the sample as a function of frequency is recorded to yield the NMR spectrum. In pulse NMR, a strong RF field (20–100 G) is applied to the sample in short bursts (~1–100 μsec) and the transient time response of the nuclear spins to the pulse recorded in the form of a voltage vs. time spectrum. However, with the advent of inexpensive minicomputers, the time response function from the pulse experiment can be digitized and then Fourier transformed to yield the steady-state frequency spectrum. Thus, in the last several years the pulse-FT method has begun to supplant the use of CW techniques in HR-NMR. Description of CW experiments can be found in standard texts[6,10,30] and are not discussed further; instead, we focus attention on FT methods.[37,38]

4.1.4.1. Pulse NMR and the FID. The effects of RF pulses on nuclear magnetization are most easily described in the rotating reference frame. The magnitude of **M** is unchanged by the transformation from laboratory coordinates. However, the H_1 field (at frequency ω) produced in a coil whose axis is perpendicular to H_0 appears as static in the transverse direction in a frame rotating at ω. Since H_1 rotates at the frequency of the rotating frame, H_1 can be arbitrarily assigned along the x' direction. Any static component of **M** in the x',y' plane (primes indicate rotating-frame coordinates) will be detected as a frequency ω in a pick-up coil along x or y in the fixed frame.[38] Consider now the effect of H_1 on **M**, when **M** initially lies along z' ($=z$). Assume that ω is equal to ω_0, the angular precession frequency of the nuclei producing **M** (i.e., an "on-resonance" RF field is applied). During the application of H_1, the torque exerted on **M** (see Fig. 10a) causes **M** to precess around the x' direction at a frequency $\omega_1 = \gamma H_1$. The angle α through which **M** rotates in a time t_p is given by

$$\alpha = \gamma H_1 t_p \quad \text{(rad)}, \tag{4.1.47}$$

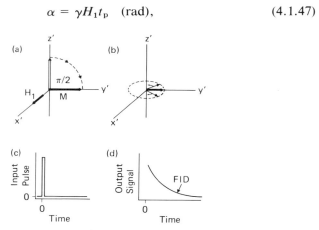

FIG. 10. (a) A 90° pulse along x' rotates **M** from the equilibrium position to the y' axis. (b) **M** decreases as magnetic moments dephase. (c) Input signal, a 90° pulse, corresponding to (a). (d) Exponential free-induction decay, corresponding to (b) (from T. C. Farrar and E. D. Becker, "Pulse and Fourier Transform NMR," Chapter 2. Academic Press, New York, 1971).

[37] For an introductory discussion see the recent books by R. J. Abraham and P. Loftus, "Proton and Carbon-13 NMR Spectroscopy." Heyden, Philadelphia, Pennsylvania, 1978; F. W. Wehrli and T. Wirthlin, "Interpretation of Carbon-13 NMR Spectra." Heyden, Philadelphia, Pennsylvania, 1978.

[38] N. Boden, in "Determination of Organic Structures by Physical Methods" (F. C. Nachod and J. J. Zuckerman, eds.), Vol. 4, Chapter 2. Academic Press, New York, 1971.

where t_p is the time H_1 is applied. A resonant RF field applied for t_p and causing **M** to precess through α degrees is termed ''an α degree pulse.'' Thus for a given magnitude H_1, if t_p is chosen so that $\alpha = \pi/2$ or a 90° pulse applied to the sample, **M** is rotated to the y' direction (Fig. 10a); if $\alpha = \pi$ or a 180° pulse is applied, **M** is rotated to $-z'$, etc.

A 90° pulse yields a maximum observable signal (induced voltage) since this produces a maximum value of M_{xy} (i.e., the component of **M** in the transverse direction), which is the magnetization to which the receiver coil responds. At the end of the pulse duration, t_p, the observed signal is termed the ''free-induction'' signal since the nuclei precess ''freely'' in the absence of H_1. The time decay of the signal intensity is termed the ''free-induction decay'' or FID[6] and is governed by the decay constant for M_{xy}, i.e., the T_2^* spin–spin relaxation time. Although the signal is measured after the rf pulse, the detector is referenced in phase to the rf by continuous application of the rf reference to the detector. Thus the detector responds to magnetization having a fixed phase relation to H_1—i.e., occurring along a fixed axis in the rotating frame (along y' in Fig. 10b). If the pulse is nonresonant ($\omega \neq \omega_0$) but still a 90° pulse on **M** (see below) then, after the pulse, **M** rotates relative to the rotating frame. The detector response displays not only the exponential decay of M_{xy} but also the interference effects as M_{xy} and the reference frequency come in and out of phase. This is illustrated by the FID of the ^{13}C resonance of methonal (proton-decoupled to yield a single resonance line) shown in Fig. 11a. Of course, for a system having multiple resonance lines arising from chemical shifts and/or spin–spin couplings the FID becomes complex as the decay of each M_{xy} produces a modulated pattern similar to that in Fig. 11a. The modulation or ringing frequencies interfere with each other to produce an interferogram. For a simple multiplet, e.g., the quartet from the proton-coupled ^{13}C spectrum of methanol (Fig. 11b) the interferogram is a regular beat pattern with periods related to the inverse of the coupling constant, J_{CH}, and the offset from the RF pulse. For the multiline spectrum, the M_{xy} interfere and also decay at different rates hence visual inspection of the FID interferogram is a useless endeavor to obtain information on the sample. However, the spectral information contained in the FID is readily extracted by Fourier transformation. Before turning to this analysis, we consider how an ''on-resonant'' pulse can be effected for each line in a spectrum (as assumed above) by an RF pulse at ω.

Consider a system having chemically shifted resonance lines that cover a frequency range of Δ Hz. Generalizing (4.1.5) we have

$$|H_{eff}| = \frac{1}{\gamma} [(\omega_i - \omega)^2 + (\gamma H_1)^2]^{1/2}, \qquad (4.1.48)$$

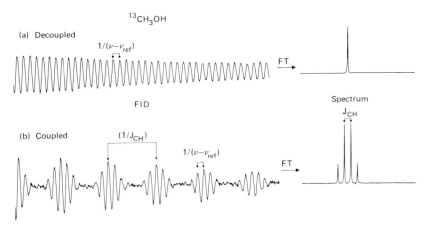

FIG. 11. (a) The proton-decoupled FID of $^{13}CH_3OH$ for resonance 100 Hz from the carrier frequency (spectrum is also shown). (b) The proton-coupled FID of $^{13}CH_3OH$ for resonance 1000 Hz from the carrier (note J coupling corresponding to the beat frequency).

which is the relationship between the precession frequency ω_i and the effective field about which the magnetization M_i precesses when H_1 is applied at ω. If H_1 is large enough to meet the condition[39]

$$\gamma H_1 \gg 2\pi\Delta',   (4.1.49)$$

where Δ' is the total range of frequencies from the RF frequency, then the first term of (4.1.48) can be ignored and

$$|H_{eff}| \cong H_1,   (4.1.50)$$

i.e., the magnetization for all the nuclei in the frequency range Δ' precess about H_1 or all M_i are rotated through the same angle α. If we combine (4.1.47) and (4.1.49), we see that for a 90° pulse, the pulse duration time must meet the condition

$$t_p \ll \tfrac{1}{4}\Delta'   \text{(sec).}   (4.1.51)$$

For $\Delta' = 4$ kHz (typical of the ^{13}C nucleus at 2.3T) $t_p \ll 60$ μsec. This clearly demonstrates the necessity for short, intense pulses. The corollary that must be attached to (4.1.51) is that t_p must be short compared with T_1 and T_2 so that relaxation processes are negligible during the pulse. The important result is that the entire range of resonance frequencies for a given type of nucleus can be excited simultaneously by a strong $\mathbf{H}_1$ field.

[39] T. C. Farrar and E. D. Becker, "Pulse and Fourier Transform NMR," Chapter 5. Academic Press, New York, 1971.

4.1.4.2. Fourier Transformation of the FID. The FID, which is a time domain function, contains all the information of the frequency domain NMR spectrum. For example, the period of the time domain signal determines the position of the resonance line (from the RF frequency) in the frequency domain; the decay time of each component of the FID is inversely proportional to the linewidth of the resonance in the frequency domain (i.e., linewidth $= 1/\pi T_2^*$); the integrated intensity of a resonance line is proportional to the initial amplitude of the envelope of its component in the FID; the peak amplitude of a resonance line is proportional to the integrated area under the envelope of its component in the FID. To convert the time domain data to the frequency domain, the FID is Fourier transformed. This is generally accomplished by digitizing the analog signal (FID) from the spectrometer, collecting the digital information in a minicomputer, and performing a discrete Fourier transform based on the Cooley–Tukey fast Fourier transform algorithm.[40]

The expression for the Fourier pair relationship between the time and frequency domains is given by

$$F(\omega) = \int_{-\infty}^{\infty} f(t)e^{-i\omega t} \, dt. \tag{4.1.52}$$

Use can be made of the complex exponential $e^{-ix} = \cos x - i \sin x$ to define sine (S) and cosine (C) transforms[39]

$$S(\omega) = 2 \int_{0}^{\infty} f(t) \sin \omega t \, dt, \tag{4.1.53}$$

$$C(\omega) = 2 \int_{0}^{\infty} f(t) \cos \omega t \, dt. \tag{4.1.54}$$

As discussed above, the functional form of the FID is an exponential, so that $f(t)$ is given by

$$f(t) = Ae^{-t/T_2} \cos (\omega - \omega_i)t, \qquad t > 0. \tag{4.1.55}$$

The Fourier transform of an exponential is a Lorentzian line in frequency space; the cosine solution is the absorption mode solution of the Bloch equations, i.e., $T_2^*/\{1 + [T_2^*(\omega - \omega_i)]^2\}$, while the sine transform is the dispersion mode solution.[7] The actual digital computation involves summing a Fourier series over a finite number of data points rather than evaluating (4.1.52). This discrete Fourier transform is defined by[39]

$$A_f = \sum_{t=0}^{N-1} X_t \exp(-2\pi i f t/N), \qquad f = 0, 1, \ldots, N - 1, \tag{4.1.56}$$

[40] J. W. Cooley and J. W. Tukey, *Math. Comput.* **19**, 297 (1965).

where A_f is the coefficient of the fth point in the frequency domain, X_t the value of the FID at time t, and N the number of data points.

4.1.4.3. Advantages of Pulse-FT Methods. The principal advantages of pulse-FT methods over CW NMR are (1) the potential for signal/noise (S/N) enhancement and (2) the greater variety of modes in which the nuclear spin system can be manipulated. Ernst and Anderson[41] have shown that for an equivalent S/N, pulse-FT methods provide a time savings compared with slow-passage frequency sweep conditions of $\sim \Delta/a$ typical linewidth, where Δ is the spectral range. Thus for the usual 100 MHz proton spectrum (i.e., one covering ~ 10 ppm or 1 kHz) and with typical linewidth of 1 Hz, the savings in time would be $\sim 10^3$. This factor arises because during the CW sweep, data are collected on a 1 Hz resonance line for only 0.001 of the sweep time; on the other hand, a strong RF pulse simultaneously excites all relevant frequencies in a few microseconds and acquisition of the FID interferogram for 1 second yields a 1 Hz resolution. In practice, CW spectra are obtained at much faster rates ($\times 10$) than slow-passage conditions and the actual time savings may be reduced to 10 or 100. Nonetheless, the savings are very significant, and in addition the FT spectra are not distorted by ringing and skewing of lineshapes found in CW spectra obtained at "normal" scanning rates.

The rapid data acquisition in pulse-FT NMR readily enables S/N to be improved by recycling the experiment a number of times. The FID from each successive pulse is digitized, coherently added, and stored in the computer. Fourier transformation is then carried out on the accumulated data. For random noise, S/N will improve as the square root of the number of accumulated FIDs. This feature of pulse-FT methods enables the spectra of nuclei such as carbon, with its sensitivity to detection $\sim 1/5700$ that of the proton, to be obtained routinely. For an abundant nucleus, the rapid data collection by FT methods enables spectra of transient or moderately short-lived species to be observed.

The rate of recycling an experiment depends on many factors, including the desired resolution and the spin relaxation times. Naturally, the question arises as to the limitations on the rate at which the spin system can be subjected to repetitive pulsing[42] and obtain optimum S/N. For ^{13}C, it is often the case that $T_2^* \ll T_1$. The FID decays rapidly, while the restoration of magnetization requires considerable time. One obvious trade-off is to use pulse widths of less than 90°. The amplitude of the FID is attenuated by $\sin \alpha$ (α = flip angle) while M_z is diminished by $M_0(1 - \cos \alpha)$.

[41] R. R. Ernst and W. A. Anderson, *Rev. Sci. Instrum.* **37**, 93 (1966).

[42] This question has been examined in detail by J. S. Waugh, *J. Mol. Spectrosc.* **35**, 298 (1970).

For small α, $\sin \alpha \gg (1 - \cos \alpha)$ and the loss of signal is recovered by the small time needed to reestablish equilibrium along H_0 and thus recycle the experiment. Another method used to obtain optimum S/N with fast recycle times is to shorten T_1 via paramagnetic relaxation agents.[43]

The application of a $\pi/2$ pulse to a sample yields the steady-state spectrum upon Fourier transformation. However, an advantage of FT NMR is that when more than one pulse is applied, the FID becomes a function of the pulse train and information not available in the steady-state spectrum can be obtained. In particular, pulse methods can be used for direct determination of relaxation parameters and for line-narrowing of resonances in solids through manipulation of the nuclear spins.

4.1.4.4. The Pulse-FT Spectrometer. Pulse-FT NMR spectrometers are manufactured by a number of analytical instrument companies. The systems range from low-field (1.4 T) single-nucleus spectrometers to high-field (~ 8 T) multinuclear instruments. In this section, we outline the requirements for a versatile pulse-FT spectrometer rather than review commercial instrumentation.

A multipurpose pulse-FT system should include the following components and associated characteristics[39]:

1. The transmitter including power amplifier should be capable of delivering short (1–10 μsec), intense (100–1000 W) bursts of RF power. The RF pulse must have rise and fall times that are short compared with the pulse width.

2. The power delivered by the transmitter must be coupled efficiently to the RF coil so that the H_1 field produced at the sample is large compared to the spectral width being examined. The H_1 field produced must be as homogeneous as possible over the sample volume; thus the sample should be confined to the volume of the RF coil. Finally, the power must be quickly dissipated (20 μsec or less) after the pulse is turned off.

3. The sample must be efficiently coupled to the receiver in order to obtain the maximum S/N from the nuclear induction signal. This requires the receiver recover quickly (2–3 μsec or less) from any overloads generated by the H_1 pulse. Hence, the transmitter and receiver should be well isolated.

4. The pulse programmer used to control pulse widths, etc. should have a stable time base and allow for precise setting of all time intervals. In most present-day commercial spectrometers use is made of a minicomputer for control of all time intervals.

5. Coupled to the spectrometer should be a versatile minicomputer that

---

[43] R. Freeman, K. G. R. Pachler, and G. N. LaMar, *J. Chem. Phys.* **55**, 4586 (1971); G. C. Levy and R. A. Komoroski, *J. Am. Chem. Soc.* **96**, 678 (1974).

controls pulse sequences and data acquisition, acts as storage for input FIDs, stores and executes programs for Fourier transformation, phase correction, and plotting of data. In essence, the minicomputer is the central unit of a pulse-FT spectrometer allowing keyboard operation (teletype) of most spectrometer features. Associated with use of a minicomputer is of course the required interface of D/A and A/D converters.

6. In addition to the above requirements, the features of any HR-NMR spectrometer such as homogeneous external DC field, facility for spinning the sample to average H_0 gradients, field frequency stabilization, phase-sensitive detection, and isolation of the spectrometer from extraneous sources of RF interference are assumed.

A simple block diagram of a pulse spectrometer based on the above discussion is given in Fig. 12a and a diagram of the associated single coil probe is given in Fig. 12b. The analytical sample resides in the RF coil L_1, which is tuned at the Larmor frequency via C_1 and is coupled to both transmitter and receiver–detection systems. L_2 and C_2 form the circuit for broad-band proton decoupling in the case of ^{13}C NMR. Often the coil is saddle-shaped and surrounds the solenoid L_1. However, more efficient decoupling can be obtained by double-tuning L_1 for both ^{13}C and ^{1}H frequencies and eliminating L_2. Finally, L_3 and C_3 comprise the tuned circuit for an external field/frequency lock system (usually ^{2}H or ^{19}F resonance is employed for the lock system). Internal lock systems in which the nucleus to be "locked on" is contained in the analytical sample (e.g., a deuterium-containing solvent) are used in most HR-NMR spectrometers and require more complex circuitry. The advantage of the internal lock is that both lock sample and analytical sample reside in the same part of H_0, thereby enabling optimization of field homogeneity over the analytical sampling by using the magnet shim coils to optimize the lock signal. In carbon NMR, where a signal often will not be detected in a single pulse, optimization of homogeneity on the lock sample is a necessity. Of course, some applications require an external lock and in these instances coil L_3 should be placed as close as possible to L_1.

An alternative to the single coil probe is the crossed-coil arrangement used in most CW spectrometers. The orthogonal arrangement of a transmitter and a receiver coil allows independent optimization of the requirements of the two systems. However, the arrangement is more complex mechanically and, because it requires a larger volume be enclosed by the transmitter coil, the utilization of the RF power from the transmitter is less efficient than in the single coil arrangement. Hence most pulse spectrometers use a single-coil probe.

In addition, to the requirements outlined in the foregoing discussion, a versatile pulse spectrometer should include the following:

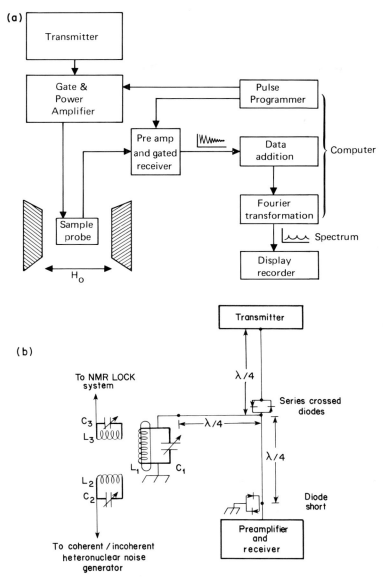

FIG. 12. (a) A block diagram of a pulse-FT NMR spectrometer. (b) The NMR sample probe. The tuned sample circuit consists of L_1 and $C_1 \cdot L_2$, which is resonant with C_2, is used to couple heteronuclear decoupling power into the sample. L_3 and C_3 make up the tuned circuit for the external NMR lock. The back-to-back diodes and quarter-wave sections of RF cable are used to optimize the sample circuitry for both intense signals of short duration (the pulse) and for very weak, slowly decaying signals (the FID signal) (from T. C. Farrar and E. D. Becker, "Pulse and Fourier Transform NMR," Chapter 3. Academic Press, New York, 1971).

1. Broad-band components (e.g., transmitter and receiver) so that a wide frequency range can be accessed, thus making it possible to observe resonance for many nuclei using the same hardware, i.e., a "multinucleus" spectrometer.

2. A probe tunable over a wide frequency range, thus enabling multinucleus observation without requiring change of probes. Of course, the probe should be Dewared to allow variable-temperature experiments.

3. Components and probe capable of delivering and sustaining, respectively, the RF powers required for experiments on solids as well as liquids and solutions.

4. Quadrature phase detection.

5. An interactive disk or other peripheral data storage system for efficient use of the computer (e.g., through use of a time-sharing capability for data processing during delay times in data acquisition) and for kinetic experiments.

4.1.4.5. Computer Requirements and Functions. As indicated in the preceding discussion, the digital computer has become an integral part of a modern NMR spectrometer. Thus in concluding this section on instrumentation we mention the functions and requirements of the computer.[44] Of first concern is the rate at which data points must be collected in order for the Fourier transformation to yield the steady-state NMR spectrum. Sampling theory requires that a data system take at least two points per cycle on a sine wave in order to represent the sine wave of a given frequency. Thus, the sampling rate must be twice the highest frequency to be observed (the so-called Nyquist frequency, N_f); e.g., to observe a 5 kHz spectral width (typical of ^{13}C at 2.3 T) requires a sampling rate of 10 kHz or once every 100 μsec. Should data (i.e., a spectral line) exist outside N_f, they are not ignored but will appear at a lower frequency $N_f - \Delta f$, where Δf is the difference in frequency of the line and the Nyquist value. This result is clearly seen by consideration of Fig. 13, which shows the sampling of three frequencies at a rate $2N_f$. For both frequencies $N_f + \Delta f$ and $N_f - \Delta f$, the computer "sees" the same information. Thus the frequency $N_f + \Delta f$ will be folded back or aliased into the spectral window at $N_f - \Delta f$. Hence in the usual experiment, the spectral width (measured from the RF pulse frequency) is selected to encompass the entire chemical shift range to avoid "fold-back" problems. Of course, noise at frequency $> N_f$ will be aliased and act to decrease S/N ratio by adding to noise already present at lower frequencies. For this reason, the FID signal is passed through an analog filter before it is digitized by the A/D converter. The filter, which is chosen to filter fre-

[44] For a review of the role of the computer in FT-NMR, see J. W. Cooper, *in* "Topics in Carbon-13 NMR Spectroscopy" (G. C. Levy, ed.), Vol. II, Chapter 7. Wiley (Interscience), New York, 1976.

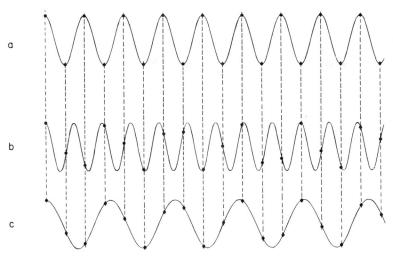

FIG. 13. (a) Sampling a sine wave at the Nyquist frequency N_f; (b) Sampling at N_f of $(N_f + \Delta f)$ (c) Sampling at N_f of $(N_f - \Delta f)$[J. W. Cooper, in "Topics in Carbon-13 NMR Spectroscopy" (G. C. Levy, ed.), Vol. II, Chapter 7. Wiley (Interscience), New York, 1976].

quencies greater than the spectral width, attenuates but does not totally eliminate the fold-back frequencies, e.g., the commonly employed four-pole Butterworth filter attenuates the frequency $1.5N_f$ tenfold.

As we have seen in a previous section, the FID contains information on the frequency separation between the transmitter frequency (carrier frequency) and that of a resonance line. Use of a single-phase detection scheme does not enable which side of the carrier frequency a resonance lines occurs to be determined. Thus if the carrier frequency is placed at the center of the spectral window, the Fourier-transformed spectrum will show all resonances correctly spaced from the carrier but all on the same side of the reference RF. Thus the aliased resonances (i.e., those with absolute frequency lower than the carrier) are not correctly spaced relative to those at higher frequency than the carrier. To circumvent this problem, the carrier frequency is placed to either high or low frequency of all the resonance lines. However, the noise from both sides of the carrier remains in the spectrum; thus, in effect S/N is reduced by $\sqrt{2}$ due to aliasing of the noise about zero frequency (the RF carrier). Since in the detection of low-sensitivity nuclei S/N is critical, most spectrometers employ a RF crystal filter that eliminates the aliased noise by a split-band technique.[45]

An alternative solution to the problem of aliased noise about zero

[45] A. A. Allerhand, R. F. Childers, and E. Oldfield, *J. Magn. Reson.* **11**, 272 (1973).

is quadrature phase detection.[44] Since positive and negative frequencies relative to the carrier do differ in phase, two phase detectors can be used to discriminate on which side of the carrier resonance lines occur. The phase detectors are shifted by 90° so that one channel collects the absorption component and the other the dispersion component of magnetization. Each channel of data is fed to an A/D and then stored in half of the computer memory (obviously, the computer requires dual A/Ds). The data are transformed, using a routine that assumes complex rather than real data, to produce a spectrum (and the associated noise) with proper frequency relationship to the carrier. Hence the RF carrier can be placed in the center of the anticipated spectrum since aliasing is avoided. In addition to the $\sqrt{2}$ increase in S/N gained by this method, the reduced spectral window ($\sim 1/2$) results in a reduction in computer sampling rates and spectrometer pulse power (since lines are closer to the carrier the power to produce an "on-resonant" pulse for all nuclei is less). The detection system is discussed in detail by Schaefer and Stejskal.[46]

The A/D converter measures the instantaneous value of the FID signal at a series of equal time intervals. The voltage measured at each time conversion is digitized by being recorded as a number of amplitude units. The minimum amplitude unit is fixed by the A/D hardware specifications of the minicomputer. Commonly, the digitizer is bipolar with ± 5 V range and analog data are converted to digital data with 12 bit (0.025%), 9 bit (0.2%), or 6 bit (1.6%) precision, depending on experiment requirements. Computer word size usually varies between 16 and 20 bits, while memory size is usually 8K (K = 1024) words for data storage and an additional 4K–8K words for storing the operating program. The size of memory and the digitizer speed (total of time for A/D conversion and addition of the digital data into memory) determine the maximum resolution possible in an experiment. Maximum digitizer rates are usually in the range 50–100 kHz. As pointed out earlier, the total chemical shift spread for ^{13}C at 2.3 T is ~ 5 kHz, requiring a 10 kHz sampling rate; if all 8K of data are used to record the FID, then sampling occurs for $8196/10,000 \sim 0.82$ seconds. Hence, the maximum digital resolution (equal to the inverse of the sampling time) upon 8K Fourier transformation is 1.25 Hz. Better resolution can be achieved with larger computer size, using a smaller spectral width if aliased lines do not overlap regions of interest, or using an interactive disk system attached to the minicomputer to do "piecemeal" transforms to simulate a larger computer core size.

After data acquisition, the computer can be used to "modify" the digi-

[46] E. O. Stejskal and J. Schaefer, *J. Magn. Reson.* **13**, 249 (1974); **14**, 160 (1974).

tized FID signal before transformation.[44] Generally, the mathematical operations include baseline correction, apodization, and digital filtering. Baseline correction is performed by determining the average value of the signal and subtracting this from all data points; this results in removal of any DC component (which contains no spectral information but will introduce a large peak at zero frequency in the spectrum). If the data acquisition time is short compared to the decay time of the analog signal (perhaps due to limited computer memory), the truncation approximates a square-wave decay and results in multiplication of the transformed spectrum by a $\sin \omega T / \omega T$ function (the transform of a square wave) where T is the acquisition time. Thus, around each sharp peak, side lobes or oscillations are created that may partially obscure the resonance line. This effect is substantially reduced by apodization, i.e., multiplying the truncated FID by a function that goes to zero more rapidly than a square wave.[39] For example, the transform of an approximately triangular apodization function is $(\sin \omega T / \omega T)^2$ and thus the side lobes fall off much more rapidly. Since noise is constant in the FID, the S/N ratio decays in the tail of the FID. If the digitized data are multiplied by a decaying exponential, the noise in the tail of the FID is reduced relative to the signal at the beginning. The noise is then reduced in the transformed spectrum; however, the total "effective" sampling time is also reduced and thus an "artificial" broadening of the resonance lines occurs related to the decay constant of the exponential. Conversely, if a S/N reduction can be tolerated, resolution can be enhanced by multiplying the FID by an increasing exponential function.

Following Fourier transformation, the computer is used to correct for phase and amplitude distortion introduced by the electronics of the spectrometer. The phase errors commonly introduced are:

(1) a zero-order or non-frequency-dependent error caused by misadjustment of the phase detector so that a mixture of absorption and dispersion modes is produced;

(2) a first-order or frequency-dependent phase error introduced by the finite length of the 90° pulse and delays interjected between the pulse and the start of receiving data to allow for receiver recovery or prevent feed through of the pulse;

(3) phase errors introduced by the filtering network.

The Fourier transform produces real and imaginary data related to the absorption (v) and dispersion (u) modes of the NMR resonance line by the expressions

$$R = v \cos \theta + u \sin \theta, \qquad I = v \sin \theta - u \cos \theta, \qquad (4.1.57)$$

where θ is the phase shift. The pure absorption mode is generated from a linear combination of sine and cosine transforms.

4.1.4.6. Other Methods. Although conventional CW and pulse-FT methods are employed in the great majority of NMR experiments, several other techniques can be used to obtain spectra. These include the methods of stochastic excitation[47] and rapid-scan spectroscopy.[48]

In the stochastic excitation method, pseudorandom noise is used to modulate the monochromatic RF, the phase, or the frequency to yield white noise excitation of a frequency range. The output noise is not random as it contains signals from resonance lines. Hence cross correlation of the input and output "noise" followed by Fourier transformation of the cross-correlation function yields the NMR spectrum. The instrumentation required has greater similarity to that of CW spectrometers than to pulse-FT instrumentation, making stochastic excitation spectroscopy easier to adapt to a CW system.

In correlation and rapid scan spectroscopy, a spectrum is scanned very rapidly in CW mode so that slow passage conditions do not hold. This produces a ringing spectrum, which is then correlated with a theoretical line or an experimental single line observed under the same rapid scan conditions to produce a normal frequency domain spectrum. The technique has similar sensitivity to FT methods but allows any region of interest to be investigated. Thus problems are avoided with large solvent peaks, giving rise to dynamic range effects, merely by not scanning through the solvent line. The method requires twice the computer memory of conventional NMR since the reference line for correlation must be stored in a block of computer memory equal to that for the ringing spectrum.

4.1.5. Measurement of NMR Parameters

The pulse-FT experiment produces a conventional frequency domain NMR spectrum from which chemical shifts and J couplings are measured by usual methods. The methods are detailed in standard texts[4,14,25,30] on NMR and as such are not described herein. Similarly, sample preparation, the effects of residual oxygen, etc., are not described; again, the reader is referred to standard texts.

The measurement of relaxation times has increased manyfold with the advent of pulse-FT spectrometers. Since relaxation rates are of current interest in studying polymer dynamics in solution, we review briefly the

---

[47] R. R. Ernst, *J. Magn. Reson.* **3**, 10 (1970).

[48] R. K. Gupta, E. Becker, and J. A. Feretti, *J. Magn. Reson.* **13**, 275 (1974).

measurement of these parameters. The usual methodology of determining NMR relaxation parameters is to apply a perturbing RF field to a spin system at equilibrium and to monitor the return to equilibrium of the component of magnetization of interest (i.e., M_z for T_1 or M_{xy} for T_2). Some of the variety of techniques by which the above experiment and variations on this experiment may be implemented using pulse-FT techniques are discussed in this section. Discussion of CW techniques can be found in Lyerla and Levy.[9]

4.1.5.1. Spin–Lattice Relaxation Time. Pulse Fourier transform methods are particularly convenient for the measurement of spin–lattice relaxation times since all resonant nuclei (of a given type) can be observed simultaneously[49] and repetition of the experiment for data accumulation in the case of low-natural-abundance nuclei is readily controlled by software programs. The usual schemes for determining T_1 are the inversion–recovery method and saturation methods.

4.1.5.1.1. INVERSION–RECOVERY (IRFT). The inversion–recovery method in its simplest form consists of a pulse sequence $\{180° - t - 90° - T\}$. A strong H_1 field is turned on for a time to rotate M_z through 180° (resulting in an *inversion* of spin population). Following the 180° pulse, M_z will return to its equilibrium value (along H_0) by the process of spin–lattice relaxation. The return to equilibrium is monitored by a 90° pulse the rotates the nonequilibrium magnetization $M_z(t)$ into the xy plane, where its FID is collected as in a conventional pulse-FT experiment. Fourier transformation thus yields a spectrum whose line intensities are proportional to $M_z(t)$. The growth curve for $M_z(t)$ as a function of t (the time between 180 and 90° pulses) is given by

$$S_t = S_\infty(1 - 2e^{-t/T_1}), \tag{4.1.58}$$

where S_∞ is the signal intensity at equilibrium. T_1 is readily determined from a semilog plot of $(S_\infty - S_t)$ vs. t. Depending on the relative value of t and the T_1 governing the resonance line, the intensity of a line may be negative, zero, or positive, as demonstrated by the spectrum of α-methyl styrene in Fig. 14.

For the case where S/N is of concern, FIDs at a given value of t are coherently added. However, before each repetition of the experiment, a waiting time T is imposed, which is of sufficient duration that complete recovery of M_z to equilibrium for each resonance line is ensured. For best results, T is chosen to be five times the longest T_1 in the compound. Since the value of S_∞ is critical to the experiment and because the delay between experiment repetitions can be long, Freeman and Hill[50] intro-

[49] R. L. Vold, J. S. Waugh, M. P. Klein, and D. E. Phelps, *J. Chem. Phys.* **48**, 3831 (1968).
[50] R. Freeman and H. D. W. Hill, *J. Chem. Phys.* **54**, 3367 (1971).

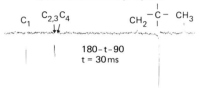

Syndiotactic α-methyl styrene

C_1 $C_{2,3}C_4$ $CH_2 - \overset{\displaystyle |}{\underset{\displaystyle |}{C}} - CH_3$

180-t-90
t = 30 ms

FIG. 14. ^{13}C-NMR spectrum of poly(α-methyl styrene) from a 180-t-90 pulse sequence, t = 30 msec.

duced a variation of the inversion–recovery sequence to minimize the effects of slow drift in spectrometer gain or resolution during a T_1 determination. The sequence is given by

$$\{T-90^\circ_\infty - T - 180^\circ - t - 90^\circ_t\}_n \tag{4.1.59}$$

and data are retained in a differential mode. The FID from a 90° pulse at equilibrium (an S_∞ value) is stored. Then the FID from a $180-t-90$ sequence is subtracted from it, resulting in an FID representative of $S_\infty - S_t$. Thus the signals obtained start with a large positive amplitude at short t and decay exponentially toward zero with increasing t, there being no inverted signals in the spectrum. In this manner, the effect of any drift in the S_∞ value is minimized.

4.1.5.1.2. PROGRESSIVE SATURATION (PSFT). When T_1 values are long (>5 sec), the waiting time T in the IRFT sequence gives rise to considerable inefficiency in acquiring data. One method of circumventing the long recovery times is to employ a progressive saturation scheme. In this experiment, the spin system is subjected to a train of 90° pulses separated by a time t. Under steady-state conditions, i.e., when a dynamic equilibrium has been established between the effects of the pulse and the relaxation during t, the observed signal amplitude after Fourier transformation of the FID is given by[50]

$$S_t = S_\infty(1 - e^{-t/T}1), \tag{4.1.60}$$

and a semilog plot of $(S_\infty - S_t)$ vs. t yields T_1 from the slope. Since the values of t are comparable to those in the IRFT experiment, this is a much faster sequence for repetitive pulsing. However, the sensitivity (or dynamic range) in the PSFT sequence is 50% that of the IRFT sequence since the signal intensity varies only between 0 and S_∞. Thus, four times as many accumulations are necessary to attain the equivalent S/N as the IRFT method. The major requirements in the PSFT experiment to determine T_1 are

1. A steady state must be established before data acquisition is initiated. Practically, this results in not acquiring the first four or five FIDs:
2. Before each 90° pulse, magnetization must be zero in the x,y plane $(M_{xy} = 0)$:
3. The magnetization should all be transverse after the 90° pulse.

Conditions (2) and (3) ensure that the observed time-dependent process is purely a T_1 phenomenon. For ^{13}C NMR, condition (2) is usually fulfilled by use of noise-modulated proton decoupling, which results in rapid decay of M_{xy}, thus avoiding the problem of refocusing of spins or echo formation.[50]

4.1.5.1.3. SATURATION–RECOVERY (SRFT). The PSFT sequence suffers from the drawback that the shortest time interval possible between pulses is controlled by the time for data acquisition which must take place during the interval t. This problem is circumvented in the SRFT scheme[51,52] in which the initial state is one of zero magnetization. Unlike the PSFT technique, where saturation refers to elimination of M_z, the SRFT method refers to saturation in the conventional sense of no net magnetization in any direction. Saturation of the spin system may be accomplished by a burst of nonselective 90° pulses[52] or by application[51] of a 90° pulse followed immediately by a field-gradient pulse along the z axis to dephase M_{xy}. The recovery of magnetization along z is then measured as a function of t, the time between establishment of the saturation condition and imposition of the monitoring 90° pulse. The equation governing signal intensity is again (4.1.60).

Numerous variations on the above sequences now exist that are aimed at circumventing the time delay in the IRFT experiment[53] and providing better dynamic range than the saturation techniques, i.e., the methods seek to optimize the S/N per unit time in the relaxation experiment. However, the method best suited for determining spin–lattice relaxation times for polymers in solution is usually the IRFT sequence or its variant the FIRFT. Because T_1s of proton-bearing carbons in polymers generally

[51] C. G. McDonald and J. S. Leigh, Jr., *J. Magn. Reson.* **9,** 358 (1973).

[52] J. L. Markley, W. J. Horsley, and M. P. Klein, *J. Chem. Phys.* **55,** 3604 (1971).

[53] D. Canet, G. C. Levy, and I. R. Peat, *J. Magn. Reson.* **18,** 199 (1975).

fall in the range 40 msec to 1 sec at ambient temperatures and at magnetic fields in the range 1.4–2.3 T, the IRFT sequence is efficient for polymers containing carbons without attached protons, the FIRFT sequence provides the best utilization of spectrometer time. In the case of solid polymers well below T_g or perhaps for polymer solutions at low temperature and at superconducting fields where τ_C is beyond the T_1 minimum, one of the saturation techniques may be preferred. Levy and Peat[54] have considered the various pulse schemes in detail, including their efficiency and the effect of imperfect 180 and 90° pulses on data accuracy.

4.1.5.2. Rotating-Frame T_1. To measure $T_{1\rho}$, the sample is first allowed to reach an equilibrium magnetization value M_0 in H_0, and then a strong RF pulse (H_1 field) is applied to the sample. After a time equivalent to a $\pi/2$ pulse, a 90° phase shift is introduced into H_1. The $\pi/2$ pulse rotates M_0 into the xy plane and the 90° phase shift then makes H_1 and M_1 (initially equal to M_0) co-linear in the rotating frame. In this situation, the spin system is termed "spin-locked" and continued application of H_1 maintains the collinearity or "forced transitory precession" of M_1 and H_1 in the rotating frame.[39] After spin-locking, the spin system begins to relax in H_1 by interaction with the lattice. This process is monitored by determining the signal amplitude M_1 as a function of the time τ H_1 is applied after the phase shift. This is accomplished by turning off H_1 after τ and opening the receiver to record the FID. Provided H_1 is much greater than the frequency width of the spectrum (so that the $\pi/2$ pulse is resonant for all nuclei), Fourier transformation of the FID from a multiline system yields a spectrum in which the intensity of each resonance line is goverend by its $T_{1\rho}$. The decay of M_1 is usually exponential and given by

$$M_1(\tau) = M_0 e^{-\tau/T_{1\rho}}. \qquad (4.1.61)$$

A semilog plot of $M_1(\tau)$ vs. τ yields $T_{1\rho}$ from the slope of the straight line. Although the relaxation is analogous to T_1, no terms involving the equilibrium magnetization in H_1, $M_1(\infty)$, appear in (4.1.61) because they are too small to be measured and hence are neglected. The $T_{1\rho}$ experiment is shown graphically in Fig. 15.

Other methods are also available to achieve the spin-locking condition.[38] One procedure involves the application of a $\pi/2$ pulse, followed immediately by a second pulse, of length τ, which is phase-shifted by 90° relative to the $\pi/2$ pulse. The advantage over the one-pulse method is that a strong H_1 field can be used for the $\pi/2$ pulse to ensure that magnetization is rotated to the xy plane for all nuclei and then a weaker H_1 field

[54] G. C. Levy and I. R. Peat, *J. Magn. Reson.* **18,** 500 (1975).

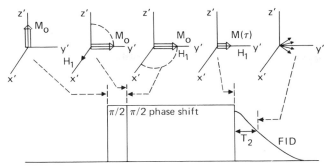

FIG. 15. Pulse sequence used to measure the rotating-frame spin–lattice relaxation time [V. J. McBrierty, *Polymer* **15**, 503 (1974)].

applied for spin-locking. This permits the amplitude of the spin-locking pulse to be varied independently of the $\pi/2$ pulse thus enabling $T_{1\rho}$ to be measured as a function of the H_1 field.

4.1.5.3. Spin–Spin Relaxation Time. Determination of the spin–spin relaxation time proceeds with more difficulty than does that of the spin–lattice relaxation time, since in general for liquids, the natural T_2 decay is masked by the much faster dispersal of M_{xy} components from different regions of the sample by inhomogeneity in H_0. As discussed earlier, the measured T_2 from spectral linewidth ($T_2 = 1/\pi \, \Delta\nu$) is thus referred to as T_2^* to indicate that it is not a true measure of T_2. However, for macromolecules in solution, correlation times are often such that the inhomogeneous contributions do not dominate the linewidth. Thus data on T_2 can be extracted from linewidths without introducing much error by subtracting out the field inhomogeneity effects. From Fig. 5, it is apparent that over a range of correlation time $\sim 10^{-12}$–10^{-7} seconds, $T_2 = T_{1\rho}$ and thus $T_{1\rho}$ measurements yield T_2. At longer correlation times, the width of the resonance line is overwhelming dominated by the natural T_2. A detailed discussion of pulse methods for determination of T_2, including echo techniques, has been given by Freeman and Hill.[55]

4.1.5.4. ^{13}C—$\{^1$H$\}$ NOE. Nuclear Overhauser enhancements for all carbons in a molecule may be determined by comparison of the integrated intensities of a spectrum obtained with continuous wide-band proton decoupling to those obtained with the gated proton decoupling scheme depicted in Fig. 16. In the latter experiment, broad-band irradiation on the protons is applied only during the ^{13}C RF pulse and subsequent data acquisition. This results in a decoupled ^{13}C spectrum but without an NOE

[55] R. Freeman and H. D. W. Hill, *in* "Dynamic NMR Spectroscopy" (L. M. Jackman and F. A. Cotton, eds.), Chapter 5. Academic Press, New York, 1975.

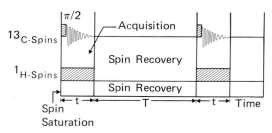

FIG. 16. Schematic of the gated decoupler pulse sequence used to obtain ^{13}C magnetic resonance spectra that are proton-decoupled but have no intensity contributions from NOE. Acquisition time is t and the added delay with the decoupler turned off is T [S. J. Opella, D. J. Nelson, and O. Jardetzky, *ACS Symposium Series* **34**, Chap. 32 (1976)].

since the enhancement builds as a T_1 process *once the decoupling field is applied*. Since the field is on only the few microseconds for the ^{13}C pulse before sampling the FID, no NOE arises. However, the NOE does build during the acquisition period. Thus the recovery time between pulses (decoupler off) must allow for decay of the non-Boltzmann polarization. During the recovery period, (4.1.24) governs the relaxation process. Full consideration of the coupled equation indicates[56] a recovery time about eight times the longest T_1 for the carbons of interest should be allowed to reestablish the Boltzmann population for purposes of recycling the experiment. The experiments are readily carried out under computer control and integrations are performed digitally. (Note that the gated-decoupler sequence is used to provide ^{13}C data for compositional analysis since it eliminates the discrepancies in intensities produced by differential NOEs.[9])

4.2. HR-NMR as a Probe of Polymer Structure

The major focus of NMR in macromolecular chemistry has been the elucidation of structural features of polymer chains. For example, the sensitivity of ^{1}H and ^{13}C chemical shifts and ^{1}H–^{1}H coupling constants to local electronic environment has allowed the identification and quantification of chain branches and configurational sequences in homopolymers and configurational and monomer sequence distributions in copolymers. In turn, these data have provided detailed insight into mechanisms of chain

---

[56] S. J. Opella, D. J. Nelson, and O. Jardetzky, *J. Chem. Phys.* **64**, 2533 (1976).

propagation and advanced the understanding of structure–property relationships for polymeric materials.[4] Studies of polymer structure usually involve measurement of chemical shifts, resonance line intensities, and coupling constants for the magnetic nuclei of a polymer (often as a function of solvent and temperature) and of model compounds closely related to the repeat unit. Computer simulation of spectra aids in elucidation of microstructure by comparison of experimental spectra to calculated spectra based on models of chain propagation. In this chapter, examples of the use of such techniques to determine polymer structure are presented.

4.2.1. Polymerization Mechanism

Many physical properties of polyethylene (PE), in particular the morphology and solid-state properties, are dependent on the nature of both the short and long chain branches. Thus, knowledge of the type and frequency of branching is a critical parameter in establishing structure–property relationships for PE. Until 1972, it was generally accepted that, based on the IR studies of Wilbourn,[57] the dominant short chain branch introduced in the high-pressure polymerization of low-density PE was ethyl. However, in 1972, Bovey et al.[58] published results of a ^{13}C NMR study of chain branching in low-density PE, which conclusively demonstrated that butyl, not ethyl, branches dominate over all other types. The 25 MHz carbon spectrum[59] of PE ($M_w/M_n = 1.12$) is shown in Fig. 17. The major resonance in the spectrum arises from the —CH$_2$— repeat unit, while the presence of the additional resonance lines is a clear indication of chain branching. (Note that the carbon spectrum covers a range of ~30 ppm relative to a proton spectral width of ~2 ppm, indicative of the high resolving power of ^{13}C NMR.)

To unravel the details of the branching, use was made[58–61] of ^{13}C chemical shift studies in linear and branched hydrocarbons[62,63] where it had been found that the ^{13}C shifts could be defined (predicted) in terms of structure-related chemical shift parameters. These semiempirical rela-

[57] A. H. Willbourn, *J. Polym. Sci.* **34,** 569 (1959).
[58] D. E. Dorman, E. P. Otocka, and F. A. Bovey, *Macromolecules* **5,** 574 (1972).
[59] F. A. Bovey, F. C. Schilling, F. L. McCrackin, and H. L. Wagner, *Macromolecules* **9,** 76 (1976).
[60] J. C. Randall, *J. Polym. Sci., Polym. Phys. Ed.* **11,** 275 (1973).
[61] M. E. A. Cudby and A. Bunn, *Polymer* **17,** 345 (1976).
[62] D. M. Grant and E. G. Paul, *J. Am. Chem. Soc.* **86,** 2984 (1964).
[63] L. P. Lindeman and J. Q. Adams, *Anal. Chem.* **43,** 1245 (1971).

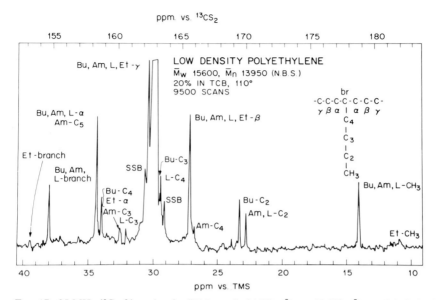

FIG. 17. 25 MHz ^{13}C of low-density PE (sample 5AS5); $\bar{M}_w$ = 15,600, $\bar{M}_n$ = 13,950; 20% in 1,2-trichlorobenzene at 110°; 9500 scans. The diagram at the upper right shows the nomenclature employed for the carbons associated with a branch. The end carbon (i.e., C_1) is designated as CH_3; Et = ethyl, Bu = n − butyl, Am = n − amyl, and L = "long" in the sense described in the text. "SSB" designates spinning side bands of the principal methylene resonance [F. A. Bovey, F. C. Schilling, F. L. McCrackin, and H. L. Wagner, *Macromolecules* **9**, 76 (1976)].

tionships allow prediction of the spectrum of a PE chain with various types of branching. In particular, for a branch in PE, designated as

$$
\begin{array}{cccccc}
\gamma & \beta & \alpha & B & \alpha & \beta & \gamma \\
-CH_2CH_2CH_2- & & -CH- & CH_2CH_2CH_2- \\
& & | \\
& & CH_2(n) \\
& & \vdots \\
& & \cdot \\
& & CH_2(2) \\
& & | \\
& & CH_3(1)
\end{array}
$$

^{13}C NMR is capable of distinguishing the branch carbon B and main-chain carbons α, β, γ to the branch site from the methylene repeat carbon. The chemical shifts for a chain branch are such that it is possible to uniquely identify the branch up through hexyl.[60] Analysis of the spectrum of PE

TABLE I. Branch Frequencies in Two Fractions of Low-Density PE[a]

Sample	5AS5		11AS2	
$\overline{M}_w$	15,600		186,000	
$\overline{M}_n$	13,950		81,790	
$[\eta]^b$	40		134	
Branch points	Per wt. av. molecule	Per 1000 CH_2	Per wt. av. molecule	Per 1000 CH_2
Ethyl	1.3	1.2	19	1.4
n-Butyl	5.8	5.2	75	5.6
n-Amyl	1.7	1.5	25	1.9
"Long"				
by ^{13}C	1.0	0.9	8.3	0.6
by viscosity	0.9	0.8	12	0.8
Total branch points (assuming ^{13}C long-branch value)	9.8	8.8	127	9.5

[a] From F. A. Bovey, F. C. Schilling, F. L. McCrackin, and H. L. Wagner, *Macromolecules* **9**, 76 (1976).

[b] Limiting viscosity number (ml gm^{-1}) in 1,2,4-trichlorobenzene at 130°.

based on the predicted shifts allows assignment of each resonance line of the spectrum in Fig. 17. When coupled with intensity data this results in the description of chain branching given in Table I. The long-branch content refers to branches hexyl or longer and is derived from the intensity of the unique resonance (32.16 ppm) of the $n - 2$ carbon of these branches. The long-branch content determined in this manner is in good agreement with estimates from intrinsic viscosity.

The dominance of *n*-butyl branches in low-density PE, as indicated by the ^{13}C data, provides strong support for the "back-biting" mechanism proposed by Roedel[64] to account for formation of the short-chain branching:

The 1,5 hydrogen transfer, which is energetically favored by the quasi-six-membered ring transition state, appears to be exclusively pre-

[64] M. J. Roedel, *J. Am. Chem. Soc.* **75**, 6110 (1953).

ferred over the 1,4 hydrogen transfer as indicated by the lack of *n*-propyl branches. Introduction of *n*-amyl branches is consistent with a backbiting mechanism during polymerization; however, the mode of introduction of ethyl branches is not clear. In this regard, recent work by Axelson *et al.*,[65] in which a large number of low-density PE samples were studied, indicates that, with respect to branch type and concentration, there is no unique LDPE. Nonetheless, *n*-butyl branches were always present and usually dominant over all others. The presence of ethyl branches in all samples suggests some details of the polymerization mechanism remain to be worked out.

Numerous examples of proton magnetic resonance as a tool to investigate polymerization mechanisms can be found in the book by Bovey[4], while additional examples of ^{13}C studies can be found in a recent book by Randall.[66]

4.2.2. Polymer Chain Stereochemistry

The physical, mechanical, and to some extent chemical properties of vinyl and diene homopolymers depend on their chain configuration[4] or tacticity, i.e., the stereochemical relationship between repeat units introduced by the pseudoasymmetric carbon. If the relative configuration of all chain units is the same (meso relation) the polymer is classed as isotactic; if the configuration of units alternates down the chain, (racemic relation) the polymer is syndiotactic; and if the stereochemical sequence is random, the polymer is termed atactic or heterotactic. (Atactic is also often used to describe a partially random polymer containing a predominance of syndiotactic-type units.) These stereochemical classifications are illustrated in Fig. 18. Clearly, characterization of the stereoregularity of a polymer is necessary for establishing property–structure relationships and developing an understanding of stereospecific polymerization. Of all the spectroscopic techniques, NMR is the most sensitive to stereochemistry and is thus amenable to determining configurational relationships in a polymer chain. In fact, the application to problems of polymer tacticity has been a dominant use of NMR in polymer science.[4,66]

Any statistical description of chain stereochemistry employs the meso and racemic relationships as a base. At the very minimum, NMR experiments allow determination of the dyad character of the chain, but usually provide data on triad distribution, i.e. the number of rr, mr (= rm) relationships. In some cases, hexad sequences have been resolved in pro-

[65] D. E. Axelson, G. C. Levy, and L. Mandelkern, *Macromolecules* **12**, 41 (1979).

[66] J. C. Randall, "Polymer Sequence Determination: Carbon-13 NMR Method." Academic Press, New York, 1977.

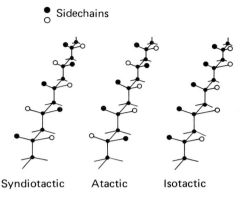

● Sidechains
○

Syndiotactic   Atactic   Isotactic

FIG. 18. Designations of chain configurations in a vinyl polymer: (a) syndiotactic, (b) atactic (irregular), and, (c) isotactic.

ton and carbon[67] studies at high field that allow for detailed tests of stereo-chemical propagation mechanisms. Once the limitation of NMR stereo-sequence resolution is established for a sample and the intensities of the units measured, the resultant description of the chain can be compared to predicted configurational content based on statistical models of propagation. The simplest model is that of a Bernoullian chain in which the stereochemistry of each unit added to the chain is purely statistical, i.e., independent of the stereochemistry at the end of the growing chain. A first-order Markov chain has the configurational content influenced by the chain end stereochemistry; tetrad data are required to test for conformity to this model. A second-order Markov chain is influenced by penultimate effects; pentad data are required to test for conformity. Higher-order Markovian and non-Markovian processes[68] are also possible and have been investigated in a few cases.[69,70]

To illustrate the sensitivity of NMR to chain configuration, several examples of poly(methylmethacrylate), PMMA, spectra are presented. In Fig. 19 are shown the ^{13}C spectra of highly isotactic and highly syndio-tactic PMMA. The isotactic spectrum displays the five resonance lines, one corresponding to each carbon of the repeat unit, expected in a chain in which the same stereochemical relationship exists between each repeat unit. The spectrum of the syndiotactic polymer shows five dominant res-

[67] K.-F. Elgert, R. Wicke, B. Stutzel, and W. Ritter, *Polymer* 16, 465 (1975).

[68] B. D. Coleman and T. G. Fox, *J. Chem. Phys.* 38, 1065 (1963).

[69] For the expressions of the stereosequence relationship to generating parameters, Bovey[4] chapter VIII should be consulted.

[70] I. R. Peat and W. F. Reynolds, *Tetrahdron Lett.* 1359 (1972).

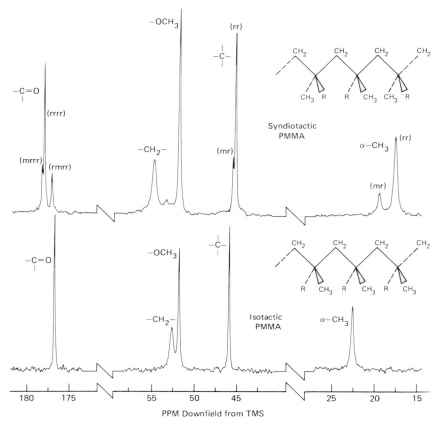

FIG. 19. The 20 MHz proton-decoupled ^{13}C spectra of isotactic and syndiotactic PMMA at 38° in pyridine-d_5 (solvent peaks eliminated). Each spectrum represents 7000 accumulated FIDs. Symbols m and r refer to meso and racemic stereochemical relationships between adjacent repeat units. Structural representation of syndiotactic and isotactic stereochemical triads are included above the respective spectrum; the symbol R represents the methyl ester sidegroup [figure from J. R. Lyerla, Jr., T. T. Horikawa, and D. E. Johnson, *J. Am. Chem. Soc.* **99**, 2463 (1977)].

onances; however, there are also several less intense resonances arising from the presence of meso relationships in the predominantly racemic chain. The carbonyl carbon, which is sensitive to pentad stereochemistry, reveals only rrrr, rmrr, and rrrm sequences, indicating the meso relationships occur largely as isolated defects.[71] The average length of

[71] J. R. Lyerla, Jr., T. T. Horikawa, and D. E. Johnson, *J. Am. Chem. Soc.* **99**, 2463 (1977).

racemic blocks $(\overline{N}_r)$ in the chain as determined from the expression $\overline{N}_r = 1 + 2(rr/mr)$ is 8.5.

Opposed to the rather simple carbon spectra of the stereoregular polymers are those for PMMA prepared by free-radical polymerization and with phenyl magnesium bromide initiator (Fig. 20). Spectral analysis gives the following scheme of carbon sensitivity to tacticity[70]: α-methyl and quaternary (triad); β-CH$_2$(tetrad); ester carbonyl (pentad); methoxy (insensitive). The differences in intensities of corresponding resonance lines in the spectra demonstrate the strong stereochemical dependency on the mode of polymerization for PMMA. In the free-radical case, the distribution of stereosequences is Bernoullian, while for the stereoblock polymer the configurational content conforms to a second-order Markov process.

The proton spectrum of PMMA usually provides equivalent information on tacticity as the carbon spectrum. However, the spectrum of the methylene protons provides an absolute measure of the polymer configu-

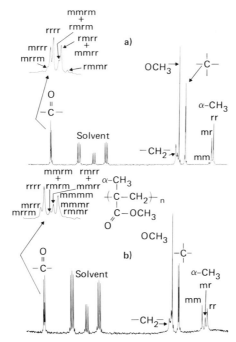

FIG. 20. The 20 MHz proton-decoupled ^{13}C spectra of 10 w/v solutions of PMMA in pyridine-d_5: (a) polymer prepared by free-radical initiation; (b) polymer prepared with phenyl magnesium bromide initiator in toluene at $-78°$C in the presence of diethyl ether. The carbonyl regions of each are shown on an expanded scale.

ration.   That is, without chemical shift evidence from model compounds, there is not an a priori basis for assigning, for example, the most shielded α-CH$_3$ proton and carbon lines to syndiotactic triads.[72]   In Fig. 21, the six possible tetrad configurational sequences are listed.   The two methylene protons are equivalent (homosteric) in rrr and mrm tetrads and appear as singlets.   In the four remaining sequences, the protons are not equivalent (hetereosteric) and appear as quartets due to ^{1}H–^{1}H spin–spin coupling. For meso tetrads, the coupling yields AX spectra while in the racemic case the spectrum is an AB quartet.   These facets of the NMR spectrum are shown in the 220 MHz ^{1}H spectra (Fig. 22) of a predominantly isotactic and a predominantly syndiotactic polymer.   Assignment of the protons to threo (t) (i.e., the proton on the opposite side of the zigzag plane to the ester group in a trans chain conformation) and erythro (e) (i.e., the proton on the same side as the ester group) is also included. Analysis of the CH$_2$ proton spectrum thus, in principle, allows definitive description of the polymer tacticity.   The PMMA situation is relatively simple because there is no J coupling between the CH$_2$ and α-CH$_3$ protons.   In monosubstituted vinyl polymers, the vicinal coupling of α- and β-methylene protons complicates the analysis, however the spectrum can usually be interpreted with aid of deuteration and high-field studies.[4] It should be noted that specific deuteration of materials represents a powerful method to study details of polymerizations.   For example, by means

[72] Schaefer has shown that for a number of monosubstituted vinyl polymers the high-field CH and CH$_2$ carbons are always isotactic triads or dyads while the high-field resonance of the α-substituent is always syndiotactic [J. Schaefer, in "Topics in Carbon-13 NMR Spectroscopy" (G. C. Levy, ed.) Vol. I, p. 155.   Wiley (Interscience), New York, 1974].

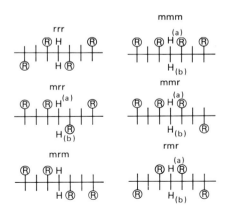

FIG. 21.   Projections of the six possible tetrad configurational sequences in a vinyl polymer chain.

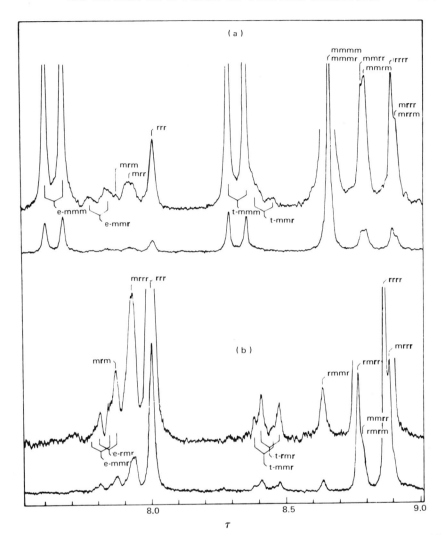

FIG. 22. The 220 MHz ^{1}H spectra of 15% w/v solutions of PMMA in chlorobenzene: (a) polymer prepared by anionic initiator (isotactic) and (b) polymer prepared by free-radical initiator (dominant syndiotactic). The β-methylene protons appear between 7.5τ and 8.5τ, while the α-methyl protons appear between 8.5τ and 9.0τ. Spectra are shown at two vertical scales. [From F. A. Bovey, *Acc. Chem. Res.* **1**, 175 (1968).]

of specific deuteration of monomers, proton NMR studies have been used to elucidate the direction of addition of monomer to the double bond during chain propagation of vinyl systems.[73,74]

4.2.3. Copolymer Sequence Distribution

The application of NMR to compositional analysis of copolymers, terpolymers, etc., is a relatively straightforward and obvious use of the technique. The particular merit of NMR is that polymer composition can usually be assayed at several resonance sites. As more complex polymers are required to meet future material specifications, the chemical shift dispersion of proton and carbon magnetic resonance will make NMR invaluable as an analytical tool in polymer chemistry. However, magnetic resonance provides not only information on the overall composition of copolymer systems, but also data on the distribution of the monomers in the chain.

It is well established that the physical and mechanical properties of copolymers have considerable dependence on the sequence distribution, i.e., the placement of comonomer units along the polymer chain. If one or both of the monomers possess an asymmetric carbon, there is also the possible influence of tacticity on copolymer properties. Thus, knowledge of these "microstructural" features is fundamental to understanding the relationship of polymer property to molecular structure. The simplest description of a copolymer is an overall composition in terms of the monomers and delineation of the chemical and stereochemical nature of the neighboring units of a monomer, i.e., knowledge of the distribution of copolymer triads.

For the general case in which both monomers have an asymmetric carbon, permutation of the sequence of the three units with the two possible tactic relationships results in the 28 triad structures of which eight are magnetically equivalent to others. Thus knowledge of the intensities of 20 triads (listed in Table II) are required for a statistical description of the copolymer. An example of such a copolymer system is methyl methacrylate (E-ester)/methacrylic acid (A-acid). The respective ^{13}C and ^1H spectra of a 35 mole% acid copolymer are shown in Fig. 23. The regions of the spectra providing the most data on microstructure are the carbonyl region of the carbon spectrum (170–190 ppm) and the α-methyl region of the proton spectrum (1–2 ppm). However, in neither spectrum is resolution sufficient to provide a *direct* compositional account of the 20 triad structures necessary to describe the chain microstructure.

[73] F. A. Bovey, *Acc. Chem. Res.* **1**, 175 (1968).
[74] F. A. Bovey, *Prog. Polym. Sci.* **3**, 29 (1971).

TABLE II.   Copolymer Triads (E and A Monomer Units)

			E-Centered			
	EmEmE		EmErE		ErErE	
EmEmA		EmErA		ErEmA		ErErA
	AmEmA		AmErA		ArErA	
			A-Centered			
	AmAmA		AmArA		ArArA	
AmAmE		AmArE		ArAmE		ArArE
	EmAmE		EmArE		ErArE	

Using a combination of specific deuteration studies, stereoregular co-polymer studies (i.e., only one type of stereochemical relationship in the chain), and variable-solvent studies, it is possible to demonstrate that the α-CH$_3$ protons are sensitive to sequence and tactic effects and that each of the 20 possible triads can be assigned to one of the six resolved α-CH$_3$ proton lines in Fig. 23.[75,76] These line assignments were employed[77] as the basis for a spectral simulation program to obtain quantitative data on microstructure using relative intensities of triad lines computer generated from copolymer statistical relationships derived by Bovey.[78] These equations are based on Bernoullian statistics for polymer tacticity and first-order Markov statistics for sequence distribution. The set of equations relates the triad intensities to three parameters: (1) the reactivity ratios of the two monomers, r_E and r_A; (2) σ_m, the probability that a monomer will add to another in a coisotactic or meso (m) fashion; and (3) the monomer feed ratio f at the start of polymerization. To account for the changing copolymer composition and monomer feed ratio with the extent of copolymerization, provision was made in the computer program to calculate triad intensities and recalculate monomer feed ratio at 1% conversion intervals. Triad line intensities at the desired degree of copolymerization were obtained by summing the line intensities over the conversion intervals. The intensity, chemical shift, and linewidth at half-height for each triad were then used as input parameters for a Lorentzian

[75] E. Klesper, W. Gronski, and A. Johnsen, in "NMR. Basic Principles and Progress" (P. Diehl, ed.), Vol. 4, p. 47.   Springer-Verlag, Berlin and New York, 1971.

[76] E. Klesper, A. Johnsen, W. Gronski, and F. W. Wehrli, *Makromol. Chem.* **176**, 1071 (1975), and references therein.

[77] J. R. Lyerla, Jr., *IBM J. Res. Dev.* **21**, 111 (1977).

[78] F. A. Bovey, *J. Polym. Sci.* **62**, 197 (1962).

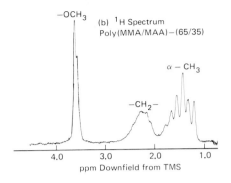

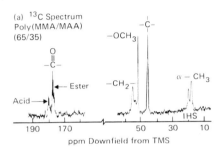

FIG. 23. (a) 20.0 MHz ^{13}C spectrum of pyridine-d_5 solution of poly(MMA-MAA), 65-35 composition, $T = 311$ K. Carbon assignments are indicated for the appropriate resonance region. Symbols: I, isotactic triad; H, heterotactic triad; S, syndiotactic triad. (b) 100 MHz ^{1}H spectrum of the same copolymer, $T = 373$ K. Proton assignments are given above the appropriate resonance regions [from J. R. Lyerla, Jr., *IBM J. Res. Dev.* **21**, 111 (1977)].

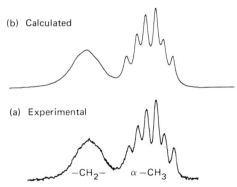

FIG. 24. (a) Experimental spectrum of the α-methyl region of the ^{1}H spectrum of poly(MMA-MAA), 52-48 composition (copolymer prepared in solution in THF). (b) Calculated spectrum [J. R. Lyerla, Jr., *IBM J. Res. Dev.* **21**, 111 (1977)].

lineshape generation program, which produced a proton spectrum that could be compared to the experimental spectrum. In Fig. 24, the result of the simulation of the proton spectrum of 48 mole% acid copolymer is depicted. The description of the chain microstructure is obtained directly from the generation parameters and is given in Table III. A test of the

TABLE IIIa. Triad Composition of Poly(MMA-MMA),[a,b] 52-48
(from Computer Analysis)[c]

Ester-centered triad	Probability[d]	Acid-centered triad	Probability
E^r E^r E	0.090	A^r A^r A	0.070
E^r E^m E	0.045	A^r A^m A	0.035
E^m E^m E	0.006	A^m A^m A	0.004
E^r E^r A	0.166	A^r A^r E	0.153
E^r E^m A	0.042	A^r A^m E	0.038
E^m E^r A	0.042	A^m A^r E	0.038
E^m E^m A	0.010	A^m A^m E	0.010
A^r E^r A	0.077	E^r A^r E	0.083
A^r E^m A	0.038	E^r A^m E	0.042
A^m E^m A	0.005	E^m A^m E	0.005

[a] Polymer prepared by solution polymerization in THF.
[b] Extent of conversion is 30%.
[c] From J. R. Lyerla, Jr., *IBM J. Res. Dev.* **21**, 111 (1977).
[d] Total probability normalized to unity.

TABLE IIIb. Final Input Parameters for Computer Simulation of Poly (MMA-MAA),
52-48 Composition[a]

Reactivity ratios	$r_E = 1.05$
	$r_A = 0.95$
Tactic probabilities	$\sigma_m^{EE} = 0.2$[b]
	$\sigma_m^{AA} = 0.2$
	$\sigma_m^{EA} = 0.2$[c]
	$\sigma_m^{AE} = 0.2$
Monomer feed ratios	$f_E = 0.5$
	$f_A = 0.5$
Linewidths[d]	Ester: 7.5 Hz[e]
	Acid: 9.0 Hz[f]

[a] Polymer prepared by solution polymerization in THF.
[b] Determined from homopolymer spectra.
[c] From α-methyl carbon data.
[d] Linewidth at half-height.
[e] From E^rE^rE resonance line of α-methyl protons.
[f] From differential relaxation rates between ester and acid.

validity of the results is provided by the degree to which they reproduce the analogue carbonyl carbon spectrum. As discussed earlier, the carbonyl of PMMA is sensitive to pentad stereochemistry, as is the carboxyl carbon in the acid. To take account of pentad tacticity and triad sequence requires the chemical shifts of 72 possible resonances. Since the measurement of all these resonances is not possible experimentally, a set of predicted shifts was derived[77] from a set of additivity parameters based on the sequence effects observed in the cotactic copolymers and chemical shifts from stereoregular homopolymers. Using these shift values, generation of the 72 line intensities via the parameters defined by the proton simulation produces the carbon spectrum in Fig. 25. The good agreement with the experimental spectrum supports the description of the copolymer obtained from the proton data.

Application of the procedure of coupling computer simulation and ¹H and ¹³C spectra to (MMA/MAA) cocopolymers has been used to demonstrate that quite different microstructures are obtained by free-radical polymerization in THF vs. toluene.[77] Triad analysis demonstrated that copolymers prepared in THF were homogeneous (i.e., the comonomer feed

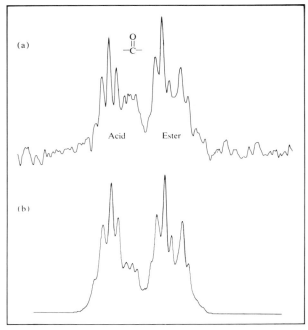

FIG. 25. (a) ¹³C spectrum of the carboxyl–carbonyl region of poly(MMA-MAA), 52-48 composition (copolymer prepared in solution in THF). (b) Computer-calculated spectrum based on triad sequence and pentad tactic effects [J. R. Lyerla, Jr., *IBM J. Res. Dev.* **21**, 111 (1977)].

determined the final composition). However, free-radical polymerization in toluene produced heterogeneous copolymers that were "blocky" in acid during the initial stages of the reaction. Additionally, the sequence distribution does not follow first-order Markov statistics, indicating the possible effects of dimerization of the acid monomer and the association of acid monomers with the carboxyl groups of growing chains in the non-polar solvent.

The discrimination of microstructural features of copolymers represents one of the most potent features of NMR—making it possibly the most valuable tool available for obtaining understanding of copolymerization mechanisms. For additional examples of copolymer applications, the reader is referred to the volumes by Bovey[4] and Diehl.[75]

4.2.4. Polymer Chain Conformation

In the previous discussions, it has been assumed that the molecules may exist in any or all of their possible conformations and that the rate of interconversion is rapid on the NMR time scale. The existence of homo- or heterostericity does not depend on any particular conformer being energetically preferred, nor upon there being a slow equilibration of conformers. However, some information can be obtained from NMR experiments on polymer chain conformation from coupling constants.

In principle, the vicinal proton coupling constant in monosubstituted vinyl polymers can be related to local conformational preferences of the polymer chain. For the all-trans chain in Fig. 26a vicinal coupling constants are designated as J_{AX} and J_{BX}. The observed proton couplings are conformationally averaged values. However, to extract data on conformation requires a number of assumptions, among which is exact staggering of conformers. This assumption then implies the gauche and trans couplings are the same in all conformers (depending only on the dihedral angle between coupling sites) and can be calculated from the Karplus[34] equation. (The freezing out of conformers at low temperatures has not been generally feasible for polymers as the rotational barriers between conformers are usually small.)[74] An additional difficulty arises from the poorer resolution obtained in polymer solutions, which makes fixing of the values of J_{HH} difficult. For these reasons, use of NMR data to obtain conformational data on polymers has not been as successful as in small organic compounds.

An example of conformational data afforded by NMR coupling constants is the study[79] of the ^{19}F and ^{1}H NMR of chlorotrifluoroethylene–isobutylene alternating copolymers. The possible conformations of the C–C bond between the CClF and CH_2 carbons of the chain are given in

[79] K. Ishigure, H. Ohashi, Y. Tabata, and K. Oshima, *Macromolecules* **9**, 290 (1976).

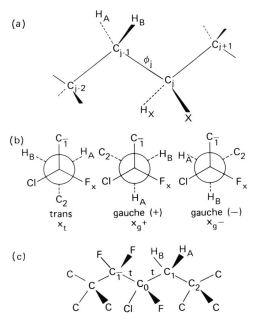

FIG. 26. (a) Designations for the vicinal ^{1}H–^{1}H coupling constants J_{AX} and J_{BX}. (b) Conformations for HF couplings in alternating copolymer of chlorotrifluoroethylene–isobutylene (c) trans conformation. [K. Ishigure, H. Ohashi, Y. Tabata, and K. Oshima, *Macromolecules* **9**, 290 (1976).]

Fig. 26b. The observed vicinal H–F coupling constants (consistent with analysis of the CH$_2$ proton spectrum) are $J_{AX} = 0.5 \pm 1$ Hz, $J_{BX} = 33.5 \pm 1$ Hz. The conformationally averaged coupling constants can be written as

$$J_{AX} = P_t J_g^{HF} + P_{g^+} J_g^{HF} + P_{g^-} J_t^{HF}$$

$$J_{BX} = P_t J_t^{HF} + P_{g^+} J_g^{HF} + P_{g^-} J_g^{HF} \qquad (4.2.1)$$

where the P_i are the probability factors for the conformation with respect to the CClF–CH$_2$ bond and J_i are the values of trans and gauche couplings. Calculation of J_g and J_t for HF vicinal couplings based on the Karplus[34] relationship indicates $J_g \ll J_t$. In fluorinated ethane derivatives $J_g^{HF} \simeq 1$–5 Hz and $J_t^{HF} \simeq 15$–30 Hz. Thus the measured values in the copolymer would suggest one of the geminal protons is consistently trans and the other predominantly gauche with respect to the vicinal F. This situation arises for both trans and gauche (−) conformers; however, two gauche (−) conformations with respect to the CClF–CH$_2$ bond are less stable than a trans conformation because of the repulsive interaction between two methyls—the "pentane effect." Hence the results of coupling constant data suggest a dominance of trans conformations (Fig. 26c).

An alternate approach[80] to correlating NMR coupling constants and conformation is to predict the coupling from conformational populations based on rotational isomeric state theory.[81] For a two-state conformational scheme (trans, gauche), and the couplings depicted in Fig. 26a J_{AX} averaged over all conformations of the jth C–C bond is

$$J_{AX} = P_{t,j}J_{AX,t} + P_{g,j}J_{AX,g} = P_{t,j}(J_{AX,t} - J_{AX,g}) + J_{AX,g}, \quad (4.2.2)$$

where the P are the probabilities of trans and gauche states and the $J_{AX,t \text{ or } g}$ the respective coupling constants. Similarly, for J_{BX}

$$J_{BX} = P_{t,j}(J_{BX,t} - J_{BX,g}) + J_{BX,g}, \quad (4.2.3)$$

resulting in

$$\Delta J = J_{AX} - J_{BX} = P_{t,j}(\Delta J_t - \Delta J_g) + \Delta J_g, \quad (4.2.4)$$

where $\Delta J_t = J_{AX,t} - J_{BX,t}$ and $\Delta J_g = J_{AX,g} - J_{BX,g}$. Assuming, as usual, that J depends only on the dihedral angle θ between the protons of concern, $\Delta J_g = -\Delta J_t$ for $\theta_t = 10°$ and $\theta_g = 110°$, then

$$\Delta J = 2\Delta J_t(2P_{t,j} - 1). \quad (4.2.5)$$

Evaluating ΔJ_t from the Karplus relation and the probabilities from rotational isomeric state theory results in values for isotactic poly(methyl acrylate) of $\Delta J = 1.9$ Hz compared with experimental values[82] of 1.3, 2.5, 1.9 Hz in $CDCl_3$, methyl formate, and benzene, respectively.

Isomeric state theory can also be used to generate conformational data to predict chemical shifts. For example, Yoon and Flory[83] have computed the proton chemical shifts of methine and aromatic protons of the central $>CHC_6H_5$ group in the meso nonad of polystyrene and in the nonads rrm_6, $mrrm_5$, m_2rrm_4, and m_3rrm_3, i.e., those containing a racemic triad isolated in a meso chain. The calculated shifts match the observed spectra both in order of occurrence of the resonances for the nonads and, approximately, in the magnitudes of the shifts between resonances. Tonelli et al.[84] have applied the rotational isomeric state model to predict the ^{13}C chemical shifts in PVC.

[80] D. Y. Yoon, U. W. Suter, P. R. Sundararajan, and P. J. Flory, *Macromolecules* **8**, 784 (1975).

[81] P. J. Flory and A. D. Williams, *J. Am. Chem. Soc.* **91**, 3118 (1969).

[82] T. Yashino, Y. Kikuchi, and J. Koniyama, *J. Phys. Chem.* **70**, 1059 (1966).

[83] D. Y. Yoon and P. J. Flory, *Macromolecules* **10**, 562 (1977).

[84] A. E. Tonelli, F. C. Schilling, W. H. Starnes, Jr., L. Shepherd, and I. M. Plitz, *Macromolecules* **12**, 78 (1979).

4.3. HR-NMR as a Probe of
Polymer Molecular Dynamics

The use of HR-NMR methods to investigate polymer chain dynamics has not been exploited to nearly the same degree as their use to elucidate chain structure. The majority of relaxation studies have been carried out on bulk samples utilizing ^{1}H broad line NMR methods.[85,86] These techniques do not generally provide motional information at individual sites in the polymer repeat unit due to insufficient resolution and the fact that the proton relaxation times in the solid are an average over all protons due to spin diffusion (see Chapter 4.4).[87] For polymers in solution, ^{1}H relaxation times can be measured for individual protons of the repeat unit[88]; however, the interpretation of the data usually requires separation of inter- and intramolecular contributions to the T_1 and T_2 processes. In this regard, interpretation of carbon relaxation parameters for carbons with directly bonded protons is less complex due to dominance by local C–H dipolar interactions. Thus, increasing interest, both from the experimental and theoretical viewpoints, has been focused on use of ^{13}C HR-NMR to study motion of macromolecules in solution. Information derived from NMR relaxation measurements complement those made by other techniques such as ESR of spin-labeled polymers,[89] neutron scattering,[90,91] fluorescence depolarization,[92] and dielectric,[93] acoustic,[94] and mechanical[95] relaxations.

In this chapter, we discuss characteristics of NMR relaxation times of high polymers and illustrate that the motional information derived is important in advancing the understanding of structure–property relation-

[85] I. Y. Slonim and A. N. Lyubimov, "The NMR of Polymers." Plenum, New York, 1970.

[86] W. P. Slichter, in "NMR. Basic Principles and Progress" (P. Diehl, ed.), Vol. 4, p. 209. Springer-Verlag, Berlin and New York, 1971.

[87] D. W. McCall, Acc. Chem. Res. **4**, 223 (1971).

[88] K. Hatada, Y. Okamoto, K. Ohta, and H. Yuki, J. Polym. Sci., Polym. Lett. Ed. **14**, 51 (1976).

[89] A. T. Bullock and G. G. Cameron, in "Structural Studies of Macromolecules by Spectroscopic Methods" (K. J. Ivin, ed.), Chapter 15. Wiley (Interscience), New York, 1976.

[90] J. S. Higgins, in "Structural Studies of Macromolecules by Spectroscopic Methods" (K. J. Ivin, ed.), Chapter 2. Wiley (Interscience), New York, 1976.

[91] G. Allen, C. J. Wright, and J. S. Higgins, Polymer **15**, 319 (1974).

[92] B. Valeur and L. Monnerie, J. Polym. Sci., Polym. Phys. Ed. **14**, 11 and 29 (1976).

[93] G. Williams and D. C. Watts, "NMR. Basic Principles and Progress" (P. Diehl, ed.), Vol. 4, p. 271. Springer-Verlag, Berlin and New York, 1971.

[94] J. H. Dunbar, A. M. North, R. A. Pethrick, and D. B. Steinhauer, J. Chem. Soc., Faraday Trans. 2, **71**, 1478 (1975).

[95] I. M. Ward, "Mechanical Properties of Solid Polymers." Wiley (Interscience), New York, 1971.

ships for polymeric materials. For the most part, the discussions concern carbon relaxation parameters. However, mention is also made of T_1 data for protons and other nuclei.

4.3.1. Chain Segmental Motion

Since the carbon relaxation times of polymers in solution, the melt and the bulk, well above T_g are dominated by $^{13}C-^1H$ intramolecular dipolar interactions, the relaxation equations (4.1.27), (4.1.34), and (4.1.46) developed for the reorientation of a C–H vector "embedded" in a rigid structure that undergoes isotropic rotational motion provide a first-order description of the motional parameters (T_1, T_2, NOE) for polymer solutions. Clearly, this is a gross simplification for a complex molecular system such as a polymer chain, where the reorientational motion of a C–H vector may depend on multiple internal[96,97] (e.g., rotations associated with conformational changes and torsional motions within a given conformation) as well as overall molecular reorientation processes. In such instances, it has become commonplace to refer to the correlation time derived from experimentally determined T_1 values as an "effective" correlation time that represents the weighted average of the correlation times for the individual motions that reorient the C–H vector of concern, i.e., $\tau_{eff} = \Sigma_k C_k \tau_k$, where the C_k are orientation- and probability-dependent coefficients.[98] Although simplistic, the model does provide insight into chain motion.

For backbone carbons of high polymers, the measured τ_{eff} is associated with chain segmental motion and thus provides information on the "local" flexibility of the chain. The correspondence of τ_{eff} with segmental motion is illustrated in Table IV, where the T_1 values for the carbons[99] of polyisobutylene (PIB) in CCl_4 are listed as a function of M_w.[100] As the T_1 results can be directly translated into τ_{eff} data (4.1.30), it is clear the effective correlation times for the two backbone carbons of the repeat unit become independent of molecular weight for $M_w \geqslant 3000$. Similar ^{13}C NMR studies on atactic polystyrene[101] and poly(ethylene glycol)[102] show that backbone τ_{eff}

[96] T. M. Connor, *Trans. Faraday Soc.* **60**, 1574 (1964).

[97] J. Schaefer, *Macromolecules* **6**, 882 (1973).

[98] J. R. Lyerla, Jr., H. M. McIntyre, and D. A. Torchia, *Macromolecules* **7**, 11 (1974).

[99] Data presented in this chapter refer, in the main, to carbons with directly bonded protons. For quaternary carbons, interpretation of results depends on proper summing of the various C–H interactions.

[100] Y. Inoue, A. Nishioka, and R. Chujo, *J. Polym. Sci., Polym. Phys. Ed.* **11**, 2237 (1973).

[101] A. Allerhand and R. K. Hailstone, *J. Chem. Phys.* **56**, 3718 (1972).

[102] J. Lindberg, I. Siren, E. Rahkamaa, and P. Tormala, *Angew. Makromol. Chem.* **50**, 187 (1976).

TABLE IV. ^{13}C Spin–Lattice Relaxation Times of PIB 25% (w/v) Solution in Carbon Tetrachloride at 40°C[a]

Molecular weight	T_1 (sec)[b]		
	—CH$_2$—	—C—	—CH$_3$
1,355	0.199	1.79	0.305
2,750	0.172	1.63	0.249
200,000	0.168	1.51	0.241
200,000[c]	0.170	1.68	0.248
700,000	0.161	1.51	0.236

[a] From Y. Inoue, A. Nishioka, and R. Chujo, *J. Polym. Sci., Polym. Phys. Ed.* **11**, 2237 (1973).

[b] Error = ±10%.

[c] Degassed sample; other samples were measured without attempting removal of dissolved oxygen.

values in these polymers also become constant in the molecular weight range 3000–10,000.   The lack of dependence on $\overline{M}_w$ and the magnitudes of the room temperature τ_{eff} values for backbone carbons (0.01–10 nsec) are clear indications that the carbon T_1 process for high polymers in solution is dominated by short-range local motions of the chain and not by macro-Brownian motion of the entire chain.

Listed in Table V are τ_{eff} values for backbone carbons for a number of polymers in solution.   There is a clear correlation between the τ_{eff} data and chain flexibility, e.g., relatively "stiff" chains such as PMMA, polystyrene, and polyvinylcarbazole are characterized by much slower reorientational times than polymers such as PE, PPO, and polyisoprene.   Also, the data demonstrate that the isotropic model for motion is not without merit, as data derived from both CH and CH$_2$ carbons (in the appropriate compounds) are in good agreement.   Apparently, there are sufficient different internal modes that segmental motion in many cases is reasonably well characterized by an isotropic walk.   Measurement of T_1 values as a function of temperature allows activation energies of skeletal motions to be estimated.   For example, in Table VI are ΔE values for several polymers derived from such data based on the assumption of an Arrhenian activation process, i.e., $\tau_{eff} = A e^{\Delta E/RT}$, where A is a preexponential factor.   The low activation energies for PE and polyethylene glycol are consistent with the absence of sidegroups that can restrict backbone motion, while the high value in PIB results from the strong steric repulsions between adjacent methyl groups.   The effect of a sidegroup on reducing backbone flexibility is readily apparent in Fig. 27, where the case of

TABLE V. τ_{eff} Values for Backbone Carbons of Various Polymers

Polymer	State	τ_{eff} (nsec)	
		CH	CH$_2$
Atactic polystyrene	Cl$_2$C=CCl$_2$ (44°)	0.57	0.50
Isotactic polystyrene	toluene-d$_8$ (35°)	0.51	—
poly acrylonitrile	DMSO (50°)	0.31	0.28
poly(vinyl chloride)	DMSO (50°)	0.39	0.34
poly isobutylene	CCl$_4$	—	0.19
poly(vinyl alcohol)	DMSO (50°)	0.43	0.31
poly(vinyl alcohol)	D$_2$O (50°)	0.32	0.23
poly(but-l-ene)	PCE(40°c)	—	.48
poly(but-l-ene sulfone)	CDCl$_3$ (40°)	23.0	22.0
poly(styrene peroxide)	CHCl$_3$ (33°)	0.20	0.20
poly(α-methyl styrene)	CDCl$_3$ (38°)	—	0.68
poly(vinylidene fluoride)	DMF (41°)	—	0.08
Poly(ethylene oxide)	neat	—	0.09
poly(propylene oxide)	CDCL$_3$ (30°)	—	0.06
poly propylene	1,2,4-trichlorobenzene (46°)	—	0.14
poly(methyl methacrylate)			
syndiotactic	Pyridine-d$_5$ (38°)	—	0.60
isotactic	Pyridine-d$_5$ (38°)	—	0.38
poly(vinyl carbazole)	CDCl$_3$ (30°)	0.95	0.95
poly(diphenylphenylene oxide)	CDCl$_3$	0.3	—
poly(dimethylphenylene oxide)	CDCl$_3$	0.3	—
poly oxymethylene	CF$_3$—CH(OH)—CF$_3$	—	0.08

TABLE VI. Activation Energies for Skeletal Motions in Polymers from ^{13}C Relaxation Data

Polymer	State	ΔE (kcal/mole)
PE	O-dichlorobenzene	2.4[a]
Polypropylene	O-dichlorobenzene	3.4[a]
PIB	O-dichlorobenzene, CDCl$_3$	4.9[a], 4.3[b]
Poly(propylene oxide)	CDCl$_3$	5.0[b]
Poly(ethylene glycol)	melt	0.36[c]
Polystyrene (atactic)	PCE	6.4[d]
PMMA (atactic)	O-dichlorobenzene	6.7[d]

[a] Y. Inoue, A. Nishioka, and R. Chujo, *J. Polym. Sci., Polym. Phys. Ed.* **11**, 2237 (1973).
[b] F. Heatley, *Polymer* **16**, 493 (1975).
[c] J. Lindberg, I. Siren, E. Rahkamaa, and P. Tormala, *Angew. Makromol. Chem.* **50**, 187 (1976).
[d] F. Heatley and A. Begum, *Polymer* **17**, 399 (1976).

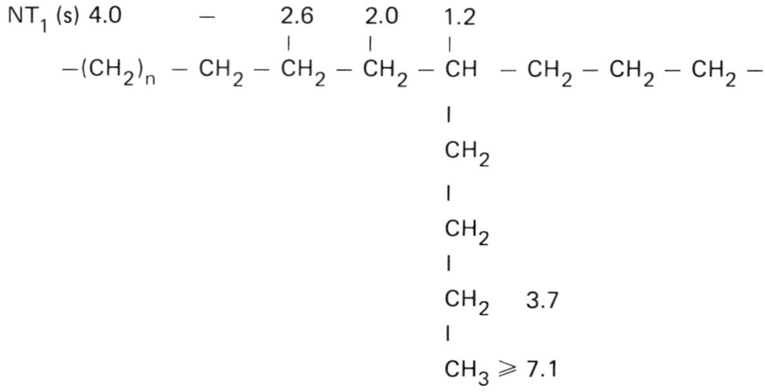

NT_1 (s) 4.0   —   2.6   2.0   1.2
$$-(CH_2)_n - CH_2 - CH_2 - CH_2 - CH - CH_2 - CH_2 - CH_2 -$$

$$| $$
$$CH_2$$
$$|$$
$$CH_2$$
$$|$$
$$CH_2 \quad 3.7$$
$$|$$
$$CH_3 \geqslant 7.1$$

FIG. 27. The ^{13}C NT_1 values in branched PE at 118°C and 67.9 MHz [D. E. Axelson, L. Mandelkern, and G. C. Levy, *Macromolecules* **10**, 557 (1977)].

branched PE is illustrated. Near the butyl branch, the backbone carbons have NT_1 ($\propto \tau_{\text{eff}}^{-1}$) values two to three times shorter than the value characterizing the linear portion of the chain.

An instructive application of ^{13}C spin–lattice relaxation to assess backbone motion is the case of 1 : 1 alternating poly(olefin sulfones). These polymers do not show a high-frequency (megahertz) dielectric loss region as is normally characteristic of segmental motions in flexible polymers having dipole components perpendicular to the chain direction.[103,104] Instead, the maximum loss is found to be in the kilohertz region (which is associated with rotary diffusional motion of polymer chains) and is found to be strongly molecular weight dependent. Thus, the dielectric data would suggest that segmental motion in these polymers is severely constrained, presumably by the sulfone units. However, this view of slow segmental motion in these polymers is inconsistent with such properties as T_g.[105] The glass-transition temperatures of poly(olefin sulfones) are similar to those for syndiotactic PMMA, which has segmental motions in solution characterized by 10^{-9}s correlation times.[71]

Recently (1977), Cais and Bovey,[105] Stockmayer *et al.*,[106] and Fawcett *et al.*[107] have examined the flexibility of the backbone of 1 : 1 alternating poly(olefin sulfones) by ^{13}C T_1 methods. In all cases, the T_1 results were

[103] T. W. Bates, K. J. Ivin, and G. Williams, *Trans. Faraday Soc.* **63**, 1976 (1967).
[104] T. W. Bates, K. J. Ivin, and G. Williams, *Trans. Faraday Soc.* **63**, 1964 (1967).
[105] R. E. Cais and F. A. Bovey, *Macromolecules* **10**, 757 (1977).
[106] W. H. Stockmayer, A. A. Jones, and T. L. Treadwell, *Macromolecules* **10**, 762 (1977).
[107] A. H. Fawcett, F. Heatley, K. J. Ivin, C. D. Stewart, and P. Watt, *Macromolecules* **10**, 765 (1977).

found to be largely independent of molecular weight and consistent with segmental motions characterized by τ_{eff} values in the 10^{-8}–10^{-9} second region. These correlation times are four orders of magnitude shorter than those from the dielectric results. Rationalization of both sets of data requires the presence of segmental motions that reorient C–H vectors without giving rise to electric dipole relaxation. Motions of this type can be postulated,[105,106] e.g., that in Fig. 28, which involves five backbone bonds (C–S–C–C–S–C). Concerted segmental transitions about two C–S bonds allows interconversion of the three conformational states ttt, g^+tg^-, and g^-tg^+. The C–C bond always remains trans and thus neighboring pairs of sulfone dipoles cancel in accord with the small dipole moments observed in polysulfones.[108] Hence the NMR relaxation data indicate that segmental motions are not as severely hindered as dielectric data would suggest; however, T_1 data on poly(but-1-ene sulfone) vs. poly(but-1-ene)[105] indicate an order of magnitude less flexibility in the sulfone chain, which may be ascribed to the constraint of backbone C–C bonds fixed in a trans conformation.

It is clear from the proton and carbon relaxation data for different stereoregular PMMAs (Table VII) that the tacticity of a polymer chain can affect T_1. The data suggest a greater degree of restriction on local motions in the syndiotactic chain. In other poly(alkyl methacrylates) the isotactic chain is also found to be more flexible, although as the bulkiness of the ester side group is increased there is a substantial reduction in backbone motion irrespective of stereochemistry. Similar studies on less stereoregular polymers have shown that longer relaxation times are ob-

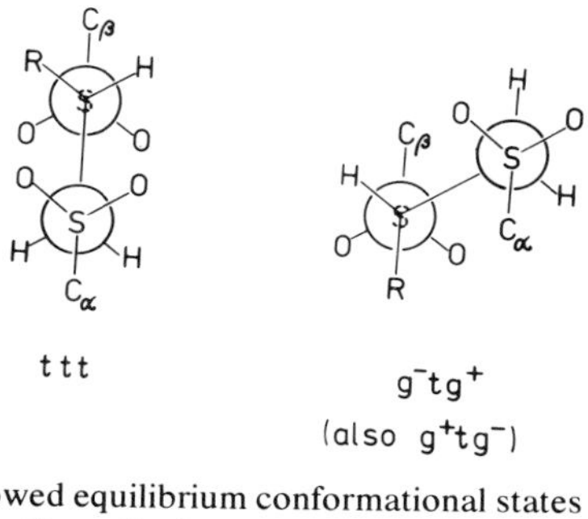

ttt

g^-tg^+

(also g^+tg^-)

FIG. 28. Proposed allowed equilibrium conformational states for poly(α-olefin sulfone)s in solution. Note that the sulfone dipoles cancel and that during the transitions ttt $\rightleftarrows$ g^-tg^+ $\rightleftarrows$ g^+tg^- there is no net reorientation of these dipoles (dielectrically inactive motions), but there is a reorientation of backbone C–H vectors (^{13}C NMR active motions) [R. E. Cais and F. A. Bovey, *Macromolecules* **10**, 757 (1977)].

[108] A. H. Fawcett and K. J. Ivin, *Polymer* **13**, 439 (1972).

TABLE VII.   Effect of Chain Tacticity on NMR Relaxation Times
in Methyl Methacrylates

Polymer	Configuration	Proton[a] T_1 values (msec)			Carbon T_1 values (msec)				
		α-CH$_3$	CH$_2$	OCH$_3$	α-CH$_3$	CH$_2$	OCH$_3$	C	C=O
PMMA	syndiotactic	177	125	541	200[b]	130	1120	1530	—[d]
	isotactic	321	206	745	377[b]	210	1500	1870	—[d]
PMMA	syndiotactic[c]				46	42	448	581	1100
	isotactic[c]				124	64	630	940	1840
PEMA[a,e]	syndiotactic	—			219	86			
	isotactic	—			419	192			
PiPMA[a,e]	syndiotactic	—			175	77			
	isotactic	—			347	129			
Pt-BMA[a,e]	syndiotactic	—			132	36			
	isotactic	—			237	74			

[a] In toluene-d_8; temperature, 110°C; data from K. Hatada, Y. Okamoto, K. Ohta, and
H. Yuki, *J. Polym. Sci., Polym. Lett. Ed.* **14**, 51 (1976).

[b] In pyridine-d_5; temperature 100°C; J. R. Lyerla, Jr., T. T. Horikawa, and D. E. Johnson,
J. Am. Chem. Soc. **99**, 2463 (1977).

[c] In pyridine-d_5; temperature, 38°C; J. R. Lyerla, Jr., T. T. Horikawa, and D. E. Johnson,
J. Am. Chem. Soc. **99**, 2463 (1977).

[d] Not measured.

[e] Data from K. Hatada, T. Kitayama, Y. Okamoto, K. Ohta, Y. Umemura, and H. Yuki,
Makromol. Chem. **178**, 617 (1977).

tained for isotactic vs. syndiotactic stereosegments of a polymer chain
(e.g., triads in PMMA and pentads in polypropylene),[109,110] again empha-
sizing the dependence of T_1 on local motions of the chain.   The differences in relaxation rate are presumably associated with the differing
rates of conformational transitions within a configurational sequence.
Relaxation times have also been shown to be dependent on solvent compatibility in triblock copolymers such as styrene/butadiene/styrene.[111]
The carbons in different resolved-chain sequences in polybutadiene (i.e.,
distribution of cis-1,4 and -1,2 units)[112] and styrene/butadiene copolymers[113] also are found to have T_1s that depend on the nature of the
neighboring units.   This composite of results clearly indicates the sensi-

[109] J. C. Randall, *J. Polym. Sci., Polym. Phys. Ed.* **14**, 1693 (1976).

[110] J. R. Lyerla, Jr., unpublished data.

[111] F. Heatley and A. Begum, *Makromol. Chem.* **178**, 1205 (1977).

[112] W. Gronski, G. Quack, N. Murayama, and K.-F. Elgert, *Makromol. Chem.* **176**, 3605
(1975).

[113] W. Gronski, N. Murayama, C. Mannewitz, and H.-J. Cantow, *Makromol. Chem.,
Suppl.* **1**, 485 (1975).

tivity of relaxation times to local chain motion and thus as a source of information on polymer chain dynamics in solution. While other techniques provide such information, the power of the NMR relaxation studies arises from the resolution of structural details of the chain (due to chemical shift sensitivity), which thereby allows molecular motion as a function of chain microstructure, temperature, solvent, etc., to be accessed at multiple sites in the repeat unit.

4.3.2. Sidechain Reorientational Motions

Determination of τ_{eff} values for sidechain carbons of polymers allows qualitative and semiquantitative information on the rotational freedom of the sidegroup to be obtained. In the effective correlation-time model, the motions that reorient C–H vectors in the backbone are presumed equally effective in reorienting C–H vectors in the sidechain. Thus, for "rigid" sidegroups (i.e., groups for which internal reorientation modes are slow relative to backbone motion), τ_{eff} for sidechain and backbone carbons would be the same. However, if the sidechain has internal reorientational modes that reorient (with respect to H_0) the C–H vectors at rates comparable to or greater than backbone motion, the τ_{eff} results will differ from backbone values. (If all motions contributing to the T_1 process are in the extreme narrowing limit, τ_{eff} value will be shorter and NT_1 values longer for the sidechains.) These effects are illustrated by the NT_1 data listed in Table VIII for several polymers having either a methyl or phenyl sidegroup. Among results of interest are those for syndiotactic PMMA and poly(α-methyl styrene) (PAMS), poly(methacrylonitrile) (PMAN), and poly(2,6-dimethyl phenylene oxide) (PMPO), all of which have comparable backbone segmental motion as judged by NT_1 data. On the other hand, the methyl groups have quite different rotational freedom, ranging from an essentially rigid group in PAMS (on the time scale of backbone motion) to a freely rotating group in PMPO.[114] Differences in steric interactions with the backbone and adjacent sidegroups appear to correlate with the various relaxation rates, e.g., strong steric interactions between sidegroups are expected for PAMS, to a lesser extent for the CH_3–ester group interactions in PMMA and for the CH_3–cyano group in PMAN, and still weaker interactions exist for PMPO.

The results for phenyl sidegroups[114,115] indicate that rotations of the ring about the chain–phenyl bond occur at rates comparable to backbone motion for poly(styrene peroxide) (PSO) and poly(2,6-diphenyl phenylene oxide) (PPPO), but that sidegroups are rigid in the case of PAMS and PS.

[114] F. Lauprêtre and L. Monnerie, *Eur. Polym. J.* **11**, 845 (1975).
[115] R. E. Cais and F. A. Bovey, *Macromolecules* **10**, 169 (1977).

TABLE VIII. NT_1 Values for Polymers with Methyl or Phenyl Sidegroup

Polymers with methyl groups			$NT_1{}^a$	
Polymer	Solvent ($T°$C)	Backbone	Methyl	Ref.
Syndiotactic PMMA	pyridine (38°)	84 (CH$_2$)^b	138	71
Isotactic PMMA	pyridine (38°)	128 (CH$_2$)	372	71
Syndiotactic poly(α-methyl styrene)	CDCl$_3$ (38°)	78 (CH$_2$)	81	110
Atactic polymethacrylonitrile	CDCl$_3$ (38°)	82 (CH$_2$)	330	110
Polypropylene	1,2,4 TCB (46°)	290 (CH)	2250	109
Poly(propylene oxide)	CDCl$_3$ (30°)	800 (CH)	3100	118
Polyisobutylene	CDCl$_3$ (30°)	280 (CH$_2$)	645	118
Poly (2,6-dimethyl phenylene oxide)	CDCl$_3$ (30°)	93 (ring)	2040	114
Poly(methacrylic acid)	H$_2$O (26°)	58 (CH$_2$)	99	d

Polymers with phenyl groups			Ring		
			C$_{2,3}$	C$_4$	
Polystyrene	CHCL$_3$ (33°)	105^c	120	110	115
Poly (α-methyl styrene)	CDCl$_3$ (38°)	78 (CH$_2$)	76	75	110
Poly(styrene peroxide)	CHCl$_3$ (33°)	240^c	390	240	115
Poly(2,6-diphenyl phenylene oxide)	CDCl$_3$ (30°)	86 (ring)	122	90	114
Polyphenylthiirane	CDCl$_3$ (25°)	196^c	223, 224	199	116

^{a} In msec.
^{b} Carbon to which data are referenced.
^{c} Average of CH/CH$_2$ data.
^{d} Data taken from J. D. Cutnell and J. A. Glasel, *J. Am. Chem. Soc.* **99**, 42 (1977).

Only the ortho and meta carbons reflect the internal motion as the para C–H vector lies along (or nearly so) the axis of reorientation—the result of the coincidence being that the motion does not reorient the para C–H vector with respect to H_0 and hence does not affect relaxation of this carbon.   The introduction of the peroxide linkage into poly(styrene) allows the phenyl group additional rotational motion as a peroxide oxygen is less sterically restrictive than a β-methylene group.   A similar result is obtained for poly(phenylthiirane) polymers.[116]

If the two-correlation time model derived by Woessner[117] for a methyl group attached to an isotropically reorienting body is assumed to be applicable for a polymer, estimates of the rotational barriers can be determined.   In this case, (4.1.17) becomes

$$J(\omega) = \frac{A\tau_{\text{eff}}}{1 + \omega^2\tau_{\text{eff}}^2} + \frac{B\tau_B}{1 + \omega^2\tau_B^2} + \frac{C\tau_C}{1 + \omega^2\tau_C^2} \qquad (4.3.1)$$

[116] R. E. Cais and F. A. Bovey, *Macromolecules* **10**, 752 (1977).
[117] D. E. Woessner, *J. Chem. Phys.* **36**, 1 (1962).

TABLE IX. Activation Energies for Methyl Carbons in PPO and PIB[118]

Sample	Activation energy (kJ/mol)		Barrier from neutron scattering (kJ/mol)[119]
	CH_2	CH_3	
PPO			
bulk	27 ± 2	10.5 ± 2	13
0.3 wt fraction in $CDCl_3$	21 ± 2	11 ± 2	
PIB			
bulk	24 ± 2	25 ± 2	24
0.3 wt fraction in $CDCl_3$	18 ± 2	18 ± 2	

where A, B, and C are geometrical factors equal to $0.125(3 \cos^2\theta - 1)^2$, $3 \sin^2\theta \cos^2\theta$, and $0.75 \sin^4\theta$, respectively, with θ the angle between the methyl C–H vector and the rotation axis and

$$\tau_B^{-1} = \tau_{\text{eff}}^{-1} + (6\tau_i)^{-1}, \qquad (4.3.2)$$

$$\tau_C^{-1} = \tau_{\text{eff}}^{-1} + 2(3\tau_i)^{-1}. \qquad (4.3.3)$$

If τ_i is the correlation time for methyl reorientation based on a stochastic diffusion process, then $\tau_B^{-1} \neq \tau_C^{-1}$ and (4.3.2) and (4.3.3) are used in (4.3.1). If τ_i is based on a jump model between three equivalent sites, (4.3.2) is also valid for τ_C^{-1}. Heatley[118] has applied (4.3.1) with $\tau_B^{-1} = \tau_C^{-1}$ to PIB and poly(propylene oxide) (PPO) T_1 data obtained as a function temperature on bulk polymers and in $CDCl_3$ solutions. T_1 data for the CH_2 carbon in PIB and the CH carbon in PPO were used to compute τ_{eff} and then methyl T_1 data used to determine τ_i. Activation energies were determined from the Arrehenius equation and are given in Table IX. The barriers are in good agreement with those obtained by neutron inelastic scattering.[119] For PPO the much lower ΔE value for τ_i than τ_{eff} suggests that methyl rotation is determined primarily by intramolecular factors. The methyl groups in PPO are well separated and thus have small interaction; in addition, reorientation does not hinder rotation of main chain bonds. Although in PIB the strong steric repulsion between neighboring methyls forces the chain bonds out of the staggered conformation by $\sim 20°$,[120] methyl rotation is still strongly hindered. From the τ_{eff} and τ_i values derived from T_1 data obtained as a function of dilution of the polymers in $CDCl_3$, Heatley concluded that methyl rotation proceeds

[118] F. Heatley, *Polymer* **16**, 493 (1975).

[119] J. S. Higgins, G. Allen, and P. N. Brier, *Polymer* **13**, 157 (1972).

[120] T. Tanaka, Y. Chatani, and H. Tadokoro, *J. Polym. Sci., Polym. Phys. Ed.* **12**, 515 (1974).

most freely by a cooperative process involving neighboring methyls and motion about chain bonds. However, analysis[121] of more recent results on higher-molecular-weight samples of PIB and at different values of H_0[122] indicate the backbone and methyl motions are decoupled. The time scale of backbone rearrangements changes by several order of magnitude between the bulk and dilute solution states, while the time scale of methyl motion remains relatively independent of the physical state. Analyses of ^{13}C T_1 data to determine the degree of motion of the phenyl ring in polystyrene[118] and PPPO[114] have also employed (4.3.1).

For side-group carbons not directly attached to the backbone and not part of a rigid sidechain, internal reorientation modes often dominate the relaxation. For example, the methoxy carbon of PMMA has a relaxation time about 10 times greater than the α-CH$_3$ in the syndiotactic polymer and five times greater than the α-CH$_3$ in the isotactic polymer.[71] This result primarily reflects the very low barrier (~ 1 kcal/mole)[91] to internal rotation of the OCH$_3$ group. For n-butyl methacrylate[123] the carbons of the butyl chain display a gradation in relaxation with the end-chain methyl T_1 about 10 times longer than the α-CH$_2$ carbon. Again, the

$$
\begin{array}{c}
CH_3\\
|\quad 75\\
+C-CH_2)_{\overline{n}}\\
|\\
C-CH_2-CH_2-CH_2-CH_3\\
O\quad\; 120\quad\; 310\quad\; 760\quad\; 1390\quad (T_1 \text{ in msec})
\end{array}
$$

data reflect the increased number of reorientational motions for carbons near the end of aliphatic chains.[98]

4.3.3. Distribution of Correlation Times

From the prior discussion, it is apparent that the "effective" correlation time for chain segmental motion derived from T_1 measurements provides a useful framework for qualitative comparison of polymer chain flexibilities. However, the obvious limitations of a single-correlation-time model in treating the relaxation results for sidechains subject to internal motion was pointed out in the previous section. The shortcomings of the model for backbone motion are illustrated by considering the motional data derived from the full range of dynamic parameters, i.e., T_1, T_2, NOE. In particular, the validity of the single-correlation-time model requires that

[121] A. A. Jones, R. P. Lubianez, M. A. Hanson, and S. L. Shostak, *J. Polym. Sci., Polym. Phys. Ed.* **16**, 1685 (1978).

[122] R. A. Komoroski and L. Mandelkern, *J. Polym. Sci., Polym. Symp.* **54**, 201 (1976).

[123] G. C. Levy, *J. Am. Chem. Soc.* **95**, 6117 (1973).

all three experimental parameters yield a consistent value of τ_{eff} for chain segmental motion. Listed in Table X are values of relaxation parameters for the backbone carbons of several polymers. The results for syndiotactic PMMA are characteristic of the data, i.e., the τ_{eff} values derived from T_1, T_2, and NOE are not the same—in this case being 0.6, 1.4, and 2.0 nsec respectively. Alternatively, based on the value of τ_{eff} derived from T_1 measurements, the T_2 and NOE values consistent with the single-correlation-time model should be 42 msec and 2.86 rather than the observed 22 msec and 2.1.

That the linewidths ($1/\pi T_2$) are larger than predicted by the single-correlation-time model based on T_1 data indicates the presence of slow modes of polymer segmental motion (characterized by long correlation times) that do not contribute to T_1 but contribute to T_2 primarily through the $J(0)$ spectral density term. As the NOE is a T_1-like process, the reduced Overhauser enhancement requires the presence of some motions associated with correlation times longer than those determining T_1 but short compared to those determining T_2. Thus, the NMR results suggest that a distribution of correlation times provides a more "realistic" description of the segmental motion—a finding perhaps satisfying to our intuitive feeling that backbone motion involve a multiplicity of reorientation modes arising from cooperative movements of chain segments of varying size. Indeed, the NMR results are in concert with studies of polymers by dielectric relaxation and mechanical loss methods, which often invoke a distribution of correlation times to describe the molecular motion.[96]

Schaefer[124,125] has formulated a description of chain motion based on a log chi square (log χ^2) distribution of isotropic correlation times, which accounts for the observed relaxation results and provides some insight into the cooperative nature of chain motion. The distribution is characterized by a width parameter p and an average correlation time $\bar{\tau}$ (as $p \to \infty$, a single correlation model is approached). The expression for the relaxation parameters are derived by substitution for the spectral densities the appropriate terms modified for the log χ^2 distribution, namely,

$$J_n(\omega) = \int_0^\infty \frac{\bar{\tau} F^{(p)}(S)(\exp_b S - 1)\, dS}{(b-1)[1 + \omega^2 \bar{\tau}^2(\exp_b S - 1)/b - 1)^2]},$$

where the distribution function $F^{(p)}(S)\, dS$ is given by

$$F^{(p)}(S)\, dS = \frac{1}{\Gamma(p)}(pS)^{p-1}\exp(-pS), \tag{4.3.5}$$

[124] J. Schaefer, in "Topics in Carbon-13 NMR Spectroscopy" (G. C. Levy, ed.), Vol. I, chap. 4. Wiley (Interscience), New York, 1974.
[125] J. Schaefer, *Macromolecules* **6**, 882 (1973).

TABLE X. Calculated ^{13}C Relaxation Parameters Based on log χ^2 Distribution of Correlation Times for Several Polymers

Polymer	State	Carbon	Experimental					Calculated		
			$T_1{}^a$	$T_2{}^a$	NOE	p^b	τ^c	T_1	T_2	NOE
Isotactic polystyrene	o-dichlorobenzene (35°)^d	CH	65	26	1.8	18	1	68	27	2.0
Isotactic PMMA	pyridine (38°)^e	CH$_2$	64	52	2.5	25	0.24	64	53	2.5
Syndiotactic PMMA	pyridine (38°)^e	CH$_2$	42	22	2.1	19	0.6	39	22	2.2
Syndiotactic α-methyl styrene	CDCl$_3$ (38°)^f	CH$_2$	39	31	2.4	30	0.57	39	32	2.3
Poly(propylene oxide)	CDCL$_3$ (21°)^g	CH	g	—	—	40	0.03	—	—	—
cis-Polyisoprene	bulk (35°)^d (amorphous)	CH	95	29	2.2	14	0.4	98	32	2.2
cis-Polyisoprene	carbon-black filled^d	CH	100	5	2.2	10	0.4	102	4	2.1
cis-Polybutadiene	bulk (35°)^d	CH	600	48	2.4	9	0.01	550	42	2.4

[a] In msec.
[b] Base, b = 1000 for all cases.
[c] In nsec.
[d] J. Schaefer, Macromolecules 6, 882 (1973); H$_0$ = 20 kG.
[e] J. R. Lyerla, Jr., T. T. Horikawa, and D. E. Johnson, J. Am. Chem. Soc. 99, 2463 (1977); H$_0$ = 18.7 kG.
[f] J. Lyerla, unpublished data; H$_0$ = 18.7 kG.
[g] F. Heatley and A. Begum, Polymer 17, 399 (1976), H$_0$ = 23.5 kG, data in graphical form.

and b is the adjustable base for the logarithmic time scale defined by

$$S = \log_b[1 + (b - 1)\tau_l/\bar{\tau}]. \qquad (4.3.6)$$

Numerical integration[126] is required to evaluate the resulting equations for T_1, T_2, and NOE as functions of b, $\bar{\tau}$, and p. The choice of the χ^2 distribution arises in the main from its asymmetry about $\bar{\tau}$ with the long-correlation-time domain having greater density than the short time domain.   Thus, allowance is made for the long-range cooperative motions that presumably occur in polymers (and affect T_2), which may be discriminated against by a normal distribution function.   The logarithmic base accommodates the wide range of correlation times encompassed by polymer motions.   In a statistical sense, p is a measure of the degrees of freedom for the system; thus, large p implies independent chain units and a single correlation time.   Alternatively, small p is characteristic of a broad distribution, while large p implies a narrow distribution.   This is illustrated in Fig. 29, where the distribution of (reduced) correlation times as a function of p is plotted.

Further description of ^{13}C relaxation in terms of the log χ^2 distribution is provided in Fig. 30, where T_1, T_2, and NOE are given as function of $\bar{\tau}$ for $p = 10$ ($b = 1000$).[127]   For the broad distribution as compared to the

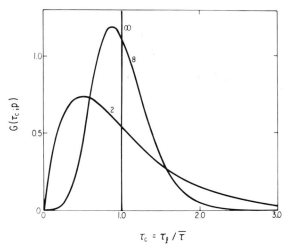

FIG. 29. A χ^2 distribution of correlation times for three values of the width parameter p. $p = \infty$ corresponds to single-correlation-time model [J. Schaefer, *Macromolecules* **6**, 882 (1973)].

[126] See footnote 20 of Lyerla *et al.*[71].

[127] For limitations on T_2, see J. R. Lyerla, Jr. and D. A. Torchia, *Biochemistry* **14**, 5175 (1975), footnote 4.

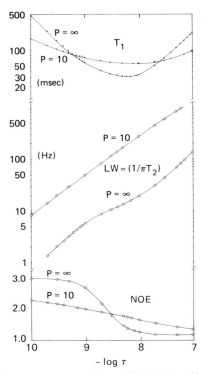

FIG. 30. NMR relaxation parameters at $H_0 = 1.8$ T: (a) T_1; (b) LW linewidth; (c) NOE plotted as a function of $\tau = \tau_c$ for a single-correlation-time model ($P = \infty$) and for a log χ^2 distribution of correlation times with $b = 1000$, $p = 10$, $\tau = \bar{\tau}$ (see text for definitions). The plots are for a methine carbon with $r_{CH} = 1.09$ Å.

single-correlation-time plots, the T_1 plot is flatter, the T_1 minimum is higher, and the T_1 vs. $\bar{\tau}$ profile is asymmetric; the NOE is not as strong a function of $\bar{\tau}$ and is reduced substantially below the maximum enhancement of 2.987; T_2 is a strong function of $\bar{\tau}$ and decreases more rapidly than for the isotropic case. The log χ^2 distribution has been applied to a number of polymers in solution and several in the bulk state. Values of p, $\bar{\tau}$, and predicted relaxation parameters are included with the measured parameters in Table X. As is evident, the experimental results can be fit within experimental error by this distribution function. The average correlation times are in accord with the flexibility of the *cis*-polybutadiene and polypropylene oxide chains and the increased rigidity of polystyrene and PMMA chains. The more interesting findings are associated with the interpretation of p as a measure of the degree of long-range cooperative motion in the polymers. A large value of p would be expected for PPO, where the "oxygen swivel joint" gives rise to high chain flexibility and presumably cooperative motions over only a few monomer units. The p values for the two tactic PMMAs suggest reorientational motion is a more cooperative phenomenon in the syndiotactic chain. The data for bulk cis-polyisoprene is not unexpected in view of the main chain double bond and the side chain methyl. The one unusual result is for bulk polybutadiene, which is characterized by the shortest $\bar{\tau}$ value but is suggested to have the highest degree of cooperative motions on the basis of the width parameter. Unlike the other polymers listed in the table, polybutadiene has no side groups off the chain, thus it can be argued that long-range cooperative motions, which would be hampered by the presence of side groups, are possible in the bulk state of the polymer. Only after more freedom is induced by a solvent are these long-range cooperative motions eliminated. Such long-range motions strongly affect T_2, of course, but not T_1.

The relaxation data for cis-polyisoprene filled with carbon black particles are also explicable by the log χ^2 model of correlation times. While the T_1 and NOE data do not change relative to an unfilled sample, the linewidths ($1/\pi T_2$) are greater by about five or six times. Schaefer showed the additional broadening does not rise from a sum of narrow lines each corresponding to a different magnetic environment by demonstrating the inability to "burn holes" in the resonance line.[128] If it is assumed that the chain entanglements resulting from filler prohibit complete motional averaging, thus giving rise to residual dipolar interactions associated with long correlation times, then the distribution function that described filled polymer would have an increased tail (i.e., smaller p) but about the same $\bar{\tau}$ relative to unfilled polymer. Indeed, a value of $p = 10$, with $\bar{\tau}$ the same as in the unfilled case, is found to describe the relaxation parameters of the filled system, lending some validity to the interpretation. (The arguments supporting the interpretation are more broad-based than outlined and the reader is referred to the original papers by Schaefer[125,128] for the complete analysis.) A similar finding was obtained in a ^{13}C relaxation study of semicrystalline cis-polyisoprene as a function of degree of crystallinity.[129] In Table XI are the relaxation parameters for the amorphous polymer and the amorphous region in two semicrystalline materials. The T_1 and NOE data are essentially unchanged but the linewidths increase considerably for all carbons. In this case, crystallinity retards complete segmental motional narrowing resulting in effects on T_2 but not T_1 and NOE since the high-frequency motions that determine these latter parameters are relatively unaffected.

The Cole–Cole distribution function[96] and the correlation function for

[128] J. Schaefer, S. H. Chin, and S. I. Weissman, *Macromolecules* **5**, 798 (1972).
[129] R. A. Komoroski, J. Maxfield, and L. Mandelkern, *Macromolecules* **10**, 545 (1977).

TABLE XI. ^{13}C Relaxation Parameters^a for Amorphous and
Semicrystalline^b cis-Polyisoprene at 0°

Carbon	Amorphous			Semicrystalline		
	$T_1{}^c$	NOE	LW^d	T_1	NOE	LW
α (=C)	2000	1.1	70	2400	1.1	108 (92)^e
β (=CH)	400	1.3	114	460	1.3	178 (142)
γ (—CH$_2$)	260	1.3	165	270	1.4	248 (197)
δ (—CH$_2$)	240	1.3	124	270	1.4	170 (146)
ϵ (—CH$_3$)	840	1.8	69	910	2.2	120 (96)

^{a} Resonance frequency 67.9 MHz, data from R. A. Komoroski, J. Maxfield, and L. Mandelkern, *Macromolecules* **10**, 545 (1977).
^{b} Degree crystallinity (1-λ) = 0.31.
^{c} In msec.
^{d} In Hz.
^{e} Values of linewidth for degree crystallinity (1-λ) = 0.12.

relaxation by conformational jumps on a diamond lattice[130] as well as log χ^2 have been used by Heatley and Begum[130a] to explain satisfactorily the T_1 and NOE results in several polymers. The Cole–Cole and log χ^2 models assume a distribution of isotropic correlation times; if this is not the case, the relaxation parameters will depend on several correlation times. In general, the distribution function that determines the contribution of each correlation time to the spectral density will depend upon the orientation of the C–H vector with respect to the rotation axes. Hence, the use of a single distribution function for different carbons in the backbone approximately "characterizes" the average chain motion.

Clearly, more work on modeling is required for detailed interpretation of polymer motions from relaxation parameters. Jones and Stockmayer[131] have used lattice models based on the occurrence of a three-bond jump on a tetrahedral lattice to examine a wide range of relaxation data. The lattice models have been successful in reproducing the experimental ^{13}C relaxation data and, as opposed to the distribution models, perhaps provide a better physical insight, as they are based on specific motions possible in linear polymers. However, in the main, the conclusions with respect to segment length are in accord with the conclusions based on distribution models. Jones et al.[132] have reviewed the results of the three-

[130] B. Valeur, J. P. Jarry, F. Geny, and L. Monnerie, *J. Polym. Sci., Polym. Phys. Ed.* **13**, 667, 675, and 2251 (1975).
[130a] F. Heatley and A. Begum, *Polymer* **17**, 399 (1976).
[131] A. A. Jones and W. H. Stockmayer, *J. Polym. Sci., Polym. Phys. Ed.* **15**, 847 (1977).
[132] A. A. Jones, G. L. Robinson, and F. E. Gerr, *ACS Symp. Ser.* **103**, 271 (1979).

TABLE XIIa. ^{13}C Spin–Lattice Relaxation Times of Poly(N-Vinyl Carbazole) (PNVC) in CDCl$_3$ at 30°C

Carbon	Carbazolyl Aromatics					Methine	
^{13}C T_1 (msec)	70	68	64	67	61	60	34

TABLE XIIb. ^{1}H Spin–Lattice Relaxation Times of Poly(N-Vinyl Carbazole) (PNVC) in CDCl$_3$ at 30°C

Aromatic[a]		Shielded aromatic[a]		Skeletal[a]	
Peak[b]	^{1}H T_1 (msec)	Peak[b]	^{1}H T_1 (msec)	Peak[b]	^{1}H T_1 (msec)
7.6	450				
6.8	330	4.9	150	3.2	90
6.3	230			1.6	70

[a] Assignments are taken from D. J. Williams, *Macromolecules* **3**, 602 (1970).
[b] Chemical shifts in ppm from TMS.

bond model. For dissolved PE, PIB, PS, and PPO, the activation energies for backbone rearrangement are 18–25 kJ, and the length of chain involved in cooperative motion is ~5–15 bonds.

4.3.4. Other Nuclei

Although carbon relaxation has been the dominant theme of this chapter, the proton T_1 data on PMMA (Table IIIa) attest to the use of proton relaxation as a guide to chain motion. With the ready measurement of relaxation data using FT methods and the use of deuterated solvents for providing an internal reference for the conventional ^{2}H field/frequency lock, it is worthwhile to determine both ^{1}H and ^{13}C T_1s for polymers. For example, both the proton and carbon chemical shifts in poly(N-vinyl carbazole) PNVC suggest a stacking-effect between carbozolyl groups in neighboring segments.[133] Specifically, there are several aromatic protons shifted to high field (ring current effect) and there is nonequivalence of carbon resonances about the pseudo-C_2 axis through the nitrogen in the side group. The ^{1}H and ^{13}C relaxation times for the resolved aromatic lines and the CH and CH$_2$ groups are given in Table XII. The NT_1 values for the carbons of the backbone and side chain are essentially equivalent, indicating that any carbazolyl side-group motion (about C–N) is much

[133] N. Tsuchihashi, M. Hatano, and J. Sohma, *Makromol. Chem.* **177**, 2739 (1976).

slower than backbone motion, thus attesting to the rigidity of the polymer in solution. By using the carbon data to determine τ_{eff}, the proton data were assessed by calculating the T_1s based on isotropic reorientation, summing $^1H-^1H$ interactions, and then comparing the calculated value to experimental data. On this basis, the small T_1 found for the highest-field aromatic protons can be ascribed to additional $^1H-^1H$ interactions arising from aromatic protons in neighboring polymer segments — a result further supporting strong steric interactions between neighboring sidegroups. The carbon ring T_1s do not show the interaction because the expected large distance between the carbons and protons on neighboring units results in a negligible contribution to carbon T_1 relative to directly bonded protons (i.e., the r_{CH}^{-6} distance dependence is the determining factor).

The ^{19}F nucleus can be employed with ^{13}C to study motion in perfluoro-polymers and materials[134] such as PVF_2. Use can also be made of ^{17}O, ^{14}N, ^{15}N, ^{33}S, ^{31}P to study heteroatom polymers. For the nuclei possessing quadupole moments, structural features such as tacticity or sequence may not be resolvable due to broad resonance lines. However, the interpretation of the relaxation data is not complicated by the need to sum over dipole–dipole interactions since the quadrupole mechanism is dominant.[18]

4.4. HR-NMR of the Solid State

In contrast to the spectral definition of a few hertz or less achievable in the liquid-state NMR spectrum of a polymer, the spectrum of a polymer solid (e.g., glassy or crystalline materials) usually consists of a single broad resonance line having a full-width at half-maximum (FWHM) of tens of kilohertz. However, this result does not imply that a solid spectrum is devoid of information. To the contrary, study of the second moment and various relaxation times of the resonance line as a function of temperature and other parameters often yields important data on the polymer such as degree of crystallinity, chain axis orientation, and degree of molecular motion. Indeed, "wide-line" NMR has been one of the primary methods of characterizing the bulk state of polymeric systems. In these studies, particular emphasis has been placed on determination of proton relaxation times to discuss dynamics in the bulk state. In Fig. 31 is shown the temperature dependence of T_1, $T_{1\rho}$, and T_2 for natural rubber. Two minima are observed in the T_1 data: the low-temperature process is ascribed to methyl group rotation, while the high-temperature process is associated with the

[134] F. A. Bovey, F. C. Schilling, T. K. Kwei, and H. L. Frisch, *Macromolecules* **10**, 559 (1977).

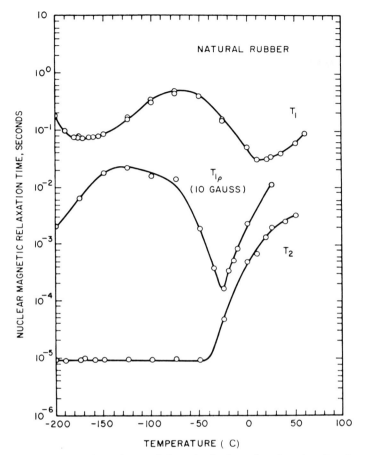

FIG. 31. Temperature dependence of the spin–lattice relaxation time T_1, the rotating frame relaxation time $T_{1\rho}$, and the spin–spin relaxation time T_2 for natural rubber. Measured at a radio frequency of 30 MHz and amplitude of 10 G. [W. P. Slichter, "NMR. Basic Principles and Progress" (P. Diehl, ed.), Vol. 4, p. 209. Springer-Verlag, Berlin and New York, 1971].

onset of large amplitude motions of backbone segments. Two minima are also seen in the $T_{1\rho}$ data, but shifted to lower temperature with respect to the T_1 minima. This result reflects the sensitivity of $T_{1\rho}$ to lower-frequency motions. The transition in T_2 occurs at about the same temperature as the major change in $T_{1\rho}$ and reflects loss of main chain motion, which averages the dipolar interactions (i.e., motions $> 10^5$ Hz). The T_1 data on PMMA (Fig. 32) show three distinct minima in temperature behavior. These correspond (from low to high temperature) to onset of

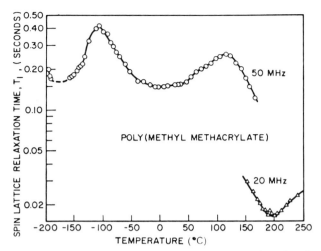

FIG. 32. Temperature dependence of the spin–lattice relaxation time in poly(methyl methacrylate), measured at 20 and 50 MHz [W. P. Slichter, "NMR. Basic Principles and Progress" (P. Diehl, ed.), Vol. 4, p. 209. Springer-Verlag, Berlin and New York, 1971].

rotation of the ester methyl group, onset of rotation of the α-methyl group, and onset of chain segmental motion.   These examples illustrate only one aspect of the utility of wide-line NMR; numerous reviews exist that catalogue further applications.[85,86,96,135]

Despite the information content of wide-line spectra, it might be logically assumed that if a HR-NMR spectrum of a solid could be obtained, new detail on the nature of the solid state might be uncovered.   Recent advances in magnetic resonance have now made possible the partial realization of this goal.[136]   For example, ^{13}C spectra having linewidths in the range 10–200 Hz has been obtained for polymers and organic solids.   In this chapter, our purpose is threefold: (1) to discuss the combination of techniques required to obtain such ^{13}C spectra; (2) to describe briefly the implementation of these techniques; and (3) to illustrate the utility of the resulting spectra in providing information on the composition of insoluble polymers, chain orientation, motion in the solid state, and other aspects of bulk polymers.

We now turn to a discussion of the effects on the NMR spectra of solids produced by the orientation dependence of the dipolar and chemical

[135] V. J. McBrierty, *Polymer* **15,** 503 (1974); D. W. McCall, *Acc. Chem. Res.* **4,** 223 (1971); W. P. Slichter, *J. Chem. Educ.* **47,** 193 (1970).

[136] J. Schaefer, E. O. Stejskal, and R. Buchdahl, *Macromolecules* **10,** 384 (1977), and references therein.

shielding interactions.   Since we are interested in the observation of ^{13}C resonance in hydrocarbon materials, we are primarily concerned with ^{13}C–^{1}H heteronuclear dipolar interactions.   However, because the discussion is general (assuming spin-$\frac{1}{2}$ nuclei) for any system composed of a "rare" or low natural abundance spin (^{13}C) and an abundant spin (^{1}H), we employ the conventional symbolism S and I for the respective rare and abundant spin species.[137]

4.4.1. The Resolution Problem: Sources of Line-Broadening and Methods for Their Removal

4.4.1.1. The I–S Dipolar Interaction. The primary cause for the broad featureless resonance lines characteristic of the NMR spectra of solids is the dipolar interaction between nuclear spins.   The interaction between two nuclear dipoles is described by a second-rank tensor $\hat{\mathbf{D}}$. Because $\hat{\mathbf{D}}$ is traceless, the isotropic average is zero and thus the interaction has no effect (to first-order) on liquid-state spectra. In the solid, the magnitude of the interaction is both orientation and distance dependent. The Hamiltonian governing the interaction is given by[26]

$$\mathscr{H}_d = \hbar^2 \sum_{i<j} \gamma_i \gamma_j \mathbf{I}_i \cdot \hat{\mathbf{D}}_{ij} \cdot \mathbf{I}_j \qquad (4.4.1)$$

where $\mathbf{I}_{i,j}$ are spin operators and $\gamma_{i,j}$ the respective magnetogyric ratios of the interacting nuclei.

The dipole interaction between two unlike spins I and S having magnetic moments μ_I and μ_S, respectively, is depicted in Fig. 33 [and given by (4.1.9)].   The magnitude of the local magnetic field h_l induced at nucleus S due to the magnet moment of nucleus I is a function of the internuclear separation R_{IS} and the angle θ this vector makes with the applied magnetic field H_0.   Whether h_l augments or detracts from H_0 depends on the orientation of μ_I and H_0 (i.e., the spin state of I); as a consequence, the resonance of the S nucleus is split into a doublet (except for $\theta = 54.7°$, where $h_l = 0$) whose members are shifted by $\pm \omega_l$ ($= \mu_S h_l$) from the resonance frequency ω_0 obtained when the dipolar interaction is zero.

In a rigid polycrystalline system, a nucleus does not experience an isolated dipole interaction but many such interactions arising from neighboring nuclei having (in most instances) different R and θ relationships with

---

[137] Although the symbols I and S appear in some equations in this chapter as spin operators or spin quantum numbers, their meaning in these instances is obvious and no ambiguity results.

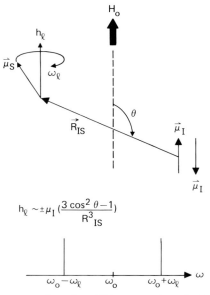

$$h_\rho \sim \pm \mu_I \frac{(3\cos^2\theta - 1)}{R^3_{IS}}$$

FIG. 33. Schematic representation of an IS heteronuclear dipole interaction. Symbols are defined in the text. The frequency spectrum is given at the bottom.

the nucleus. The result is a Gaussian-shaped resonance line centered at ω_0 and characterized by a second moment given by[138]

$$\langle \Delta\omega^2 \rangle_{SI} = \frac{1}{3} \gamma_I^2 \gamma_S^2 \hbar^2 I(I + 1) \cdot \frac{1}{N} \sum_{j,k} \frac{(1 - 3\cos^2\theta_{jk})^2}{r_{jk}^6} \qquad (4.4.2)$$

Since the rigid lattice linewidth for a carbon bonded to a hydrogen is $\sim 10-40$ kHz, it is clear that the width of the resonance line (except at very large applied fields) encompasses the entire range of carbon chemical shifts associated with diamagnetic organic structures. Of course, the linewidth is reduced if specific molecular motions are present to partially average[139] the dipolar interaction by imparting a time dependence to θ. As described in Section 4.1.3.1 in solution of the liquid state, rapid molecular tumbling results in isotropic averaging of $\langle 3\cos^2\theta - 1 \rangle$ and hence a vanishing value of h_l. The highly resolved spectra that are obtained reflect chemical shift and scalar (nondipolar) spin–spin coupling interac-

---

[138] C. P. Slichter, "Principles of Magnetic Resonance," p. 71. Springer-Verlag, Berlin and New York, 1978. Because S is a "rare" spin, we neglect the $\langle \Delta\omega^2 \rangle_{SS}$ contribution.

[139] For example, methyl group reorientation about the C_3 axis.

tions. To approach equivalent resolution in the solid state thus requires a method to force $h_l \rightarrow 0$. We now examine this situation for the carbon–proton case.

The truncated or secular dipolar Hamiltonian $\mathcal{H}_d$ for two species I and S may be written as[140]

$$\mathcal{H}_d = \mathcal{H}_{II}^0 + \mathcal{H}_{IS}^0 + \mathcal{H}_{SS}^0, \qquad (4.4.3a)$$

where

$$\mathcal{H}_{II}^0 = \frac{\gamma_I^2 \hbar^2}{2} \sum_{i<j}^{N_I} \sum^{N_I} r_{ij}^{-3}(3\cos^2\theta_{ij} - 1)(\mathbf{I}_i \cdot \mathbf{I}_j - 3I_{iz}I_{jz}), \qquad (4.4.3b)$$

$$\mathcal{H}_{IS}^0 = -\gamma_I\gamma_S\hbar^2 \sum_i^{N_I} \sum_k^{N_S} r_{ik}^{-3}(3\cos^2\theta_{ik} - 1)I_{iz}I_{kz}, \qquad (4.4.3c)$$

$$\mathcal{H}_{SS}^0 = \frac{\gamma_S^2 \hbar^2}{2} \sum_{k<m}^{N_S} \sum^{N_S} r_{km}^{-3}(3\cos^2\theta_{km} - 1)(\mathbf{I}_k \cdot \mathbf{I}_m - 3I_{kz}I_{mz}). \qquad (4.4.3d)$$

The expressions for the various dipolar interactions have been written as a product of a spatial term and a spin term to facilitate discussion. As mentioned above, molecular tumbling isotropically averages the spatial term to zero in the liquid state. In a somewhat analogous manner, i.e., by operating on the angular dependence of the interaction, dipolar broadening can also be removed in the solid state. If a solid sample is rotated at frequency ω_r about an axis making an angle β with respect to H_0 and angle α with respect to the internuclear vector connecting interacting dipoles, it is straightforward to decompose θ into α, β, and $\phi = \omega_r t$. This results in an expression for the local dipolar field given by[141]

$$h_l = \pm\mu_j/R_{ij}^3[(3\cos^2\alpha - 1)(3\cos^2\beta - 1) + \tfrac{3}{2}\sin 2\alpha \sin 2\beta \cos\omega_r t + \tfrac{3}{2}\sin^2\alpha \sin^2\beta \cos^2\omega_r t]. \qquad (4.4.4)$$

From (4.4.4) it is evident that h_l consists of a static contribution and a fluctuating contribution. If β is chosen to be 54.7°, the so-called magic angle, the time-independent term vanishes and the spectrum consists of a line at ω_0 (the Larmor frequency) having sidebands at $\omega_0 \pm n\omega_r$. The dipolar broadening is completely removed from the central line and is manifest only in the appearance of side bands. For fast spinning rates, the intensity of the side bands becomes essentially zero and the spectrum consists

[140] C. P. Slichter, "Principles of Magnetic Resonance," Chapter 3. Springer-Verlag, Berlin and New York, 1978.
[141] B. Schneider, D. Doskocilová, and H. Pivcová, *Magn. Reson. Chem. Biol. Proc. Int. Summer Sch., 1971* p. 127 (1971). Note the definitions of α and β are interchanged here to be consistent with use of β in Eq. 4.4.6.

only of the narrowed line at ω_0. It is important to note that $h_l \rightarrow 0$ irrespective of the initial orientation of R and H_0; thus all dipolar interactions in the molecular ensemble are removed.

We now inquire as to the rotational velocities required for dipolar line-narrowing by magic-angle spinning. Averaging of the dipolar Hamiltonian, which is in essence the desired end result of the spinning experiment, requires the rotation period of the sample be short compared to the spin–spin relaxation time or dephasing time T_2 of the individual spins. Since T_2 is approximately the inverse of the linewidth, it is apparent that, for effective removal of dipolar broadening, $\omega_r/2\pi$ should be large compared to the dipolar linewidth. For proton–proton and carbon–proton dipolar interactions, this requires spinning speeds in the range 10–50 kHz. While spinning rates as high as 15 kHz have been achieved,[142] the upper limit is impractical as most materials used to confine samples disintegrate due to centrifugal forces. Thus, use of the MAS technique to remove dipolar broadening, while successful in removing homonuclear interactions for phosphorous and nuclei of smaller magnetogyric ratio,[26] is of limited utility for proton homonuclear and heteronuclear interactions.

An alternative approach to removing the dipolar interaction that is particularly effective for the C–H case is to operate on the spin part of the Hamiltonian (4.4.3) rather than the spatial term. If we focus on Fig. 33, we see that one method of reducing the I–S dipolar term is to modulate the orientation of the I spins with respect to H_0. By forcing the spins to change states at a rate large compared to the frequency of the I–S interaction, h_l is reduced to zero. This is accomplished by applying RF irradiation at the I spin frequency of sufficient amplitude to "decouple" the interaction.[143] (In essence, this results in coherent averaging of the dipolar Hamiltonian in spin space as opposed to the incoherent averaging in molecular space due to Brownian motion in the liquid state.) As discussed in Section 4.1.3.6, such double-resonance techniques are standard in carbon magnetic resonance since high-resolution spectra in solution are usually obtained by decoupling the ^{13}C–^1H scalar spin–spin interactions (J coupling). However, the RF power required (assuming the same coupling efficiency with the probe) is considerably greater to remove the 10–40 kHz dipolar coupling as compared to the 100–200 Hz scalar coupling. For this reason, the removal of the I–S dipolar interaction by RF irradiation has been referred to as "dipolar" decoupling[136] (DD).

[142] K. W. Zilm, D. W. Alderman, and D. M. Grant, *J. Magn. Reson.* **30**, 563 (1978). These authors cite that rates of 21 kHz have been obtained by J. W. Beams [*Rev. Sci. Instrum.* **8**, 795 (1937)] using H_2 as the driving gas.

[143] F. Bloch, *Phys. Rev.* **111**, 841 (1958).

Because of the 1.1% natural abundance of the ^{13}C isotope, the probability of having directly bonded two carbons carrying a magnetic moment is $\sim 10^{-4}$. Since the dipolar interaction attenuates rapidly (inverse third power) with distance between dipoles, and carbon has a small magnetogyric ratio relative to the proton, the homonuclear dipolar term $\mathcal{H}_{SS}^0$ of (4.4.3d) is a negligible source of line broadening for carbons relative to the $\mathcal{H}_{IS}^0$ interaction. Thus, high-power proton decoupling is particularly effective in reducing the linewidths of ^{13}C resonance lines in the solid state.

The third term in (4.4.3) is the homonuclear dipolar interaction for the I spins. In the absence of decoupling, exchange interactions that induce rapid "flip-flop" motion of the I spins can reduce the I–S broadening from its full rigid-lattice value. In this manner, the magnitude of I–I dipolar interaction can affect the width of the S resonance. The details of this spin exchange narrowing, which is analogous to classical motional narrowing of resonance lines, is beyond the scope of this chapter and the interested reader is referred to the literature.[144]

The effect of proton dipolar decoupling on the carbon resonance of a solid is illustrated in Fig. 34 for the glassy state of poly-

(a) ^{13}C - NMR Spectrum of Lexan in Glassy State

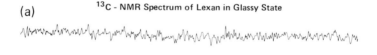

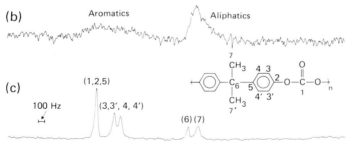

FIG. 34. ^{13}C NMR spectra of polycarbonate at 27°C: (a) spectrum obtained under FT-NMR conditions normal for high resolution on polymer solutions; (b) spectrum obtained under DD/CP conditions; (c) spectrum obtained under DD/CP/MAS conditions. All spectra represent FT of 2000 FID accumulations. Resonance line assignments are given in the figure.

[144] See, for example, G. Sinning, M. Mehring, and A. Pines, *Chem. Phys. Lett.* **43**, 382 (1976), and references therein. The *I–I* flip–flop term, if nonzero, results in the necessity of having the *I* decoupling field large compared to the flip–flop rate for *I* spins to decouple the *I–S* interaction. See M. Mehring, "NMR Basic Principles and Progress" (P. Diehl, E. Fluck, and R. Kosfeld, eds.), Vol. 11, p. 153. Springer-Verlag, Berlin and New York, 1976.

dioxydiphenylpropane-carbonate (the polycarbonate of bisphenol-A). The spectrum in Fig. 34a (or lack thereof) was obtained under conditions of proton decoupling considered normal for measuring carbon spectra in solution, while the spectrum in Fig. 34b was obtained using a decoupling field of ~32 kHz (7.5 G)—ten times that used in Fig. 34a. Removal of C–H dipolar interactions allows resonances from the aliphatic carbons to be distinguished from the aromatic carbons. However, there remains a residual broadening of the resonances that inhibits greater resolution. We now examine the nature of this remaining broadening. [It should be noted that the apparent lack of even a broad line spectrum in Fig. 34a results from two factors: (1) the low-abundance of the ^{13}C isotope and (2) the relatively small number of free induction decays (FID) accumulated for a case where intensity is dispersed over a kilohertz frequency range. Of course, the expected long spin–lattice relaxation times in the solid make repolarization of the carbon spins, and thereby addition of free induction decays for signal/noise improvement, an inefficient process. We return to this point in later discussion.]

4.4.1.2. Chemical Shift Anisotropy. As discussed earlier in this section, the chemical shift is a second-rank tensor that, because of rapid molecular tumbling, is averaged to its isotropic value, i.e., $\frac{1}{3}$ the trace of the principal elements, in solution. In a polycrystalline or amorphous solid, the orientation dependence of the chemical shift is not averaged out and as a consequence can result in broadening of a resonance line. In particular, unless the electronic shielding of a nucleus is spherically symmetric, the chemical shift of the nucleus will depend on the orientation of the principal axes of the shielding tensor with respect to the magnetic field direction. The NMR spectrum of a powder then consists of a superposition of lines encompassing the frequency range between the maximum and minimum values of the shielding—in essence, a powder pattern. The calculated lineshapes for an isotropic probability distribution (powder) of tensor orientations for uniaxially symmetric and nonaxially symmetric tensors are presented in Fig. 35.[145] The principal values of the shielding tensor $\hat{\sigma}$ can be obtained directly from the spectrum as indicated. In Fig. 35 are the respective ^{13}C spectra of solid benzene (at $-40°C$) and a highly cyrstalline PE obtained under dipolar-decoupled conditions. At $-40°C$, the rapid reorientation of the benzene ring about the sixfold axis reduces the shielding tensor to one of axial symmetry. The nonaxial symmetry of the shielding in PE is quite apparent. Knowledge of the principal elements of $\hat{\sigma}$ and of the effect of temperature (and thereby molecular motion) on the shape and width of the powder pattern

[145] A. Pines, M. G. Gibby, and J. S. Waugh, *J. Chem. Phys.* **59**, 569 (1973).

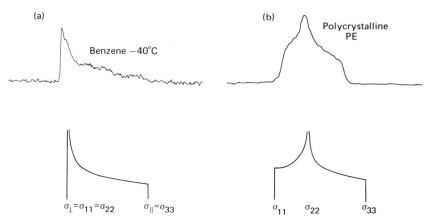

FIG. 35. Calculated [N. Bloembergen and T. J. Rowland, *Acta Metall.* **1**, 731 (1953)] absorption lineshapes and representative ^{13}C spectra for polycrystalline samples: (a) an axially symmetric chemical shielding tensor $\sigma_\parallel$ and $\sigma_\perp$ denote shielding parallel and perpendicular to the symmetry axis), e.g., the spectrum of frozen benzene at $-40°C$; (b) a general shielding tensor, e.g., the spectrum of polycrystalline polyethylene.

provides a source of important information on a molecule [145]—a point we return to in later discussion. Benzene and PE (if chain ends are ignored) represent systems in which there is only one carbon resonance; in the more general case, where a molecule or polymer repeat unit possesses several magnetically inequivalent carbons, the anisotropy in the chemical shielding will usually give rise to an NMR spectrum composed of overlapping powder patterns, which make it virtually impossible to extract structural information in any straightforward manner. This is precisely the case for the spectrum of the polycarbonate given in Fig. 34b. To obtain a more highly resolved spectrum requires the averaging of the various carbon shielding tensors in the solid state.

The required averaging can be induced by utilizing the magic-angle spinning technique introduced in connection with the dipolar interaction. The chemical shift of a nucleus i with principal shielding elements σ_{i1}, σ_{i2}, σ_{i3} and with direction cosines between the principal axes of $\hat{\sigma}$ and $\mathbf{H}_0$, λ_{i1}, λ_{i2}, λ_{i3} is given by[26]

$$\sigma_{izz} = \lambda_{i1}^2 \sigma_{i1} + \lambda_{i2}^2 \sigma_{i2} + \lambda_{i3}^2 \sigma_{i3}. \qquad (4.4.5)$$

If the solid sample in which i is resident is rotated at angular velocity ω_r about an axis inclined at an angle β to $\mathbf{H}_0$ and at angles ϕ_{i1}, ϕ_{i2}, ϕ_{i3} to the principal axes of $\hat{\sigma}$, the direction cosines become

$$\sum_{k=1}^{3} \lambda_{ik} = \cos\beta \cos\phi_{ik} + \sin\beta \sin\phi_{ik} \cos(\omega_r t + \chi_{ik}), \qquad (4.4.6)$$

where χ_{ik} is the azimuthal angle of the kth principal axis of $\hat{\sigma}$ at $t = 0$. Substitution of (4.4.6) into (4.4.5) and taking the time average yields

$$\bar{\sigma}_{izz} = \tfrac{1}{2}\sin^2\beta(\sigma_{i1} + \sigma_{i2} + \sigma_{i3}) + \tfrac{1}{2}(3\cos^2\beta - 1)\sum_{k=1}^{3}\sigma_{ik}\cos^2\phi_{ik}. \quad (4.4.7)$$

If β is the magic angle (54.7°), (4.4.7) reduces to

$$\bar{\sigma}_{izz} = \tfrac{1}{3}(\sigma_{i1} + \sigma_{i2} + \sigma_{i3}), \quad (4.4.8)$$

which is the average value obtained in fluids. The result holds for all nuclei in the sample irrespective of the initial orientation of the principal axes of $\hat{\sigma}$ with respect to $\mathbf{H}_0$ and to the spinning axis. To illustrate the effect of spinning, Fig. 36 depicts the chemical shift vs. time plot for a nu-

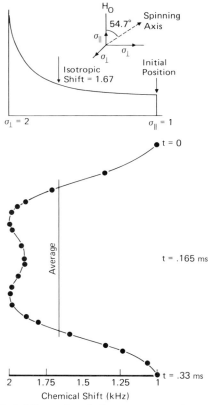

FIG. 36. Plot of chemical shift vs. time during one rotor cycle at a spinning rate of 3 kHz for a carbon with an axially symmetric shift tensor having principal values $\sigma_{11} = \sigma_{22} = 2$ kHz, $\sigma_{33} = 1$ kHz. Data calculated using Eqs. (4.4.5) and (4.4.6) for magic-angle spinning (see diagram), initial tensor orientation of σ_{33} parallel to H_0, and azimuthal angles of 0° for σ_{22} and 90° for σ_{11}.

cleus with an axially symmetric shielding tensor during one full rotation of the sample about an axis making the magic angle with H_0. The average value of σ_{izz} obtained from the plot is $\frac{1}{3}(\sigma_{i1} + \sigma_{i2} + \sigma_{i3})$.

As in the case of the dipolar interaction, the question again arises as to how rapidly the sample must be rotated to reduce the broadening due to chemical shift anisotropy. In the first approximation, the spinning frequency should be fast compared to the anisotropy expressed in frequency units or equivalently to the frequency range encompassed by the powder pattern. Typical shift anisotropy values for aliphatic carbons are 15–50 ppm, while for aromatic and carbonyl carbons values of 120–200 ppm are observed.[146] These data reflect the larger directional variation in electron density (and hence shielding) associated with multiple bonds; this accounts for the broader resonance observed for the ring carbons as compared to the aliphatic carbons in Fig. 34b. At an applied field of 1.4 T, the anisotropy for aromatic and carbonyl carbons translates to ~ 1.8–3 kHz. Thus spinning speeds greater than 3 kHz would lead to removal of the broadening. In Fig. 34c is the spectrum of polycarbonate obtained employing a combination of dipolar decoupling and magic-angle spinning at 4 kHz. The spectrum is sufficiently resolved that the majority of carbons in the repeat unit appear as individual resonance lines, thus making it possible to derive detailed data on bulk polymers. However, it is clear the resonance lines are not as narrow as in a solution spectrum. We discuss the sources of residual broadening and the limitations to resolution in bulk polymers in Section 4.4.4.3.

4.4.2. The Sensitivity Problem

4.4.2.1. General Considerations and Definitions.

We have seen in the previous section that central to the success of the dipolar decoupling and MAS schemes in narrowing carbon resonances was the fact that homonuclear dipolar interactions between carbons could be ignored because of the low isotopic abundance. Of course, the facility with which carbon spectra can be observed depends strongly on the application of digital methods of signal averaging to overcome the "rare" spin sensitivity problem. Acquisition of n free-induction decays (FID) results in a S/N enhancement of $\sqrt{n}$. As is well-known, the rate at which the FID following a $\pi/2$ RF pulse can be accumulated is usually not limited by the spectrometer hardware nor the signal acquisition time, but by the rate at which carbon magnetization M_0^C can be reestablished along H_0 by spin–lattice relaxation processes (see Chapter 4.1). The efficiency of T_1 processes depends on the spectral density of motions at or near the Larmor frequency of the observed nucleus (i.e., the megahertz region). As we have

[146] A. Pines, M. G. Gibby, and J. S. Waugh, *Chem. Phys. Lett.* **15**, 373 (1972).

seen, for polymers in solution or the melt, carbon relaxation times are relatively short—being a few hundred milliseconds at ambient temperatures for carbons with attached protons and several seconds for carbons without directly bound protons.[147] This occurs because the molecular motion produces a high density of the appropriate fluctuating local field needed to couple the spin system and lattice and thus induce relaxation. In contrast, motion at high frequencies in solids is usually of small amplitude, thereby resulting in a substantially reduced spectral density in the Larmor frequency region.[148] As such, the T_1 values for carbons in solids may vary between a few seconds to minutes and even hours, e.g., the carbon spin–lattice relaxation time for crystalline polyethylene is ~700 sec at ambient temperature.[149] Clearly, obtaining a spectrum can be time consuming if indeed not prohibitive in the case of multiline spectra. The situation is further aggravated by the fact that linewidths in the solid spectrum may be expected to be ~10–50 Hz as compared to 1–5 Hz widths obtained for polymers in the melt or in solution. The distribution of spectral intensity over a wider frequency range manifests itself in requiring a greater number of data acquisitions on the solid sample relative to the liquid sample to obtain a spectrum having the same "apparent" S/N (here we define S/N as peak height, irrespective of width, divided by RMS noise).

As indicated in the above discussion, the carbon T_1s in solids depend to some extent on the physical state of the system. For example, in Figs. 37a and c are shown the ^{13}C spectra for a polycrystalline sample of the material PHBA (polyhydroxy benzoic acid) and for a glassy sample of polycarbonate (Lexan[167a]), respectively. Both spectra were obtained using conventional Fourier transform techniques and $\pi/2$ repetitive pulsing with delay times of 3.5 seconds. Clearly, motion in the amorphous sample results in reasonably short carbon relaxation times T_1^C (although the smaller amplitude for the lines assigned to carbons without directly bound protons indicates full repolarization does not take place for these carbons in the repetition interval), while in the crystalline sample, the lack of a spectrum suggests reduced molecular motion, quite long relaxation times, and little repolarization between pulses. Fortunately, the situation for obtaining spectra of crystalline systems is not as difficult as this result would suggest. We now turn to some brief comments of proton T_1 characteristics and on spin–lattice relaxation times in the rotating frame.

[147] J. Schaefer, in "Structural Studies of Macromolecules by Spectroscopic Methods" (K. J. Ivin, ed.), Chapter 11. Wiley, New York, 1976.

[148] A. Abragam, "The Principles of Nuclear Magnetism," Chapter IX. Oxford Univ. Press, London and New York, 1961.

[149] D. L. VanderHart, *Macromolecules*, **12**, 1232 (1979).

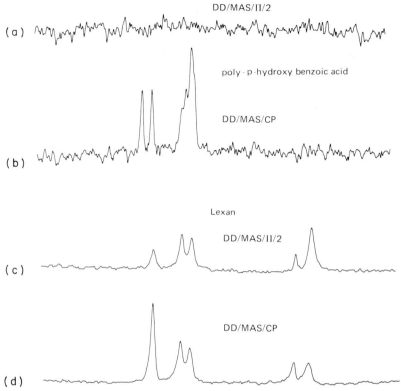

DD/MAS/II/2

(a)

poly - p -hydroxy benzoic acid

DD/MAS/CP

(b)

Lexan

DD/MAS/II/2

(c)

DD/MAS/CP

(d)

FIG. 37. DD/MAS ^{13}C-NMR spectra of polycarbonate and PHBA. All spectra were taken with a 3.5 sec experimental repeat time and represent 1000 FID accumulations. In each case, one spectrum was recorded with 90° pulses and one with CP (CP time = 1.5 msec).

4.4.2.2. Proton T₁s in Polymers. In the case $T_1 \gg T_2$, the coupling within the spin system is much stronger than with the lattice; hence, the spin system comes to internal equilibrium more rapidly than it reaches equilibrium with the lattice. As a consequence, if a spin system is perturbed from equilibrium with the lattice, e.g., by a $\pi/2$ pulse, the excess energy put into the spin system by the RF field can remain for a time long compared to the time for the spin system to establish internal equilibrium. However, energy transfer to the lattice is an efficient process near paramagnetic impurities, lattice imperfections or near molecular segments that are in rapid motion (e.g., rotating methyl groups or amorphous regions in a semicrystalline polymer).[135] Even small concentrations of these relaxation sites in a material can strongly affect the overall rate of spin–lattice relaxation because the excess spin energy can be transferred to these preferred relaxation sites, where in turn it is transmitted to the lattice. In this manner, the entire spin system comes to equilibrium with the lattice.

It is outside the scope of this discussion to explore the details and complexities of spin diffusion. Put simply, it is similar to any diffusion process in that it involves transport driven by a concentration gradient. In this instance, it is spin magnetization that is transported by a spatial gradient in the magnetization.[150] The magnetization does not diffuse by physical movement of the nuclei or electrons having magnetic moments but by mutual spin–flips of like moments via the exchange term of the dipolar Hamiltonian.[151] The coefficient of spin diffusion is typically of the order 10^{-12} cm²/sec and depends on the distance between nearest like neighbors and the probability of flip–flop transitions.[148] The important aspect of the process for our present purposes is the distance dependence, which enters the expression for a diffusion-controlled[152] spin–lattice relaxation process as the inverse third power, i.e., $1/T_1 \propto r^{-3}$. For this reason, the proton T_1 of a solid when spin diffusion is the operative relaxation mechanism will be much shorter than the corresponding ^{13}C T_1 owing to the reduced proximity of ^{13}C moments. Indeed, it is the more efficient spin diffusion process between the amorphous and crystalline regions that enables the protons in the crystalline regions of semicrystalline PE to relax in ~ 1–5 sec at ambient conditions[153] while the carbon T_1 process for the crystalline region requires $\sim 10^3$ sec.

4.4.2.3. Relaxation in the Rotating Frame. In Chapter 4.1 we discussed spin–lattice relaxation in the rotating frame and the process of spin-locking. At this point it is instructive to examine the consequences of the spin-locking process in the context of the Boltzmann expression. At equilibrium in H_0, the observed magnetization $M_0{}^1 \propto \exp(2\mu_1 H_0 / kT_S)$, where T_S is the temperature of the spin system,[154] which is equal to that of

---

[150] N. Bloembergen, "Nuclear Magnetic Relaxation." Benjamin, New York, 1961.

[151] The exchange term is contained in the $I_i \cdot I_j$ and $I_k \cdot I_m$ terms of the Hamiltonians given in Eqs. (4.4.3b) and (4.4.3d).

[152] M. Goldman, "Spin Temperature and NMR in Solids," Chapter 3. Oxford Univ. Press, London and New York, 1970.

[153] These arguments are simplistic; clearly consideration must be given to how "good" the thermal contact is between the two regions (i.e., the size of the interfacial regions), whether all the crystalline region makes the same thermal contact with the amorphous region (size of the crystallites is important), etc.

[154] The population distribution between the two spin states is conveniently described by the concept of a spin temperature. For a Boltzmann distribution $T_S = T_L$; however, for a non-Boltzmann distribution $T_S \neq T_L$ but approaches T_L in the spin–lattice relaxation time. See Reference 152 for details.

the lattice T_L. After the spin-locking sequence, we have initially $M_1^I = M_0^I \propto \exp(2\mu_1 H_{1I}/kT_S)$. Since $H_{1I} = 10^{-3}H_0$, T_S must be $10^{-3}T_L$ for the condition to hold. In the language of spin thermodynamics, the result of the spin-locking sequence is to cool the I spins to near 0K. Of course, except at very low temperatures, there is in essence no change in the temperature of the sample because the heat capacity of the spin system is negligible relative to that of the lattice. The approach toward equilibrium in H_{1I} then corresponds to a heating of the spin system.

This prologue on proton relaxation times and spin-locking now puts us in a position to understand the solution to the ^{13}C sensitivity problem in solids proposed and demonstrated by Pines, Gibby, and Waugh.[145]

4.4.2.4. Cross Polarization. Based on the above discussion, it is clear that often advantage could be obtained in the signal-averaging process if the ^{13}C experiment could be recycled in the time required for proton repolarization ($\sim 5 \times T_1^H$) rather than for carbon repolarization ($\sim 5 \times T_1^C$). For this advantage to be realized requires the existence of mechanisms to establish carbon polarization via the protons despite the fact that in the static DC field, the large difference in resonance frequency of ^{13}C and ^{1}H results in essentially isolated spin systems. A solution to the problem of coupling the two spin systems was first recognized by Hahn.[155] This involves the application of two resonant alternating RF fields H_{1I} and H_{1S} to the respective I and S spins such that

$$\gamma_I H_{1I} = \gamma_S H_{1S}. \tag{4.4.9}$$

This condition, known as the Hartmann–Hahn condition,[155] corresponds to making the Zeeman splittings of the I spins quantized along H_{1I} in the I spin rotating frame equal to the Zeeman splitting of the S spins quantized along H_{1S} in the S spin rotating frame. With Zeeman levels matched, the spin systems can exchange energy via the coupling produced by the I–S dipolar interaction.[156] A classical description of the process is that the precession of M_1^I about H_{1I} produces a sinusoidally oscillating component of the dipolar field along the direction of H_0. If the Hartmann–Hahn condition is matched, this frequency is correct to induce transitions of the S spins relative to H_{1S}. This method of polarizing the S spins is termed cross polarization (CP).

One of the variety of experimental methods[145] for establishing contact

[155] S. R. Hartmann and E. L. Hahn, *Phys. Rev.* **128**, 2042 (1962).

[156] Note, the secular part of the *IS* dipolar Hamiltonian (4.4.3c) contains no flip–flip term, as appears (4.4.3b) and (4.4.3d) due to the frequency mismatch of *I*- and *S*-spin systems in H_0; thus, the effect of the Hartmann–Hahn experiment is to make the *IS* flip–flop term secular.

of I and S spin systems and then observing the S spin polarization is represented by the following sequence:

(1) establish I spin polarization in H_0,
(2) spin-lock I spins along H_{1I},
(3) contact I and S spins with a Hartmann–Hahn match,
(4) turn off H_{1S} and record S spin FID,
(5) repeat sequence.

In understanding the CP process it is instructive to discuss the steps of
this sequence from a simple thermodynamic picture. Consider the I and
S spin systems to be two thermodynamic reservoirs that are in thermal
contact with the lattice (see Fig. 38). Once the I spins are in equilibrium
with the lattice (requiring a time $\sim 5T_1^I$), the magnetization is given by
(4.1.7), which can be rewritten as

$$M_0^I = \beta_L C_I H_0 \qquad (4.4.10)$$

where C_I is the Curie constant $(= N_I \gamma_I^2 \hbar^2 I(I + 1)/3)$. The relation $\beta = 1/kT$
defines an inverse spin temperature that for the I spins at the end of step 1
is equal to the inverse of the lattice temperature β_L. Application of the
spin-locking sequence (step 2) effectively cools (see previous discussion)
the I spins to an initial rotating-frame temperature

$$\beta_0^I = (H_0/H_{1I})\beta_L. \qquad (4.4.11)$$

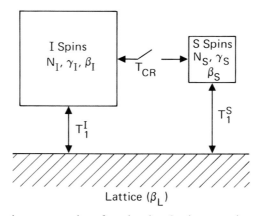

FIG. 38. Schematic representation of an abundant I spin reservoir and a rare S spin reservoir, coupled to the lattice as indicated by their spin–lattice relaxation times T_1^I and T_1^S.
Coupling between the spin reservoirs is represented by T_{CR}, the cross-relaxation time. The
coupling can be induced by a cross-polarization process in the rotating frame. Figure
similar to those of G. Sinning, M. Mehring, and A. Pines [*Chem. Phys. Lett.* **43**, 382
(1976)] and A. Pines, M. G. Gibby, and J. S. Waugh [*J. Chem. Phys.* **59**, 569 (1973)].

The S spins, which are assumed unpolarized (i.e., $T_1^S \gg T_1^I$), are brought into contact with the I spins (step 3) and the two-spin systems come rapidly (in a time $\sim T_{CR}$) to equilibrium at a common spin temperature β_1. Because the spin–lattice relaxation times are long, the two spin reservoirs are essentially isolated from the lattice during the cross-polarization process and thus the total energy in the combined spin systems is conserved. Invoking conservation, we find[145]

$$\beta_1 C_I H_{1I}^2 + \beta_1 C_S H_{1S}^2 = \beta_0^I C_I H_{1I}^2, \qquad (4.4.12)$$

which allows us to determine β_1 as

$$\beta_1 = \beta_0^I (1 + \epsilon)^{-1}, \qquad (4.4.13)$$

where $\epsilon = S(S + 1)N_S / I(I + 1)N_I$. For the $^1H/^{13}C$ case, $\epsilon \sim 0.01$ and thus $(1 + \epsilon)^{-1}$ can be approximated by $(1 - \epsilon)$. The ^{13}C magnetization now established in the rotating frame after one cross-polarization contact is

$$M_1^S = \beta_1 C_S H_{1S} = \beta_1^0 (1 - \epsilon) C_S H_{1S}. \qquad (4.4.14)$$

Rewriting (4.4.14) using (4.4.11) and invoking the Hartmann–Hahn condition results in

$$M_1^S = \beta_L C_S H_0 (\gamma_I / \gamma_S)(1 - \epsilon). \qquad (4.4.15)$$

Thus the carbon polarization obtained in the CP process is about four times that established in H_0 (i.e., $M_0^S = \beta_L C_S H_0$).

After CP the contact between I and S spin systems is broken by removal of the H_{1S} field and the FID of M_1^S is observed (step 4). The S spins then return to an unpolarized state. The field H_{1I} remains on during the recording of the FID to eliminate I–S dipolar broadening. (The I magnetization does not vanish in this instance but decoupling is achieved since the axis of magnetization is perpendicular to H_0.)

The transfer of polarization between I and S spin systems is controlled by a cross-relaxation time constant T_{CR}. The rate (T_{CR}^{-1}) at which I and S spins reach thermal equilibrium in the double-rotating frame depends, in part, on the S spin second moment due to I–S dipolar interactions [see (4.4.2)]. The geometric dependence and motional dependence of this interaction suggests that (1) different types of S spins (e.g., carbons having directly bound protons as opposed to nonproton bearing carbons), (2) rigid groups as opposed to rapidly reorienting groups (e.g., methyl groups), and (3) different orientations with respect to H_0 of the same I–S spin pair (e.g., different portions of an anisotropy pattern) will reach full equilibrium [see (4.4.15)] at different rates. Thus determination of T_{CR}^{-1} values by interrupting the CP step after a time τ and observing the S spectrum as a function of τ provides, in principle, a valuable source of information on motion and orientation in solids.[145]

From (4.4.13), it is apparent that the initial CP contact results in only a small heating of the I spins. Thus, ideally, the remaining I magnetization locked in the rotating frame is

$$M_1^{\mathrm{I}} = \beta_{\mathrm{L}} C_{\mathrm{I}} H_0 (1 - \epsilon) = M_0^{\mathrm{I}} (1 - \epsilon). \qquad (4.4.16)$$

If $T_{1\rho}^{\mathrm{I}} > T_{\mathrm{CR}}$ (a necessary condition for cross polarization to occur, for otherwise the I magnetization will decay before it can be transferred to the S spins), it is possible to repeat the contact and observe the sequence (steps 3 and 4 above) before requiring I spin repolarization (step 1). If spin–lattice relaxation can be neglected, the S spin magnetization obtained from the nth contact will be[145]

$$M_1^{\mathrm{S}} = (\gamma_{\mathrm{I}}/\gamma_{\mathrm{S}})(1 - \epsilon)^n M_0^{\mathrm{S}}. \qquad (4.4.17)$$

Coaddition of the S spin's FID from each of the multiple-contact experiments yields, upon ultimate Fourier transformation, a spectrum (from one spin-lock period) having a total sensitivity enhancement even greater than 4×. For the interested reader, detailed analyses of the sensitivity considerations and of the CP process itself are presented by Pines et al.[145]

Although the development of a solution to the rare spin sensitivity problem as described in this section proceeded with an investigation of cross polarization on the basis of a more rapid recycle time for experiments, it is clear that the sensitivity enhancement obtained by cross polarization is also an important basis for utilization of the technique. As an example, in Figs. 37b and d are the spectra of PHBA and polycarbonate using the same recycle time as in Fig. 37a,c but employing cross polarization in conjunction with MAS and DD. A spectrum is now observed for PHBA and the intensities of the polycarbonate spectrum are quantitative. The first illustrations that combination of DD/CP/MAS techniques could be used to obtain high-resolution ^{13}C spectra of solids are due to Schaefer, Stejskal, and Buchdahl,[136,157,158] who reported a series of results on glassy polymers. We now turn to a brief discussion of experimental implementation of the DD/CP/MAS techniques followed by a discussion of their utility in studying macromolecules.

4.4.3. Experimental Implementation of DD/CP/MAS

In order for the reader to gain some insight into the instrumentation required for DD/CP/MAS experiments, a brief description of the spec-

---

[157] J. Schaefer, E. O. Stejskal, and R. Buchdahl, *Macromolecules* **8**, 291 (1975).
[158] J. Schaefer and E. O. Stejskal, *J. Am. Chem. Soc.* **98**, 1031 (1976).

trometer we have employed[159,160] to obtain HR-NMR spectra in solids is given in this section. Particular emphasis is placed on the MAS apparatus, which is designed for variable-temperature operation.[161]

4.4.3.1. Dipolar Decoupling/Cross Polarization. The basic spectrometer is a Nicolet TT-14 pulse-FT system operating at 1.4 T with resonance frequencies of 15.087, 60.0, and 56.4 MHz for ^{13}C, ^{1}H, and ^{19}F, respectively. The magnet is a Varian HA-60 electromagnet with a 1-$\frac{5}{8}$ in. polegap. The spectrometer has been modified for DD/CP by (1) constructing a probe capable of withstanding the RF power levels required; (2) adding external gating circuitry (for spin-locking, etc.) and RF amplifiers; and (3) rerouting some of the timing logic. The probe (10 mm i.d.) is a variation of the double-tuned single-coil design reported by Stoll, Vega, and Vaughan.[162] The coil is tuned for ^{13}C resonance and either ^{1}H or ^{19}F resonance. The probe has a "straight-through" design of Dewar and an external deuterium (D_2O) lock system. Broadband (0.25–105 MHz) 100 W RF power amplifiers (ENI-3100L) are employed to amplify the basic output of the spectrometer's ^{13}C and ^{1}H channels or the low-level ^{19}F output derived from a frequency synthesizer. With this arrangement, it is possible to achieve a Hartmann–Hahn CP match condition up to ~55 kHz for both the ^{13}C–{^{1}H} experiment and the ^{13}C–{^{19}F} experiment. If CP is not required, the proton decoupling network is capable of sustaining a H_1 field of 26 G at 100 W for dipolar decoupling, while the ^{19}F network yields a 22 G H_1 field at 100 W. A two-level decoupling scheme was added to the spectrometer in order to apply a continuous low-level (~1 G) H_{1H} field to the sample and then gate-on the strong H_{1H} field for short periods. When spectra are observed by $\pi/2$ pulses rather than CP, the weak decoupling field saturates the protons, thus sustaining any nuclear Overhauser enhancement (NOE)[163] present; application of the strong decoupling field during observation of the FID provides for dipolar decoupling. To remove artifacts in the CP spectrum that can arise from the long RF pulses in the observe channel, data are collected using a scheme[164] of alternating the proton spin temperature.

[159] C. A. Fyfe, J. R. Lyerla, W. Volksen, and C. S. Yannoni, *Macromolecules* **12**, 757 (1979).

[160] C. A. Fyfe, J. R. Lyerla, and C. S. Yannoni, *J. Am. Chem. Soc.* **100**, 5635 (1978).

[161] C. A. Fyfe, H. Mossbrugger, and C. S. Yannoni, *J. Magn. Reson.* **36**, 61 (1979).

[162] M. E. Stoll, A. J. Vega, and R. W. Vaughan, *Rev. Sci. Instrum.* **48**, 800 (1977).

[163] From the purely operational viewpoint, the existence of an NOE provides increased S/N. However, the two-level decoupling scheme also allows measurement of the NOE and thus additional insight into the molecular dynamics of the system under investigation (see Chapter 4.1).

[164] E. O. Stejskal and J. Schaefer, *J. Magn. Reson.* **18**, 560 (1975).

4.4.3.2. Magic-Angle Spinning. Two main techniques have been used to produce the high spinning rates required for MAS—one in which the rotors are supported by gas bearings and one in which rotors are supported by an axle assembly. The latter design, due to Lowe,[165] usually employs a cylindrical rotor that spins on a horizontal phosphor-bronze axle. The rotor is driven by compressed gas directed from a jet onto peripheral vanes on the rotor's exterior. The gas bearing design, pioneered by Beams[142] and adapted to NMR by Andrew,[26] requires a stator/rotor assembly with the rotors having a conical underface and a cylindrical stack (see Fig. 39). Compressed gas enters the stator through an inlet tube and is directed through inclined jets onto the rotor's flutes.[26] The rotor is thus supported and driven by the gas. The advantages of the Lowe design are the ease of loading the sample cylinder into the RF coil of the probe, ease of reaching the magic angle, and ease of adapting the assembly to variable-temperature operation. Disadvantages of the design include rapid wear of the rotor (a polymer) by the axle, sensitivity of spinning to rotor balance, and the small filling factor of the RF coil volume by the sample. The advantages of the gas bearing design are reduced wear, less sensitivity to rotor balance, and a good filling factor as the cylinder stack fits precisely into the coil (the cone of the rotor is outside the coil), which is canted at the magic angle rather than normal to the DC field.[166] Disadvantages of this design include difficulty in sample loading (often the coil itself must be removed), difficulty in the fine adjustment of

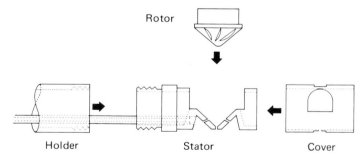

FIG. 39. Schematic representation of the component parts of the spinner assembly and their relationship (see text for description) [figure from C. A. Fyfe, H. Mossbrugger, and C. S. Yannoni, *J. Magn. Reson.* **36**, 61 (1979)].

[165] I. J. Lowe, *Phys. Rev. Lett.* **2**, 285 (1959).
[166] At this angle, the amplitude of the perpendicular component of H_{1l} is $0.816 H_{1l}$. This results in a voltage induced by transverse magnetization reduced by the same factor. The loss in S/N is compensated (under MAS conditions) by the better filling factor relative to a perpendicular arrangement.

the magic angle condition, and difficulty in adapting the design to variable temperature operation.

The spinning assembly we have employed is a new design[161] whose characteristics represent something of a compromise of the two above-described techniques. The general philosophy behind the design is to provide for (1) ease of sample loading and operation using conventional designs of probes, and (2) temperature variation. The importance of a variable-temperature capability is obvious if features such as barriers to reorientational modes, effects of phase transitions on chain mobility, the nature of secondary relaxations, etc. are to be determined. Additionally, the efficiency of the double-resonance experiment depends on $T_{1\rho}^I$ being of sufficient duration to allow complete polarization transfer and T_1^I being sufficiently short to allow rapid repolarization of the I spins. Since the two relaxation times are functions of temperature, a variable-temperature spinning capability allows for an optimally efficient[167] double-resonance experiment to be performed.

The spinner assembly is shown schematically in Fig. 39. The apparatus consists of four parts: holder, stator, rotor, and cover. The stator is threaded to the holder and the gas line (in the case of ambient temperature experiments) attached with a press fit. The rotor is inserted into the stator cup and the cover sleeve fitted over the stator. The general arrangement of stator cup and rotor is similar to that of Andrew.[26] The gas flow enters the conical "donut-shaped" cavity behind the stator cup (which is sealed to the outside by the sleeve), is directed toward the rotor through three small tangential holes (two of which are shown in the diagram) drilled through the stator cup, and impinges on the rotor. The arrangement of rotor flutes and stator holes is such that a clockwise movement of the rotor is produced. The exact shapes of these two components and the size and angles of inclination of the holes for the driving gas have to be carefully optimized as the apparatus makes a very tight fit into the probe insert and thus exit gas flow becomes important in stability considerations. The spinning axis is at right angles to the holder and stator assembly.

Spinner assemblies have been constructed from three materials: Kel-F, Delrin, and machinable boron nitride (BN).[167a] For observation of hydrocarbon materials at ambient and low temperature, the Kel-F assembly is

---

[167] The desired ratio of $T_{1\rho}^I/T_1^I$ being as close to unity as possible. Of course, account must be taken of the other factors which affect the recycle time such as the time required to collect data for the desired resolution (i.e., digital resolution), etc.

[167a] Teflon and Delrin are registered trademarks of the E. I. duPont Nemours & Co. (Inc.); Lexan is a registered trademark of the General Electric Co.; Kel-F is a registered trademark of the 3M Co.

used. It displays suitable mechanical properties and does not interfere with the carbon spectrum since the resonances of the carbons in the Kel-F are >10 kHz in width due to the unremoved C–F dipolar interactions. To observe fluorocarbon materials at ambient and low temperature by C–F dipolar decoupling/CP/MAS, the Delrin assembly is used since unremoved C–H dipolar interactions broaden the Delrin signal beyond detection. Finally, to obtain high-temperature spectra, the BN system, which is transparent in ^{13}C-NMR, has been employed. Samples in the form of powders are loaded into hollow rotors (volume ~65 μl), which are fitted with threaded caps to confine the sample. In some cases, the polymer under study was itself machined in the form of a rotor and used as the analytical sample. The rotors have a cone diameter of 8 mm and experience centripetal acceleration on the perimeter of $\sim0.4 \times 10^6$ G when rotating at 5 kHz.[168]

Since variable-temperature operation down to at least liquid nitrogen temperature is desirable, we have utilized helium as the propellant gas. At ambient temperature and 15 psi pressure, spinning rates >5 kHz are readily achieved. (The maximum theoretical rate for an 8 mm diameter rotor with helium as the propellant is ~38 kHz at 0°C.)[169] These rates are obtained on rotors without flutes. As fluting a rotor is a complicated machining operation, the ability to utilize smooth rotors is a desirable feature. Temperatures below ambient are achieved by mixing helium at ambient temperature with helium precooled by flow through a heat exchanger immersed in liquid nitrogen. The mixture of gas is transferred by a stainless steel Dewar (over which the spinning assembly fits) directly into the stator cup, resulting in spinning and cooling. Temperatures above ambient are obtained by preheating the helium at entry into the transfer Dewar. To the present, we have operated over a temperature range of -195 to $+100$°C.

For an electromagnet, there are two alternative ways in which the magic-angle arrangement can be achieved by continuous adjustment using the spinner design. The correct angle can be obtained by inserting the holder vertically into the probe (with the rotor spinning around a horizontal axis) and rotating it until the correct angle is reached, or by inserting the holder horizontally into the probe (also mounted horizontally) with the rotor spinning around a vertical axis and then rotating the holder until the correct angle is reached. Although the first arrangement in reaching the angle corresponds to a conventional RF coil/H_0 field geometry, the

[168] Despite numerous start-ups and "crashes," the Kel-F and Delrin rotors exhibit little wear after several hundred hours of operation.

[169] Determined from the maximum linear velocity of sound in helium and the diameter of the rotor periphery [$\omega = (v^2/r^2)^{1/2}$].

variable-temperature hardware is more easily arranged with the second geometry and, as such, is used on our spectrometer. In both arrangements, the exact angle chosen is reached by continuous adjustment of the rotation angle, which is controlled by a goniometer head attached to the top or front of the probe body. This gives considerable fine control over the exact angle chosen in an experiment.

In addition to the spinner assembly detailed here, several other variations of Andrew-type spinners have recently appeared in the literature.[141,142,170] Also commercial instrumentation to carry out DD/CP/MAS experiments is now entering the marketplace.[171]

4.4.4. Applications of ^{13}C NMR to the Solid State of Macromolecules

4.4.4.1. Structure and Composition.

In Fig. 40 is shown the ^{13}C spectrum obtained from a polycrystalline sample of the homopolymer of *p*-hydroxybenzoic acid (PHBA).[172] The spectrum is resolved into three distinct resonances of 1 : 1 : 5 relative intensities with the resonance of fivefold intensity having partially resolved components. The various carbons of the repeat unit are readily assigned to specific resonances on the basis of model compound studies and cross-relaxation studies (see

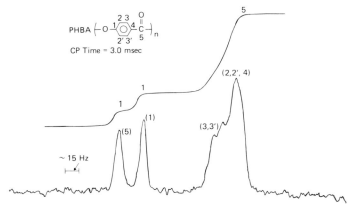

Fig. 40. DD/CP/MAS ^{13}C NMR spectrum of PHBA at 27°C. Spectrum obtained from 2K FT of 1500 FID accumulations with a CP time of 2.0 msec and an experiment repetition time of 3.5 sec. Line assignments are given in the figure and the spectral integration is displayed.

[170] R. G. Pembleton, L. M. Ryan, and B. C. Gerstein, *Rev. Sci. Instrum.* **48,** 1286 (1977).
[171] *Chemical and Engineering News,* October 16, p. 23 (1978).
[172] J. Economy, R. S. Storm, V. I. Matkovich, S. G. Cottis and B. E. Nowak, *J. Polym. Sci., Polym. Chem. Ed.* **14,** 2207 (1976).

below). Clearly, the spectrum represents a quantitative "fingerprint" of the polymer in the bulk state and makes possible the study of molecular dynamics at individual sites in the molecule. The PHBA polymer is insoluble in all common solvents and therefore not subject to solution characterization. This capability to analyze insoluble and cross-linked polymers promises to be one of the more useful aspects of the solid-state NMR techniques.

Another important facet of the PHBA spectrum is the appearance of separate resonances for the C-3 and C-3' carbons. The assignment of the resonances is definitive, based on the sequence of cross-relaxation spectra displayed in Fig. 41. As explained in Section 4.4.2.4, the rate of cross polarization depends primarily on the magnitude of the static C—H dipolar interaction experienced by a carbon. Thus, in the first approximation, carbons with directly bonded protons polarize more rapidly than those without direct interactions.[173] At CP times of 0.02 and 0.05 msec, only resonances from the two sets of ortho ring carbons are present in the spectrum. The 1:1:2 intensity pattern and the value of the chemical shifts allow the two downfield resonances in the two spectra to be accorded to the C-3 and C-3' carbons. At longer contact times, the region between the lines of the 2,2' and 3,3' carbons begins to develop intensity and at about the same rate as the line assigned to C-1. The development of line asymmetry in the 120–135 ppm region is clearly due to C-4 and demonstrates that this carbon resonates at 125.4 ppm, downfield of carbons 2 and 2'. The spectra at longer CP times are also in agreement with the proposed assignment of the carboxyl carbon in that the cross-polarization rate of the C-5 resonance is the slowest, as is expected on the basis of a larger distance to protons.

The nonequivalency of the C-3,3' ring carbons is direct evidence that the motion of the ring is confined to small-amplitude oscillations about an asymmetric equilibrium position. The spectrum has been examined in the temperature range −155 to +80°C and the separation of the C-3,3' carbons remains constant at 51 Hz. The 80°C spectrum allows a minimum barrier to rotational averaging of >17 kcal/mole to be calculated. We are currently attempting to obtain spectra at higher temperatures using the BN spinning assembly in order to observe the coalescence of the two lines and determine activation parameters. However, dielectric

---

[173] Also, the number of interactions is important, as well as effects of molecular motion. For example, a methyl group would cross-polarize more rapidly than a methylene or methine in a rigid-lattice situation; however, if the methyl group undergoes rapid reorientation about the C_3 axis, partial averaging of the dipolar interaction will occur, which may result in the methyl carbon being polarized more slowly than methylene or methine carbons despite the greater number of C—H interactions.

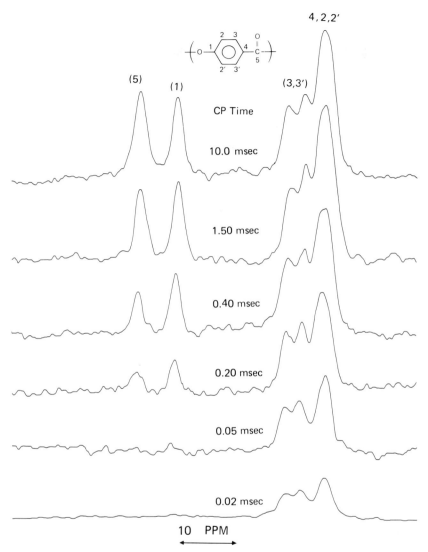

FIG. 41. Magic-angle CP ^{13}C NMR spectrum of PHBA at 27°C as a function of CP contact time (in msec). All spectra obtained from 2K FT of 1000 FID accumulations at a 3.5 sec experiment repetition time [figure from C. A. Fyfe, J. R. Lyerla, W. Volksen, and C. S. Yannoni, *Macromolecules* **12,** 757 (1979)].

data[172] on the polymer suggest that motion sufficient to cause averaging of the C-3,3′ chemical shifts may not occur until 325–350°C.

The proposed structure of PHBA is a double helix with the two chains having a reversed head-to-tail order,[172] resulting in a highly rigid polymer chain. The NMR data supports a rigid structure in several ways:

(1) The nonequivalent ^{13}C chemical shifts of the 3,3′ carbons indicate the limited torsional oscillations of the aromatic ring.

(2) The fact that the linewidths and shift separations in the spectrum of PHBA at − 155°C are the same as at + 80°C suggests that temperature reduction does not produce any dispersion in chemical shift due to nonaveraging conformations nor any significant change in lattice motions since the marginal decoupling field (6.5 G) still removes proton broadening effects at the low temperature. Both these conclusions support a highly rigid polymer chain at ambient temperature.

(3) The proton spectrum of PHBA, which is characterized by a linewidth of ~6 G and T_1 values of 3–4 sec at 25°C, is also unchanged at − 155°C.

(4) As pointed out in Section 4.4.2.4, it is not possible to obtain a spectrum of PHBA with DD, MAS, and $\pi/2$ repetitive pulses on the carbon spin system using as much as 20 sec recycle times and 8000 FID accumulations. This result suggests that the carbons have long T_1 values, which implies only small spectral density in the 15 MHz region.

The magic-angle CP spectra obtained from 25 mg powder samples of two insoluble methoxy derivatives of PHBA are shown in Fig. 42a,b. The perturbations in the electron shieldings of PHBA caused by the methoxy substituents result in extremely well-resolved solid-state spectra. By applying chemical shift substituent parameters for a methoxy group introduced onto an aromatic ring[174] to the chemical shifts of PHBA, it is possible to tentatively assign the resonance lines in Fig. 42 to specific carbons of the repeat units. The observed line intensities indicate that at the chosen value of cross-polarization time, the spectra are essentially quantitative. The apparent nonequivalency of the methoxy carbons and 3,3′ ring carbons in the dimethoxy polymer indicates restricted reorientation of the aromatic ring and thereby the polymer backbone. Because the monomethoxy polymer would be expected to have a repeat unit composed entirely of nonequivalent carbons, it is not possible to make any qualitative conclusions regarding motion without the benefit of relaxation studies.

[174] J. B. Stothers, "Carbon-13 NMR Spectroscopy," p. 197. Academic Press, New York, 1972.

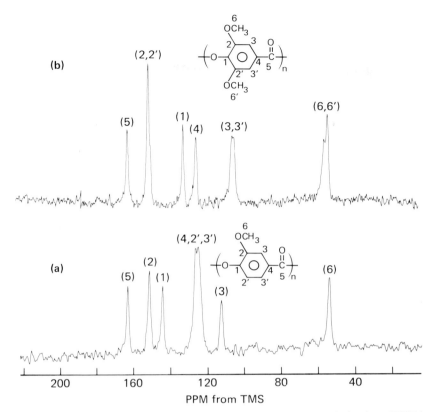

FIG. 42. (a) Magic-angle CP ^{13}C NMR spectrum of a monomethoxy derivative of PHBA at 27°C. Spectrum obtained from 2K FT of 21,900 FID accumulations with a CP contact time of 2.5 msec and experiment repetition time of 3.0 sec. (b) Magic-angle CP ^{13}C NMR spectrum of a dimethoxy derivative of PHBA at 27°C. Spectrum obtained from 4K FT of 19,000 FID accumulations with other conditions as in (a) [figure from C. A. Fyfe, J. R. Lyerla, W. Volksen and C. S. Yannoni, *Macromolecules* **12**, 757 (1979)].

The nonequivalency of ortho ring carbons observed in the aromatic polyesters is also observed in noncrystalline materials when the aromatic ring is incorporated in the backbone of a polymer.[136,175] For example, Garroway *et al.*[175] have carried out variable-temperature MAS experiments on the epoxy polymer obtained by curing the diglycidyl ether of bisphenol A(DGEBA) with piperidine. The 100° temperature range covered by the experiments is sufficient that restricted rotation, coalescence,

[175] A. N. Garroway, W. B. Moniz, and H. A. Resing, *ACS Symp. Ser.* **103**, 67 (1979).

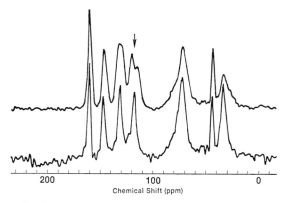

FIG. 43. ^{13}C DD/CP/MAS spectra of piperidine cured epoxy at $-36°$ (top) and 58°C (bottom). The splitting in the aromatic ring carbons ($-36°$C) denoted by the arrow is 5.5 ppm [figure from A. N. Garroway, W. B. Moniz, and H. A. Resing, *ACS Symp. Ser.* **103,** 67 (1979)].

and motional averaging of the ortho ring carbons in DGEBA that are three bonds removed from the methylene are observed (see Fig. 43).

The resolution attained in the solid state by employing DD/CP/MAS techniques is sufficient to warrant using NMR to determine composition of polymer blends and copolymers. In Fig. 44, the ^{13}C spectrum of one of the compatible blends of poly(phenylene oxide)/poly(styrene) (PPO/PS)

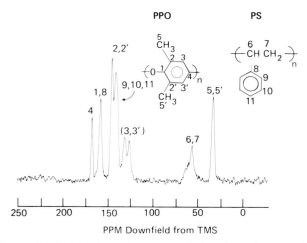

FIG. 44. The DD/CP/MAS ^{13}C NMR spectrum of a blend of polystyrene/polyphenylene oxide at 27°C. Assignments are indicated.

is presented. Comparison of the intensity of the methyl carbon resonance of PPO to that of the backbone carbons of PS is an obvious basis for
quantitative analysis. As in the PHBA spectrum, one set of ortho ring
carbons is nonequivalent in PPO—again, reflecting the restricted rotation
in the solid state since the carbons are equivalent in solution.[136] Aside
from the analytical aspect, the degree of resolution obtained on these
blends may allow insight to be gained into the nature of the interactions
responsible for compatibility of the two homopolymers, e.g., from
variable-temperature heteronuclear cross-relaxation and rotating-frame
spin–lattice relaxation studies as a function of blend composition. An
example of copolymer analysis is provided by the spectra of two insoluble p-
hydroxybenzoic acid/biphenylene terephthalate copolymers given in Fig.
45. The monomer feed ratios for the compounds were $2:1$ and $1:2$ molar
ratios respectively. Although the spectra show considerable overlap
from the many aromatic resonances, the lines corresponding to the carboxyl carbons and the aromatic carbons bonded to oxygen are resolved.
The resonance of the carbons bonded to oxygen in the HBA units differs

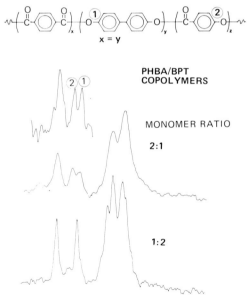

FIG. 45. Magic-angle CP spectra for $2:1$ and $1:2$ HBA/BPT copolymers at 27°C. Spectra
were obtained with CP times of 3 msec and recycle times of 2.5 sec. The spectra are the result of 8000 accumulations. The inset spectrum for the $2:1$ copolymer is one obtained with
better setting of the magic-angle spinning condition [figure from C. A. Fyfe, J. R. Lyerla, W.
Volksen, and C. S. Yannoni, *Macromolecules* **12,** 757 (1979)].

in chemical shift from the corresponding carbons in the BPT units. For the 2:1 copolymer the two resonance lines would be of equal intensity since the BPT unit has two equivalent carbons of this type, while in the 1:2 copolymer the ratio of the resonance lines would be 1:4. Clearly, these are the approximate ratios observed for these two resonance lines in Fig. 45, demonstrating that, under appropriate cross-polarization conditions, the solid-state spectra can be used for quantitative compositional analysis on such copolyesters. Unfortunately, the degree of resolution is not sufficient in the spectra to allow sequence data to be obtained.

4.4.4.2. Orientation. Aromatic copolyesters of HBA with ethylene terephthalate are examples of polymers that display liquid-crystalline order—in this case, thermotropic liquid-crystalline[176] behavior is exhibited (for certain compositions) in the melt. Mesomorphic order in polymers is of current interest because of the technological implications with respect to ultrahigh-strength/modulus materials.[177] Proton magnetic resonance has been employed to study the nature of low-molecular-weight mesophases for many years.[178] The use of DD/CP methods to observe ^{13}C resonances in mesophases appears to be capable of providing even greater detail on mesomorphic systems, since dipolar decoupling removes nonaveraged C–H dipolar interactions arising from the orientation of the molecules in the magnetic field.[179–181] For example, the ^{13}C spectra of the model thermotropic liquid crystal p-methoxy benzylidene-p'-n-butylaniline (MBBA) in the nematic state as a function of cross-polarization time are shown in Fig. 46. The chemical shifts observed between isotropic and nematic states are typical of a molecule whose long axis (about which motion is rapid) is preferentially aligned along the magnetic field direction.[179] Thus, the aromatic carbons, having their maximum shielding component perpendicular to the long axis,[146] experience downfield shifts in the nematic state relative to the isotropic state. The side-chain carbons are more shielded in the long-axis direction and thus experience upfield shifts; however, the shifts are much less than for the aromatic carbons owing to the smaller shielding anisotropies for the aliphatics and to a lesser degree of ordering in the aliphatic tail. The

[176] W. J. Jackson, Jr. and H. F. Kuhfuss, *J. Polym. Sci., Polym. Chem. Ed.* **14,** 2043 (1976).

[177] A. Blumstein, "Liquid Crystalline Order in Polymers." Academic Press, New York, 1978.

[178] G. R. Luckhurst, *in* "Liquid Crystals and Plastic Crystals" (G. W. Gray and P. A. Winsor, ed.), Vol. II, Chapter 7. Ellis Horwood Ltd., Chichester, U.K.

[179] A. Pines and J. J. Chang, *Phys. Rev. A* **10,** 946 (1974).

[180] A. Pines and J. J. Chang, *J. Am. Chem. Soc.* **96,** 5590 (1974).

[181] A. Pines, D. J. Rueben, and S. Allison, *Phys. Rev. Lett.* **33,** 1002 (1974).

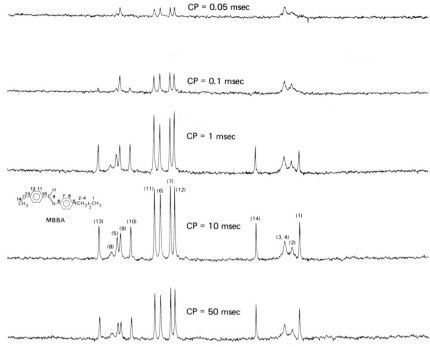

FIG. 46. ^{13}C NMR spectra of MBBA in the nematic phase at 22°C as a function of cross-polarization time. Resonance line assignments are indicated in the spectrum obtained with CP = 10 msec. Each spectrum represents the FT of 400 FID accumulations. Results are reported on a normalized scale.

dynamics of polarization build-up indicate that the methoxy and methyl carbons have quite reduced residual C–H dipolar interactions as compared to the carbons of the aromatic core. This is, no doubt, the result of chain flexibility and internal rotation about C_3 axes. Orientation data on the degree of alignment of the long axis with H_0 are also available in the spectra. In the case of smectic-A mesophases, the DD/CP spectra are strongly dependent on orientation of the "director axis" and the field. A parallel orientation can be established by cooling in the magnetic field through a nematic into a smectic phase. Once the smectic phase is formed under the influence of H_0, the high viscosity of the smectic phase, unlike the nematic, prevents realignment of the long axis along H_0 when the sample is rotated, thus producing a spectral orientation dependence characteristic of a uniaxial crystal.[179–181] The extension of these techniques to polymeric liquid-crystal compounds seems straightforward and offers the potential to learn about the degree of ordering in such systems, the composition of coexisting phases, and molecular motion.

Obviously, the spectra of the liquid-crystalline systems are obtained without MAS, since rapid spinning would destroy the orientation data. Indeed, as was alluded to in Section 4.4.1, the improvement in resolution gained by magic-angle spinning is at the expense of the information contained in the chemical shift anisotropy. The significance of the directional dependence of the chemical shift in elucidating polymer structure is evidenced by the spectra of PE obtained under dipolar-decoupled conditions (Fig. 47). The principal elements of the shielding tensor (51.4, 38.9, 12.9 ppm from TMS) have been assigned with respect to the geometry of an all-trans polymer chain.[182,183] Based on these assignments, the carbons of a PE chain in which the chain axis is aligned along H_0 will resonate at 12.9 ppm. As demonstrated by VanderHart,[184] this provides a method of monitoring the chain orientation induced in a PE sample by processes such as drawing. The ^{13}C spectrum of an oriented PE sample in which the sample was configured in the spectrometer such that the draw direction is parallel to H_0 is given in Fig. 48a. The bulk of intensity from the sample (which was prepared by extrusion at 2400 atm and 130°C through a conical dye to produce a draw ratio $\sim 12:1$)[184] is centered at

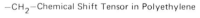

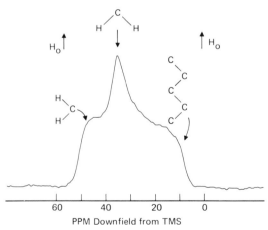

FIG. 47. The DD/CP ^{13}C spectrum of polycrystalline PE. The three elements of the chemical shielding tensor and their assignment in an all-trans chain are indicated.

[182] J. Urbina and J. S. Waugh, *Proc. Natl. Acad. Sci. U.S.A.* **71**, 5062 (1974).
[183] D. L. VanderHart, *J. Chem. Phys.* **64**, 830 (1976).
[184] D. L. VanderHart, *J. Magn. Reson.* **24**, 467 (1976).

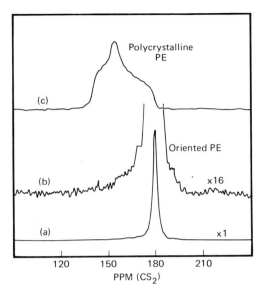

FIG. 48. The DD/CP ^{13}C NMR spectra of (a) fiber-oriented PE, draw direction fixed parallel to H_0; (b) as in a but amplification $\times$ 16; (c) a polycrystalline sample of PE [figure similar to D. L. VanderHart, *Macromolecules* **12**, 1232 (1979)].

12.9 ppm, consistent with a high degree of chain orientation in the draw direction. Examination of the spectrum under higher amplification (Fig. 48b) shows that intensity is spread over the entire 38 ppm chemical shift range encompassed by polycrystalline PE, indicative of the presence of misaligned chains in the oriented sample.

Spectra of the oriented PE sample as a function of temperature[184] are given in Fig. 49. Several features are of interest:

(1) As the temperature is raised, there is a significant decrease in the width of the downfield shoulder on the central resonance line. This result indicates that the misaligned chains show greater motional averaging than the oriented chains, as might be expected on the basis of poorer packing in the regions of misaligned chains.

(2) At high amplification, satellites are present in the spectra that arise from ^{13}C–^{13}C dipolar couplings. Because of the sample's high degree of orientation, the resonances occur at discrete positions such that the satellite splitting is given by

$$\Delta = 1.5\gamma_C^2\hbar^2(1 - 3\cos^2\theta_{ij})/R_{ij}^3. \qquad (4.4.18)$$

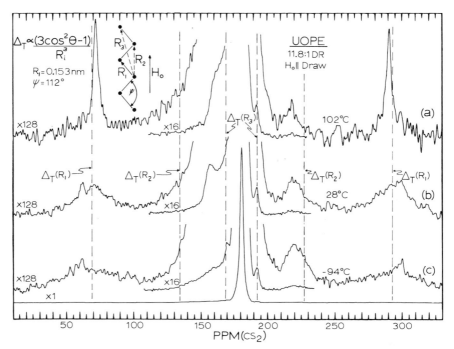

FIG. 49. ^{13}C cross-polarization spectra of ultraoriented linear polyethylene with an 11.8:1 draw ratio. The draw direction is parallel to the magnetic field. The three pairs of $^{13}C-^{13}C$ dipolar satellite positions based on $R_1 = 0.153$ nm and $\psi = 112°$ are given by the dashed lines. Temperatures and spectral amplification factors are also given [figure from D. L. VanderHart, *J. Magn. Reson.* **24**, 467 (1976)].

Calculation of the satellite splittings[184] based on an all-trans PE chain indicates the three strongest interactions (marked by the dashed lines in Fig. 49) arise from pairs of carbons on the same chain, which are one, two, and three bonds separated. Only for the directly bonded carbons is the pair of satellites observed, the downfield members of the two- and three-bond interactions being obscured by the resonances from misaligned chains. The breadth of the satellite resonances arises from the presence of misaligned chains, and the narrowing that occurs at 102°C is most easily interpreted as resulting from rotational averaging of θ_{ij} about the chain axis for those chains slightly misaligned relative to H_0.

The restricted reorientational freedom in the crystalline regions of PE results in spin–lattice relaxation times of ~700 sec,[149] while CH_2 carbons of PE chains in amorphous material have T_1s ranging from ~0.3 to

3 sec.[185]   Advantage can be taken of this differential in relaxation rates to study only the noncrystalline regions of PE. VanderHart[149] has shown that by performing a 180–t–90 pulse sequence with t = 10 sec and an experiment recycle time of 10 sec, it is possible to null and saturate all carbon intensity except that originating from carbons having $T_1 \sim 3$ sec or less. Thus, for the four spectra of drawn LPE shown in Fig. 50, the observed intensity originates almost exclusively from PE chains in non-crystalline regions of the samples. (The spectra have been normalized to represent the appropriate amount of noncrystalline material in the sample.)   A distinguishing feature between the spectra is the relative amount of aligned vs. misaligned chains.   Clearly, the two cold-drawn samples (draw ratio 15) have a much higher concentration of mobile chains that are aligned in the draw direction (represented by the major resonance at 12.9 ppm) than the high-temperature-drawn sample (draw ratio 12).   This may be due to annealing processes that can take place in the higher-temperature drawing and result in more perfect packing of chains and a concomitant reduction in chain mobility.   In the cold-drawn samples, larger concentrations of imperfections and voids may be frozen

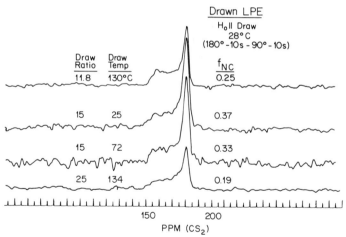

FIG. 50. Proton-decoupled ¹³C NMR spectra from the noncrystalline regions of linear polyethylene as a function of draw ratio and temperature. The noncrystalline fraction of material is indicated by f_{NC}. The pulse sequence used is indicated. Spectra are given on a scale normalized to f_{NC} [figure courtesy of D. L. VanderHart, National Bureau of Standards].

[185] R. A. Komoroski, J. Maxfield, F. Sakaguchi, and L. Mandelkern, *Macromolecules* **10**, 550 (1977).

in, resulting in more volume for aligned chains to undergo rotation about the chain axis, and thus accounting for the relative intensities observed in the spectra of Fig. 50.

4.4.4.3. MAS of Polyethylene and Fluoropolymers.

The DD/CP/ MAS spectrum of semicrystalline PE [159] (~65% crystallinity) is shown in Fig. 51 (top). It is obvious that the resonance line is asymmetric on the high-field side. Observation of the spectra as a function of time between 90° pulses using DD/MAS conditions shows the presence of two peaks separated by ~2.3 ppm. It is apparent the high-field resonance line has a much shorter carbon T_1 (having attained equilibrium intensity in 2 sec) and a greater width (40–50 vs. 10 Hz) than the low-field resonance line. These results are consistent with the assignment of the low-field line to the crystalline regions of PE and the high-field line to the non-crystalline regions. In particular, the long T_1 of the low-field line is consistent with the limited mobility of chains in the crystalline regions relative to noncrystalline regions.

The 10 Hz linewidth of the peak from the crystalline region of PE is in

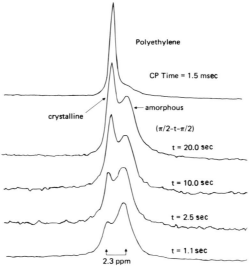

FIG. 51. Magic-angle CP and $\pi/2$ ^{13}C NMR spectra of semicrystalline polyethylene at 27°C. The spectra obtained with a $\pi/2-t-\pi/2$ pulse sequence used the indicated values of t. The number of FID accumulations varied from 2000 to 4500 but the spectra are displayed on a normalized scale. The spectrum obtained from the CP experiment used a CP contact time of 1.5 msec and an experiment recycle time of 2 sec. The spectrum is the result of 400 FID accumulations [figure from C. A. Fyfe, J. R. Lyerla, W. Volksen, and C. S. Yannoni, *Macromolecules* **12**, 757 (1979)].

accord with the linewidths ($\sim$15 Hz) observed for the highly crystalline aromatic polyesters discussed earlier. The 40–50 Hz width of the peak from noncrystalline regions is similar to the 50–100 Hz linewidths observed in glassy systems.[136] One consequence of the chain rigidity in crystalline regions may be a substantial reduction of motions in the 10^4–10^5 Hz region, as well as in the megahertz region, relative to that for glasses. Motions in the 10^4–10^5 frequency region do not result in complete motional averaging but instead yield a dipolar broadening that is neither removable by realistically achievable proton-decoupling fields nor with currently achievable magic-angle spinning speeds. Thus, a reduction of these motions in the aromatic polyesters and crystalline PE may, in part, account for the narrower resonance lines relative to glassy systems.[159]

If a greater homogeneity of the local environment experienced by each repeat unit can be associated with crystal habit, then the high degree of crystallinity of these aromatic polyesters may result in substantial reduction of line broadening from dispersions of isotropic chemical shifts. Such dispersions are associated with the steric (in vinyl polymers) and conformational isomerism common in the glassy state.[136] If reduced shift dispersion is indeed the primary source of the line narrowing in crystalline polymers relative to glasses, it would support the proposal[136] that it may not be possible to narrow lines in glassy polymers below 40–50 Hz.

The chemical shift between the resonance maxima in PE presumably arises from constraint to trans bond conformations in the crystalline region as opposed to trans/gauche conformations in the amorphous regions. Assuming that the chemical shift difference (6.4 ppm) between the crystalline peak in PE and the resonance of cyclohexane, which exists in an all gauche conformation, can be attributed entirely to conformational effects, Earl and VanderHart[186] have concluded that the 2.3 ppm shift between resonances for the crystalline and noncrystalline regions is 0.6 ppm less than would be expected for a fraction of gauche states based on thermal considerations.[187] Although speculative, constraints on noncrystalline chains biasing the equilibrium population of trans/gauche toward trans would account for the difference. Such a result would also be in accord with the linewidth observed in the noncrystalline region. Motional effects account for only about one-half the total width; chemical shift dispersion contributes about another 50% to the width.[186] The dispersion can be attributed to variations in trans/gauche population ratios

---

[186] W. Earl and D. L. VanderHart, *Macromolecules* **12**, 762 (1979).

[187] At 300°K and using a barrier of 600 cal for t-g conformational jump, a fraction of gauche states equal to 0.42 is calculated.

in various chains or chain segments. These initial results suggest that more detailed investigations, including variable-temperature studies, are warranted to determine the potential of solid state NMR in elucidating the nature of semicrystalline PE.

From the discussion on PE it is apparent that the static as well as the magic-angle spinning double-resonance experiments provide useful information. However, we have seen earlier (e.g., Fig. 34b) that overlap of the shielding anisotropies usually precludes detailed analysis of orientational effects in multiline spectra. Fortunately, because the chemical shift anisotropy is totally static in origin, sample spinning at a rate much less than the anisotropy is sufficient to completely displace the spectral density from zero frequency and remove the line broadening due to this source.[188] Of course, sidebands are prominent in slow-spinning experiments when the chemical shift anisotropies are large; however, the intensity pattern of the spinning sidebands reflect the magnitude of the shift anisotropy and by proper analysis this information can be recovered by measuring spectra at a few spinning frequencies.[188,189] The utility of the technique has been demonstrated by Stejskal et al. for PMMA.[188] An alternative approach to obtain chemical shift anisotropy data in complex systems is to isotopically enrich with ^{13}C at sites of interest. Enrichment levels are required that produce a S/N sufficient to reduce the signal intensity from nonenriched sites to the noise level yet do not introduce problems from ^{13}C–^{13}C dipolar broadening. In less complex systems, variable-angle spinning can be used to determine the elements of the shielding tensor. For example, in the static ^{13}C spectrum of cis-1,4 polybutadiene (CPBD) at $-150°C$ only two elements (Fig. 52a) of the shielding tensor $\hat{\sigma}$ are resolved. The third overlaps the —CH_2— pattern, thus making it impossible to fully assign the tensor. However, variable-angle spinning allows the full assignment to be made through the angular functional dependence of the spectrum (see Fig. 52b). At the magic-angle (Fig. 52a) the spectrum collapses to a line with FWHM of 50 Hz. With the sample spinning at an angle of 87°, the $P_2(\cos \theta)$ functional dependence predicts a spectrum, relative to the static case, reduced in width by 0.492 and with the principal elements of the shielding tensor reversed with respect to the isotropic shift.[144] On this basis, the principal values of $\hat{\sigma}$ calculated from the data of Fig. 52b are $\sigma_{11} = 236 \pm 4$, $\sigma_{22} = 115 \pm 4$, $\sigma_{33} = 35 \pm 4$. Assuming a similar orientation of the shielding tensor as in

[188] E. O. Stejskal, J. Schaefer, and R. A. McKay, J. Magn. Reson. **25**, 569 (1977).

[189] E. Lippmaa, M. Alla, and T. Tuherm, in "Magnetic Resonance and Related Phenomena" (H. Brunner, K. H. Hausser and D. Schweitzer, eds.), p. 113. Heidelberg-Geneva, 1976.

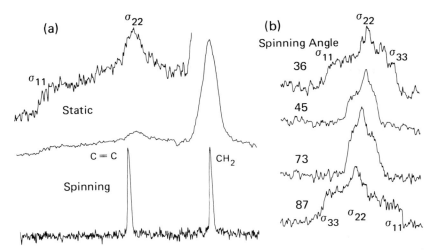

FIG. 52. (a) DD/CP ^{13}C NMR spectra of *cis*-1,4 polybutadiene at −150°C: static and spinning at the magic-angle. For the static spectrum, two elements of the shielding tensor are indicated for the vinyl carbon in the 4 × vertical expansion (inset). (b) DD/CP spectra as a function of spinning angle. Tensor elements are indicated.

ethylene ($\sigma_{11} = 236$, $\sigma_{22} = 124$, $\sigma_{33} = 27$),[190] these elements correspond to shielding approximately perpendicular (but in plane) to the double-bond direction, parallel to the bond direction, and perpendicular to the plane of the bond, respectively. The isotropic shift obtained from these values is 128.7 ppm as compared to 129.7 determined from the spectrum in the elastomeric state and the MAS spectrum at −150. [Note that, in the MAS spectrum, both resonances despite large anisotropy differences have the same FWHM and side-bands are absent from the spectrum demonstrating the efficiency of the spinning device (described in the experimental section) at low temperature.] A strong driving force for observing chemical shift anisotropies in polymers is that the manner in which the shielding tensor undergoes averaging may provide insight as to the nature of motions (i.e., along chain axis, perpendicular to chain axis, etc.) that enter a polymer chain as a function of temperature (and thereby physical state).

Fluoropolymers are subject to analysis by DD/CP/MAS techniques by using the ^{19}F nuclei to cross-polarize the ^{13}C spins. Figure 53 displays the ^{13}C spectrum of a highly crystalline Teflon at −120°C. The spectrum of the spinning sample is reduced about fivefold in width relative to the non-

[190] K. W. Zilm, R. T. Conlin, D. M. Grant, and J. Michl, *J. Am. Chem. Soc.* **100**, 8038 (1978).

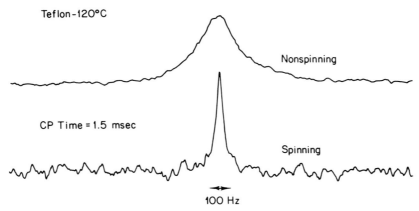

FIG. 53. Static and spinning fluorine-decoupled CP ^{13}C NMR spectra of highly crystalline poly(tetrafluoroethylene) at − 120°C.

spinning sample. An anisotropy pattern was not obtained, which may be due to the large anisotropy of the fluorine resonance[191] coupled with a marginal decoupling field. The combination of these factors results in incomplete removal of C−F dipolar interactions and thus failure to define the shape of the anisotropy pattern (expected to be a nonaxially symmetric pattern).

The DD/CP/MAS spectrum of poly(chlorotrifluorethylene) (CTFE) displays the expected two-carbon resonances at ambient temperature and above (Fig. 54). As the temperature is reduced, the upfield resonance, which is assigned to the CFCl carbon, begins to broaden substantially and ultimately disappears into the noise. This broadening apparently results from nonaveraged C−Cl dipolar interactions. Although 3 kHz MAS is sufficient to partially average this interaction, the fact that chlorine usually has a large quadrupole coupling constant[192] in an organic system such as CTFE results in the chlorine nuclear moments being fixed at some orientation in the molecular axis system rather than along the Zeeman field. The moments follow the mechanical rotation of the sample to a degree dependent on the ratio of the Zeeman and quadrupole interaction energies of the nucleus,[193] resulting in line broadening and line splitting. At temperatures near T_g (~40), motion in the polymer backbone is apparently sufficient to cause the chlorine T_1 to be short enough to induce

[191] M. Mehring, R. G. Griffin, and J. S. Waugh, *J. Chem. Phys.* **55,** 746 (1971).
[192] F. A. Cotton and C. B. Harris, *Proc. Natl. Acad. Sci. U.S.A.* **65,** 12 (1966).
[193] E. Lippmaa, M. Alla, and E. Kundla, *18th Exp. NMR Conf., 1977.*

"self-decoupling" of the C–Cl dipolar interaction.[194]   At temperatures well below T_g, motional averaging is severely reduced and thus the observed broadening.

4.4.4.4. Dynamics in the Solid State. In principle, resolution of individual carbon resonances in bulk polymers allows relaxation experiments to be performed that can be interpreted in terms of main- and side-chain motions in the solid.   In addition to the spin–lattice relaxation time in the Zeeman field, the spin–spin relaxation time and nuclear Overhauser enhancement, other parameters providing data on polymer dynamics include the proton and carbon spin–lattice relaxation times in the rotating-frame $T_{1\rho}$, the cross-relaxation time T_{CR}, and proton relaxation in

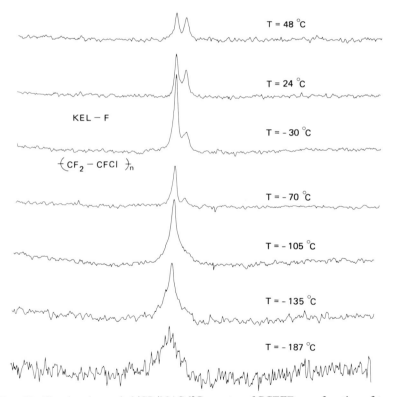

FIG. 54. Fluorine-decoupled/CP/MAS ^{13}C spectra of PCTFE as a function of temperature.

[194] A. Abragam, "The Principles of Nuclear Magnetism," Chapter VIII. Oxford Univ. Press, London and New York, 1961.

the dipolar field.   Schaefer and Stejskal[136,157,195] have carried out pio-
neering work in exploring polymer dynamics using solid-state NMR techniques.   Measurement of ^{13}C T_1 values in glassy PMMA at ambient temperature reveals that the α-CH$_3$ carbon relaxes in <0.1 sec, the ester
methyl and methylene carbons in ~1 sec, and the two nonprotonated
(carbonyl and quaternary) carbons in ~10 sec.   These results[157] are consistent with the onset of internal reorientation of α-CH$_3$ at this temperature, relatively unhindered internal rotation of the side-chain methyl, and
restriction of any large-scale motion by the main chain (T_g ~ 105°C).   The
presence of an approximately full chemical shift anisotropy for the carbonyl carbon in the nonspinning spectrum of PMMA at room temperature
is direct proof that the ester side group as a whole does not engage in extensive internal rotational reorientations comparable to the internal motions of the ester-methyl (or ester-methoxy) and α-methyl carbons.

The NOE values determined for the three proton-bearing carbons were
all found to be about 2.0, which is less than maximum (3.0) but far from
the minimum (1.15) that might be expected for a rigid system.   This result argues for the presence of a broad distribution of correlation times associated with a variety of segmental rotational and torsional motions.
The presence of a distribution of correlation times tends to level the
NOEs of most carbons to about the same value even though these
carbons may still have significantly different T_1s.   Unlike the situation for
polymer systems well above T_g, the distribution of correlation
times[71,196,197] for solid PMMA probably has a tail in the direction of short
correlation times (see Chapter 4.3).   In other words, below T_g, long-range
cooperative motions (necessarily associated with long correlation times)
are discriminated against, while many short-range, high-frequency
(10^4–10^6 Hz) torsional motions can apparently still persist, even for the
main chain, and even well below T_g.   Thus, deviations from a symmetrical, narrow distribution of correlation times are most likely to be toward
the short-correlation time region.   The net result is that a substantial
NOE can still exist for carbons in a solid, glassy polymer.   Naturally the
high-frequency motions resulting in the NOE are also effective contribu-
tors to the spin–lattice relaxation times of the various carbons.   This is
the reason the methylene-carbon T_1 in PMMA has a short T_1 as compared
to those found in highly crystalline materials.

Long ^{13}C T_1s make measurement of T_1 and NOE by the usual methods
(e.g., 180°–t–90° sequence) tedious if not prohibitive because of the long

[195] J. Schaefer and E. O. Stejskal, *in* "Topics in Carbon-13 NMR Spectroscopy" (G. C.
Levy, ed.), Vol. III, 283 (1979).
[196] J. Schaefer, *Macromolecules* **6**, 882 (1973).
[197] J. R. Lyerla, Jr. and D. A. Torchia, *Biochemistry* **14**, 5175 (1975).

delay times required for carbon repolarization. Recently, Torchia[198] has developed a method of determining T_1 that allows repolarization by the more rapid CP process. Additionally, the method appears to be well-suited for T_1 measurements in two-phase systems. Thus, measurement of T_1 should be possible for most polymers; on the other hand, it is expected that the interpretation of the results will require considerable development of theory and models.

While the T_1 and NOE parameters are sensitive to polymer motions in the 15–100 MHz region, it is main-chain motions in the 15–100 kHz region that are likely to be important in determining the mechanical properties of polymers below T_g. As described earlier, the kilohertz region of the frequency spectrum is accessible by measurement of the carbon rotating-frame relaxation time $T_{1\rho}^C$. Schaefer et al.[136] have measured this relaxation time for a number of glassy polymers using the experimental sequence depicted in Fig. 55. Briefly, carbon magnetization is established by the Hartmann–Hahn CP procedure, contact between the ^1H and ^{13}C spin systems destroyed (i.e., H_{1H} turned off), the carbons held in their rotating frame for a variable time, H_{1C} turned off, and the carbon magnetization sampled under dipolar decoupled conditions. $T_{1\rho}^C$ is then extracted from a semilog plot of carbon intensity vs. hold time after breaking CP contact. An example of spectra of a relaxation sequence for both

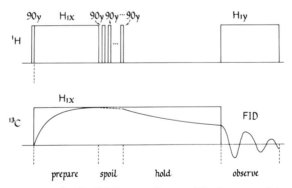

FIG. 55. Pulse sequence for the ^{13}C $T_{1\rho}$ experiment. The first part of the experiment is a matched Hartmann–Hahn CP preparation of carbon magnetization. The second part (the "spoil" sequence of 90° pulses separated by ^1H T_2s) is usually omitted; it is included only when searching for possible effects of transients in the ^1H channel on the ^{13}C channel. The third part of the experiment is the variable time the carbons are held in their rotating frame without CP contact with the protons. This is followed in the fourth part by the observation of the carbon free induction decay with resonant dipolar decoupling of the protons [figure from J. Schaefer, E. O. Stejskal, and R. Buchdahl, *Macromolecules* **10**, 384 (1977)].

[198] D. A. Torchia, *J. Magn. Reson.* **30**, 613 (1978).

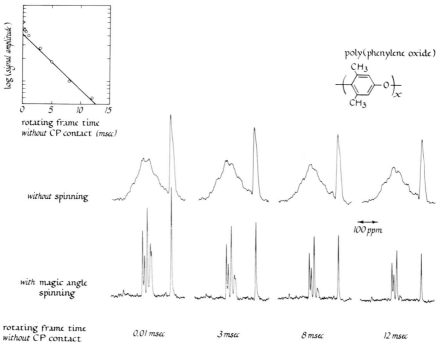

FIG. 56. Cross-polarization ^{13}C-NMR spectra of poly(phenylene oxide), with and without magic-angle spinning, as a function of the time the carbon magnetization was held in the rotating frame without CP contact. The insert shows the $T_{1\rho}^C$ relaxation behavior of the protonated aromatic-carbon doublet [figure from J. Schaefer, E. O. Stejskal, and R. Buchdahl, *Macromolecules* **10**, 384 (1977)].

spinning and nonspinning cases is given in Fig. 56 for poly(phenylene oxide).

The interpretation of carbon $T_{1\rho}$ data is complicated by the fact that spin–spin (cross-relaxation) processes as well as rotating frame spin–lattice processes may contribute to the relaxation. This arises because the proton dipolar state is strongly coupled to the lattice. During the period in which carbon magnetization decays with H_{1H} off, carbon polarization can decay by motional processes ($T_{1\rho}^C$) or by polarization transfer to the proton dipolar state, then to the lattice via the proton dipolar spin–lattice process (T_{1D}). Schaefer *et al.*,[195] VanderHart and Garroway,[199] and Garroway *et al.*[175,200] have examined the problem in detail. While

[199] D. L. VanderHart and A. N. Garroway, *J. Chem. Phys.* **71**, 2773 (1979).

[200] A. N. Garroway, W. B. Moniz, and H. A. Resing, *Faraday Soc. Symp.* **13**, Chap. VI (1979).

there remains some degree of controversy as to interpretation, at present it seems that so long as the rotating-frame field is much greater than the proton local dipolar field, a system with motion in the kilohertz region will have $T_{1\rho}$ determined mainly by spin–lattice processes. According to Schaefer et al.[195] this will be the case for many glassy polymers and biopolymers at room temperature and probably plastic crystals near the melting point. On the other hand, crystalline polymers (for example, PE) and small organic molecules well below the melting point will show little effect of spin–lattice processes on $T_{1\rho}^C$. It is possible that heavily cross-linked polymers will form an intermediate case.

In a comprehensive paper on carbon $T_{1\rho}$ in glassy systems, Schaefer et al.[136] examined seven systems in considerable detail. The relaxation data were evaluated in terms of spin–lattice processes and interpreted in terms of distinct main- and side-chain motions in the 10–50 kHz regime (note, for example, the much shorter $T_{1\rho}^C$ value for the protonated ring carbons in PPO (Fig. 56) relative to the side-group methyl). The data allowed conclusions to be drawn as to the short-range nature of certain low-frequency side-group motions (e.g., the ester side-group motion in PMMA), and the long-range cooperative nature of some main-chain motions (e.g., polycarbonate and polysulfone). Interestingly, a direct correlation was shown between the ratio of cross-polarization and rotating-frame relaxation times ($T_{CR}/T_{1\rho}^C$) for the main chain carbons of a polymer with the toughness or impact strength of the polymer.

Although much work remains to be done to fully assess the value of the various carbon relaxation times in understanding the mechanical properties of polymers, the initial results appear promising even if ambiguous in some instances.

4.4.4.5. Biopolymers. Although in this chapter we have concentrated on the application of solid-state NMR techniques to synthetic macromolecules, the techniques are also appropriate to the study of biopolyers.[201] For example, Torchia et al.[202] have utilized differences in the ^{13}C spectra of deoxyhemoglobin S obtained by scalar decoupling, (proton) dipolar decoupling, and cross polarization to study the intracellular gelation of deoxyhemoglobin S. The gelation is thought to be responsible for the "sickling" deformation of the cell. Only isotropically mobile hemoglobin molecules (correlation time, $\tau_c \lesssim 10^{-6}$ sec) are detected under scalar-decoupling conditions. By contrast, dipolar decoupling detects both mo-

[201] D. A. Torchia and D. L. VanderHart, in "Topics in Carbon-13 NMR Spectroscopy" (G. C. Levy, ed.), Vol. III, 325 (1979).

[202] J. W. H. Sutherland, W. Egan, A. N. Schechter, and D. A. Torchia, Biochemistry 18, 1797 (1979).

bile and motionally restricted molecules. Comparison of the two types of spectra thus allows discrimination of the fractional amounts of deoxy-S monomer and "polymer."[203]

A typical sequence of decoupled spectra for hemoglobin S in solution is presented in Fig. 57. Figure 57a,b represents the oxy and deoxy forms obtained using low-power (scalar) proton decoupling. Clearly, there is a decrease (42%) in total intensity in the deoxy relative to the oxygenated form—a result that differs from the hemoglobin A case, where the spectra of both oxygenated and deoxygenated forms have the same total intensity

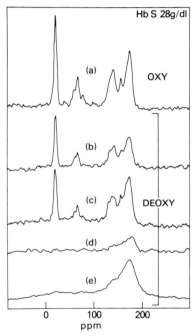

FIG. 57. Comparison of spectra of a 28 g/dl preparation of hemoglobin S at 37°C. (a) Oxygenated sample, $90°-t$ pulse sequence, $t = 2$ sec, scalar decoupled; (b) deoxygenated sample, $90°-t$ pulse sequence, $t = 2$ sec, scalar decoupled; (c) deoxygenated sample, $90°-t$ pulse sequence, $t = 2$ sec, dipolar decoupled; (d) spectrum (c) minus spectrum (b); (e) deoxygenated sample, proton-enhanced, 1 msec Hartmann–Hahn matched contract, 2 sec repetition time dipolar decoupled. 4096 transients were accumulated in each case. Digital line broadenings of 20 Hz in (a), (b), and (c) and 50 Hz in (d) and (e) were employed to enhance sensitivity. The chemical shift scale is relative to external CS_2 [figure from J. W. H. Sutherland, W. Egan, A. N. Schechter, and D. A. Torchia, *Biochemistry* **18**, 1797 (1979)].

[203] In this context, the word polymer refers to an aggregated or highly associated state, not a long chain molecule with covalent bonds.

under scalar-decoupling conditions.[202] Figure 57c is the spectrum of deoxy-S obtained using dipolar decoupling conditions. An increase in intensity is apparent and the difference spectrum between the scalar and dipolar decoupled spectra is presented in Fig. 57d. The residual intensity in the aliphatic region demonstrates the presence of motionally restricted molecules in the deoxy form, which may be considered to arise from gels or hemoglobin aggregates.

The cross-polarized spectrum of deoxy-S, which represents only intensity from carbons with nonvanishing static dipolar interactions (i.e., nonisotropically mobile molecules), is shown in Fig. 57e. Again, this result is in contrast to the deoxy-A case, where no CP spectrum is observed. The width of the signal (~ 150 ppm) observed in the carbonyl–aromatic region of the CP spectrum is typical of a chemical shift anisotropy for a crystalline solid. Spin–lattice relaxation measurements indicate a reduction of the backbone rotational correlation time by 10^4 upon gelation ($10^{-8} \rightarrow 10^{-4}$ sec). Thus, the set of results is consistent with a two-phase model for the state of deoxyhemoglobin S in solution—a mobile monomeric phase and an immobilized gel phase. Studies of hemoglobin gelation in cells by the same NMR techniques indicate that as much as 80% of the hemoglobin is aggregated in deoxygenated sickled erythrocytes.

Jelinski and Torchia[204] and Torchia and VanderHart[205] have used double-resonance techniques to investigate the mobility of the collagen peptide backbone in solution and the fibrillar state. Specifically, collagen samples in which glycine (every third residue in the chain) was enriched to $\sim 50\%$ at either the carbonyl or the methylene carbon were prepared and NMR parameters such as linewidths, T_1 values, NOE, and CP enhancements were determined under a variety of physical conditions. The results indicate that the collagen molecule in solution undergoes torsional reorientation, as well as rodlike reorientation about its long axis. The NMR data on the glycyl residues provides strong evidence that rapid axial motion ($\tau_c \sim 10^{-6}$) persists in the fibrillar state. However, the angular range of reorientation is restricted to $\sim 30°$.

The examples of ^{13}C NMR spectra of bulk polymers cited in this chapter clearly demonstrate that spectra of solids in which resonance linewidths approach those of liquids can be obtained by combining resolution (DD) and sensitivity (CP) enhancement techniques with magic-angle spinning. The results represent only a portion of reported ^{13}C studies of bulk macromolecules. In addition to the ^{13}C data, investigations utilizing ^{29}Si, ^{15}N, ^{31}P, and ^{19}F nuclei are beginning to be reported. Also a number

[204] L. W. Jelinski and D. A. Torchia, *J. Mol. Biol.* **133**, 45 (1979).
[205] D. A. Torchia and D. L. VanderHart, *J. Mol. Biol.* **104**, 315 (1976).

of investigators have reported high-resolution proton and fluorine spectra of bulk polymers by use of multiple-pulse NMR techniques. These procedures involve the repetitive application of periodic and cyclic pulse sequences to modify the time development of the proton or fluorine spin system. Details can be found in the books by Mehring[144] and Haeberlen.[206] The initial results of high-resolution NMR in solids suggest considerable potential for elucidating structure and dynamics in the solid state. However, the ultimate utility of these methods for defining the nature of semicrystalline polymers, etc., awaits the results of much more investigation.

4.5. Summary

As discussed in Chapters 4.2–4.4, NMR spectroscopy has wide-ranging application in the investigation of the structure and dynamics of macromolecules. The advent of pulse-FT techniques and the rapid development of ^{13}C NMR in the last 10 years have led to increased utilization of magnetic resonance by polymer chemists and physicists to probe the details of molecular structure/property relationships. The recent introduction of techniques for high-resolution NMR in the solid state holds the promise of a wealth of new information on bulk systems. Thus, it ap-

TABLE XIII. HR-NMR Parameters and the Features of Macromolecules

Parameters	Types of information derived from high-resolution spectra
Structural	
Chemical shifts	Primary structure of homopolymers, elucidation
Spin–spin coupling constants	of chain stereochemistry and conformation,
Resonance line intensities	elucidation of polymerization mechanisms, de-
Chemical shift anisotropy	termination of rotational barriers, determination of copolymer composition and sequence, determination of chain orientation
Dynamical	
Resonance linewidths	Determination of rates of molecular reorientation
Spin–lattice relaxation times (T_1)	and rotational barriers, elucidation of chemical
Spin–spin relaxation times (T_2)	exchange processes, data on the density of
Rotating frame relaxation times $(T_{1\rho})$	high- (MHz) and low-frequency (kHz) segmen-
Nuclear Overhauser factors	tal and side-chain motions of polymers, deter-
Cross-polarization times	mination of degree of crystallinity
Chemical shift anisotropy	

[206] U. Haeberlen, *Adv. Magn. Reson. Suppl.* 1 (1976).

pears that NMR spectroscopy will continue, as it has since its discovery, to be an analytical technique for the study of macromolecular systems. In concluding, we summarize in Table XIII the types of information on polymers that can be derived from NMR studies.

Acknowledgments

The author would like to express his appreciation to Dr. Frank Bovey (Bell Telephone Laboratories, Murray Hill, New Jersey), Dr. Jacob Schaefer (Monsanto Chemical Co., St. Louis, Missouri), Dr. David VanderHart (NBS, Washington, D.C.), Dr. Alan Garroway (NRL, Washington, D.C.), and Dr. Dennis Torchia (NIH, Bethesda, Maryland) for communicating recent work in advance of publication and for helpful discussions. Also, the author would like to express his sincere appreciation to his colleagues Dr. Colin Fyfe (Univ. of Guelph, Guelph, Ontario, Canada) and Dr. Costantino Yannoni (IBM, San Jose, California) for their painstaking efforts in introducing the nuances of solid NMR and for their many valuable discussions during the course of our collaborative efforts. The inventive technical assistance of R. D. Kendrick in modifying the commercial spectrometer and in adding new "twists" to the experiments is gratefully acknowledged. Finally, the author would like to thank Julie Countryman for her expert help in preparing this manuscript and for her good humor in persevering through many revisions.

5. PROBE AND LABEL TECHNIQUES

5.1 Positron Annihilation†

By J. R. Stevens

5.1.1. Introduction

Positrons can be used as very sensitive microprobes of the physical properties of polymeric matter. In an extremely short time an energetic positron in condensed matter loses all of its kinetic energy by collisions with electrons and ions. The thermalized positron becomes trapped, becomes bound to an electron to form positronium (Ps) or becomes part of a chemical complex with a polymer chain. In any event after a period of time much longer than the thermalization time, the positron annihilates with an electron with the resultant conversion of their masses to energetic photons whose energies, momenta, and times of emission may be measured with high precision. These observable characteristics of the annihilation process, which manifest its quantum electrodynamic nature, depend upon the history of the positron in the medium and upon the state of the positron–many-electron system at the time of annihilation.

Experimental results obtained represent statistical averages of many consecutive trials performed with a single thermalized positron. Each such positron is virtually at rest in a sample fixed in the laboratory and utilizes its charge and spin properties in sensing, through the overlap of wavefunctions, the local electron density. Ultimately the positron annihilates with a nearby electron whose momentum is carried away by the annihilating photons. The annihilation characteristics therefore give information about the density and momentum distribution of valence electrons in the sample.

In this review attention will be given to the positron method applied to the study of the physics of polymers. The use of this method to study the chemical properties of organic liquids and solids is discussed, for ex-

---

† See also: Vol. 5B (Nuclear Physics) of this Series, Sections 2.5.4.5; 2.5.5.4; 2.8.2.2.2.

ample, by Goldanskii,[1] Ache,[2] Ito and Tabata,[3] Tao,[4,5] Green,[6] Merrigan et al.,[7] Thosar et al.,[8] Kajcsos et al.,[9] and Goldanskii and Shantarovich.[10] However, insofar as the positron method elucidating the chemical properties of polymers is concerned, very little has been done. What has been done has been mainly associated with polymerization and degradation processes in irradiated matter. This aspect will be discussed briefly.

An extensive bibliography is provided that includes several articles of historical[11-16] as well as of fundamental interest.[14,16,17]

5.1.2. Positrons and Positronium

5.1.2.1. The Positron and Free Positron Annihilation.
The positron is the well-known antiparticle of the electron, carrying a positive charge equal in magnitude to the negative charge carried by the electron. It was predicted from Dirac's study of the relativistic quantum mechanics of the electron[11] and discovered in cosmic ray showers by Anderson.[12]

Quantum electrodynamics is capable of providing an explanation for all of the observed properties of the annihilation of slow positron–electron pairs.[17-20] Briefly, if these conjugate particles annihilate with their spins antiparallel, that is, a singlet [1]S interaction, two-photon emission almost

[1] V. I. Goldanskii, At. Energy Rev. **6**, 3 (1968).

[2] H. J. Ache, Angew. Chem., Int. Ed. Engl. **11**, 179 (1972); Angew. Chem., Int. Ed. Engl. **84**, 234 (1972); Adv. Chem. Ser. **175**, 1 (1979).

[3] Y. Ito and Y. Tabata, Chem. Phys. Lett. **15**, 584 (1972).

[4] S. J. Tao, Appl. Phys. **3**, 1 (1974).

[5] S. J. Tao, Appl. Phys. **10**, 67 (1976).

[6] J. H. Green, in "Radiochemistry" (A. G. Maddock, ed.), p. 251. Butterworth, London, 1972.

[7] J. A. Merrigan, J. H. Green, and S. Tao, Phys. Methods Chem. Part 3D, p. 501 (1972).

[8] B. V. Thosar, R. G. Lagu, V. G. Kulkarni, and G. Chandra, Phys. Status Solidi **55**, 415 (1973).

[9] Z. S. Kajcsos, J. Dezsi, and D. Horvath, Appl. Phys. **5**, 53 (1974).

[10] V. I. Goldanskii and V. P. Shantarovich, Appl. Phys. **3**, 335 (1974).

[11] P. A. M. Dirac, Proc. Cambridge Philos. Soc. **26**, 361 (1930).

[12] C. D. Anderson, Phys. Rev. **43**, 491 (1933).

[13] A. E. Ruark, Phys. Rev. **68**, 278 (1945).

[14] A. Ore, Univ. Bergen, Arbok, Naturritensk. Rekke No. 9 (1949).

[15] M. Deutsch, Prog. Nucl. Phys. **3**, 131 (1953).

[16] R. A. Ferrell, Rev. Mod. Phys. **28**, 308 (1956).

[17] P. R. Wallace, Solid State Phys. **10**, 1 (1960).

[18] A. I. Akhiezer and V. B. Berestetskii, "Quantum Electrodynamics." Wiley (Interscience), New York, 1965.

[19] V. B. Berestetskii, E. M. Lifshitz, and L. P. Pitaevskii, "Relativistic Quantum Theory." Pergamon, Oxford, 1971.

[20] R. N. West, Adv. Phys. **22**, 263 (1973).

always occurs. Each of the two photons has energy $m_0c^2 = 0.511$ MeV (conservation of energy) and they emerge from the site of the annihilation event 180° apart (conservation of momentum). Because of parity considerations (the annihilating pair must have odd parity) two-photon emission is forbidden for collisions where the particles meet with their spins parallel, that is triplet ^3S interactions. In this case three-photon emission is the most probable. Since the z-component of the total angular momentum of the singlet state ($j = 0$) is zero and since the z-component of the total angular momentum of the triplet state ($j = 1$) is 0 or ± 1, the statistical ratio of singlet to triplet interactions is $1:3$. The ratio of singlet to triplet annihilation rates is approximately $1115:1$.[21] Therefore, the ratio of probabilities of two-photon to three-photon annihilations resulting from free positron interactions is $1115:3$ or $372:1$.

In experimental studies the positron is most frequently produced in the radioactive decay of a suitable isotope within or adjacent to the specimen to be studied. Depending upon its energy and the density of the specimen it usually penetrates up to several tenths of a millimeter thermalizing in <10 psec,[22,23] and possibly (comparing the positron with the case of an electron) in <1 psec.[24]

The case of the positron losing its energy in a molecular substance should be analogous to that for an energetic electron. In what follows, information is drawn from an excellent review by Hunt[24] on early events in radiation chemistry. Initially the energy of the positron is reduced very rapidly ($dE/dt \simeq 10^{16}$ eV sec^{-1}) through inelastic collisions, which produce an avalanche of ionization products including secondary electrons. These secondaries have energies in the range 0–50 eV, most of which is deposited in the medium. These initial ionizations and excited states are not formed homogeneously throughout the substance but instead are concentrated in relatively small volumes, which in liquids are called spurs. In molecular substances the positron slows down from a few electron volts to thermal energy mainly through positron–phonon interactions. Thermalized positrons not bound to any one particular molecular segment (M^+) will be bound to the bulk molecular solid, most probably in traps of varying potential energy depth. In polar polymers these traps are deeper than in nonpolar polymers by at least a factor of two[25] and include interstitial cavities distributed among the polymer chains. The mean size of the cavity in which a positron is trapped would determine its mean lifetime.

[21] A. Ore and J. L. Powell, *Phys. Rev.* **75**, 1969 (1949).
[22] P. Kubica and A. T. Stewart, to be published.
[23] A. Perkins and J. P. Carbotte, *Phys. Rev. B* **1**, 101 (1970).
[24] J. W. Hunt, *Adv. Radiat. Chem.* **5**, 185 (1976).
[25] J. R. Miller, *J. Chem. Phys.* **56**, 5173 (1972); *Chem. Phys. Lett* **22**, 180 (1973).

This lifetime is found experimentally to be in the range 0.1–0.9 nsec for "free" positrons in polymeric solids and is a function of the local electron density.

5.1.2.2. Positronium and Bound-State Annihilation. Positronium (Ps), the bound-state of the electron and the positron, can be theoretically described in terms of the electromagnetic field and the electron–positron field. Although the existence of the ground state of Ps was first demonstrated in 1951 by Deutsch,[26] the first observation of the 2430 Å, 2P → 1S emission line was not made until 1975.[27] The ground states of Ps are the singlet ^1S state or parapositronium (p-Ps) and the triplet ^3S state, or orthopositronium (o-Ps).

The energetics of Ps formation in molecular solids such as polymers can be semiquantitatively discussed in terms of the Ore gap model, a concept originating from the corresponding problem in gases.[14] A schematic that plots the probability of Ps formation against the kinetic energy of the moderating positron is shown in Fig. 1 and illustrates the Ore gap. The bind-

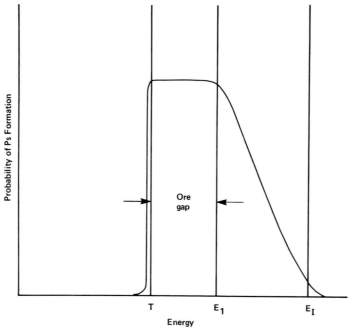

FIG. 1. Formation probability of Ps as a function of positron energy.

[26] M. Deutsch, *Phys. Rev.* **82,** 455 (1951).
[27] K. F. Canter, A. P. Mills, and S. Berko, *Phys. Rev. Lett.* **34,** 177 (1975).

ing energy of the Ps atom is 6.8 eV in free space, but may be smaller in matter. Since, according to this model, the formation of Ps requires that some of the positron's kinetic energy be used to remove an electron from a molecule in order to form this bound state, the positron must have kinetic energy E of at least $(E_I - 6.8)$ eV, labeled the threshold energy T in Fig. 1, where E_I is the ionization energy of the molecule. If $E > E_I$, inelastic scattering with ionization will be more probable[17] and if $E > E_1$, the lowest electronic excitation energy, this excitation will compete with Ps formation. Thus Ps formation is most probable for the range of positron kinetic energies

$$E_1 > E > T,$$

since the positron only undergoes elastic scattering in this range called the Ore gap. If we assume that a high-energy positron has equal probability of deexciting to any energy level below E_I so that the distribution of E below E_I is constant, the probability P of Ps formation should be in the range

$$\frac{E_I - T}{E_I} > P > \frac{E_1 - T}{E_1}.$$

The foregoing analysis, strictly speaking, applies to a gas. For condensed matter, the binding energies of the positron and Ps to the liquid or solid should be incorporated.[16] Also for long-chain hydrocarbons, an abundance of excitation levels, including vibrational modes, the formation of positron and molecule complexes, and impurities will complicate this simple picture. Nevertheless, the Ore gap model has served as a very useful first approximation even in polymeric solids.[1,28]

Mogensen[29] has proposed that a sizable fraction of Ps is formed through the capture by a positron of an electron found in ionization debris in spurs formed by the moderating positron, rather than being formed according to the Ore gap model. This proposal was made to explain experimental results on Ps decay, particularly in liquids. In a condensed medium of static dielectric constant ϵ there is a critical distance r_c, between the charges of an ion pair, at which the thermal kinetic energy kT of the pair is the same as its electric potential energy (i.e., as related to the attractive force between the pair). Thus

$$r_c = e^2/4\pi\epsilon_0\epsilon \, kT, \tag{5.1.1}$$

where ϵ_0 is the dielectric constant of free space. The Ps yield from spurs

[28] W. Brandt, *Appl. Phys.* **5**, 1 (1974).
[29] O. E. Mogensen, *J. Chem. Phys.* **60**, 998 (1974).

can be estimated as

$$P = f_0 P_0 [1 - \exp(-r_c / \bar{r})], \qquad (5.1.2)$$

where $\bar{r}$ is the charge separation, P_0 the total relative Ps yield, and f_0 the fraction of positrons that can potentially form Ps by capturing spur electrons in competition with the capture process by positron–molecule scattering in the Ore gap. Therefore, the significance of r_c, the Onsager radius, is that whenever the members of an ion pair separate to a distance exceeding r_c, dissociation may be considered complete in the sense that $1/e$ of the charges escape recombination with other charges in the spur.

The relevance of these two models, the Ore gap model and the spur model, in accounting for experimental observations in polymeric matter will be discussed in Section 5.1.6.

The theory of Ps decay has been reviewed by Stroscio[30] and the agreement between theory and experiment continues to be improved,[31] although a recent experimental determination of the annihilation rate of o-Ps[32] in vacuum is in disagreement with the theoretical value and with values obtained from experiments in gases. One-quarter of the positrons in Ps are formed in the singlet state p-Ps and $\frac{3}{4}$ are formed in the triplet state o-Ps. However, positrons in Ps will not necessarily annihilate with the p-Ps mean lifetime of 125 psec or the o-Ps mean lifetime of 138 nsec since various processes that occur during collisions of the neutral Ps atom with the molecules of the system may affect the annihilation rate.

Goldanskii[33] suggests that if Ps experiences a polarization "swelling" as it interacts with the molecules of the medium the lifetime of p-Ps, for example, would increase. This introduces the possibility that the annihilation rate involving the positron and some surrounding "lattice" electron having a para-spin orientation (antiparallel) relative to the positron would become much greater than the self-annihilation rate of p-Ps. The former process is called "pick-off."

Annihilation of o-Ps by pick-off decreases its mean lifetime by two orders of magnitude even in the absence of polarization swelling and predominates in polymeric solids. This annihilation mode is very important in the observation of density or free volume changes in polymers since the annihilation occurs with an electron in the surrounding medium rather than with an electron bound to a Ps atom or positron complex. Any

[30] M. A. Stroscio, *Phys. Lett. C* **22**, 215 (1975).

[31] W. E. Caswell, L. P. Lepage, and J. Sapirstein, *Phys. Rev. Lett.* **38**, 488 (1977).

[32] D. W. Gidley, A. Rich, P. W. Zitzewitz, K. A. Marko, and A. Rich, *Phys. Rev. Lett.* **40**, 737 (1978).

[33] V. I. Goldanskii, *in* "Positron Annihilation" (A. T. Stewart and L. O. Roellig, eds.), p. 224. Academic Press, New York, 1967.

process that shortens the mean lifetime of the positron in o-Ps is called "quenching." In addition to pick-off, these processes include ortho–para conversion and two-photon annihilation resulting from chemical reactions of o-Ps.

5.1.3. Theoretical Considerations

The cross section for two-gamma annihilation of a free positron and a free electron was calculated by Dirac[11] as

$$\sigma_{2\gamma} = \frac{\pi r_0^2}{\gamma + 1} \left\{ \frac{\gamma^2 + 4\gamma + 1}{\gamma^2 - 1} \ln[\gamma + (\gamma^2 + 1)^{1/2}] - \frac{\gamma + 3}{(\gamma^2 + 1)^{1/2}} \right\}, \quad (5.1.3)$$

where $\gamma = 1/(1 - \beta^2)$ and $\beta = v/c$. In the case of $v/c \ll 1$ Eq. (5.1.3) reduces to

$$\sigma_{2\gamma} = \pi r_0^2 c/v,$$

which is inversely proportional to the electron's velocity v. The parameter $r_0 = e^2/mc^2 = 2.8 \times 10^{-15}$ m is the classical radius of the electron or positron.

Based upon Dirac's result [Eq. (5.1.3)], $\sigma_{2\gamma}$ is found for large v to be negligible in comparison with the cross sections for ionization and excitation. In fact as long as these processes are energetically possible, then they should predominate over annihilation. Thus to a first-order approximation, positron annihilation is considered to take place in a medium containing η_e electrons per cm³ with a rate

$$\lambda_{2\gamma} = \sigma_{2\gamma} v \eta_e = \pi r_0^2 c \eta_e. \quad (5.1.4)$$

It is important to note that η_e is the average number density of electrons at the position of the positron in any spin state (i.e., it is the square of the Coulomb wavefunction evaluated at contact). This result is obtained by recognizing that Dirac's $\sigma_{2\gamma}$ involves averaging over random spin orientations and, since the probability of a triplet state annihilating into two photons is zero, the probability of a singlet state is increased to four times the spin-averaged value or

$$\lambda_{2\gamma} = 4\pi r_0^2 c \eta_e',$$

where η_e' is the average number density of electrons in the medium at the position of a positron in a singlet state relative to an electron. Equation (5.1.4) is obtained by remembering that $\eta_e' = \frac{1}{4}\eta_e$.

The ground-state Ps wavefunctions have the approximate form

$$\psi(r) = (\pi a^3)^{-1/2} \exp(-r/a),$$

where r is the relative position coordinate, $a = 2\hbar^2/m_0 e^2$ is the Bohr radius of Ps (twice that for hydrogen), and $m_0/2$ is the reduced mass. Since the electron and positron momenta in the ground state are of the order of a^{-1} we can regard annihilation as approximating that of a free positron and a free electron of zero momenta but definite spin orientation.

The decay rate of the ground state of p-Ps corrected to order α, the Sommerfeld fine-structure constant, in radiative and Coulomb effects is[30]

$$\lambda_{2\gamma} = \frac{\alpha^5 mc^2}{2\hbar} [1 - (\alpha/\pi)(5 - \tfrac{1}{4}\pi^2)] = 0.79854 \times 10^{10} \quad \text{sec}^{-1},$$

which is in good agreement with the result from a measurement of $\lambda_{2\gamma}$ in argon.[34]

The decay rate of o-Ps, $\lambda_{3\gamma}$, has been measured in gases. The two results $0.7262 \times 10^7 \text{ sec}^{-1}$[35] and $0.7275 \times 10^7 \text{ sec}^{-1}$[36] are 2% higher than the most recent theoretical result, $0.70386 \times 10^7 \text{ sec}^{-1}$.[31] However Gidley et al.[32] have reported a value in vacuum of $0.709 \times 10^7 \text{ sec}^{-1}$ for $\lambda_{3\gamma}$, considerably closer to the above theoretical value but still in disagreement. Further work is needed on this problem.

The pick-off annihilation rate depends on the degree of overlap of the positron component of the Ps wavefunction with the "lattice" wavefunction. The spin-averaged pick-off annihilation rate λ_p is given by

$$\lambda_p = \pi r_0^2 c \int \eta_{Ps}^+(\mathbf{r}) \eta_L(\mathbf{r}) \, d^3r, \qquad (5.1.5)$$

where η_{Ps}^+ is the density of the positron bound in a Ps atom and η_L is the density of the electrons bound to molecules in the lattice L.

Brandt et al.[37] used a simple lattice cellular model of free volume and excluded volume in evaluating Eq. (5.1.5.). Subsequently, Brandt and Spirn[38] and Brandt and Fahs[39] extended this model to include the effects of lattice vibrations on λ_p. This model neglects the mutual Ps and lattice polarization, which could be important in polymeric solids, and assumes a regular array of cells of volume v. A molecule is placed at the center of every cell such that in each cell a volume v_0 is excluded, leaving free volume $v_f = v - v_0$ surrounding the spherically symmetric molecule. A Ps atom would be found delocalized in the free-volume portion of the cell moving in a potential energy distribution, which increases rapidly at the boundary of each molecule. This model could be considered as a good

[34] E. D. Theriot, R. H. Beers, and V. W. Hughes, Phys. Rev. Lett. 18, 767 (1967).
[35] R. H. Beers and V. W. Hughes, Bull. Am. Phys. Soc. [2] 13, 633 (1968).
[36] P. G. Coleman and T. C. Griffith, J. Phys. 36, 2155 (1973).
[37] W. Brandt, S. Berko, and W. W. Walker, Phys. Rev. 120, 1289 (1960).
[38] W. Brandt and I. Spirn, Phys. Rev. 142, 231 (1966).
[39] W. Brandt and J. H. Fahs, Phys. Rev. B 2, 1425 (1970).

first approximation of a polymer phase, a monomer segment replacing each molecule in the model. A polymer phase would be better represented if this model did not require that every cell contain a monomer segment; however, this would complicate the evaluation of Eq. (5.1.5). The result predicted for λ_p by this model is temperature dependent through the mean-squared amplitude of lattice vibrations associated with each cell and through the free volume. The pick-off lifetime $\tau_p = \lambda_p^{-1}$ is predicted correctly (Fig. 2[38]) to initially increase with temperature (change in free volume), reach a maximum or shoulder because of the smearing of the electron density distribution caused by increased thermal motion at higher temperatures, and increase further with temperature increase to

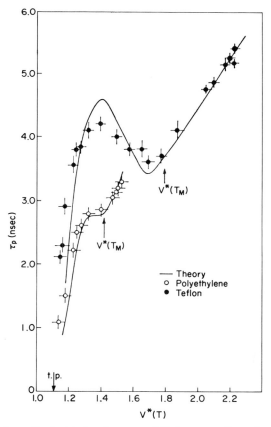

FIG. 2. Comparison of calculated and measured lifetimes τ_p. Reproduced by permission of the American Institute of Physics from Brandt and Spirn.[38]

still higher temperatures. In Fig. 2, the solid curves pertain to a cylindrical cell geometry and v^*, the reduced cell volume, is given by

$$v^*(T) = 1 + v_f/v_0 = v/v_0.$$

In fitting the experimental data obtained for positrons annihilating in a series of low molecular weight paraffins,[40] polyethylene (PE), and polytetrafluoroethylene (PTFE)[38] assumptions were made about the height of the potential energy barrier U_0 and the radius r_0 of the molecule. This theory does not attempt to account for absolute pick-off annihilation rates.

5.1.4. Experimental Techniques

There are three experimental techniques now in general use for measuring the times of emission, the energies, and the momenta of positron–electron annihilation photons. The two-photon annihilation event is most frequently studied in polymers. Information is provided about the lifetimes of positrons annihilating in a substance and two components of the electron momentum distribution in the substance. The components of the electron momentum distribution are obtained by measuring the Doppler-broadened energy spectrum of the annihilation photon and the two-photon angular correlation. Quantitative and semiquantitative evidence for Ps formation can also be obtained from angular correlation studies.

5.1.4.1. Lifetimes. Positron lifetime studies make use of radioactive sources that emit a gamma photon simultaneously (< 10 psec) with a positron. Detection of the gamma photon is the zero time reference for a lifetime measurement.

A fast-slow, delayed coincidence system is used that incorporates time-to-pulse height conversion and multichannel analysis and storage. A typical system is shown in Fig. 3. The time-to-pulse height converter (TAC) is activated and deactivated by positron "birth" and "death" (delayed) signals processed from the plastic scintillators through the photomultipliers and the fast timing discriminators. The TAC then puts out a pulse proportional in height to the time difference between the birth and death signals. Dynode signals from each of the photomultipliers are passed through slow side channels and used to improve timing precision and to ensure that only TAC pulses of related birth and death signals are coincidence gated to the multichannel analyzer. The stored spectrum of pulse heights consists of a convolution of the real lifetime spectrum of the annihilated positrons with the time resolution function of the system. A typical lifetime spectrum is shown in Fig. 4.

[40] W. Brandt and J. Wilkenfeld, *Phys. Rev. B* **12**, 2759 (1975).

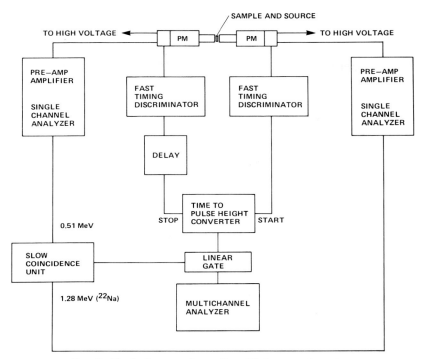

FIG. 3. Block diagram of a positron lifetime spectrometer.

Components for a suitable positron lifetime spectrometer are commercially available. Time resolutions of 250–300 psec (FWHM) can be obtained with careful control of the ambient temperature and humidity. Unfortunately these spectrometers are count-rate limited. Above a count rate of about 10 ksec⁻¹ (a few microcuries) the probability of random events being recorded is dramatically increased, even with pile-up rejection electronics, and so prolonged experiments are normally required to obtain the statistics necessary for accurate deconvolution of the lifetime spectrum. Over long time periods electronic drift, especially that which leads to an asymmetric time resolution function and an uncertainty in the zero of time, becomes a serious problem.

Crisp et al.[41] have proposed a modified spectrometer incorporating a digital stabilizer in which any drift in zero time is common to both the time resolution function and the delayed lifetime spectrum. In their system the time resolution function that arises from the simultaneous detection of

[41] V. H. C. Crisp, I. K. MacKenzie, and R. N. West, *J. Phys. E* **6**, 1191 (1973).

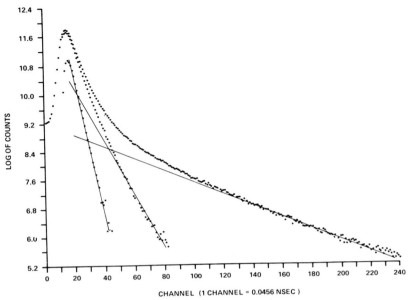

FIG. 4. Typical positron lifetime spectrum (PE at 393 K). Only τ_1 (= 0.15 nsec), I_1 (= 0.24), τ_4 (= 2.39 nsec), and I_4 (= 0.19) are shown resolved. The two intermediate components (Table II, Ref. 63) are not drawn as resolved for the sake of clarity.

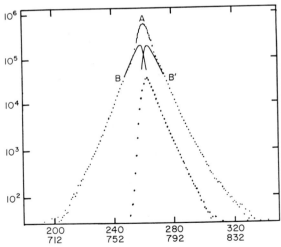

FIG. 5. Prompt and delayed spectra, ^{22}Na and indium metal. The prompt curve is the spectrum of time-pulse height converter before routing. Scale, 1 channel: 39 psec. Reproduced by permission of the Institute of Physics from Crisp et al.[41]

the two annihilation photons by the collinear detectors is stored along with a symmetrical delayed spectrum in the first half of a 1024 channel memory (Fig. 5). A routing technique is used to store a conventional delayed spectrum in the second half of the memory. Slow side channels are not required in this technique, which reduces the time necessary to obtain statistical accuracy by an order of magnitude. Recently Myllyla and Karras[42] have constructed and tested a positron lifetime spectrometer with "state of the art" electronics and claim significant improvements in time resolution, counting rate capability, and temperature stability.

5.1.4.2. Angular Correlation. Conservation of momentum requires that the photons from two-photon annihilation be emitted in opposite directions. If the electron–positron pair is at rest then the angle between the photon directions is exactly 180°. However, if the annihilating pair has a net linear momentum **p** prior to annihilation then the angle between the photon directions deviates from 180° by a small angle, which is given (in the small-angle approximation) by[20,43]

$$\delta\theta = p_z/m_0c \quad \text{rad.}$$

Assuming that the positron in the pair is at rest, p_z is the electron momentum transverse to the direction that would have been taken by the two photons if they had been emitted from an annihilating pair at rest. The experimentally observed range for $\delta\theta$ is between 0 and 50 mrad.

For angular correlation measurements the positron source selected need not emit a gamma photon coincident with the positron unless lifetime measurements are made concurrently. The plastic or NaI scintillation detectors are shielded from a direct view of the source, which is usually located at a fixed distance above the specimen surface attached to a large lead shield, which limits the cone of positrons emitted and generally reduces background problems.

A typical experimental arrangement for the measurement of the angular correlation of annihilating photons is shown in Fig. 6. The collimator geometry defines the instrumental angular resolution (<0.2 mrad). The coincident (<50 nsec) counting rate from the two detectors is measured as a function of the displacement of one detector with respect to a reference line through the region of the sample being probed and the second detector. For good statistics and to minimize errors due to electronic drift, the angular distributions of annihilation photons are usually a function of cycling many times through the angular range. Source strengths between 0.01 and 1 Ci are used. The scintillators are mounted in a way in

[42] R. Myllyla, *Nucl. Instrum. Meth.* **148**, 267 (1978).
[43] H. Weisberg, Ph.D. Thesis, Brandeis University, Waltham, Massachusetts (1965).

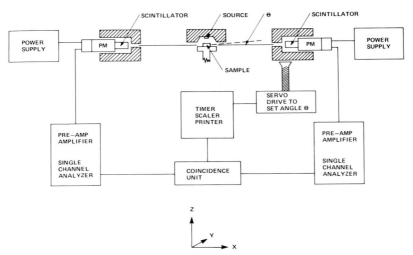

FIG. 6. Block diagram of an angular correlation spectrometer.

which the long dimension is parallel to the long counting slits; i.e., one dimension of the counting slit is made very much larger than the width of a typical angular distribution. Thus the detectors are insensitive to the other two components of momentum.

Figure 7 shows the momentum distribution $N(p)$ of positron–electron pairs annihilating in PTFE (Teflon).[44] $N(p)$ is the probability that the mo-

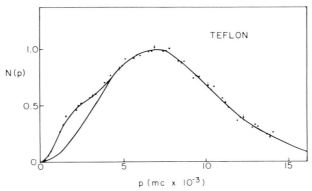

FIG. 7. The momentum distribution of positron–electron pairs annihilating in Teflon. Reproduced by permission of the National Research Council, Canada from Kerr.[44]

[44] D. P. Kerr, *Can. J. Phys.* **52**, 935 (1974).

mentum of the annihilating pair is in the approximate range $m_0c(\delta\theta)$ to $m_0c(\delta\theta + d\theta)$. In Fig. 7 two components are resolved and will be discussed in Section 5.1.6.1.

In performing angular correlation experiments care should be taken to ensure that any self-absorption occurring in the sample is approximately equal at all angles and to eliminate distortion of the momentum distribution as a result of the diffraction of the annihilating photons by the collimating system. This can be accomplished by rotating the sample through a small angle of about 40 mrad.[45] Also care should be taken to recognize spurious temperature dependences of the angular distribution when using shielded sources external to the sample. These effects are eliminated if internal sources are used.[45]

Recently[46,47] two groups have announced developments in constructing two-dimensional counting systems that would permit measurement of the p_z and p_y components. Berko et al.[46] have used 64 conventional NaI scintillation crystals to construct and test two two-dimensional detectors, which provide for the detection of all 1024 cross correlations (32×32). Their sample to detector plane distance is 10 m. Douglas and Stewart[47] are using PbBi converters in constructing their two-dimensional angular correlation system. The application of these two-dimensional angular correlation techniques to the study of polymers may be difficult since for sources separate from the specimen the positrons must be heavily focused to form a point source on the specimen. Radiation damage could result.

5.1.4.3. Doppler Broadening. Any momentum component p'_x of the annihilating electron–positron pair gives rise to a Doppler broadening of the annihilation line at 0.511 MeV. The Doppler shift is related in first order to the longitudinal momentum component through

$$E = m_0c^2 \pm cp'_x,$$

in which E are the energies ($\pm$) of the annihilation photons in the laboratory frame of reference. Assuming that the positron is at rest, $p'_x = \frac{1}{2}p_x$, where p_x is the x-component of the electron momentum. Therefore

$$E = m_0c^2 \pm cp_x/2,$$

and the Doppler shift $\delta E = \pm cp_x/2$ is about 1.5 keV for 10 eV electrons. For an isotropic medium such as a polymeric glass all the momentum

[45] J. L. Campbell, T. E. Jackman, I. K. MacKenzie, C. W. Schulte, and C. G. White, Nucl. Instrum. Methods 116, 369 (1974).
[46] S. Berko, M. Hoghgooie, and J. J. Mader, Phys. Lett. 63A, 335 (1977).
[47] R. J. Douglas and A. T. Stewart, Proc. Int. Conf. Positron Annihilation, 4th, Helsingor, 1976 H14 (1976).

components are equal. Therefore,

$$p_x = p_z = m_0 c(\delta\theta)$$

and

$$\delta E = \pm\tfrac{1}{2}m_0 c^2(\delta\theta) = \pm 255.5(\delta\theta) \quad \text{keV},$$

where $\delta\theta$, the angular correlation, is in radians.

Lithium-drifted germanium detectors [Ge(Li)] and hyperpure germanium detectors are used to detect and measure the energies of annihilation photons and are commercially available with energy resolution $\delta E = 1.2$ keV. Inserting this value in the above equation for δE, a value of about 5 mrad is obtained for $\delta\theta$, which is an order of magnitude poorer than the angular resolution for the angular correlation technique. Furthermore, Ge(Li) detectors must be stored and operated at liquid N_2 temperatures to decrease thermal conductivity due to a low energy gap (0.67 eV) and to prevent the migration of Li atoms from the depletion region. In spite of these problems, data can be rapidly accumulated with only one detector and specimens and conditions that are not amenable to angular correlation measurements can be studied.

The experimental set-up for a high-resolution, high count rate Ge(Li)

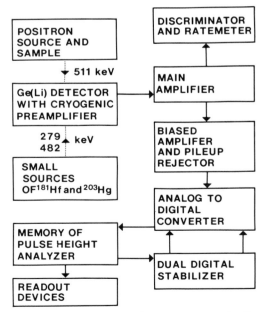

FIG. 8. Block diagram of a Ge(Li) high-resolution spectrometer for Doppler-broadened energy profile measurements.

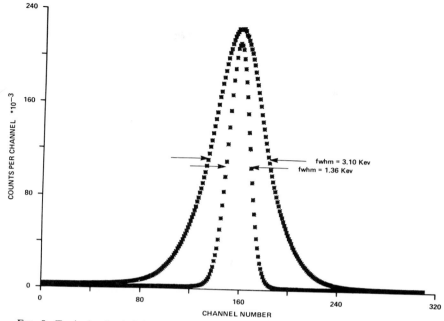

FIG. 9. Typical pulse-height spectra recorded in Ge(Li) detector showing the resolution function at 511 keV and the Doppler-broadened energy spectrum of the annihilation photons.

spectrometer is shown in Fig. 8.   The energy resolution and a typical Doppler-broadened annihilation energy spectrum are shown in Fig. 9.   In Fig. 8 the biased amplifier permits expansion of the energy profile for closer examination and the pile-up rejector permits the use of high count rates (up to 50 k sec^{-1}).   The dual digital spectrum stabilizer is used to prevent small zero and gain shifts from broadening the energy spectrum and requires two reference sources for operation.   The highly stable pulser is used to monitor analog to digital converter (ADC) stability.

Lynn et al.[48] have recently developed a double Ge(Li) detector system that measures the energy difference between the annihilation photons in two-gamma annihilation.   The energy difference ΔE determines p'_x since $\Delta E \simeq \mathbf{c} \cdot \mathbf{p}$, where $\mathbf{c}$ is the velocity of light in the direction of the higher-energy annihilation photon and $\mathbf{p}$ is the momentum of the electron–positron pair.   This technique makes the very high momentum components more accessible experimentally.

[48] K. G. Lynn, J. R. MacDonald, R. A. Boie, L. C. Feldman, J. D. Gabbe, M. F. Tobbins, E. Bonderup, and J. Golovchenko, *Phys. Rev. Lett.* **38**, 241 (1977).

5.1.4.4. Multiparameter Methods. Using two-parameter multichannel analysis, lifetime techniques have been combined with angular correlation or Doppler-broadening techniques.

McGervey and Walters[49] combined angular correlation and lifetime measurements to correlate the three components in the lifetime distribution of positrons annihilating in PTFE with $\delta\theta$. Their results will be discussed in Section 5.1.6.

Sen and MacKenzie[50] combined Doppler broadening and lifetime measurements to produce simultaneous energy shift and lifetime information for each recorded annihilation event. Spectra for PE and PTFE were obtained.

Still in the developmental stage, Fluss and Smedskjaer[51] are developing a computer-based data analysis system for simultaneous collection of data from all three techniques. So far their system has only been tested to collect, store, and analyze data from lifetime and Doppler-broadening techniques.

The combination of the double Ge(Li) detector technique of Lynn et al.[48] with the angular correlation technique and a source of a few microcuries should be investigated by workers applying the positron method to polymers.

5.1.4.5. Positron Sources. Table I[45] summarizes the decay characteristics of five positron-emitting radionuclides that have been used in studies on condensed matter. By far the most commonly used source for positron experiments in polymeric substances has been ^{22}Na. It is readily obtainable and inexpensive. Sodium-22 sources are conventionally prepared by permitting a few droplets of radioactive sodium salt solution to evaporated on thin ($\simeq 0.02$ mm) mica, plastic, or metal foils, which are then covered by similar foils and sealed. The effect of positrons annihilating either in the source material or in the covering material has usually been neglected. It has also been assumed that positrons probe the bulk of the specimen rather than its surface. Sometimes the source has been sandwiched between pieces of the specimen after being directly deposited. This eliminates the problem of annihilation in the covering but care must be taken to guard against degradation of the sample by the ^{22}Na carrier. If the ^{22}Na source preparation method developed by Herlach and Maier[52] for metals could be adopted for polymers the problem of degradation could also be overcome. In this technique carrier free ^{22}Na in the form of

[49] J. D. McGervey and V. F. Walters, Phys. Rev. B 2, 2421 (1970).

[50] P. Sen and I. K. MacKenzie, Nucl. Instrum. Methods 141, 293 (1977).

[51] M. J. Fluss, L. C. Smedskjaer, M. K. Chason, D. G. Legnini, and R. W. Siegel, J. Nucl. Mater. 69, 586 (1978).

[52] D. Herlach and K. Maier, Appl. Phys. 11, 199 (1976).

TABLE I. Decay Data of Positron Emitters

Radionuclide	Half-life	Maximum energy (keV)	Intensity of 511 keV (%)	Principal γ rays		Lifetime studies[b]
				Energy	Intensity[a] (%)	
^{58}Co	71 days	470	80	810	99	Yes
^{22}Na	2.6 years	545	180	1275	100	Yes
^{64}Cu	12.8 hours	660	38	1340	0.5	No
^{44}Ti/^{44}Sc	48 years	1470	188	1159	100	Yes
^{68}Ge/^{68}Ga	274 days	1880	176	1078	3.5	No

[a] Intensities are expressed as percentage per decay.

[b] An emitter is suitable for lifetime studies if an intermediate nuclear level having lifetime of the order of a few picoseconds is strongly populated.

^{22}NaCl is evaporated in high vacuum onto the specimen. A 2000 Å thick layer of a metal is then evaporated onto the specimen coating the source. The source is then sealed using electron beam welding. For polymers the source could be sealed with an epoxy and a specimen sandwich formed by using another identical piece of the specimen. Recently a method of ion implantation of ^{22}Na has been suggested.[52a]

Campbell et al.[45] have recently reviewed the sources utilized in positron annihilation condensed-matter research and have concluded after a systematic study of five positron emitters (Table I) that the commonly used radionuclides may give rise to systematic errors that would cast doubt on the reliability of much of the research data in the literature. Figure 10 reproduces the lifetime spectra for positrons annihilating in cobalt recorded with external sources ^{22}Na and ^{44}Ti and an internal ^{58}Co source. Aluminum and Kapton [a polyimide resin in the form of a film (Dupont)] were used as source covers for the external sources. Cobalt-58 was produced by the (γ, n) reaction using the 30 MeV bremsstrahlung from an electron beam accelerator. On the basis of their research these authors recommend ^{68}Ge/^{68}Ga and ^{44}Ti/^{44}Sc sources, which both emit high-energy positrons. ^{68}Ge/^{68}Ga is available and relatively inexpensive. Unfortunately it cannot be used for lifetime studies. Titanium-44 is produced by the ^{45}Sc(p,2n)^{44}Ti reaction, is expensive, and is commercially available only in limited quantities. Lambrecht and Lynn[52b] are investigating alternative production methods.

[52a] M. J. Fluss and L. C. Smedskjar, Appl. Phys. 18, 305 (1979).

[52b] R. M. Lambrecht and K. G. Lynn, Proc. Int. Conf. Positron Annihilation, 4th, Helsingor, 1976 H17 (1976).

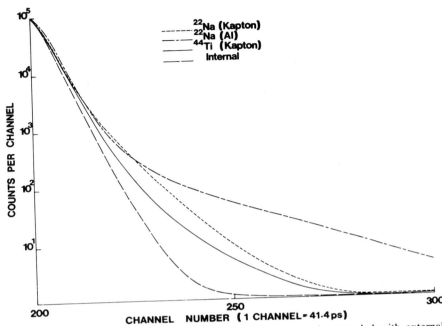

FIG. 10. Lifetime spectra for positrons annihilating in cobalt, recorded with external sources ^{22}Na and ^{44}Ti and an internal ^{58}Co source. Reproduced by permission of the North-Holland Publishing Co. from Campbell *et al.*[45]

5.1.5. Methods of Data Analysis

Along with the source problem mentioned above the other continuing problem in using positrons as a microprobe of the physical properties of polymeric substances is the analysis of the data obtained. Exceptional electronic stability, excellent counting statistics, reliable deconvolution analysis, and theoretical models that correctly relate the parameters observed to the characteristics of the annihilation process in molecular solids are essential for meaningful results.

5.1.5.1. Lifetime Spectra Analysis. A typical annihilation lifetime spectrum obtained from positrons annihilating in polymers is considered to consist of at least three exponentially decaying components each characterized by a mean lifetime and a probability for annihilation with that particular mean lifetime (i.e., intensity). The instrumental time resolution function is convoluted with the complex exponential decay and the resulting spectrum is perturbed due to background radiation and statistical fluctuations. In order to unfold the exponential decay components reliably, the zero of time and the shape of the time resolution function must be established with high precision. Fluctuations in the zero of time over the duration of the long counting period (usually several hours), often required to obtain statistical accuracy, will make meaningful analysis of short-lived components impossible. Further, if there are any source annihilations or nonlinearities in the time calibration these should be known to better than a few percent and corrected for.

It is generally the case that the time resolution is defined by the full width at half maximum (FWHM) of a "prompt" peak derived from detecting the simultaneous 1.17 and 1.33 MeV gamma photons from ^{60}Co using the same experimental settings for the electronics as is used in the experiment. Zero time is determined by the centroid of that peak.

Lichtenberger et al.[53] have proposed an analytical model of the experimental counting distribution that permits the instrumental time resolution to be unfolded and the entire counting distribution analyzed in cases where the prompt resolution function can be approximated by a double-sided exponential. (The shape of the prompt peak from their spectrometer justified this approximation.) Using the double-sided exponential makes it possible to evaluate the convolution integral analytically. An algorithm that combines the best features of the gradient search and the Taylor series search is used to obtain a least-squares estimation of the nonlinear parameters. The model was tested on Monte Carlo simulated data, and since the correct parameters are known, a realistic estimation of the errors associated with the analysis is obtained.

Kirkegaard and Eldrup[54] use a least-squares fitting technique but assume that the time resolution function can be represented by a sum of Gaussian functions. In this case the evaluation of the convolution integral gives a nonanalytical expression involving error functions. These authors have determined an instrumental resolution function for their system by analyzing the lifetime spectra for positrons annihilating in some standard having a well-established single lifetime. (The need for standard materials and calibration procedures in positron annihilation lifetime spectroscopy has been pointed out by MacKenzie et al.[55]) The resolution function so determined, approximated by a sum of Gaussian functions, is being used in their analysis since they find that it differs from the ^{60}Co prompt curve for their lifetime system. In fact, as Eldrup et al.[56]

[53] P. C. Lichtenberger, J. R. Stevens, and T. D. Newton, *Can. J. Phys.* **50**, 345 (1972).
[54] P. Kirkegaard and M. Eldrup, *Comput. Phys. Commun.* **7**, 401 (1974).
[55] I. K. MacKenzie, N. Thrane, and P. Sen, *Proc. Int. Conf. Positron Annihilation, 4th, Helsingor, 1976* H1 (1976).
[56] M. Eldrup, Y. M. Huang, and B. T. A. McKee, *Appl. Phys.* **15**, 65 (1978).

point out, even small uncertainties ($\simeq 1\%$) in the width of the time resolution function used in an analysis may cause uncertainties in the resulting
lifetimes and intensities several times the statistical uncertainties for a
spectrum of 10^6 counts. Thus a large number of counts does not always
guarantee an accurate analysis. It is worth noting here that if there is
little interest in the short lifetimes and associated intensities the analysis
could exclude the peak region and reduce the effect of uncertainties in the
shape of the time resolution function. However, accurate corrections for
source annihilations and background must have been made or neglecting
the peak region will make matters worse.

The technique of Crisp *et al.*[41] referred to in Section 5.1.4.1 lends itself
to the analysis of relative mean lifetimes by measuring the time displacement between the centroid of the delayed spectrum and the prompt spectrum.[57] As pointed out already the accumulation of a statistically viable
symmetric lifetime spectrum can occur in about 15 minutes, greatly reducing the time required for an experimental run and indicating the possibility of the positron lifetime technique being used as an analytical tool.

In the analysis of positron lifetime spectra, the salient need is for a deconvolution routine that can reliably deal with at least eight unknown
independent parameters. This assumes high electronic stability and adequate statistics.

5.1.5.2. Angular Correlation Analysis. Let $N_z(\theta)$ be the rate of annihilation into the momentum region between $m_0c(\delta\theta)$ and $m_0c(\delta\theta + d\theta)$.
Then[58]

$$N_z(\theta) = A \sum_i \int_{-\infty}^{\infty} \int \lambda_{2\gamma_i}[p_x, p_y, m_0c(\delta\theta)] \, dp_x \, dp_y,$$

where A is a normalization constant and the sum is over all occupied electron states. This integral is, at the moment, intractable for organic
molecular solids. However, studies[44,59] have shown that for organic
materials both the broad and narrow (p-Ps) components of angular distributions of momenta can be well represented by Gaussian functions. The
same nonlinear least-squares fitting routines used in unfolding positron
lifetime spectra[53,54] can be used to separate out the Gaussian components
in an angular correlation spectrum. The same uncertainties also exist including the problem of defining and unfolding the instrument resolution
function. Kirkegaard and Mogensen[60] report success with such a pro-

[57] Z. Bay, *Phys. Rev.* **72**, 419 (1960).
[58] S. Berko and J. S. Plaskett, *Phys. Rev.* **112**, 1877 (1958).
[59] J. H. Hadley and F. H. Hsu, *Chem. Phys. Lett.* **12**, 291 (1971).
[60] P. Kirkegaard and O. E. Mogensen, *Proc. Int. Conf. Positron Annihilation, 4th,
Helsignor, 1976* H12 (1976).

gram if care is taken in correlating the parameters obtained to physical quantities.

Difference spectra can also be used to compare angular correlation distributions under slightly different circumstances.

5.1.5.3. Doppler-Broadening Analysis. In analyzing the Doppler-broadened energy profile for the annihilation photon the analysis techniques described in Section 5.1.5.2 can be used, although the instrument resolution function is a factor of 10 worse. Nevertheless one can take advantage of the symmetry of the profile and of the effective stability of these Ge(Li) systems and devise parameters of good statistical precision for comparison purposes. MacKenzie et al.[61] discuss such parameters as the first and second moments of the profile (i.e., mean and variance) and a line shape parameter, which is the ratio of the sums of counts over a fixed group of channels in the center of the line and symmetrically placed groups on the wings of the line. Furthermore, it should be possible to measure the resolution function and the unknown simultaneously so that the operating resolution function is known. Sources such as ^{103}Ru (497 keV), or ^{7}Be (477 keV) with energies below the annihilation energy could be run at the same time as the annihilation run and not distort the unknown spectrum.

5.1.6. The Positron and Positronium in Polymers

Two current problems in using the positron as an experimental probe have already been discussed. These are the positron source problem (Section 5.1.4.5) and the deconvolution problem, especially with respect to positron lifetime spectra (Section 5.1.5.1). A third serious problem has to do with the preparation and characterization of the polymer sample used. Only recently[40,62-67] have workers started to use well-characterized and carefully prepared samples. At the very least the density and the number-average and weight-average molecular weights should be known and reported. It is also important that the percentage of crystallinity for crystalline polymers be known from at least two different methods. If, in the future, this information and information on sample preparation do not accompany results reported in the literature then the

[61] I. K. MacKenzie, P. C. Lichtenberger, and J. L. Campbell, *Can. J. Phys.* **52**, 1389 (1974).

[62] J. R. Stevens and A. C. Mao, *J. Appl. Phys.* **41**, 4273 (1970).

[63] J. R. Stevens and P. C. Lichtenberger, *Phys. Rev. Lett.* **29**, 166 (1972).

[64] J. R. Stevens and R. M. Rowe, *J. Appl. Phys.* **44**, 4328 (1973).

[65] D. H. D. West, V. J. McBrierty, and C. F. G. Delaney, *Appl. Phys.* **7**, 171 (1975).

[66] S. Y. Chuang and S. Y. Tao, *J. Appl. Phys.* **44**, 5171 (1973).

[67] A. E. Hamielec, M. Eldrup, O. Mogensen, and P. Jansen, *J. Macromol. Sci., Rev. Macromol. Chem.* **9**, 305 (1973).

contribution that those results make to the advancement of knowledge in this field will be minimal.

Having recognized these problems what do we believe we presently know and do not know about positrons annihilating in polymeric solids?

5.1.6.1. Experimental Observations. From lifetime studies it appears that positrons annihilate in polymers with at least three mean lifetimes[40,44,63,64,68] and quite possibly four, especially in crystalline polymers.[44,63] These lifetimes range from about 100 psec to 5 nsec. We know that the longest of these lifetimes is temperature,[38,62,64,69] pressure,[70] and molecular-weight dependent,[65,66] and that it reflects the glass transition[62] and melting.[38] These dependences arise as a result of the positron annihilating in domains of lower electron densities or "free volume" sites found throughout the polymer and especially concentrated around the ends of long molecular chains. We also know that in polyethylene (PE) and in polytetrafluoroethylene (PTFE) the probability that a positron will annihilate with the longest mean lifetime decreases with increasing external electric field approaching a limit asymptotically.[40,63,71] Finally, we know that the probabilities (intensities) and the lifetimes of some of the positron annihilation modes depend on the morphology of the sample under study.[40,63,64,72,73]

The various modes of positron annihilation in polymers and the mechanisms by which Ps is formed are still not completely understood; nor is the relation of these modes to the details of the polymer morphology.

A discussion of some of the above points is in order.

Table II is a summary of positron lifetime data from the literature, where at least three lifetime components have been identified. In three cases, two for PE and one for PTFE, the short component identified in the three component analyses has been resolved into two components: an operation that requires precise knowledge of the shape of the time resolution function, the zero of time, background, and source corrections.

The polymers listed in Table II in addition to PE and PTFE to which reference has already been made are poly(vinyl chloride) (PVC); polyamide (PA), commercially referred to as Nylon 6; polypropylene (PP); poly(methyl methacrylate) (PMMA), commercially referred to as Plexiglass or Lucite; polystyrene (PS); polybutadiene (PB); and polyisobutyl-

[68] M. Bertolaccini, A. Bisi, G. Gambarini, and L. Zappa, *J. Phys. C* **7**, 3827 (1974).

[69] B. C. Groseclose and G. D. Loper, *Phys. Rev. A* **137**, 939 (1965).

[70] R. K. Wilson, P. O. Johnson, and R. Stump, *Phys. Rev.* **124**, 2091 (1963).

[71] M. Bertolaccini, A. Bisi, G. Gambarini, G. Padovini, and L. Zappa, *Appl. Phys.* **12**, 93 (1977).

[72] J. R. Stevens and M. J. Edwards, *J. Polym. Sci. C* **30**, 297 (1970).

[73] S. J. Tao and J. H. Green, *Proc. Phys. Soc., London* **85**, 463 (1965).

TABLE II. Summary of Positron Lifetime Data

Polymer	Cryst. (%)	$\bar{M}_w$	$\bar{M}_w/\bar{M}_n$	Positron lifetimes (nsec)				Positron intensities				Temp. (°K)	Ref.
				τ_1	τ_2	τ_3	τ_4	I_1	I_2	I_3	I_4		
PE	46			0.31	0.65	2.61		0.63	0.10	0.27		293[a]	40
	51			0.33	0.81	2.67		0.67	0.09	0.24		293	40
	53			0.31	0.87	2.54		0.69	0.09	0.22		293	40
	69			0.32	0.80	2.53		0.72	0.10	0.18		293	40
	80			0.34	0.75	2.54		0.73	0.09	0.18		293	40
	75			0.15	0.35	0.66	2.39	0.24	0.47	0.10	0.19	293	40
	—	41,700	3.7	0.1	0.34	1.22	2.67	0.72		0.11	0.17	293	63
PTFE	60			0.31	1.02	4.17		0.66	0.18	0.16		293	44
	50			0.32	0.92	3.75		0.66	0.08	0.20		293	40
PMMA	—	atactic		0.23	0.39	1.23	4.33	0.21	0.47	0.18	0.14	293	40
NYLON	0			0.33	0.62	1.94		0.62	0.11	0.27		293	44
PP				0.28	0.56	1.61		0.45	0.21	0.18		293	40
		7,000		0.30	1.24	2.38			0.12	0.15		295	68
		8,900		0.30	0.82	2.14			0.10	0.21		295	66
		16,000		0.26	0.68	2.14			0.15	0.21		295	66
		300,000	4.9	0.28	0.81	2.20			0.11	0.19		295	66
		420,000	5.7	0.26	0.76	2.13			0.14	0.19		295	66
		433,000	7.2	0.27	0.72	2.16			0.14	0.20		295	66
		580,000	3.9	0.23	0.51	2.08			0.24	0.21		295	66
		870,000	5.4	0.24	0.59	1.93			0.18	0.19		295	66
PS	0	atactic		0.24	0.54	2.07		0.26	0.27	0.33		293	68
PVC	—			0.31	0.52	1.88		0.56	0.33	0.05		293	68
PB	—	368,000	4.4	0.17	0.47	2.53		0.22	0.25	0.53		273	64
PIB	0	1,540,000	2.6	0.17	0.46	2.18		0.19	0.38	0.43		273	64

[a] 293°K was assumed if a temperature was not reported.

ene (PIB). The first five polymers—PE, PTFE, PVC, PA, and PP—are partly crystalline. Commercial PVC usually has a low crystallinity of about 10% (syndiotactic) and commercial PP usually has a crystallinity of about 50% (isotactic). In solution, PVC and PMMA have dipole moments per monomer unit, respectively, 1.39 D (in tetrahydrofuran) and 1.35 D (in benzene),[74] and have relatively high dielectric loss factors in the bulk (tan $\delta \simeq 0.01$ at 1 MHz, 25°C[74]). PA exhibits a dielectric loss of the same order of magnitude. These solid polymers are polar. PS could be classed as weakly polar and the remainder as nonpolar. For instance, PS has a dipole moment per monomer unit of 0.36 D (atactic in toluene).[74] All have dielectric constants in the range of 2 to 3 over a wide range of frequencies, with the polar polymers having the highest values. These dielectric properties are of course temperature and frequency dependent.

It is now fairly certain that the longest positron lifetime in polymers is that of o-Ps annihilating by pick-off, τ_p (Section 5.1.3). The variation of τ_p with temperature has already been discussed in Fig. 2[38] (Section 5.1.3). This variation was discussed in terms of a simple cell model in which τ_p depends on the free volume [through the reduced volume $v^*(T)$] and on the reduced temperature $T^* = T/\theta_D$, where θ_D is the Debye temperature. The overlap of the positron part of the o-Ps wavefunction with the wavefunction of a "lattice" electron will increase (shorter lifetime) or decrease (longer lifetime) according to whether the free volume cell within which the o-Ps resides decreases or increases (possibly coalescing with other packets) with temperature, pressure, or different morphologies. There is a lower limit to the volume of a cell, which is compatible with a stable Ps atom. Brandt and Wilkenfeld[40] have used $v^*_{cr} = 0.23$ for this limit.

A similar simple cell model is used by Gibbs and DiMarzio[75] as a basis for their classic theory of the nature of the glass transition and the glassy state. The glass transition in polymeric glasses is associated with the onset of a coordinated segmental motion about the polymer chain axis as the temperature of the glass is raised.

Essentially the total free volume in a polymeric solid can be divided into a set of identical packets equal in number to the total number of chain segments. These packets do not remain with individual segments on a one to one basis but are distributed within the solid with increased concentration at polymer chain ends and almost zero concentration in the chain-folded crystalline regions.

As well as explaining the temperature and pressure[70] dependences this

[74] J. Brandrup and E. H. Immergut, eds., "Polymer Handbook," 2nd ed. Wiley, New York, 1975.
[75] J. H. Gibbs and E. A. DiMarzio, *J. Chem. Phys.* **28**, 373 (1958).

simple model can be used to understand why τ_p is sensitive to the glass transition[62,64,67,69] and to first-order phase changes.[69] Hamielec et al.[67] report results on the lifetimes and intensities of positrons annihilating in characterized commercial grade polymers. However, because of the preliminary nature of their work and the fact that their lifetime spectra were analyzed into two components only (as did Stevens and Mao[62]), no further comments will be made.

Experimental evidence exists that indicates that the lifetimes and intensities of positrons in polymers are sensitive to the morphology of the sample.[63,72,73,76,76a] Spherulites exist in untreated bulk crystalline polymers between which there are amorphous (randomly coiled chains) regions and within which there are chain-folded regions of long-range order interspersed with loops and cilia. In addition, some crystalline polymers contain long-range order in regions where the polymer chains are extended and aligned over distances much longer than the chain fold period (150–200 Å). Noncrystalline polymers are largely amorphous but may contain microscopic regions of short-range order.

Many kinds of defects are found in polymeric solids. These are discussed by Wunderlich.[77] Briefly, there are many "empty" microscopic regions within these solids, which would be occupied for perfect order. These vacancies cannot be filled as readily as in metals, for example, since the continuity of each polymer chain (conformation) and the necessity of interstitial or substitutional atoms to be bound to it are of overriding importance. Theories of chain disorder are couched in terms such as kinks, jogs, chain torsion, ends, and folds.

There is evidence that τ_p and its related intensity I_p (τ_3 and I_3 or τ_4 and I_4 in Table II depending on which τ is the longest) are associated with annihilation in the amorphous regions of a polymer and that the intermediate lifetimes and intensities could be associated with annihilation in the crystalline regions.[63,72,73] On this assumption, Stevens and Lichtenberger[63] have been able to obtain excellent agreement between values of the crystallinity of a PE sample (Table II) as determined by positron annihilation compared with X-ray diffraction and differential scanning calorimetry. These authors attribute the τ_2 component to positrons annihilating in the chain-folded regions and τ_3 to the defects (cilia and loops) in these regions.

[76] G. Chandra, V. G. Kulkarni, R. G. Lagu, A. V. Patankar, and B. V. Thosar, *Proc. Nucl. Phys. & Solid State Phys. Symp. Bombay 1965,* p. 210 Atom. Energy Establ. (1965).

[76a] Yu. V. Zelenev and A. I. Filipev, *Sov. Phys. Solid State* **18**, 787 (1976).

[77] B. Wunderlich, "Macromolecular Physics," Vol. 1. Academic Press, New York, 1973.

Referring to Table II the results of Kerr[44] and Brandt and Wilkenfeld[40] could also likely be used to predict the crystallinity of their PE samples if their shortest-lifetime component had been resolved into two components. Kerr's sample[44] must have been highly crystalline. For the PE samples of Brandt and Wilkenfeld[40] note how I_3 decreases with increasing crystallinity and how I_1, part of which should be associated with positrons annihilating in crystalline regions, increases with increasing crystallinity. The same conclusions could be drawn for the highly crystalline PP series,[66] and the other PTFE samples.[40] For Kerr's PTFE sample[44] the ratio $(I_2 + I_3)/(I_2 + I_3 + I_4)$ is 0.82 and the sample could be said to be highly crystalline (82%).

All of the samples discussed so far with respect to percentage of crystallinity determination have been nonpolar. However, we might be able to reach the same conclusions about the data reported for the highly polar PA.[68] Bertolaccini et al.[68] have reported intensity data for the three longest-lived components and acknowledge the existence of at least one more component with lifetime ≤ 0.2 nsec. For PA their intensities total to 0.849. If it is assumed that the intensity of the short component is missing, then this PA sample would be 79% crystalline.

Unfortunately we cannot confirm any of the above conjectures or account for the anomalous PVC data (highly crystalline?) unless the percentage crystallinities are known or the lifetime spectra are resolved into four components or both.

For PMMA, PA, PB, and PIB the values of I_3 tend to be higher than for the crystalline polymers. To account for the I_2 intensities, Stevens and Rowe[64] suggested that positrons were being trapped and annihilating in regions of short-range order that tend to reduce in size as the temperature is increased.

5.1.6.2. Modes of Annihilation in Polymers. In the previous section the longest positron lifetime component for positrons annihilating in polymeric solids was attributed to the decay of o-Ps by pick-off. The assignment of this annihilation mode has been supported by combined lifetime and angular correlation measurements on PE[44] and PTFE.[44,49] The narrow peak in the momentum distribution (i.e., the low-momentum component in Fig. 7) results from the self-annihilation with rate λ_s of p-Ps whose intensity is $\sim \frac{1}{3}$ the intensity of o-Ps. Evidence of a narrow peak centered on 180° or a low-momentum peak is taken as evidence of the formation of Ps. However, in angular correlation experiments on solid benzene, solid cyclohexane, and solid methane the intensity of the narrow-momentum peak is reported by DeBlonde et al.[78] to be very weak or ab-

[78] G. DeBlonde, S. Y. Chuang, B. G. Hogg, D. P. Kerr, and D. M. Miller, Can. J. Phys. 50, 1619 (1972).

sent even though these substances exhibited a long lifetime component. These results were confirmed by Chuang and Tao[79] for solid nitrobenzene in benzene. This leads to a conjecture[29] that positrons may become trapped and annihilate with long lifetime (2–5 nsec) without forming Ps. Goldanskii[33] has suggested that p-Ps may swell under the influence of polarization forces in its locality. Annihilation of the positron by pick-off with rate λ_p could then predominate ($\lambda_p \gg \lambda_s$) and therefore no significant low-momentum component would be seen. Bertolaccini et al.[68] have shown experimentally that Ps does "swell" or "relax" at least in PE, PS, PVC, and PA. Preliminary work in our laboratory has confirmed these results for PS and PE. This work is continuing. Brandt and Wilkenfeld[40] have presented a theoretical argument supported by experimental evidence that the width of the low-momentum component should increase as λ_p increases with decreasing temperature. The results of DeBlonde et al.[78] and Chuang and Tao[79] (from low-temperature experiments) are best explained by a combination of the last two of these three proposals, eliminating the first one.[29] It is concluded, therefore, that all long lifetimes of positrons annihilating in organic solids are a result of the pick-off decay of o-Ps.

The shortest lifetimes (0.1–0.3 nsec) for positrons in polymers are most likely those for positrons annihilating within a para-bound state such as in p-Ps or in some chemical complex. Brandt and Wilkenfeld[40] propose that the intensity of the shortest-lived component is due to p-Ps self-decay and the annihilation of positrons in the bulk but not trapped in A^+ centers.[28] As has already been suggested, their shortest-lived component should be resolvable into two components, thus isolating p-Ps decay as the annihilation mode with the shortest lifetime. If p-Ps annihilates by pick-off because of polarization swelling as has been shown,[68] then it may not have the shortest lifetime.

The annihilation mode that results in positron lifetimes in polymers in the intermediate range (0.3–0.9 nsec) has been assigned, at least in part, to positrons trapped in the substance and annihilating without forming Ps.[40,63] The pick-off annihilation of p-Ps could also result in lifetimes in this range. Whether most lifetimes in this range result from positrons annihilating in some form of trap and whether annihilation without Ps formation occurs only in ordered regions in the polymer are questions not completely answered. Certainly the problem of positron trapping in polymeric solids, especially as it relates to Mogensen's spur model of Ps formation,[29] needs further investigation.

Hunt[24] points out that there is a high probability that electrons formed

[79] S. Y. Chuang and S. Tao, Appl. Phys. 11, 247 (1976).

in the wake of a moderating positron will be trapped (in 10^{-12}–10^{-14} sec) before recombination. Although the positron and the electron cases are not analogous, close to thermalization there are appropriate sites in the polymer microstructure for positron trapping and the time before trapping should be of the same order as for electrons. In polar organic glasses the electron traps appear to be much deeper (1.5–2.5 eV) than in nonpolar organic glasses (0.7 eV).[25] The experimental evidence that supports this conclusion also supports a proposal that electrons tunnel quantum mechanically (in competition with diffusion) to acceptor sites. Miller[25] calculates on the basis of square-well penetration and the specific tunneling rate due to Kauzmann[80] that

$$r(t) = r_0 + \frac{15 + log_{10} t}{0.443(V_0 - E_0)^{1/2}} \quad \text{Å}, \qquad (5.1.6)$$

where $r(t)$ is the distance an electron may tunnel if it has time t (sec). Since V_0 is the depth of the trap and E_0 is the ground-state energy of the electron, then $V_0 - E_0$ is the binding energy of the electron in the trap. The factor r_0 corrects for the finite radius of the trapped electron and electron acceptor and is assumed to be 4–5 Å. According to Eq. (5.1.6) an electron could tunnel 14 Å in 10^{-11} sec. This theory assumes that the tunneling of the electron from its trap to the acceptor is perfectly "resonant" (no symmetry, Franck–Condon restrictions, etc.). This assumption requires that the tunneling rates are independent of the nature of the acceptor, provided the electron affinity of the acceptor is greater than $V_0 - E_0$, the binding energy of the trapped electron. It is therefore conceivable that an electron could tunnel to a thermalized positron acceptor trapped in an organic glass (amorphous) with Ps resulting. More information is required to determine whether this mechanism would pertain for crystalline regions of the polymer, since if diffusion dominates the positron needs to be in a sufficiently open structure in order to capture an electron to form Ps.

Mogensen[29] has proposed the spur model to account for the formation of Ps. He proposes[81] that this is likely the only appropriate model, especially in liquids, and argues against the Ore gap model. Until there is more experimental support from both positron annihilation studies and studies of early events (<1 psec) in radiation chemistry, these models at least should be considered as competing. Maddock et al.[82] support this view from their studies of positrons annihilating in aqueous solutions. The very early trapping of low energy (<0.2 eV) electrons originating

[80] W. Kauzmann, "Quantum Chemistry," p. 188. Academic Press, New York, 1957.
[81] O. E. Mogensen, Appl. Phys. 6, 315 (1975).
[82] A. G. Maddock, J. C. Abbe, and A. Haessler, Chem. Phys. Lett. 47, 314 (1977).

from a spur in an organic solid could very well be an important feature in the formation of Ps where positrons with energies below T (Fig. 1) are involved. Also the Ore gap model has been very useful and refinements to it that incorporate positron, electron, and Ps binding energies should be made.[16]

Two further points on modes of annihilation in polymeric solids should be noted. First, as pointed out by Brandt,[28] there is now considerable support for the suggestion that Ps can escape from insulators and annihilate according to conditions it encounters outside the solid. Second, if positron lifetimes and intensities in polymers are to reflect the morphology then the probability for positron annihilation by pick-off from o-Ps and the probability for annihilation from traps must be, respectively, proportional to the fractions of amorphous and crystalline regions in the solid.

5.1.6.3. Electric Field Effects. Several investigators[40,63,71] have noted a change in the intensity of Ps formation in PE and PTFE with increasing applied electric field. Brandt and Feibus[83] have derived a comprehensive yield function in an attempt to explain these changes. Brandt and Wilkenfeld[40] have tested this theory, which incorporates the Ore gap model, by studying the annihilation of positrons in a series of normal paraffins and polyethylenes. They observe a decrease in the Ps yield $P_0(\mathscr{E})$ with increasing electric field $\mathscr{E}$ in nonpolar PE but not in polar PMMA. Their results are reproduced in Fig. 11, where a Ps yield function

$$\Phi = P_0(\mathscr{E}) \exp(v_a^* / v_{cr}^* - 1)$$

is plotted vs. an applied electric field function $X = s_0 \mathscr{E}$, where $s_0 = |d\Phi/d\mathscr{E}|_{\mathscr{E}=0}$. The theory[83] is the solid curve. An applied field enhances the scatter of positrons out of the Ore gap decreasing $P_0(\mathscr{E})$. If positrons with energies less than T (Fig. 1) are not trapped, there is a possibility that they could be scattered up into the Ore gap, thus increasing $P_0(\mathscr{E})$. The former effect dominates at low fields, the latter at high fields. The steady-state value of Φ near 0.4 implies that the fraction $0.4/0.66 = 0.6$ of all positrons below the Ore gap are trapped and cannot be accelerated by the electric field back into the Ore gap.[28]

The results of Stevens and Lichtenberger[63] agree qualitatively with those of Brandt and Wilkenfeld.[40] Intensity values derived from a three-component fit of their data[63] fall above the curve in Fig. 11 as indicated by the small arrows for a few high-field points. The polymer chain alignment effects observed by these authors as the electric field was increased need further study.

[83] W. Brandt and H. Feibus, *Phys. Rev.* **174**, 454 (1968).

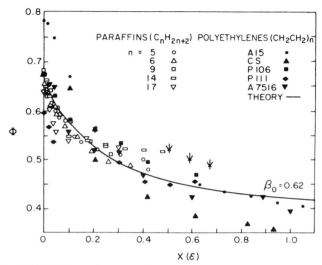

FIG. 11. Ps yield for the homologous series of paraffins and polyethylenes as a function of $\mathscr{E}$ in terms of the reduced variable X. Reproduced by permission of the American Institute of Physics from Brandt and Wilkenfeld.[40]

A recent study by Bertolaccini et al.[71] indicates that $P_0(\mathscr{E})$ increases with time after an electric field is applied to the PE sample. At 50 kV cm^{-1} an equilibrium yield was obtained after 100 hours, and at 100 kV cm^{-1} 50 hours were required. Any future studies of $P_0(\mathscr{E})$ in PE must certainly take these results into account. In the experiment already reported[63] the applied field was increased in small increments and held at each value for at least 14 hours, the lifetime spectrum being accumulated over the last 11 hours. It may well be that because of this time dependence for $\mathscr{E}$ all earlier measurements are not possible to interpret, although the work of Brandt and Wilkenfeld[40] is very convincing.

5.1.6.4. Polymerization. Ito et al. have analyzed the mechanism of solid state polymerization of dimethyl itaconate[84] and acrylamide.[85] Gamma photons from a ^{60}Co source at 30°C were used to initiate the polymerization. They observed a decrease in λ_p during the polymerization induction period (increase in free volume) and a decrease in I_p (increased trapping of positrons). These authors recommend the use of positron annihilation in the study of solid-state polymerization.

[84] Y. Ito, S. Katsura, and Y. Tabata, J. Polym. Sci., Part A-2 9, 1525 (1971).
[85] Y. Ito, K. Okuda, and Y. Tabata, Bull. Chem. Soc. J. 44, 1764 (1971).

5.1.7. Conclusion

Tremendous interest exists in the possibilities of using positrons to probe the microstructure of polymeric matter. However, the problems of source, deconvolution, and sample preparation discussed in the text must be resolved before progress can be made in this field.

Further research is required to firmly establish the mechanisms by which Ps is formed and to develop more fully the dependences of positron lifetimes and intensities on such polymer characteristics as free volume, average dipole field, and morphology. The mechanisms by which positrons are trapped in polymers and the manner of annihilation from traps must also be more fully understood.

Acknowledgment

Acknowledgment is given to many helpful discussions with I. K. MacKenzie. The suggestion at the end of Section 5.1.5.3 of simultaneously obtaining the resolution function and the Doppler-broadened energy spectrum of the unknown is his.

5.2. Fluorescence Probe Methods†

By L. Lawrence Chapoy and Donald B. DuPré

5.2.1. Introduction

Traditionally fluorescence spectroscopy has been utilized either (1) as an analytic tool in which a component under analysis could if necessary be converted into a fluorescent derivative, the emission intensity of which would then be measured and the concentration determined with the help of a calibration curve, or (2) as a structural tool in which the details of the spectrum, the lifetime, and the quantum yield can lead to information regarding the molecular structure, energy levels, and transitions of the fluorescent compound itself. During recent years, some of the more subtle aspects of the fluorescence phenomenon have been creatively employed to gain insight into the structure of polymeric materials. Closely related to this application in the area of polymer science is the use of the fluorescence phenomenon in ascertaining structural details of micelles, biological membranes, and liquid crystals. The nature of these experimental techniques and the interpretation of the results will be the subject of this chapter. Also of interest will be problems of instrumentation and criteria for choosing suitable guest or probe molecules for the host that is to be studied. Probes that are chemically incorporated will be referred to as *labels*. In this case one must also consider the strategy of the chemist in attaching the fluorescent label to the polymer in question. General references that may be of interest are Hercules,[1] Pesce *et al.*,[2] Becker,[3] and Nishijima.[4]

The experimental results presented here do not constitute an exhaus-

[1] D. M. Hercules, ed., "Fluorescence and Phosphorescence Analysis." Wiley (Interscience), New York, 1966.

[2] A. J. Pesce, C.-G. Rosén, and T. L. Pasby, eds., "Fluorescence Spectroscopy." Dekker, New York, 1971.

[3] R. S. Becker, ed., "Theory and Interpretation of Fluorescence and Phosphorescence." Wiley (Interscience), New York, 1969.

[4] Y. Nishijima, *J. Polym. Sci., Part C* **31**, 353 (1970).

† See also Vol. 13 (Spectroscopy) of this Series, Sections 3.1.4.6 and 5.2.3.6.

METHODS OF EXPERIMENTAL PHYSICS, VOL. 16A

tive review but merely serve to illustrate the broad areas where the technique can be utilized.

5.2.2. The Fluorescence Phenomenon

Fluorescence is a form of emission spectroscopy and as such consists of at least two quantum-mechanical steps: an absorptive excitation and a subsequent emission. The fact that it is a multi-step process makes it considerably more complicated than various types of absorption spectroscopy, but these very complications give rise to additional possibilities of extracting extra information about the host system not available from the single-step absorptive methods.

To proceed, one must consider the fluorescent phenomenon in detail so that all of the competing processes can be appreciated in terms of how they may be exploited to yield structural information about the host. A flow scheme containing the processes of interest here is shown in Fig. 1.

Upon absorption of radiation in the proper region of the spectrum the ground-state singlet electronic state S_0 is raised to a higher energy level S_n, where n will depend on the wavelength employed. Each $S_0 \rightarrow S_n$ absorption will have a different transition moment vector associated with it and as such the spectrum for an ordered array of molecules will be polarized. The absolute spatial assignment of the transition moment oscillators in the molecular geometry is a difficult undertaking but has been of great interest. Such assignments are accomplished by using single crystals of known structure[5] or by embedding the molecule of interest in an ordered, transparent polymer film[6,7] or liquid crystals.[8] For anisotropic molecules a secondary "hand and glove" or guest-host type of ordering is achieved in these systems. Since these are electronic transitions, one is concerned with the visible and ultraviolet portions of the electromagnetic spectrum.

The transitions $S_n \rightarrow S_1$ occurs very rapidly by vibrational decay, and it is of no special importance other than that an intramolecular energy transfer occurs and the energy is transferred to a new vector representing the $S_0 \rightarrow S_1$ transition moment oscillator.

The $S_1 \rightarrow S_0$ transition is what is commonly referred to as *fluorescence*. Since the molecules fall to higher vibrational levels in the S_0 level, the emission is red-shifted. Competing with this process are radiationless

[5] D. S. McClure, *Solid State Phys.* **8**, 1 (1959).

[6] E. W. Thulstrup and J. H. Eggers, *Chem. Phys. Lett.* **1**, 690 (1968).

[7] N. S. Gangahedkar, A. V. Namjoshi, P. S. Tamhane, and N. K. Chaudhuri, *J. Chem. Phys.* **60**, 2584 (1974).

[8] E. Sackmann, *J. Am. Chem. Soc.* **90**, 3568 (1968).

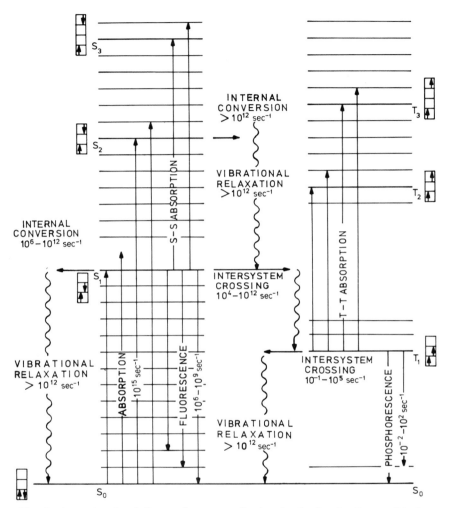

FIG. 1. A quantum level diagram for a generalized molecule showing the possible decay paths of an excited state.

transitions to the ground state. The relative magnitude of the rate constants for these decay processes determines the *quantum yield* or the efficiency of recovery of the excitation energy as radiation in a particular region of the spectrum. The transition $S_1 \rightarrow T_1$, where T_1 is the triplet state, is strictly speaking spin forbidden, but possible under certain circumstances. The $T_1 \rightarrow S_0$ transition, commonly known as *phosphorescence,* is also spin forbidden, resulting in very long lifetimes and a propensity to *quench,* a term denoting the dissipation of excitation energy through nonradiative mechanisms. In concentrated solution the situation is further complicated by the possibility of (1) reabsorption of emission, leading to decreased net emission intensity, (2) intermolecular energy transfer controlled by the square of the cosine of the angle between adjacent molecules and the reciprocal sixth power of the distance between the moment oscillators,[9,10] and (3) *excimer formation* caused, for example, by the presence of a monoexcited dimer as in pyrene solutions[11] or as in the case of isotactic polystyrene.[12]

5.2.3. Fluorescence Quenching Techniques

5.2.3.1. Theory of Quenching. As pointed out in Section 5.2.2, considerable energy is lost from the excited state via paths that do not lead to radiative emission. This concept can be quantified in terms of the *quantum yield:*

$$q = \frac{\text{number of photons emitted}}{\text{number of photons absorbed}}. \tag{5.2.1}$$

The concentration of excited states $S_1(t)$ is depopulated following a pulsed excitation by competing processes that are approximately first-order in nature; that is,

$$S_1(t) = S_1(0) \exp(-\Sigma K_i t), \tag{5.2.2}$$

where the rate constants K_1, K_2, and K_3 reflect the degree of fluorescent emission, intersystem crossing to the triplet state, and radiationless transition to the ground state, respectively. The triplet state is very long-lived, so that molecules from this state will usually revert to the ground state via radiationless transitions and not via phosphorescence except under special conditions. Hence K_2 and K_3 processes both involve quenching. In the absence of quenching, the intrinsic lifetime of the excited state τ_0 will be equal to K_1^{-1}, but generally the lifetime $\tau = (\Sigma K_i)^{-1}$.

The fluorescent quantum yield will then be a function of the competing rate constants, which depopulate the excited state:

$$q = K_1/(K_1 + K_2 + K_3), \tag{5.2.3}$$

giving

$$q = \tau/\tau_0. \tag{5.2.4}$$

[9] F. Perrin, *Ann. Phys. (Paris)* [10] **12**, 169 (1929).
[10] T. Förster, *Ann. Phys. (Leipzig)* [6] **2**, 55 (1948).
[11] J. B. Briks and L. G. Christophorou, *Spectrochim. Acta* **19**, 401 (1963).
[12] S. S. Yanari, F. A. Bovey, and R. Lumry, *Nature (London)* **200**, 242 (1963).

This leads to the fact that the effective excited state lifetime is proportional to the quantum yield, since those individual molecules that take a longer time to emit are quenched before they can do so. This is expressed quantitatively by the well-known Stern–Volmer equation[13]:

$$1/q = 1 + k[Q]\tau_0, \tag{5.2.5}$$

where the degree of quenching depends on the concentration of a quencher $[Q]$.

The molecular mechanism of quenching can be quite involved and the interpretations thereof somewhat varied. Collisional quenching occurs as a consequence of a collision between a molecule in an excited state S_1 and a quencher Q. It is sometimes possible for a long-lived complex to form that does not fluoresce. This can happen, for example, if the excited state S_1 is dissipated to heat via vibrational coupling in the complex. Alternately, a collision may promote intersystem crossing to the triplet state T_1, which is more easily quenched. This is primarily due to the longer lifetime of the triplet, but is also because of the lower energy of these states, which makes vibrational coupling with the ground state more probable. Molecules that can act as quenchers in this way often perturb the excited state through the presence of a dipole, a quadrupole, or a paramagnetic spin (examples, molecular oxygen, alkyl halides, or paramagnetic metal chelates). Other quenching mechanisms are (1) the formation of electron transfer complexes, (2) the formation of nonfluorescing complexes via hydrogen bonds or van der Waals interactions, (3) changes in ionization caused by pH changes that lead to nonfluorescing species.

Polar solvents or environments can lead to increased quenching because of the dipole perturbation as noted above, giving what are referred to as hydrophobic probes.[14] On the other hand one must consider the fact that polar solvents can in some cases shift the character of the transition from $n \to \pi^*$ to $\pi \to \pi^*$. The terms n and π designate ground-state molecular orbitals, while the term π^* refers to a molecular orbital occupied in the excited state. Since $\pi\pi^*$ states have shorter lifetimes, the possibility for intersystem crossing and hence the degree of quenching is reduced, giving so-called hydrophilic probes.[15] The possibility thus exists for using both hydrophilic and hydrophobic probes, although most experiments have been confined to the latter.

5.2.3.2. Information Obtained Using Quenching Techniques. The fact that the quantum efficiency is sensitive to a number of the above

[13] O. Stern and M. Volmer, *Phys. Z.* **20**, 183 (1919).
[14] G. M. Edelman and W. O. McClure, *Acc. Chem. Res.* **1**, 65 (1968).
[15] N. Mataga, Y. Kaifu, and M. Koizumi, *Bull. Chem. Soc. Jpn.* **29**, 373 (1956).

factors can be utilized to infer the nature of the *local environment* of a guest fluorescent molecule.[16] The effects noted necessarily monitor the local environment, as all are extremely short-range phenomena. Also the brevity of the lifetime relative to the diffusion constant ensures that one will not observe a completely time-averaged picture of the medium. The concept of the local environment is a very important one, since on the molecular level, polymers and micelles appear to be very heterogeneous systems. One may encounter local regimes that are amorphous, ordered, crystalline, hydrophilic, hydrophobic, ionic, polar, specifically solvated, etc., all within a single system. A full understanding of these systems requires a picture of all of these local environments and not an integrated average picture over the whole system. Specifically one can ascertain the polarity of a local environment, and the accessibility of the site. Also the transfer of electronic excitation can occur between an energy donor and an energy-receptive chromophore. Since the probability for transfer and hence the fluorescent intensity is a strong function of the distance, one has in effect a spectroscopic ruler.[17]

5.2.3.2.1. PROTEIN STRUCTURE. A review of the strategy and possibilities of this technique has been given by Edelman and McClure.[14] Briefly, a variety of hydrophobic probes are available for which the quantum yield increases dramatically as they are displaced from an aqueous to a hydrophobic environment. For proteins, the hydrophobic portions are usually buried inside the three-dimensional structure, giving only limited access to the probe. Only those probes in the hydrophobic environment contribute, however, to the observable intensity. The intensity is thus very sensitive to denaturation of the protein or even slight conformational changes (connected with enzyme activity, for example), which can alter the accessibility of hydrophobic regions to fluorescent probes. That very subtle conformational differences can be detected is shown by recent work on poly(L-lysine) using 2-p-toluidinyl-naphthalene-6-sulfonate (TNS), a probe often used in studies of this type.[18] Different proteins show vastly different quantum yields for the same probe, indicating their compositional and conformational differences. Different probes on the other hand react differently to the same protein because of specific steric accessibility to the available hydrophobic regions. For a more unambiguous interpretation of results, more must be learned about the nature of the interaction and the exact placement of the probe with the polypeptide. Progress in this direction has been made for the case of TNS.[19] The pos-

[16] B. L. Van Duuren, *Chem. Rev.* **63**, 325 (1963).
[17] L. Stryer, *Science* **162**, 526 (1968).
[18] G. Witz and B. L. Van Duuren, *J. Phys. Chem.* **77**, 648 (1973).
[19] C. F. Beyer, L. C. Craig, and W. A. Gibbons, *Biochemistry* **11**, 4920 (1972).

sibility for the labeling of specific amino acid residues in the polypeptide also exists.[20,21]

5.2.3.2.2. BINDING EQUILIBRIA. Binding equilibria are often of great importance for the understanding of various biochemical processes. Specially labeled fluorescent compounds, which are chemically analogous to those naturally participating in a binding process, can be prepared having the property that the quantum yield is greatly affected by the degree of binding.[22,23] The fraction of bound probes can then be estimated spectroscopically as a function of some desired parameter such as pH or ionic strength. Alternatively, the quantum yield of a conventional probe can be studied as a function of added reagents, which may be biochemically active[24] or act as drugs.[25]

In the realm of synthetic polyelectrolytes, the quenching method has been used to study the binding of fluorescent small mobile ions and the nature of the ionic atmosphere in the vicinity of a polyion.[26]

5.2.3.2.3. STRUCTURE AND DYNAMICS OF SYNTHETIC POLYMERS. The quenching technique does not appear to have been used extensively in the area of synthetic polymers. Segmental motion has been investigated using a specially labeled poly(4-vinyl pyridine) containing both fluorescing and quenching labels.[27]

In a novel technique, segmental motion was observed by the increase in fluorescent intensity following the photoinduced isomerization of specially prepared polymers.[28] This technique does not involve quenching as such, but is included in this section since it involves the observation of intensity changes that give information regarding the behavior of the host system. Conformational structure and transitions of a synthetic polyelectrolyte have been investigated with a hydrophobic probe as a function of degree of neutralization and solvent composition.[29] The quenching of phosphorescence of a labeled polymer by molecular oxygen as a function of temperature is a similar concept, which reflects segmental mobility and its effect on the accessibility of oxygen to quench the triplet state.[30]

[20] R. F. Steiner, A. Lunasin, and C. Horan, *Biochim. Biophys. Acta* **336**, 407 (1974).
[21] C.-W. Wu, L. R. Yarbrough, and F. Y.-H. Wu, *Biochemistry* **15**, 2863 (1976).
[22] M. Martinez-Carrion and M. A. Raftery, *Biochem. Biophys. Res. Commun.* **55**, 1156 (1973).
[23] B. R. Dean and R. B. Homer, *Biochim. Biophys. Acta* **322**, 141 (1973).
[24] H. N. Aithal, V. K. Kalra, and A. F. Brodie, *Biochemistry* **13**, 171 (1974).
[25] G. Sudlow, D. J. Birkett, and D. N. Wade, *Mol. Pharmacol.* **9**, 649 (1973).
[26] I. A. Taha and H. Morawetz, *J. Am. Chem. Soc.* **93**, 829 (1971).
[27] Yu. E. Kirsh, N. R. Pavlova, and V. A. Kabanov, *Eur. Polym. J.* **11**, 495 (1975).
[28] D. T.-L. Chen and H. Morawetz, *Macromolecules* **9**, 463 (1976).
[29] J.-C. Fenyo, C. Braud, J. Beaumais, and G. Muller, *J. Polym. Sci., Part B* **13**, 669 (1975).
[30] A. C. Somersall, E. Dan, and J. E. Guillet, *Macromolecules* **7**, 233 (1974).

5.2.3.2.4. APPLICATION TO BIOLOGICAL MEMBRANES AND MICELLES. There has been considerable biophysical interest in the application of the fluorescent probe-quenching technique to explore the structure of micelles and lipid bilayers. The nature of the binding of the hydrophobic probe, 1-anilonaphthalene-8-sulphonate (ANS), has been studied in detail.[31,32] It is thought to lie at the interface of the membrane with the sulfonate portion in the polar head group environment. The observed fluorescent intensity is sensitive to the surface charge and potential of the membrane as achieved by varying the pH or the divalent cationic environment. Pyrene in contrast is a nonpolar probe that lies in the hydrophobic portion of the bilayer. The exact nature of its placement has been the subject of recent investigations.[33-35] The pyrene molecule in the hydrophobic phase of the membrane can be quenched by the presence of a suitable molecule in the aqueous phase.[33,36] The extent to which this occurs is a measure of the degree with which the aqueous phase extends between the polar head groups in close proximity to the hydrophobic layer. In addition, changes in pyrene emission intensity as a function of temperature are thought to reflect structural transitions in the hydrophobic portion of the membrane.[36] For concentrated pyrene solutions, self-quenching will occur because of excimer formation following collisions between a pyrene molecule in the excited state and one in the ground state. This diffusion-controlled process[37] when considered as a two-dimensional problem[34,38] gives the diffusion coefficient of pyrene in the hydrophobic region as 3×10^{-8} cm²/sec, indicating that the hydrophobic portion of the membrane is very fluid from the standpoint of small-molecule diffusion. The mechanism of active membrane transport of lactose has been studied using a specially labeled galactoside that inhibits lactose transport but is not itself actively transported. The observed increase in fluorescence emission indicates that the galactoside is transferred to a hydrophobic environment in the membrane.[39]

Positive correlations have been found between local anesthetic activity

[31] M. T. Flanagan and T. R. Hesketh, *Biochim. Biophys. Acta* **298**, 535 (1973).

[32] R. A. Badley, H. Schneider, and W. G. Martin, *Biochem. Biophys. Res. Commun.* **49**, 1292 (1972).

[33] M. A. J. Rodgers, M. E. Da Silva, and E. Wheeler, *Chem. Phys. Lett.* **43**, 587 (1976).

[34] J. M. Vanderkooi, S. Fischkoff, M. Andrich, F. Podo, and C. S. Owen, *J. Chem. Phys.* **63**, 3661 (1975).

[35] A. C. McLaughlin, F. Podo, and K. Blaise, *Biochim. Biophys. Acta* **330**, 109 (1973).

[36] S. Cheng, J. K. Thomas, and C. F. Kulpa, *Biochemistry* **13**, 1135 (1974).

[37] W. R. Ware and T. L. Nemzek, *Chem. Phys. Lett.* **23**, 557 (1973).

[38] C. S. Owen, *J. Chem. Phys.* **62**, 3204 (1975).

[39] J. P. Reeves, E. Shechter, R. Weil, and H. R. Kaback, *Proc. Natl. Acad. Sci., U.S.A.* **70**, 2722 (1973).

and ANS fluorescence.[40] While local anesthetics enhance fluorescence in ANS-labeled membranes, the nitroxide-labeled anesthetic analogs cause a severe quenching. Since nitroxide quenching is thought to be effective at a distance of 4–6 Å, the activity of local anesthetics can thus be shown to be a membrane-related phenomenon.[41] Local anesthetics containing aromatic amines have also been shown to be quenchers in a manner similar to that of the nitroxide-labeled anesthetics.[42]

5.2.4. Fluorescence Polarization Techniques

The nature of the polarized fluorescence spectrum lends itself more readily than the quenching techniques to a precise mathematical treatment leading to well-defined physical quantities such as the rotational diffusion constant and the molecular orientational distribution function. First, however, let us examine qualitatively the origins of polarized fluorescent emission. The extent of the partially polarized fluorescent emission is detected by observing an angular dependence of emission intensity when viewed through a rotatable analyzer. This difference is expressed in terms of the degree of polarization or the emission anisotropy, which will be quantitatively defined in Section 5.2.4.1.1.3). Polarization of fluorescent emission occurs because of the photoselection process: the probability of absorption is proportional to the square of the cosine of the angle between the plane of polarization of the exciting radiation and the transition moment oscillator of an active molecule. The emitted radiation from this photoselected subset of molecules is in effect the projections of the emission oscillators of the subset onto the principal axis of the analyzer. Since the subset is anisotropic by virtue of the selection rule for absorption, the emission is polarized. Broadening of the molecular orientational distribution function toward a more random state, a finite angle between the absorption and emission oscillators, and rotational diffusion due to Brownian motion within the lifetime of the excited state all tend to decrease the observed polarization or emission anisotropy. The existence of a nonzero angle between the absorption and emission oscillators will be expressed in terms of the maximum obtainable value of the polarization in a random glass, where all rotational motion is restricted. Depolarization due to Brownian motion will be expressed as a dimensionless number relating the rotational relaxation time to the excited-state lifetime.

---

[40] J. Vanderkooi and A. Martonosi, in "Probes of Structure and Function of Macromolecules and Membranes" (B. Chance, C. Lee, and J. K. Blasie, eds.), Vol. 1, pp. 293–301. Academic Press, New York, 1971.

[41] D. D. Koblin, S. A. Kaufmann, and H. H. Wang, *Biochem. Biophys. Res. Commun.* **53**, 1077 (1973).

[42] D. D. Koblin, W. D. Pace, and H. H. Wang, *Arch. Biochem. Biophys.* **171**, 176 (1975).

In this regard it is of most importance to note that any experimental variables that lead to changes in quenching as noted in the previous section will also alter the excited-state lifetime. The degree of polarization can thus change as a result of spectroscopic causes that do not reflect any changes in either the molecular orientational distribution function or the rapidity of the Brownian motion occurring, i.e., in the physics of the host system. The strategy of the following section will therefore be to construct a formalism isolating depolarization effects into those caused by (1) noncollinearity of absorption and emission oscillators, (2) the molecular orientational distribution function, (3) rotational Brownian motion of the excited molecules during the lifetime of the excitation.

5.2.4.1. Theory of Fluorescence Depolarization. 5.2.4.1.1. GEO-METRIC CONSIDERATIONS. *5.2.4.1.1.1. Instrumental geometry.* Let **P** and **A** be vectors describing the orientation in the spectrometer of the polarization of incident beam and exit analyzer, respectively. That is, **P** and **A** may be represented by vectors

$$\mathbf{P} = \begin{pmatrix} P_x \\ P_y \\ P_z \end{pmatrix}, \qquad \mathbf{A} = \begin{pmatrix} A_x \\ A_y \\ A_z \end{pmatrix}, \tag{5.2.6}$$

referenced to the laboratory frame O-xyz where $\Sigma P_i = \Sigma A_i = 1$.

We assume that incident radiation propagates from the left along the axis, as in Fig. 2. The design of most fluorescent spectrometers is done with an appreciation of orthogonality with either right-angle or straight-through observation and polars parallel to some symmetry axis of the specimen. The straight-through geometry is experimentally difficult to

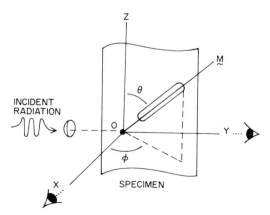

FIG. 2. A fluorescent probe placed in the laboratory coordinate system showing the viewing axes.

work with because of the large ratio of excitation to emission intensity. Special filters are required. We limit our discussion to these cases here although an experiment with **P** and **A** off the chosen axes of Fig. 2 could also be described.

5.2.4.1.1.2. Probe (guest molecule) geometry. The observed fluorescence intensity depends on the component of the incident light that is effective in exciting the absorption oscillator of the molecule and the component of the emission oscillator radiation that eventually emerges from the analyzer. Actually it is the square of these components that is involved. In the fluorescence experiment, radiative capture and reemission may be complicated by the factors listed above, and so the geometry and mobility of the active molecule must be considered in more detail. Figure 3 illustrates the general situation at the site of a cylindrically symmetric fluorescent probe molecule embedded in a medium of similar geometry. *It is an axiom of spectroscopic probe techniques that the presence of small quantities of orientational probes does not perturb the structure to be monitored and that the induced alignment of such guest molecules adequately reflects the order of the host.*

Letting **a** and **e** be unit vectors along the absorption and emission moments, respectively, Fig. 3 defines polar angles α and ϵ of their positions with respect to the long axis of the molecule. The azimuthal angles β and

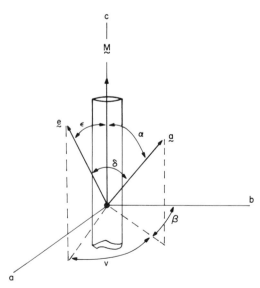

FIG. 3. The location of the emission and absorption oscillators in the molecular coordinate system relative to the molecular axis.

$\beta + v$ are random, though connected, variables under the supposed cylindrical symmetry of the molecule. There are, however, significant degrees of freedom except when α or ϵ equals zero. δ is the intramolecular angle of energy transfer between the absorption and emission dipoles, where $\cos \delta = \mathbf{a} \cdot \mathbf{e}$, and may be regarded as a fixed, structural property of the molecule or as some effective thermal average thereof in the lifetime of the excitation. For an idealized case, $\delta \equiv 0°$ for excitation to the first excited state S_1, and $\delta \equiv 90°$ for excitation to the second excited state S_2. This is because the absorption oscillators for the $S_0 \rightarrow S_1$ and $S_0 \rightarrow S_2$ transitions will often lie at right angles to each other, while the emission will most always emanate from the $S_1 \rightarrow S_0$ transition. The latter case thus leads to an effective intramolecular rotation of the plane of polarization during the excited-state lifetime. These idealized cases are seldom if ever achieved in practice and therefore should not be assumed without careful experimental consideration. $\mathbf{a}$ and $\mathbf{e}$ may be most simply written for our purpose as row vectors in the molecular frame:

$$\mathbf{a} = (\sin \alpha \sin \beta, \sin \alpha \cos \beta, \cos \alpha),$$

$$\mathbf{e} = [\sin \epsilon \sin (\beta + v), \sin \epsilon \cos(\beta + v), \cos \epsilon].$$

(5.2.7)

It is important to remember that the molecular coordinate frame of Fig. 3 is tilted with respect to that of the laboratory. Later a matrix transformation will serve to bring the information of the two very different physical levels into coincidence.

5.2.4.1.1.3. Sample (host) geometry (uniaxial symmetry). In this section, the uniaxial case is treated in a general way to allow for the three contributions to depolarization that will be discussed. Historically the polarization effects due to Brownian motion and orientation have been treated as separate cases because users of the technique were interested in either mobility in random systems, e.g., segmental mobility in polymer solutions, or anisotropy in rigid systems, e.g., orientation in drawn fibers. The growing interest in studying liquid crystalline systems that are ordered but highly fluid leads to the desirability of examining both effects simultaneously.[43] For uniaxial systems the development is tractable and will be presented here in some detail. Note that the uniaxially ordered immobile case and the random mobile case are obtainable under suitable limits of the rotational relaxation time and the distribution function.

We consider a macroscopically aligned sample situated such that the major symmetry axes of the specimen are parallel to the laboratory-based coordinate frame o-xyz. Alignment in the sample may be achieved by mechanical orientation, the application of external electric or magnetic

[43] L. L. Chapoy and D. B. DuPré, *J. Chem. Phys.* **69**(2), 519 (1978).

fields, or in some cases by sympathetic attachment of the medium to prepared surfaces. The alignment is not perfect due to thermal disruption of
the disposition of molecules about the stress field. A representative molecule whose unique symmetry axis (long axis) is denoted by **M** can be located in this coordinate system by polar and azimuthal angles (θ, ϕ) as in
Fig. 2. In discussing the orientational order of a collection of such molecules, it is convenient to introduce a distribution function $f(\theta, \phi)$ relating
the vectors **M** to the macroscopic coordinates of the specimen. The function is a measure of the relative density of the molecular vectors **M** on the
surface of a unit sphere surrounding the sample. Alternatively, $f(\theta, \phi)$
$\sin \theta \, d\theta \, d\phi$ may be regarded as the fraction of molecules with their long
axes found within the solid angle $\sin \theta \, d\theta \, d\phi$ and is thus normalized to
unity through

$$\int_0^{2\pi} \int_0^{\pi} f(\theta, \phi) \sin \theta \, d\theta \, d\phi = 1. \qquad (5.2.8)$$

For samples possessing uniaxial symmetry it is sufficient to deal with an
abbreviated orientation distribution $f(\theta)$. In this case the angle ϕ is an unimportant random variable. Normalization requires that

$$\int_{-1}^{1} f(\theta) \, d(\cos \theta) = 1, \qquad (5.2.9)$$

and the statement of uniaxial symmetry is that $f(\theta) = f(-\theta)$. Furthermore, $f(\theta) = f(\theta + \pi)$, as the distribution is invariant to a 180° rotation of
the molecules. That is, the anisotropic molecular distribution is not polarized. (The direction of **M** in the stress field is equivalent to $-\mathbf{M}$ as far
as physical properties are concerned.)

In research directed toward the understanding of anisotropic materials
it is the function $f(\theta)$ that is sought as it completely describes the long-
range and anisotropic orientational order that gives these materials their
unique physical properties. The significance of $f(\theta)$ is to these ordered
systems as the single-particle distribution function is to "normal" liquids.
It is equally as difficult to obtain. The function, however, may be reproduced with accuracy sufficient for most purposes by considering the following expansion[44-47] in terms of the complete set of Legendre polyno-

[44] S. Nomura, H. Kawai, I. Kimura, and M. Kagiyama, *J. Polym. Sci., Part A-2* **8**, 383 (1970).
[45] C. R. Desper and I. Kimura, *J. Applied Phys.* **38**, 4225 (1967).
[46] M. Abramovitz and I. A. Stegun, "Handbook of Mathematical Functions," pp. 332–341. Natl.Bur. Stand. Washington, D.C., 1965.
[47] E. B. Priestley, P. J. Wojtowicz, and P. Sheng, eds., "Introduction to Liquid Crystals." Plenum, New York, 1975.

mials $P_l(\cos\theta)$:

$$f(\theta) = \sum_{l=0}^{\infty} c_l P_l(\cos\theta), \tag{5.2.10}$$

where the coefficients c_l are given by

$$c_l = \frac{2l+1}{2} \int_{-1}^{1} P_l(\cos\theta) f(\theta)\, d(\cos\theta) = \frac{2l+1}{2} \langle P_l(\cos\theta) \rangle. \tag{5.2.11}$$

The two symmetry conditions on $f(\theta)$ amount to $f(\theta) = f(\pi - \theta)$, which eliminates the odd terms in the expansion. Hence,

$$f(\theta) = \sum_{\substack{l=0 \\ \text{even}}}^{\infty} \frac{2l+1}{2} \langle P_l(\cos\theta) \rangle P_l(\cos\theta). \tag{5.2.12}$$

The first few averages can be written explicitly in terms of the lowest moments of the distribution, i.e. $\overline{\cos^2\theta}$ and $\overline{\cos^4\theta}$ as

$$\langle P_0 \rangle = 1, \tag{5.2.13a}$$

$$\langle P_2 \rangle = \tfrac{1}{2}(3\,\overline{\cos^2\theta} - 1), \tag{5.2.13b}$$

$$\langle P_4 \rangle = \tfrac{1}{8}(35\,\overline{\cos^4\theta} - 30\,\overline{\cos^2\theta} + 3). \tag{5.2.13c}$$

[The bar indicates the same type of statistical average as in Eq. (5.2.11) and will be used interchangeably with $\langle \ldots \rangle$, where space may be conserved.] The average of the second Legendre polynomial $\langle P_2 \rangle = S$ is most usually reported as the *order parameter*[48] or *orientation factor*[49] but it is seen that $\langle P_4 \rangle$ contains further information about orientational order present and should in fact be more sensitive to molecular fluctuations because of the higher powers of the deviation angle. $\langle P_4 \rangle$ could therefore be termed the *hyperorder parameter*. For perfect order along the Z axis, $\overline{\cos^2\theta} = \overline{\cos^4\theta} = 1$, so that $S = \langle P_2 \rangle = 1$ and $\langle P_4 \rangle = 1$. For complete disorder (random distribution in the isotropic state), $\overline{\cos^2\theta} = \tfrac{1}{3}$ and $\overline{\cos^4\theta} = \tfrac{1}{5}$, so that $\langle P_2 \rangle = \langle P_4 \rangle = 0$. Intermediate order corresponds to S values between 0 and 1. The behavior of $\langle P_4 \rangle$ in intermediate conditions is restricted by a mathematical inequality (the Schwarz inequality): $\overline{\cos^2\theta} \geq$

[48] A. Saupe, *Z. Naturforsch., Teil A* **19**, 161 (1964); *Angew. Chem., Int. Ed. Engl.* **7**, 97 (1968).

[49] P. H. Hermans, Contribution to *"The Physics of Fibers,"* pp. 195ff. Elsevier, Amsterdam, 1946.

$\overline{\cos^4\theta} \geq [\overline{\cos^2\theta}]^2$. Examining Eq. (5.2.13) it is seen that $\langle P_4 \rangle$ may be negative for values of $\langle P_2 \rangle \leq 0.6.$†

Although X-ray scattering[52] can in principle measure all the moments of the distribution $f(\theta)$, in practice it is limited by the lack of definite repeating crystalline order in oriented amorphous or partially crystalline polymers and in liquid crystals. Optical birefringence, infrared, visible, and ultraviolet dichroic measurements, and magnetic resonance techniques are the most frequently cited means of determining the degree of order (see also Chapter 13, this volume). The optical methods are inherently limited to yield only a measure of $\overline{\cos^2\theta}$. Both the second and fourth moments are available, however, from suitable analysis of magnetic resonance data and polarized Raman and fluorescent emission spectra of active sites (embedded probes or sensitive molecular elements) in the array. Hence a description of $f(\theta)$ up to the third term of the series Eq. (5.2.12) is possible.

Some workers have proposed model forms for the distribution function with one or more adjustable parameters used to fit data of some experimental property sensitive to an average over $f(\theta)$. These attempts are open to the criticism of the reasonableness and uniqueness of the chosen approximation to $f(\theta)$. That is, more than one model function, perhaps of widely differing physical significance, can be made to fit the data. If one is satisfied with the necessary truncation of the series, Eq. (5.2.12) is the best unambiguous approximation to the orientational distribution available.

Intensity of fluorescent emission. The fluorescence intensity may now be given by (45)

$$I_{ij} = \frac{1}{(2\pi)^2} \int_0^{2\pi} \int_0^{2\pi} \int_0^{\pi} M_{ai}^2 M_{ej}^2 \, f(\theta) \sin\theta \, d\theta \, d\beta \, d\phi, \qquad (5.2.14)$$

where I_{ij} is the intensity of light emitted that is observed emerging through a polarization analyzer placed directly along laboratory symmetry axis j, resulting from light absorbed from an excitation source polarized directly

[50] E. B. Priestley and P. S. Pershan, *Mol. Cryst. Liq. Cryst.* **23**, 369 (1973).

[51] S. Jen, N. A. Clark, P. S. Pershan, and E. B. Priestley, *Phys. Rev. Lett.* **31**, 1552 (1973).

[52] I. M. Ward, ed., "Structure and Properties of Oriented Polymers," Chapters 1 and 5. Appl. Sci. Publ. Ltd., London, 1975.

† Negative values of S are not possible for the arrangement considered here where the preferred direction of alignment of the long axes of the molecules is taken along Z. Negative values of S would correspond to molecular alignment in the XY plane of the specimen. This could occur, for example, with an electromagnetic field applied along Z only if the molecules possess negative dielectric or diamagnetic susceptibility anisotropies, which we assume is not the case here. The physical meaning of negative $\langle P_4 \rangle$ values has been discussed elsewhere.[50,51]

along laboratory axis i. Here M^2_{ai} and M^2_{ej} are the squares of the absorption and emission components projected onto the laboratory axes i and j ($i, j = x, y, z$), respectively, and $f(\theta)$ is the angular distribution function for the case of uniaxial symmetry. Specifically,

$$M^2_{ai} = [\mathbf{a} \cdot 0 \cdot \mathbf{P}_i]^2 \tag{5.2.15}$$

$$M^2_{ej} = [\mathbf{e} \cdot 0 \cdot \mathbf{A}_j]^2 \tag{5.2.16}$$

where 0 is the rotational transformation matrix required to bring the molecular and laboratory coordinate frame into coincidence[53]:

$$0 = (0_{ij}) = \begin{pmatrix} \cos\theta\cos\phi & \cos\theta\sin\phi & -\sin\theta \\ -\sin\phi & \cos\phi & 0 \\ \sin\theta\cos\phi & \sin\theta\sin\phi & \cos\theta \end{pmatrix}. \tag{5.2.17}$$

In the absence of molecular motion, the emission occurs as a fully correlated two-step process and hence the average is performed over the product of the sequence.

A physical description of the two-step fluorescence process can be easily seen in the form of Eqs. (5.2.14)–(5.2.17). In Eq. (5.2.15) the rotational matrix 0 places the incident beam in coincidence with the long axis of the fluorescent molecule. In the event that the absorption moment is not parallel to $\mathbf{M}$, the next operation of Eq. (5.2.15) brings the radiation field to the proper component of the absorption moment and a selective excitation results. The emission step is essentially the reverse. Equation (5.2.16) takes the emission field to the molecular long axis and brings out a selected component in the laboratory frame (along axis j). Instrumental, concentration, and volume factors have been incorporated into the definition of the I_{ij}, which are thus reduced intensities. At this point it is the effect of the geometric disposition of fluorescent molecules that we seek to clarify.

All the information on the fluorescence intensities may be displayed in a 3×3 matrix $\mathbf{I}$, whose elements are given by Eqs. (5.2.14)–(5.2.17). The matrix $\mathbf{I}$ is not in general symmetric and none of the elements are necessarily equal. The sum $\sum_{i,j=1}^{3} I_{ij} = 1$, however. In certain special cases, which correspond to the more idealized probe geometries, simplifications occur. For example, if $\mathbf{a}$ and $\mathbf{e}$ are collinear (though not necessarily parallel to $\mathbf{M}$) matrix $\mathbf{I}$ is symmetric ($I_{ij} = I_{ji}$).

We consider first the form of $\mathbf{I}$ when $\mathbf{a}$ and $\mathbf{e}$ are collinear ($\delta = 0°$) and parallel ($\alpha = 0°$) or perpendicular ($\alpha = 90°$) to the long molecular axis. These are the most frequently assumed cases in experimental work but as pointed out above represent idealizations to the physics of the system.

[53] J. H. Nobbs, D. I. Bower, I. M. Ward, and D. Patterson, *Polymer* **15**, 287 (1974).

The influence on the depolarization when **a** and **e** are at a fixed nonzero angle with respect to one another and the effect of Brownian motion is then discussed.

Fluorescent intensity matrices for special systems. The simplest case occurs where **a** and **e** are collinear ($\delta = 0°$) and parallel to **M** ($\alpha = \epsilon = 0°$). The matrix elements I_{ij} may be generated almost by inspection.

In a large collection of molecules there is an equal probability of finding a molecule at each azimuthal angle ϕ. When the angle appears in Eq. (5.2.14) it can be averaged therefore as a random variable. Polar dispositions of molecules about Z are, however, governed by the anisotropic distribution function $f(\theta)$, information about which is our primary concern. Notice that when $\alpha = 0°$ rotation about the long molecular axis is unimportant. Now vertically polarized incident light will be absorbed by the molecule in proportion to $\cos^2\theta$. If the subsequent emission is viewed through an analyzer also of vertical orientation, another $\cos^2\theta$ will be included in the integral of Eq. (5.2.14). Hence $I_{33} = I_{zz} = \overline{\cos^4\theta}$, after integrating over all ϕ. If the same emergent light were viewed with the analyzer along the y axis, a factor of $\sin^2\theta \sin^2\phi$ would appear in the integral instead of the last $\cos^2\theta$ and $I_{32} = I_{zy} = \frac{1}{2}(\overline{\cos^2\theta} - \overline{\cos^4\theta})$. Similarly all elements of I can be generated from simple geometric considerations of the squares of projections along all combinations of the polarizers on the axes xyz. Table I presents a summary of the requisite geometric factors whose averaged combinations yield the following intensity matrix:

$$
\mathbf{I} = \begin{pmatrix} \frac{3}{8}(1 - 2\,\overline{\cos^2\theta} + \overline{\cos^4\theta}) & \frac{1}{8}(1 - 2\,\overline{\cos^2\theta} + \overline{\cos^4\theta}) & \frac{1}{2}(\overline{\cos^2\theta} - \overline{\cos^4\theta}) \\ \frac{1}{8}(1 - 2\,\overline{\cos^2\theta} + \overline{\cos^4\theta}) & \frac{3}{8}(1 - 2\,\overline{\cos^2\theta} + \overline{\cos^4\theta}) & \frac{1}{2}(\overline{\cos^2\theta} - \overline{\cos^4\theta}) \\ \frac{1}{2}(\overline{\cos^2\theta} - \overline{\cos^4\theta}) & \frac{1}{2}(\overline{\cos^2\theta} - \overline{\cos^4\theta}) & \overline{\cos^4\theta} \end{pmatrix}
$$

$$\alpha = 0°, \qquad \delta = 0°. \quad (5.2.18)$$

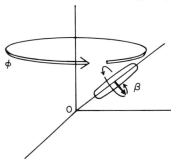

FIG. 4. A fluorescent probe placed in the laboratory coordinate system with the absorption moment located at 90° to the molecular axis. Rotational degrees of freedom for a fixed polar angle are indicated.

TABLE I. Moment Geometric Factors in the Laboratory Frame ($\delta = 0°$)

Moment	$\alpha = 0°$	$\alpha = 90°$
$M_{az}^2 = M_{ez}^2$	$\cos^2\theta$	$\sin^2\theta\,\cos^2\beta$
$M_{ay}^2 = M_{ey}^2$	$\sin^2\theta\,\sin^2\theta$	$\cos^2\theta\,\cos^2\beta\,\sin^2\phi + \sin^2\beta\,\cos^2\phi$ $+\ 2\cos\theta\,\sin\beta\,\cos\beta\,\sin\phi\,\cos\phi$
$M_{ax}^2 = M_{ex}^2$	$\sin^2\theta\,\cos^2\phi$	$\cos^2\theta\,\cos^2\beta\,\cos^2\phi + \sin^2\beta\,\sin^2\phi$ $-\ 2\cos\theta\,\sin\beta\,\cos\beta\,\sin\phi\,\cos\phi$

In this case the matrix is symmetric ($I_{ij} = I_{ji}$) and $I_{11} = I_{22}$, $I_{13} = I_{23}$, $I_{11} = 3I_{12}$.

In the event the absorption or emission moments are not parallel to **M**, rotation about the long axis of the molecule has a significant effect on the fluorescent intensity. Consider the probe geometry of Fig. 4, where $\alpha = 90°$, $\delta = 0°$, another important special case. It is seen that projections of **a** and **e** on the major axis will now also depend on the angle β. The intensity matrix for the case $\alpha = 90°$, $\delta = 0°$, is somewhat more complicated but can be readily obtained to give a result analogous to Eq. (5.2.18) by using the suitable factors recorded in Table I.

In the case of random orientation of immobile molecules (such as in a glass) the **I** matrices reduce to

$$\mathbf{I} = \tfrac{1}{15}\begin{pmatrix} 3 & 1 & 1 \\ 1 & 3 & 1 \\ 1 & 1 & 3 \end{pmatrix} \quad \begin{array}{l} \text{for both} \quad \alpha = 0°,\quad \alpha = 90° \\ \text{and } \delta = 0°. \end{array} \tag{5.2.19}$$

Depolarization due to the noncoincidence of absorption and emission oscillators. The absorption and emission moments of a fluorescent probe molecule are not necessarily parallel although they are frequently assumed to be so. If there is theoretical or spectroscopic reason to believe a useful probe does not possess this simplicity, the appropriate intensity matrix for a fixed, nonzero δ should be generated from Eqs. (5.2.14)–(5.2.17). The general case will not be presented here as the equations for the I_{ij} elements are cumbersome. The situation is less complicated, however, if one of the vectors **a** and **e** is parallel to **M**. We have developed the case where $\alpha = 0$ and $\epsilon = \delta \neq 0°$. Equation (5.2.20) is the result.

Fluorescent intensity matrix
$\delta = \cos^{-1}(\mathbf{a} \cdot \mathbf{e}) \neq 0;\ \mathbf{a}\|\mathbf{M}$

$$\mathbf{I} = \begin{pmatrix}
\tfrac{3}{8}\cos^2\delta\,\overline{\sin^4\theta} + \tfrac{1}{16}\sin^2\delta\,(3\,\overline{\sin^2\theta\,\cos^2\theta} + \overline{\sin^2\theta}) & \tfrac{1}{8}\cos^2\delta\,\overline{\sin^4\theta} + \tfrac{1}{16}\sin^2\delta\,(\overline{\sin^2\theta\,\cos^2\theta} + 3\,\overline{\sin^2\theta}) & \tfrac{1}{2}\cos^2\delta\,\overline{\sin^2\theta\,\cos^2\theta} + \tfrac{1}{4}\sin^2\delta\,\overline{\sin^4\theta} \\
\tfrac{1}{8}\cos^2\delta\,\overline{\sin^4\theta} + \tfrac{1}{16}\sin^2\delta\,(\overline{\sin^2\theta\,\cos^2\theta} + 3\,\overline{\sin^2\theta}) & \tfrac{3}{8}\cos^2\delta\,\overline{\sin^4\theta} + \tfrac{1}{16}\sin^2\delta\,(3\,\overline{\sin^2\theta\,\cos^2\theta} + \overline{\sin^2\theta}) & \tfrac{1}{2}\cos^2\delta\,\overline{\sin^2\theta\,\cos^2\theta} + \tfrac{1}{4}\sin^2\delta\,\overline{\sin^4\theta} \\
\tfrac{1}{2}\cos^2\delta\,\overline{\sin^2\theta\,\cos^2\theta} + \tfrac{1}{4}\sin^2\delta\,(\overline{\cos^2\theta} + \overline{\cos^4\theta}) & \tfrac{1}{2}\cos^2\delta\,\overline{\sin^2\theta\,\cos^2\theta} + \tfrac{1}{4}\sin^2\delta\,(\overline{\cos^2\theta} + \overline{\cos^4\theta}) & \cos^2\delta\,\overline{\cos^4\theta} + \tfrac{1}{2}\sin^2\delta\,\overline{\sin^2\theta\,\cos^2\theta}
\end{pmatrix} \tag{5.2.20}$$

Note that this is an *intra*molecular energy transfer, which leads to depolarization. Depolarization can also occur as the result of an *inter*-molecular energy transfer[54] or excimer formation.[55] These latter effects occur in concentrated solution thus requiring that the probes be dilute with respect to the host (in practice $< 10^{-3}$ moles/liter), or not in close proximity in the case of a labeled polymer.

Degree of polarization and emission anisotropy. The extent of the partially polarized fluorescent emission is frequently quoted in terms of one of two parameters measured in the right-angle geometry: p, the *degree of polarization*, or r, the *emission anisotropy*, defined through[56,57]

$$p = \frac{I_\parallel - I_\perp}{I_\parallel + I_\perp} = \frac{I_{zz} - I_{zy}}{I_{zz} + I_{zy}}, \tag{5.2.21}$$

$$r = \frac{I_\parallel - I_\perp}{I_\parallel + 2I_\perp} = \frac{I_{zz} - I_{zy}}{I_{zz} + 2I_{zy}}, \tag{5.2.22}$$

where $I_\parallel$, $I_\perp$ are the measured intensities with the polars parallel and perpendicular, respectively. A correction for the inherent polarization bias of the optics and electronics of the instrumentation must be made in the specification of these intensity ratios.[58]

The denominator of r is the trace of the emission matrix, thus expressing the total intensity. The quantity r therefore has a certain physical appeal and in some cases is mathematically preferable. The measures are of course related and it can be shown that, in general,

$$\frac{1}{p} - \frac{1}{3} = \frac{2}{3}r^{-1}. \tag{5.2.23}$$

For a randomly oriented collection of immobile molecules whose absorption and emission moments are parallel, matrix Eq. (5.2.19) gives $p_0 = \frac{1}{2}$, $r_0 = \frac{2}{5}$, where the subscript denotes this special case. These values are obtained irrespective of the angle **a** or **e** makes with **M** so long as they remain collinear and immobile during the lifetime of the fluorescence process. If **a** and **e** form a fixed, nonzero intramolecular angle δ, the polarization of emission is reduced and it can be shown from matrix Eq. (5.2.20) that

$$p_0 = \frac{3 \cos^2\delta - 1}{3 + \cos^2\delta}, \qquad r_0 = \frac{3 \cos^2\delta - 1}{5}. \tag{5.2.24}$$

[54] G. Weber, *Trans. Faraday Soc.* **50**, 552 (1954).

[55] M. Yokoyama, T. Tamamura, M. Atsumi, M. Yoshimura, Y. Shirota, and H. Mikawa, *Macromolecules* **8**, 101 (1975).

[56] G. Weber, *in* "Fluorescence and Phosphorescence Analysis" (D. M. Hercules, ed.), Chapter 8. Wiley (Interscience), New York, 1966.

[57] M. Shinitzky, A. C. Dinnoux, C. Gitler, and G. Weber, *Biochemistry* **10**, 2106 (1971).

[58] T. Azumi and S. P. McGlynn, *J. Chem. Phys.* **57**, 2413 (1962).

Equations (5.2.24) are therefore *intrinsic* polarization and anisotropy factors of the molecule, which must be measured in the isotropic phase in such a way as to eliminate the contribution of thermal motions that also act to reduce the polarization. This can be accomplished either by quenching (in the thermal sense of the term) the specimen to freeze out the molecular motions or by decreasing the excited state lifetime through the addition of a radiative quencher as described by the Stern–Volmer equation, Eq. (5.2.5), such that ratio of the rotational relaxation time to the excited-state lifetime becomes $\gg 1$. (See Section 5.2.4.2.1.)

The range of values of p_0 and r_0 is

$$-\tfrac{1}{3} \le p_0 \le \tfrac{1}{2}, \qquad \tfrac{1}{5} \le r_0 \le \tfrac{2}{5}, \tag{5.2.25}$$

as δ may vary between 0 and $\pi/2$.

Equation (5.2.24) may be put into a neater form where a multiplicative factor resembling the second Legendre polynomial appears if we write:

$$\frac{1}{p_0} - \frac{1}{3} = \frac{5}{3}\left(\frac{2}{3\cos^2\delta - 1}\right), \tag{5.2.26}$$

$$r_0 = \frac{2}{5}\left(\frac{3\cos^2\delta - 1}{2}\right). \tag{5.2.27}$$

If the distribution of molecular orientations is not random but otherwise static (as would be the case for a liquid crystal quenched from a temperature within its mesomorphic range or an oriented polymer below its glass transition temperature T_g), similar equations obtain. For example, if the absorption axis of the molecule is along **M**, i.e., $\alpha = 0°$, $\delta \ne 0°$.

$$\frac{1}{p} - \frac{1}{3} = \left(\frac{1}{p} - \frac{1}{3}\right)_0 \left(\frac{2}{3\cos^2\delta - 1}\right), \tag{5.2.28}$$

where

$$\left(\frac{1}{p} - \frac{1}{3}\right)_0 = \frac{4}{3}\frac{\overline{\cos^2\theta}}{(3\overline{\cos^4\theta} - \overline{\cos^2\theta})}, \tag{5.2.29}$$

the subscript referring to the measurement one would obtain in the immobile, anisotropic medium if the intramolecular angle δ were zero. It is this latter quantity that gives the information we seek about the moments of the distribution $f(\theta)$. [Equation (5.2.29) reduces to the constant factor $\tfrac{4}{3}$ of Eq. (5.2.26) for the isotropic distribution.] Another experiment is necessary, however, to separate $\overline{\cos^2\theta}$ from $\overline{\cos^4\theta}$ in Eq. (5.2.29). A measurement of the absorption dichroic ratio $D = \langle M_{az}^2\rangle / \langle M_{ax}^2\rangle$ would be sufficient as

$$\overline{\cos^2\theta} = D/D + 2, \qquad \alpha = 0°. \tag{5.2.30}$$

Choosing another fluorescence geometry will also provide a second expression for the moments, which can be simultaneously solved along with Eq. (5.2.29). If we define a new degree of polarization p' measured in the right-angle geometry with the incident polarization lying now in the xy plane, i.e.,

$$p' = \frac{I_{xz} - I_{xy}}{I_{xz} + I_{xy}}, \tag{5.2.31}$$

then

$$\frac{1}{p'} - \frac{1}{3} = \left(\frac{1}{p'} - \frac{1}{3}\right)_0 \left(\frac{2}{3 \cos^2\delta - 1}\right), \tag{5.2.32}$$

where

$$\left(\frac{1}{p'} - \frac{1}{3}\right)_0$$

$$= \frac{2}{3} \frac{(5 - 6 \overline{\cos^2\theta} + \overline{\cos^4\theta}) + (6 \overline{\cos^2\theta} - 3 \overline{\cos^4\theta} - 3) \cos^2\delta}{(6 \overline{\cos^2\theta} - 5 \overline{\cos^4\theta} - 1)}. \tag{5.2.33}$$

The intramolecular vector angle δ does not factor out completely as in Eqs. (5.2.28) and (5.2.29) but $\cos^2\delta$ is a number intrinsic to the molecule and can be obtained from measurements on the random "glassy" state.

The depolarizing effect of rotational brownian motion. Molecular disorder and the intramolecular transfer of energy between nonparallel absorption and emission moments have been shown to destroy, at least partially, the polarization of the incident field.

Considerations were developed above under the explicit assumption that the experiment monitors a statistical average over a large collection of molecules statically oriented throughout the lifetime of the fluorescence process. Rotational diffusion of the fluorescent probe will result in a further degradation of the polarization. In this section we describe the complications of rotational relaxation.

Due to the anisotropic nature of the absorption process, molecules in certain orientations with respect to the incident beam polarization are more likely to become excited. Polarized illumination therefore has the effect of producing an oriented population of molecules within the medium even when the initial distribution of all molecular axes is random. That is, inherent in the physics of the experiment, we always have to do with a subset of all molecules present in the illuminated volume, those not eliminated through unfavorable orientations in the initial absorption step. The process may be referred to as *photoselection*. In an anisotropic sample photoselection occurs in an already ordered condition of the molecules so that initial anisotropic and isotropic molecular organizations are still distinguishable.

Diffusion complications may be introduced in terms of a time-dependent orientational distribution function of the photoselected population, which is governed by a rotational diffusion equation.[59] The time evolution of the intensity elements, Eq. (5.2.14), can thus be formed and converted by averaging over time to steady-state intensities, which are measured in the usual fluorescent experiment under conditions of continuous illumination.[43]

The expression that results[43] for p, similar in form to Eq. (5.2.28), is

$$\frac{1}{p} - \frac{1}{3} = \frac{8}{3} \frac{\overline{\cos^2\theta}}{(3 \overline{\cos^4\theta} - \overline{\cos^2\theta})} (\tau/\tau_R + 1) \cdot [(3\cos^2\delta - 1)]^{-1}$$

(5.2.34)

where τ_R is the rotational relaxation time of the fluorescent molecule.

If the emission is instantaneous ($\tau = 0$) or the rotational diffusion highly hindered ($\tau_R \gg \tau$), Eq. (5.2.34) reduces to Eqs. (5.2.28) and (5.2.29). On the other hand if rotational motion of the probe relaxes out much quicker than the lifetime of the excitation ($\tau_R \ll \tau$), Eq. (5.2.34) diverges as

$$I_{\parallel} \simeq I_{\perp} \cdot I_n .$$

In this limit no information about the fourth moment of the distribution is obtainable. In this case the fluorescence experiment supplies no more information than an absorption measurement but is complicated by the possible noncoincidence of absorption and emission oscillators of the probe molecules.

In the limit of random orientation of order (isotropic fluid), Eq. (5.2.34) becomes

$$\frac{1}{p} - \frac{1}{3} = \left(\frac{1}{p_0} - \frac{1}{3}\right)\left(1 + \frac{\tau}{\tau_R}\right).$$

(5.2.35)

Equation (5.2.35) was originally derived by Perrin[60] to describe depolarization due to Brownian motion of a rotating, spherical molecule.† Equa-

[59] K. A. Valiev and L. D. Eskin, *Opt. Spectrosc. (USSR)* **12**, 429 (1962).
[60] F. Perrin, *J. Phys.* **7**, 390 (1926).
[61] G. Weber, *Adv. Protein Chem.* **8**, 415 (1953).

† The absence of the multiplicative factor of 3 in the form of Perrin's equation as in Weber[61] is due to the difference in the definition of the rotational relaxation time.

tion (5.2.34) describes the more general situation, such as that found in liquid crystals and uniaxially oriented polymers, where an anisotropic molecular organization is present even before photoselection.

If we treat the probe molecule as a sphere obeying Stokes' law for rotation, the friction coefficient $\zeta = 8\pi\eta a^3$, where a is the radius of the effective volume of the molecule and η is the viscosity of the medium in which it rotates, then $\tau_R^{-1} = kT/V\eta$ and Eq. (5.2.36) may then be written as

$$\frac{1}{p} - \frac{1}{3} = \left(\frac{1}{p_0} - \frac{1}{3}\right)\left(1 + \frac{kT}{V\eta}\,\tau\right), \qquad (5.2.36)$$

where V is the molecular volume. A plot of $(1/p - \frac{1}{3})$ vs. T/η can give information about the molecular volume, local viscosity ("microviscosity"), or the emission lifetime of the probe in solution, if two of these quantities are available from other sources. (The complication of rotational relaxation is not necessarily the nuisance it was implied to be above.)

5.2.4.2. Information Obtained from Polarized Fluorescent Emission. 5.2.4.2.1. THE DETERMINATION OF p_0. In as much as the quantitative interpretation of fluorescence polarization data can never be better than the determination of p_0, extreme care must be given to this aspect of the experiment, even though it provides in essence no useful information in itself. A perusal of the literature might lead one to believe that not enough attention has been given to this matter, and the interpretation of some results may be realistically questioned on these grounds. The idealized value of 0.5 is never obtained, the best values being reported as approximately 0.45. There are several ways in which p_0 can be obtained experimentally. Using Eq. (5.2.36), a plot of $(1/p - \frac{1}{3})$ vs. T/η will give $(1/p_0 - \frac{1}{3})$ from the intercept in the limit of $T/\eta \to 0$. In the event that the intensity changes with temperature, a correction of relative intensity must be made in the T/η axis in as much as intensity changes relate directly to the excited state lifetime [cf. Eq. (5.2.4)].

In practice one obtains an absorption spectrum of a dilute solution of the probe molecule. The strongly allowed low-energy (long wavelength) absorption represents the $S_0 \to S_1$ electronic transition, which is as previously noted preferred for fluorescence polarization measurements. This wavelength, λ_{ex}, is used for the excitation energy in the fluorescent experiment. For anthracene, $\lambda_{ex} = 365$ nm, and a red-shifted emission spectrum as shown in Fig. 5 is obtained. Fluorescence intensity measurements are performed after adjusting the emission monochromator to an intense emission band. If the strongest band lies too close to the scattering peak or to a Raman peak originating from the solvent, the next strongest should be chosen. For anthracene the peak at 422 nm is chosen

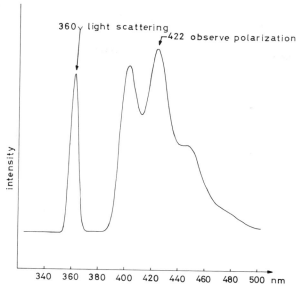

FIG. 5. The emission spectrum of polystyrene end-terminated with anthracene when excited at 360 nm in dimethylphthalate.

as λ_{em}. Since p_0 is on the order of 0.1 for anthracene, extreme care must be taken in measuring the intensities (see Section 5.2.7). Performing the experiment on a dilute solution of polystyrene that is end-terminated with anthracene in dimethylphthalate one obtains a plot as shown in Fig. 6.[62]

The linearity of this plot and the development of Eq. (5.2.36) are based on the rotational diffusion of a sphere. For anisotropic rotational motion, a nonlinear plot can result, leading to difficulties in the extrapolation. Alternatively using the Stern–Volmer equation (5.2.5), one can extrapolate the polarization to zero emission intensity upon the addition of a quenching species.[63] Since quenching results in a shortening of the lifetime, Eq. (5.2.36) also predicts that $1/p - \frac{1}{3}$ will also give $1/p_0 - \frac{1}{3}$ as the intercept in the limit of $\tau \to 0$. Alternatively one might choose to incorporate the probe into a glassy polymer matrix and make a single measurement of the polarization at $T < T_g$. This may be a risky procedure, however, if the glassy polymer is in a nonequilibrium specific-volume state. Great care must be taken in annealing the glass to bring about volume relaxation and the subsequent reduction of free volume in the sample. It

[62] L. L. Chapoy, *Proc.—Int. Congr. Rheol., 7th, 1976* p. 184 (1976).
[63] K. Brown and I. Soutar, *Eur. Polym. J.* **10**, 433 (1974).

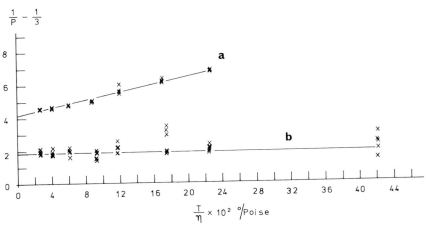

FIG. 6. A Perrin plot for (a) polystyrene end-terminated with anthracene and (b) 4-dimethylamino-4'-nitrostilbene (DS) both in dimethylphthalate.

has been shown that excess free volume in the glass and volume relaxation can affect the polarization of a dissolved probe.[64] Literature values of p_0 should be used with caution. While it has been shown, for example, that p_0 for fluorescein is not sensitive to solvent,[65] it has been our experience in work with anthracene that p_0 for this probe is sensitive to both solvent and chemical attachment to polystyrene. It would seem that more work of this type should be undertaken with the intention of developing a literature source of well-characterized probes.

5.2.4.2.2. THE DETERMINATION OF ORIENTATION. Orientation is a very important phenomenon within the scope of polymer physics. Polymers can be consciously oriented as in the case of the uniaxially drawing of synthetic monofilament yarns or as in the preparation of biaxially oriented films. Alternatively, orientation follows as a by-product of polymer processing inasmuch as strains induced from constrained flows do not have sufficient time to fully relax as the material is cooled below T_g. This gives rise to residual orientations that are kinetically frozen-in in the immobile glassy state.[66] For one-dimensional flows such as fiber drawing or pipe extrusion, one might reasonably expect to have uniaxially orientation in the draw direction or long direction of the extruded article ($\theta \to 0°$), while for plastometer characterization and possibly during some types of compression-molding one might expect that squeeze film flow

[64] L. L. Chapoy, *Chem. Scr.* **2**, 35 (1972).

[65] D. Biddle, *Ark. Kemi* **29**, 543 (1968).

[66] H. Heron, S. Pedersen, and L. L. Chapoy, *Rheol. Acta* **15**, 379 (1976).

would lead to a two-dimensional random planar orientation ($\theta \to 90°$).[67] For the injection-molding of simple geometries one might also be able to accurately estimate the geometry and symmetry of the flow and resulting orientation. Orientations have been shown to greatly affect the mechanical properties of injection-molded materials.[66] For the injection-molding of nontrivial shapes and vacuum-forming of intricate parts, the flows will be complicated and it will not be possible to specify the symmetry for an arbitrary volume element of the part in question. This latter problem may be thought of as the general case for the analysis of orientation.

For partially crystalline polymers it has been shown that incorporated probes will generally enter the amorphous regions of the polymer making the fluorescence technique specific to those regions.[68,69] This is in contrast to other methods of characterizing orientation such as X-ray diffraction, which gives information about the crystalline regions, or optical birefringence, which gives information that is an average over both phases. For anthracene dissolved in polyethylene, however, two types of emission have been reported, which result from crystalline and amorphous interstitial environments.[70] This effect was used to investigate small structural changes in the polyethylene upon mechanical deformation.[71]

As we have noted, fluorescence depolarization is one of the few available methods that allows an estimation of $\overline{\cos^4\theta}$, a term necessary for a more subtle determination of the molecular orientational distribution function $f(\theta)$, as discussed in Section 5.2.4.1.1.3.

5.2.4.2.2.1. Uniaxial Orientation.[53,72] Uniaxial orientation is a special case having one principal material axis. This case can be rigorously treated for the fluorescence polarization experiment as shown in Section 5.2.4.1.1.3, so that one can obtain the orientational distribution function $f(\theta)$ to second order. This can be done with either (1) polarized fluorescent emission in two geometries or (2) a dichroic absorption measurement and a polarized emission measurement, to give the necessary parameters, i.e., $\overline{\cos^2\theta}$ and $\overline{\cos^4\theta}$. It has, however, not been widely appreciated that these two parameters are related to the first two coefficients of the expansion of $f(\theta)$.

[67] G. Brindley, J. M. Davies, and K. Walters, *J. Non-Newtonian Fluid Mech.* **1**, 19 (1976).

[68] H. P. Frank and H. Lehner, *J. Polym. Sci., Part C* **31**, 193 (1970).

[69] Y. Nishijima, Y. Onogi, R. Yamazaki, and K. Kawakami, *Rep. Prog. Polym. Phys. Jpn.* **11**, 407 (1968).

[70] G. P. Egórov and E. G. Moisya, *J. Polym. Sci., Part C* **16**, 2031 (1967).

[71] Yu, S. Lipatov, E. G. Moisya, P. N. Logvinenko, and G. Ya. Mengeres, *J. Appl. Polym. Sci.* **20**, 115 (1976).

[72] S. Hibi, M. Maeda, H. Kubota, and T. Miura, *Polymer* **18**, 137 (1977).

This kind of orientation will occur in the case of a drawn monofilament or an extruded profile. In the limit of very high degrees of orientation one can also point to monodomain samples of nematic liquid crystals.[47] These materials have been the subject of much recent study. The development of rigid aromatic polyamides having liquid crystalline properties and capable of forming superstrength fibers[73,74] promises to open a whole new field of material science.

The determination of the second and fourth orientational parameters $\langle P_2 \rangle$ and $\langle P_4 \rangle$ measured throughout the liquid crystalline temperature range of oriented P-methoxybenzylidene-p'-n-butylaniline (MBBA) using a fluorescent probe, has been chosen as an illustrative example.[75] The technique is, however, general and is equally suited for use on highly oriented uniaxially aligned polymers. The probe employed was 4-dimethylamino-4'-nitrostilbene (DS), shown in Fig. (7). Note that the chemical structure is both highly elongated and stiff. This makes it suitable for use as an orientational probe. λ_{ex} for DS was 445 nm, making it suitable for use in MBBA, which is transparent above about 380 nm. λ_{em} was shifted to 570 nm, which was desirable in view of the copious light scattering from the liquid crystal. The approximately zero slope in the Perrin plot for DS, Fig. 6, implies that $\tau/V \to 0$ and that it thus may be reasonable to ignore Brownian motion in this system. $\langle P_2 \rangle$ and $\langle P_4 \rangle$ are measured by (1) absorption dichroism and emission dichroism Eqs. (5.2.28) and (5.2.30), respectively, and (2) two emission geometries, Eqs. (5.2.28) and (5.2.32). These results are shown in Fig. 8. In the isotropic liquid the polarization approaches that which is predicted from the Perrin plot Fig. 6, Eq. (5.2.26), thus partially justifying the assumption that Brownian motion is not relevant in this system.

5.2.4.2.2.2. Biaxial Orientation. The special case of biaxial orientation will occur for certain injection-molding geometries[66] as well as for extruded films that are stretched in both the extrusion and lateral directions. The orientational distribution function for this case can be developed in a way similar to that for uniaxial orientation. However, one must retain the aximuthal angle ϕ in the distribution function, with $f(\theta, \phi)$ normalized

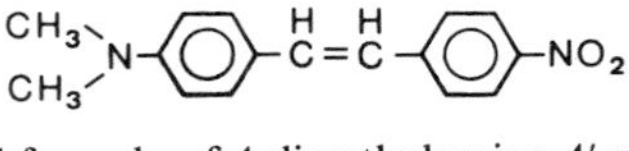

FIG. 7. Chemical formula of 4-dimethylamino-4'-nitrostilbene (DS).

[73] G. Alfonso, E. Bianchi, A. Ciferri, S. Russo, F. Salaris, and B. Valenti, *Polym. Prepr., Am. Chem. Soc. Div. Polym. Chem.* **18**, 179 (1977).

[74] P. W. Morgan, *Polym. Prepr., Am. Chem. Soc., Div. Polym. Chem.* **18**, 131 (1977).

[75] L. L. Chapoy, D. B. DuPré, and E. T. Samulski, *In* "Liquid Crystal Ordered Fluids." (J. F. Johnson and R. S. Porter, eds.), Vol. 3, pp. 177–189. Plenum, New York, 1978.

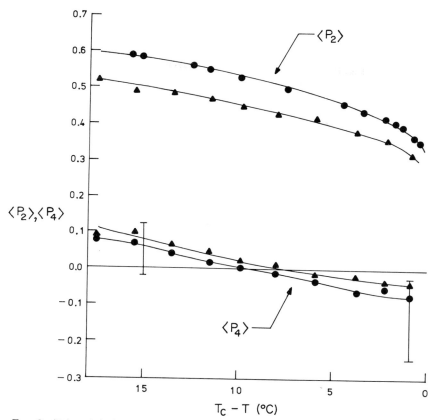

FIG. 8. $\langle P_2 \rangle$ and $\langle P_4 \rangle$ for p-methoxybenzylidene-p'-n-butylaniline (MBBA) doped with DS. Triangles are from the simultaneous solution of emission data. Circles are from a combination of fluorescent emission and optical dichroism data.

as in Eq. (8). This function may be represented as a series expansion, in this case involving the spherical harmonics,[76] in an analogous fashion to the Legendre polynomial expansion of the uniaxial case. In addition to order parameters $\bar{P}_2$, $\bar{P}_4$ involving averages in the polar angle θ, there are now mixed orientation factors describing the correlation or coupling between degrees of order in both polar and azimuthal angles. It has been pointed out,[77] for example, that the first such correlation coefficient

[76] L. L. Chapoy and D. B. DuPré, to be published.
[77] S. Nomura, H. Kawai, I. Kimura, and N. Kagiyama, *J. Polym. Sci.*, Part A-2 **5**, 479 (1967).

$3\langle\sin^2\theta \cos 2\phi\rangle$ is a more general measure of biaxial orientation than the factor $\langle\cos 2\phi\rangle$ usually quoted in optical birefringence studies.[78]

The measurement of the expansion coefficients for the biaxial systems requires a limited number of right-angle rotations of the sample with respect to the laboratory axes.[45,76] A continuous rotation[79] of the sample is, however, not necessary[45] to specify biaxial orientational order.

5.2.4.2.2.3. Generalized Orientations.[79] The most encompassing technique for studying orientation involves a goniometer device in which the sample can be rotated through the solid angle while the parallel and perpendicular components of the emission are simultaneously recorded.[80] The angular dependence of the emission can then be curve-fitted to a model distribution function. The uniqueness of these distribution functions can, however, be questioned. The fact that the distribution function can be written as an expansion of its moments for cases of simple symmetries does not appear to be widely appreciated. Most data obtained are in fact restricted to cases involving simple uniaxial and biaxial orientations. The method would, however, appear to be tempting to study generalized orientations resulting from complex flows.

A comparison of orientation information from birefringence and fluorescence techniques enables one to compare the orientation in the crystalline and amorphous phases in, for example, poly(vinyl alcohol) films.[81] The birefringence data give average information for the whole sample, while the fluorescence data apply only to the amorphous phase. The correlation of draw ratio with orientation has been examined for poly(ethylene terephthalate) fibers[82] and polypropylene films.[83]

5.2.4.2.3. STUDIES INVOLVING MOBILITY. Given information regarding the angle between the emission and absorption oscillators for the probe in question (expressed as p_0) and the orientational distribution function $f(\theta)$ for the probes, one can study the effects of Brownian motion or rotational diffusion on the emission polarization. These effects are described by Eqs. (5.2.34)–(5.2.35). Most work of this type has, however, been done on random systems, i.e., $f(\theta)$ equal to a constant, using the Perrin equation (5.2.35), which expresses the depolarization by a dimensionless time group expressed as the ratio of the excited-state lifetime τ to the rotational relaxation time τ_R. For probe studies, one is generally interested in determining the effect of external variables on τ_R. The life-

[78] R. S. Stein, *J. Polym. Sci.* **31**, 335 (1958).

[79] Y. Nishijima, Y. Onogi, and T. Asai, *J. Polym. Sci., Part C* **15**, 237 (1966).

[80] Y. Nishijima, Y. Onogi, and T. Asai, *Rep. Prog. Polym. Phys. Jpn.* **11**, 391 (1968).

[81] Y. Nishijima, Y. Onogi, R. Yamazaki, and K. Kawakami, *Rep. Prog. Polym. Phys. Jpn.* **11**, 407 (1968).

[82] G. E. McGraw, *Am. Chem. Soc., Div. Org. Coat. Plast. Chem., Pap.* **32**, 77 (1972).

[83] Y. Nishijima, Y. Onogi, and R. Yamazaki, *Rep. Prog. Polym. Phys. Jpn.* **11**, 415 (1968).

time τ must thus be determined independently as a spectroscopic time scale for the experiment so that the equation may be solved explicitly for τ_R. Since the determination of τ is difficult, one often considers τ_R/τ as a function of external variables, implicitly assuming τ to be constant. This in turn implies that the degree of quenching is maintained during the course of the experiment, an assumption that can be checked by examining the quantum yield q, Eq. (5.2.4). According to the Stokes–Einstein relationship, the rotational relaxation time τ_R can be split into the product of an effective volume V related to the size of the entity that is depolarizing the probe through Brownian motion on the time scale of interest multiplied by an effective viscosity η, which expresses the local frictional resistance to motion that the probe entity experiences. The magnitude of the excited-state lifetime dictates that depolarization will reflect diffusion processes and segmental motion. It is very important to keep the concentration of probes dilute, both to avoid gross structural perturbations of the sample and to diminish the formation of excimers, the polarization properties of which are complex and not sufficiently understood at this point to use in mobility studies.

By determining Perrin plots, Fig. 6, for a series of polystyrenes, which have been end-terminated with anthracene, having various molecular weights one can thus calculate τ_R/τ for the terminal segment as shown in Fig. 9. Note the relatively large error in τ_R/τ in spite of the precise data in the Perrin plot, Fig. 6.[62]

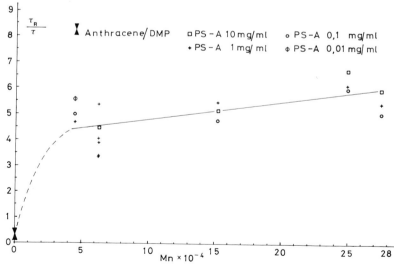

FIG. 9. The reduced rotational relaxation time τ_R/τ for terminal segment behavior as a function of molecular weight. All measurements are performed in dilute solution in dimethylphthalate.

5.2.4.2.3.1. Free Probes. For a free probe dissolved in a matrix, V will be approximately equal to the molecular volume of the probe and η will be the total frictional drag that can be focused on a point within the matrix. For a simple low molecular weight fluid, this frictional drag will be expressed by Stokes' law incorporating the viscosity of the liquid, and thus the microscopic and macroscopic viscosity will be identical. For a more complex system such as an entangled polymer solution, the shear viscosity will be large and non-Newtonian and determined primarily by the entanglement density or concentration of network junctions. The local viscosity affecting the probe on the other hand will be determined primarily by short-range frictional drag exerted by both chain segments and solvent. In contrast to the result experienced from a measurement of the shear viscosity, the local viscosity will be largely oblivious to the presence of net work junctions and thus the apparent frictional drag will be only slightly greater than that for the pure solvent. The local viscosity is then a measure of the Stokes' friction, which would be relevant when considering the diffusion of a small molecule within a polymeric matrix. This is in fact a very general problem occurring in such diverse situations as the diffusion of low molecular weight additives in rubbers and plastics, the transport of water through polymer membranes, and the transport of monomer to growing chains during polymerization reactions.

The local viscosity for polymer solutions increases in steps at the concentration ranges 20–30 and 60–70%[84] indicating the development of internal structure. The limiting law for the local viscosity in dilute solutions is found to have second-power dependence on polymer concentration. This behavior is obeyed at concentrations much higher than one would experience in dealing with the macroscopic viscosity[85,86] and has been subjected to theoretical analysis.[87] In agreement with the intuitive concepts presented above, the local viscosity is found to be Newtonian. Non-Newtonian behavior for the macroscopic viscosity is associated with changes in entanglement density, which have little effect on the local friction properties.[88] The local viscosity is also found to be independent of molecular weight above some critical value.[89] Again, the large increase in macroscopic viscosity observed for high molecular weight polymers is associated with the formation of a comprehensive network, which has little effect on the local friction properties. On the other hand, the change

[84] Y. Nishijima and Y. Mito, *Rep. Prog. Polym. Phys. Jpn.* **10,** 139 (1967).

[85] J. D. Hostetler and J. L. Schrag, *J. Polym. Sci., Part A-2* **10,** 2281 (1972).

[86] L. L. Chapoy, *Chem. Scr.* **2,** 38 (1972).

[87] R. M. Mazo, *Biopolymers* **15,** 507 (1976).

[88] S. Claesson and H. Odani, *Discuss. Faraday Soc.* **49,** 268 (1970).

[89] Y. Nishijima and M. Saito, *Rep. Prog. Polym. Phys. Jpn.* **10,** 135 (1967).

in free volume at T_g is an effect that leads to a change in the local friction factor, giving a marked change in polarization associated with T_g.[64]

5.2.4.2.3.2. *Labeled Systems.*[90-97] A fluorescent label is obtained when a fluorophore probe is attached covalently or tightly bound to a polymer chain or some other larger organizational entity. The depolarization due to Brownian motion in such an experiment is considerably more difficult to interpret, since usually only a part of the total entity is involved in the motion.

If a polymer behaved as a rigid rod one could treat the depolarization as a problem of rigid body mechanics considering the rod to be rotating in a damping fluid. Since polymers, in general (even helical structures), are flexible chains one is in effect observing some average over a spectrum of normal modes whose frequencies are sufficiently high to lead to depolarization. This implies that the active segment will be short, making the technique sensitive to local phenomena. The chemistry of attachment must show sufficient specificity so that only a single type of segment mobility is observed. In addition, the label must be chosen with a judicious choice of absorption and emission oscillators such that chain motions will effectively depolarize the emission.[94] Tumbling of the entire molecule must be taken into account as a possible source of depolarization.[93] The relaxation time for tumbling, τ_D, is given by the Kirkwood–Riseman equation

$$\tau_D = 2M\eta_0[\eta]/3RT, \qquad (5.2.37)$$

where M is the molecular weight, η_0 the solvent viscosity, $[\eta]$ the intrinsic viscosity, R the gas constant, and T the absolute temperature.

If Perrin behavior is obtained for the depolarization data, the volume of the active segment can be obtained from the T/η plot. This implies that the matrix viscosity is primarily responsible for regulating the segmental mobility. Indications have been obtained, however, that internal energy barriers to segmental mobility can also be observed to play a role.[98] Such

[90] B. Valeur, J. P. Jarry, F. Geny, and L. Monnerie, *J. Polym. Sci., Part A-2* **13**, 667 (1975).
[91] B. Valeur, L. Monnerie, and J. P. Jarry, *J. Polym. Sci., Part A-2* **13**, 675 (1975).
[92] B. Valeur, J. P. Jarry, F. Geny, and L. Monnerie, *J. Polym. Sci., Part A-2* **13**, 2251 (1975).
[93] B. Valeur and L. Monnerie, *J. Polym. Sci., Part A-2* **14**, 11 (1976).
[94] B. Valeur and L. Monnerie, *J. Polym. Sci., Part A-2* **14**, 29 (1976).
[95] A. North and I. Soutar, *J. Chem. Soc., Trans. Faraday Soc. 1* **68**, 1101 (1972).
[96] J. P. Bentz, J. P. Beyl, G. Beinert, and G. Weill, *Eur. Polym. J.* **11**, 711 (1975).
[97] S. S. Skorokhodov, M. G. Krakovkjak, E. V. Anufrieva, and N. S. Shelekhov, *J. Polym. Sci., Part C* **42**, 1229 (1973).
[98] D. Biddle and T. Nordström, *Ark. Kemi* **32**, 359 (1970).

effects have been taken into account in the internal viscosity model[99] or the damped torsional oscillator model.[100]

Owing to the extreme sensitivity of the technique, polymers having a molecular weight of 100,000 can be labeled on the terminal segment, i.e., containing only a single label per chain, and still be observed at concentrations of 0.01 weight percent polymer. As noted, this enables one to specifically observe terminal segment mobility under isolated chain conditions.[62]

The segmental volume associated with the depolarization is strongly determined by the conformational structure of the polymer as demonstrated by the appreciable changes in polarization accompanying the denaturation of the helical structure of specially labeled polylysine.[101]

5.2.4.2.3.3. Application to Biological Membranes and Micelles.[32,36,102] There has been a growing interest, within the sphere of biophysics, in studying the microviscosity in the hydrophobic portion of micelles and lipid bilayers.

In this case one selects a suitable nonpolar hydrocarbon probe, which becomes embedded in the hydrophobic portion of the membrane. Depolarization studies utilizing the Perrin equation enable one to calculate the local viscosity experienced by the tumbling probe.[57,103]

The compositional dependence of the microviscosity can be studied as well as the occurrence of phase transitions of the hydrocarbon tails in the hydrophobic phase.[104-106] Differences in microviscosities have been detected in normal lymphocytes and malignant lymphoma cells.[107]

5.2.5. Energy Transfer Probes

The understanding of excimer behavior at this writing is not very complete. However, a great amount of research is currently underway because of the close association of excimer formation with the capability of generating photocurrents. These concepts have also been used to estimate the sequence distribution in a synthetic copolymer containing one fluorescent monomer.[108] Additionally it has been found that excimer formation

[99] A. Peterlin, *J. Polym. Sci., Part B* **10,** 101 (1972).

[100] A. V. Tobolsky and D. B. DuPré, *J. Polym. Sci., Part A-2* 1177 (1968).

[101] T. Iio, Y. Iwashita, and H. Watanabe, *Bull. Chem. Soc. Jpn.* **45,** 2206 (1972).

[102] G. W. Pohl, *Z. Naturforsch., Teil E* **31,** 575 (1976).

[103] G. W. Stubbs, B. J. Litman, and Y. Barenholz, *Biochemistry* **15,** 2766 (1976).

[104] M. P. Andrich and J. M. Vanderkooi, *Biochemistry* **15,** 1257 (1976).

[105] M. Shinitzky and Y. Barenholz, *J. Biol. Chem.* **249,** 2652 (1974).

[106] Y. Suurkuusk, B. R. Lentz, Y. Barenholz, R. L. Biltonen, and T. E. Thompson, *Biochemistry* **15,** 1393 (1976).

[107] M. Shinitzky and M. Inbar, *J. Mol. Biol.* **85,** 603 (1974).

is reduced at temperatures sufficiently low such that favorable encounters are rendered improbable during the lifetime of the excited state. Monomer fluorescence then results. This concept has been exploited to detect relaxation processes in the vicinity of T_g for polystyrene.[109] The reciprocal sixth-power dependence on distance for nonradiative energy transfer has recently been employed as a spectroscopic ruler to estimate, for example, the distance between the calcium binding site and the specificity pocket for the enzyme bovine trypsin.[110] Similar studies have been performed on thermolysin.[111]

5.2.6. Probe Selection

There is a large number of criteria that must be simultaneously satisfied if a probe is to be regarded as suitable. The probe must be soluble and absorb at a wavelength for which the medium is transparent. It must possess a reasonably large molar extinction coefficient and good quantum yield so that high levels of emission can be achieved at low concentrations. The emission should be sufficiently red-shifted so that Raman peaks from the medium and the Rayleigh scattering intensity, which can be severe for polymeric systems, will not interfere with the fluorescence signal of interest. The probe should not undergo chemical reaction on the time scale of the experiment. This includes photochemistry, which even if reversible may cause time dependence in the characteristics of the emission. Since the systems of interest in the current connection will be very nonideal compared to spectroscopic studies in the literature, one must exercise care before assuming any values for spectroscopic constants found in this type of work. Spectroscopic studies, for example, often exercise great precaution to exclude oxygen, which can be an efficient radiative quencher. For mobility studies, one must have a high value of p_0 in order that τ_R can be ascertained with reasonable accuracy. In addition, probe choice must be made with a judicious choice of oscillator placement relative to the labeling site or orientation axis. For orientation studies a sufficiently high aspect ratio must be present to guarantee that the probe will align properly and reflect orientation in the sample. For free probes, one

[108] Y. Nishijima, M. Yamamoto, and S. Kumagawa, *Rep. Prog. Polym. Phys. Jpn.* **9,** 469 (1966).

[109] C. W. Frank, *Macromolecules* **8,** 305 (1975).

[110] D. W. Darnall, F. Abbott, J. E. Gomez, and E. R. Birnbaum, *Biochemistry* **15,** 5017 (1976).

[111] V. G. Berner, D. W. Darnall, and E. R. Birnbaum, *Biochem. Biophys. Res. Commun.* **66,** 763 (1975).

must consider carefully the chemical affinity of the probe so that one can rationalize its local environment in the system. Recently, for example, a labeled lecithin-type probe has been developed for use as a hydrophobic membrane probe.[112-114] For all applications, it is critical that the presence of the probe does not cause any disruption of the system. This can be achieved by comparing measurements with those obtained by other techniques. In this regard, it should be noted that a fluorescent spin probe has been developed[115] which should enable results of fluorescence measurements to be directly compared and complemented with results of electron spin resonance. The preparation of useful labeled compounds can be very problematical since in addition to all of the above-mentioned factors, one must be able to chemically couple the probe to the system of interest. This requires that the reaction proceeds under mild condition in a very selective way without the occurrence of side reactions so that one knows the exact structure of the labeled compound.

Generally this requires placing a reactive function on an existing fluorophore so that suitable coupling can be achieved. Placement of the reactive function on the labeling reagent must also occur highly specifically. One can, thus, be put in the position of having to make compromises and select a fluorophore such as anthracene for which the chemistry is well known and relatively simple with regard to the 9 position (i.e., the carbon in the middle ring) while the p_0 value is relatively low. Copolymers can be prepared with 9-vinyl anthracene.[94,98] Alternatively, 9-methyl anthracene can be used as a chain-transfer agent[63] or 9-chloro methylanthracene as an anionic terminating agent[62] to give an end-capped polymer. Fluorocein isothiocyanate can be used to couple a fluorocein moiety to a hydroxyl group,[116] which has been shown to be a very suitable label in aqueous media.[65,117] Poly-ene labels can be introduced in polyvinyl chloride by the action of ultraviolet light, i.e., a partial degradation.[118] The employment of such labels should be accompanied with caution as ketones are simultaneously formed, thus complicating the interpretation of the resulting luminescence.[72]

[112] J. A. Monti, S. T. Christian, and W. A. Shaw, *Annu. Meet. Am. Soc. Biol. Chem.* Abstract 740, p. 1505 (1976).

[113] W. Stoffel and G. Michaelis, *Hoppe-Seyler's. Z. Physiol. Chem.* **357,** 7 (1976).

[114] W. Stoffel and G. Michaelis, *Hoppe-Seyler's Z. Physiol. Chem.* **357,** 21 (1976).

[115] R. A. Long and J. C. Hsia, *Can. J. Biochem.* **51,** 876 (1973).

[116] I. A. Wolff, P. R. Watson, and C. E. Rist, *J. Am. Chem. Soc.* **75,** 4897 (1953).

[117] D. Biddle and S. Pardhan, *Ark. Kemi* **32,** 349 (1970).

[118] Y. Nishijima, Y. Onagi, and M. Ogawa, *Rep. Prog. Polym. Phys. Jpn.* **9,** 465 (1966).

5.2.7. Instrumentation

A wide variety of good quality commercial ultraviolet and fluorescence spectrophotometers are available for use in realizing the experimental techniques described in the previous sections. An ultraviolet spectrum is essential for determining the wavelength region for the various $S_0 \rightarrow S_n$ electronic transitions of the probe. Generally one will want to excite the probes in the lowest-energy transition $S_0 \rightarrow S_1$, so that no intramolecular depolarization will occur. Quartz cuvettes are available that are transparent over the range 200–1000 nm. The fluorescent spectrum is essential to determine the wavelength of the emissions for the excitation wavelength as determined above. Given the emission spectrum one can then monitor changes in emission intensity or polarization at a suitable wavelength as a function of some experimental parameter. Fluorimeters can have two monochromators for use with the excitation beam and the emitted radiation. Alternatively the excitation monochromator can be replaced by a bandpass filter, which will allow the passage of a spectral emission line from a low-pressure mercury lamp. This arrangement results in higher illumination energies at principally 365 and 436 nm at the expense of wavelength flexibility. For intensity measurements the emission monochromator may be replaced with a suitable filter arrangement as well. Care must be taken in chosing an emission wavelength to exclude both Raman and Rayleigh scattering from the signal being monitored. Viewing usually occurs at 90° to avoid complications arising from incomplete exclusion of the exciting beam. While specialized devices can be constructed for making polarization measurements,[119] commercial instruments can be equipped with rotatable polarizers and analyzers. Since the measurement of fluorescent emission is an absolute determination and not a relative compensated one as in the case of double-beam absorption spectroscopy, extreme instrumental stability is required. Better commercial instruments express emission intensity in a reference mode relative to instantaneous lamp intensity so that fluctuations from this source are eliminated. A typical commercial double monochromated instrument, the Perkin–Elmer MPF-2A, which we have used with good advantage, is schematically shown in Fig. 10. Pseudo-double-beam instruments are now available so that solvent contributions can be eliminated directly without extensive calibration. Determination of emission intensities can be improved with the use of digital meters, which can often be directly coupled to the fluorimeter. High-frequency noise (> 10 Hz) can be eliminated with capacitive filters. Low-frequency noise is more diffi-

[119] G. Weber, *J. Opt. Soc. Am.* **46**, 962 (1956).

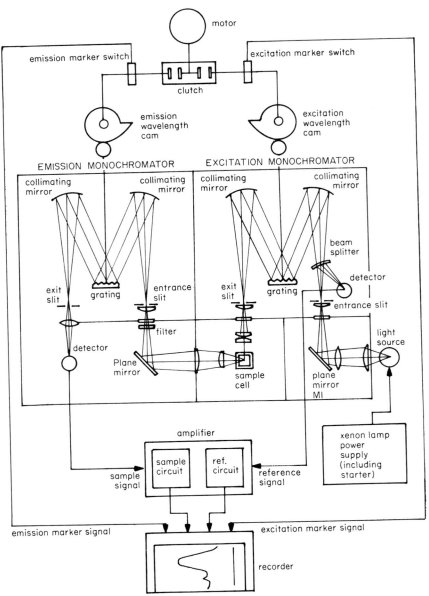

FIG. 10. A schematic diagram of the Perkin–Elmer MPF-2A fluorescence spectrophotometer.

cult to eliminate, but can be handled by signal averaging on the real-time domain either with an on-line computer or with a hard-wired device such as the Solartron 1860 Computing Digital Voltmeter, which can evaluate a noisy time-dependent signal $i(t)$ as

$$\langle i \rangle = \frac{1}{t} \int_0^t i(t) \, dt, \tag{5.2.38}$$

which becomes constant in the limit $t \to \infty$.

Fluorescent lifetimes are measured with difficulty owing to their brevity in the nanosecond range. Commercial equipment such as the Nanosecond Spectral Source System manufactured by TRW Instruments has been available for some time but the area is under rapid development and numerous components are available. Techniques generally rely on pulsing lamps with a pulse width of comparable duration compared to the lifetime. To obtain the lifetime, the decay curve must be deconvoluted from the effects of the finite width of the exciting pulse and the finite response times of the components.[120-124] Picosecond lasers might be used to good advantage in this area. The use of single-photon counting enables one to measure both the polarization and excited state lifetime in a simple experiment, thus avoiding many problems of data interpretation.[92,93125] A comprehensive review appears to be lacking at this writing.

Acknowledgment

The authors would like to thank Statens tekniskvidenskabelige Forskningsråd, Contract 516-6569.K-339, for partial support of this work.

[120] D. K. Wong and A. M. Halpern, *Photochem. Photobiol.* **24**, 609 (1976).
[121] L. S. Forster, *Photochem. Photobiol.* **23**, 445 (1976).
[122] P. R. Hartig, K. Sauer, C. C. Lo, and B. Leskovar, *Rev. Sci. Instrum.* **47**, 1122 (1976).
[123] L. J. Cline Love and L. A. Shaver, *Anal. Chem.* **48**, 364A (1976).
[124] M. A. West and G. S. Beddard, *Int. Lab.*, p. 61 (May–June 1976).
[125] E. D. Cehelnik, R. B. Cundall, J. R. Lockwood, and T. F. Palmer, *J. Chem. Soc., Trans. Faraday Soc. 2* **70**, 244 (1974).

5.3. Paramagnetic Probe Techniques*,†

By Philip L. Kumler

The use of paramagnetic spin probes, as a technique for studying both structure and dynamics in high polymers, is receiving increasing attention. A paramagnetic species is incorporated into the polymer of interest, either by mechanical methods (spin-probe experiment) or by covalently attaching the paramagnetic center to the polymer backbone (spin-label experiment). Electron spin resonance (ESR) spectroscopy can then be used to study the dynamics of the paramagnetic center. The observed ESR spectra of the doped polymer will then reflect details of the structure and dynamics of the polymeric phase in which the probe is incorporated.

5.3.1. ESR Spectroscopy—Theory

Electron spin resonance (ESR), also known as electron paramagnetic resonance (EPR), is receiving increasing interest as a tool for the study of both structure and dynamics in macromolecular systems. ESR has proven to be a powerful technique for the spectroscopic study of any system that possesses a net electron spin angular momentum. Typical examples of systems that are amenable to study by ESR include:

(a) free radicals produced by physical† or chemical means in the solid, liquid, or gaseous states,

(b) biradicals, molecules containing two unpaired electrons at sufficient separation such that the unpaired electrons only weakly interact,

(c) triplet state molecules, which include ground-state triplet molecules as well as triplet states produced by thermal or photochemical excitation,

(d) conduction electrons in metals,

(e) semiconductors,

(f) most transition metal ions.

5.3.1.1. Fundamental Principles. The basic phenomenon detected in an ESR experiment is the transition between Zeeman levels of a paramag-

---

* See also Vol. 2B (Electronic Methods) of this Series, Chapter 9.7.

† This aspect is described in Chapter 14 (this volume, Part C).

netic system contained in a static magnetic field. Transitions between two Zeeman levels can be induced by an electromagnetic field of the appropriate frequency ν, in which $h\nu$ corresponds to the energy level separation ΔE:

$$\Delta E = h\nu. \tag{5.3.1}$$

A rigorous description of the electron spin resonance phenomenon requires a quantum-mechanical treatment,[1] but the fundamental principles may be realized by classical physics. A spinning electron acts as a circular current and creates magnetic dipoles; the total magnetic dipole is the sum of the orbital magnetic dipole (due to the motion of the electron around a nucleus) and the spin magnetic dipole (from the spinning of the electron around its own axis). There is normally only a very small contribution to the net dipole due to the former. The magnetic dipole moment $\boldsymbol{\mu}$, a vector quantity, expresses the interaction of a magnetic dipole with a magnetic field.

5.3.1.2. The Resonance Phenomenon. In a magnetic field $\mathbf{H}$ the position of minimum energy is when the dipoles are aligned in a parallel fashion with the external field. To achieve any other orientation of the magnetic dipoles with respect to the field $\mathbf{H}$ requires an input of energy E such that

$$E = -\boldsymbol{\mu} \cdot \mathbf{H} = -\mu H \cos(\mu, H) = -\mu_z H, \tag{5.3.2}$$

where E is the amount of work required, $\boldsymbol{\mu}$ the magnetic dipole moment, $\mathbf{H}$ the applied magnetic field, μ the magnitude of the magnetic moment $\boldsymbol{\mu}$, H the magnitude of the magnetic field $\mathbf{H}$, $\cos(\mu, H)$ the cosine of the angle between $\boldsymbol{\mu}$ and $\mathbf{H}$, and μ_z the projection of $\boldsymbol{\mu}$ along the z axis (the direction of H).

The spin angular momentum ($\mathbf{P}$) of an electron is always proportional to the magnetic moment $\boldsymbol{\mu}$:

$$\boldsymbol{\mu} = \gamma\mathbf{P}, \tag{5.3.3}$$

where γ is the gyromagnetic (or magnetogyric) ratio. This proportionality constant (γ) can also be defined via classical mechanics as

$$\gamma = q/2mc, \tag{5.3.4}$$

where q and m are the charge and mass, respectively, of a particle rotating in a circular path, and c is the speed of light. For an isolated electron, q is the electron charge ($-e$), and γ allows the conversion of angular momentum to magnetic moment. This treatment is, however, only appli-

[1] J. E. Wertz and J. R. Bolton, "Electron Spin Resonance: Elementary Theory and Practical Applications." McGraw-Hill, New York, 1972.

cable to systems possessing only pure orbital angular momentum. For all other cases, γ can be defined by

$$\gamma = -ge/2mc, \tag{5.3.5}$$

where g is a factor characteristic of the electron under consideration; for a free electron $g = 2.00232$.

From quantum mechanics it can be shown that the component of $\mathbf{P}$ along a particular axis, i.e., P_z, can have only two discrete values, defined by

$$P_z = M_s(h/2\pi), \tag{5.3.6}$$

where M_s is the spin quantum number with allowed values of $\pm \tfrac{1}{2}$. Combinations of Eqs. (5.3.3)–(5.3.6) lead to

$$\mu_z = -g(eh/4\pi mc)M_s = -g\beta M_s, \tag{5.3.7}$$

where $\beta = eh/4\pi mc$ and is called the Bohr magneton. Substitution of Eq. (5.3.7) into Eqs. (5.3.1) and (5.3.2) leads to an equation defining the electron Zeeman energy:

$$E = h\nu = g\beta H M_s = \pm \tfrac{1}{2} g\beta H. \tag{5.3.8}$$

Thus, in an applied magnetic field the electron Zeeman levels are no longer degenerate, and the splitting is a function of the magnetic field strength, as shown schematically in Fig. 1.

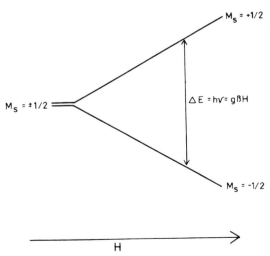

FIG. 1. The splitting of electron Zeeman levels in a magnetic field.

TABLE I. Field, Frequency, and Wavelength for Typical ESR Experiments

Band	ν (MHz)	λ (cm)	H (G)
X	9,500	3	3,400
K	36,000	0.8	13,000

The fundamental prerequisite for resonant absorption to occur is that the applied electromagnetic radiation must equal the energy level separation between two Zeeman levels, that is,

$$\Delta E = h\nu = g\beta H_r, \qquad (5.3.9)$$

where the frequency is expressed in hertz and H_r is the magnetic field strength at which resonance occurs. This equation implies that M_s [see Eq. (5.3.8)] changes by a factor of one unit (from $M_s = +\frac{1}{2}$ to $M_s = -\frac{1}{2}$). The g factor in Eq. (5.3.9) is a constant ($g_e = 2.00232$) characteristic for a free electron if H_r is the magnetic field at the electron. For a collection of unpaired electrons, the ratio of the number of electrons in the $M_s = +\frac{1}{2}$ state ($n_{1/2}$) to those in the $M_s = -\frac{1}{2}$ state ($n_{-1/2}$) is

$$n_{1/2}/n_{-1/2} = \exp(-\Delta E/kT) = \exp(-g\beta H/kT). \qquad (5.3.10)$$

Equation (5.3.8) predicts a simple proportionality between the frequency of irradiation ν and the field strength at resonance H. In principle, resonant absorption could be detected either by fixing the frequency and varying the magnetic field or by fixing the magnetic field and varying the frequency. In a typical ESR experiment, the frequency is fixed and the energy levels are *tuned* by varying the field H. The frequencies normally employed in ESR experiments are in the microwave region and typical combinations of frequency, wavelength, and field strength are shown in Table I.

5.3.1.3. Relaxation and Linewidths. An ESR absorption line has a finite linewidth; the two major processes contributing to line broadening are *secular broadening* and *lifetime broadening*. Paramagnetic centers interact chemically and magnetically with each other and with their environment; such interactions lead to finite linewidths for ESR absorption lines. Linewidth analysis can give important information about time-dependent phenomena.

Any process that generates varying local magnetic fields can lead to line broadening (such processes are often referred to as secular broadening). Secular broadening can be either homogeneous or inhomogeneous. If the local magnetic field experienced varies as a function of time, the result is homogeneous broadening. If, however, the local magnetic field experienced by paramagnetic centers varies from center to center but remains constant as a function of time, then inhomogeneous broadening results.

A different source of line broadening is lifetime broadening, which has a purely quantum-mechanical origin. A particular spin state will have a finite lifetime (Δt) and thus a finite linewidth (ΔE) as a result of the Heisenberg uncertainty principle:

$$\Delta E \,\Delta t \sim h/2\pi. \qquad (5.3.11)$$

It can also be shown that the linewidth ΔH is related as follows:

$$\Delta H \sim t/g\beta \,\Delta t = 1/\gamma_e \,\Delta t, \qquad (5.3.12)$$

where γ is the gyromagnetic ratio of the electron and ΔH the half-width of the absorption line. An increase in the rate of transition between two spin states will decrease Δt and hence increase the linewidth ΔH; such broadening, which affects all members of a line system equally, is an additional example of homogeneous line broadening.

Magnetic dipoles can undergo a change in their spin state (M_s) by interaction of the spin system with the surrounding environment. The Δt corresponding to the rate of this energy exchange is related to a time T_1, the so-called *spin lattice relaxation time*; T_1 is characteristic of the mean lifetime of a particular spin state. The width of an ESR line is normally, however, much greater than one predicts from consideration of T_1 alone. A different relaxation time T_2 is based on the width of a normalized line. Assuming that microwave power levels are sufficiently low to avoid line-broadening effects from saturation, then it is true that

$$\frac{1}{T_2} = \kappa\gamma_e\Gamma, \qquad (5.3.13)$$

where Γ is the half-width at half-height of an ESR line, γ_e is the gyromagnetic ratio, and κ is a constant dependent on lineshape ($\kappa = 1$ for Lorentzian-shaped lines, and $\kappa = (\pi \ln 2)^{1/2}$ for Gaussian-shaped lines).

The relaxation time T_2 includes both lifetime broadening (characterized by T_1) and all other broadening processes. These other processes can be factored by considering a new relaxation time T_2' (the so-called spin–spin relaxation time) such that

$$\frac{1}{T_2} = \frac{1}{T_2'} + \frac{1}{2T_1}. \qquad (5.3.14)$$

The factor of 2 in Eq. (5.3.14) accounts for the fact that the average spin lifetime is $2T_1$.

5.3.1.4. *g* Factors. The *g* factor in Eq. (5.3.9) is characteristic of a free electron in the absence of an external magnetic field. However, in a typical ESR experiment, internal magnetic fields will be generated by the external field, and these fields may add to or subtract from the external field. It is thus convenient to redefine H_r in Eq. (5.3.9) as the external applied magnetic field, and allow the g factor to vary to account for differences in internal magnetization, leading to an effective g value,

$$g_{eff} = h\nu/\beta H_r. \qquad (5.3.15)$$

This g factor then becomes characteristic of the particular species in which the unpaired electron is located and is of value in determining the nature of the paramagnetic species responsible for a particular signal. While many g values may be quite similar to that for a free spin (2.00232), considerable variation in g values is also observed; some typical ranges for g values are shown in Table II.

The g factor is independent of field direction only in certain cases; such would be the case if the electron were located in the center of a site of octahedral, tetrahedral, or cubic symmetry. Magnetically isotropic behavior is also possible in dilute solutions of low viscosity; in such cases the isotropic behavior is a result of rapid, random reorientation of the paramagnetic centers.

For most molecules the g factors are anisotropic; that is, the H_r (and thus the g value) is a function of the orientation of the molecule with respect to the external magnetic field. If the principal axes in the paramagnetic molecule are called X, Y, and Z, then a g factor for the field H along the X axis of the molecule (that is, g_{XX}) can be defined. In the absence of symmetry, $g_{XX} \neq g_{YY} \neq g_{ZZ}$ and three distinct g values are char-

TABLE II.   Representative g Values

Entity	g Value
Free electron	2.00232
Nitroxides	2.0050–2.0060
·CH$_3$	2.00255
·CH$_2$OH	2.0033
·CH$_2$CHO	2.0045
R—O—O·	2.01–2.02
R—S·	2.02–2.06
Cu^{2+} complexes	2.0–2.4
Fe^{3+} (low-spin) complexes	1.4–3.1
Fe^{3+} (high-spin) complexes	2.0–9.7
DPPH (α,α'-diphenyl-β-picryl hydrazyl)	2.0037
Naphthalene anion	2.00276
Perylene anion	2.00267
Tetracene cation	2.00260

acteristic of the system. If, however, the molecule possesses a threefold (or higher) axis of symmetry (taken to be the Z axis), then X and Y are equivalent and only two g factors are observed. The g factor parallel to the symmetry axis (g_{ZZ}) is called g parallel ($g_\parallel$), whereas the g factor for H in the XY plane (perpendicular to Z) is called g perpendicular ($g_\perp$).

As mentioned above, paramagnetic species in dilute solutions of low viscosity exhibit a single, isotropic g factor that results from an averaging process such that

$$g = (g_{XX} + g_{YY} + g_{ZZ})/3. \tag{5.3.16}$$

For the resultant g value to be truly isotropic, the rate of reorientation must be sufficiently fast; such conditions are normally satisfied only with low-molecular-weight materials in low-viscosity solutions. In many high-molecular-weight materials, line broadening due to g factor anisotropy can and does occur (see below).

5.3.1.5. Nuclear Hyperfine Interactions. In addition to local magnetic fields induced by the external field (leading to a variation in g factors as discussed above), most paramagnetic molecules of interest possess permanent local magnetic fields that are not induced by an external field. These local magnetic fields are due to the presence of other magnetic nuclei in the molecule, and the interaction of the unpaired electron with these localized magnetic moments leads to nuclear hyperfine interactions. This hyperfine interaction may also be isotropic or anisotropic, and will normally lead to hyperfine splitting (hfs). Because of hfs ESR spectra often appear as a series of characteristic lines or patterns rather than a single absorption line. Much detailed information concerning the structure of the paramagnetic species can be deduced by careful analysis of the splitting patterns. Both protons and neutrons possess nuclear spin and resultant nuclear magnetic moments. Thus many nuclei possess an intrinsic spin angular momentum, characterized by the spin quantum number I which may have values of $\frac{1}{2}$, 1, $1\frac{1}{2}$, 2, $2\frac{1}{2}$, 3, ... The components of the nuclear spin angular momentum vector are quantized, identical to the corresponding electron spin vectors, and restricted to values of M_I ranging from $-I$ to $+I$ in unit increments, resulting in $2I + 1$ components in an arbitrary direction. For all nuclei for which the atomic mass and atomic number are both even, $I = 0$. If the atomic number is odd and the atomic mass is even, I is an integer; if the atomic mass is odd, I will be a half-integer ($\frac{1}{2}$, $\frac{3}{2}$, $\frac{5}{2}$, . . .).

The field H_r in the resonance equation (5.3.9) is the external magnetic field experienced by the unpaired electron. Any local magnetic fields (H_{loc}) present will add vectorially to the external field, resulting in the effective field (H_{eff}) actually experienced by the electron:

$$H_{\text{eff}} = H_{\text{ext}} + H_{\text{loc}}. \qquad (5.3.17)$$

The multiplicity of nuclear spin states is given by $(2I + 1)$, where I is the nuclear spin quantum number. For single electrons interacting with any nucleus for which $I = 0$, the resonance absorption will be a single line. However, for interaction of an electron with a nucleus possessing a nonzero nuclear spin, the absorption signal will occur as a set of $(2I + 1)$ lines of equal intensity. The simplest example of this nuclear hyperfine interaction is the case of the hydrogen atom (one electron and one proton; $I = \frac{1}{2}$ for H). In this case, M_I has two allowed values ($\pm\frac{1}{2}$) and we expect a pair of lines of equal intensity. There are two values of the external magnetic field at which resonance will occur, corresponding to the two possible local fields (due to the proton) experienced by the single electron, that is,

$$H_R = (H' \pm a/2) = (H' - aM_I). \qquad (5.3.18)$$

The term $a/2$ is the magnitude of the local magnetic field and H' is the resonant magnetic field when $a = 0$. A schematic representation of the expected spectrum (shown as the first derivative of the absorption curve) and the energy level splitting as a function of external field (at constant frequency) is shown in Fig. 2. The interval (a) in the spectrum is a measure (normally expressed in gauss) of the magnitude of the splitting and is called the hfs constant. Each of the electron Zeeman levels is split into two levels by the nuclear hyperfine interaction. The only allowed transitions are those corresponding to $\Delta M_I = 0$ and $\Delta M_s = \pm 1$. Note that the allowed transitions (represented by arrows in Fig. 2) all possess the same energy difference $h\nu$.

For a single electron interacting with a nucleus in which $I = 1$, a three-line (equal intensity) spectrum is predicted. A schematic representation of the Zeeman splitting and derivative spectrum is shown in Fig. 3; there will be three allowed values of the resonant field H_r. Such a spectrum would be characteristic of a deuterium atom or a suitably substituted nitroxide (see Fig. 6). More complex splitting patterns result if the electron interacts with (is magnetically coupled with) more than one magnetic nuclei, and will not be treated here.

The nuclear hyperfine interaction is a result of the electron and nuclear dipoles interacting and is treated in some detail in recent monographs.[1,2]

[2] L. J. Berliner, ed., "Spin Labeling: Theory and Applications." Academic Press, New York, 1976.

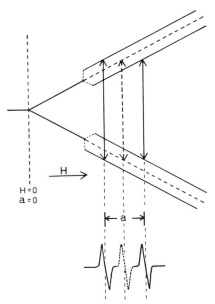

FIG. 2. Observed energy levels for interaction of one electron with one nucleus having $I = \frac{1}{2}$ as a function of magnetic field at a constant frequency. The dotted transition would be observed if a were zero.

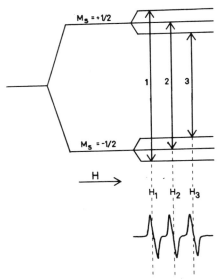

FIG. 3. Observed energy levels for interaction of a single electron with one nucleus having $I = 1$, at constant field. The expected derivative spectrum is shown (at constant frequency) at the bottom.

5.3.1.6. Double-Resonance Techniques. Double-resonance techniques have not been applied extensively to ESR as they have been to NMR, and have received very little use in studies of synthetic polymeric materials. There are two distinct double-resonance ESR experiments and both involve irradiating a paramagnetic nucleus with an electromagnetic field at more than one frequency.

In the electron–nuclear double-resonance technique (ENDOR) the sample is irradiated at the appropriate ESR frequency and simultaneously irradiated at a frequency in the rf range that will induce transitions between nuclear spin levels. This technique was first demonstrated in 1956[3] and has been utilized in the study of biological macromolecules.[1,4] In a typical experiment, the magnetic field is "locked" on one component of a hyperfine pattern; the microwave power level is then increased to cause partial saturation of the ESR line. By means of a separate coil around the sample an intense rf field is applied and swept through a range appropriate to the nucleus under consideration. At two different values of the rf frequency the intensity of the ESR line will increase. From such experiments very accurate values of the hyperfine coupling constant may be determined; in addition the mean value of these two frequencies will be approximately equal to the NMR frequency of the nucleus in the particular applied magnetic field, which will greatly facilitate the identification of the particular nucleus.

A different double-resonance technique, electron–electron double-resonance (ELDOR),[5] also involves the partial saturation of a particular ESR hyperfine line followed by application of a second microwave frequency. When this second frequency satisfies the resonance condition for a second transition, the intensity of the first signal is decreased. This technique is used primarily to study the mechanism of relaxation processes and has proven especially useful for the study of nitroxides in liquids[6]; recent work by Dorio and Chien[7] has, however, extended the ELDOR technique to the study of nitroxide motion in polymeric solids (polystyrene and polypropylene).

5.3.1.7. Effect of Tumbling Rates. As briefly mentioned above, the rate of molecular reorientation can have a very dramatic effect on the observed ESR spectrum. At very rapid tumbling frequencies the anisotropic contributions to both the g factor and the hfs are averaged out.

[3] G. Feher, *Phys. Rev.* **103**, 384 (1956).
[4] H. M. Swartz, J. R. Bolton, and D. C. Borg. eds., "Biological Applications of Electron Spin Resonance." Wiley (Interscience), New York, 1972.
[5] J. S. Hyde, J. C. W. Chien, and J. H. Freed, *J. Chem. Phys.* **48**, 4211 (1968).
[6] E. Stetter, H.-M. Vieth, and K. H. Hausser, *J. Magn. Reson.* **23**, 493 (1976).
[7] M. M. Dorio and J. C. W. Chien, *Macromolecules* **8**, 734 (1975).

Thus, paramagnetic molecules can be used as sensitive indicators of the dynamics of the local environment. Rates of rotational reorientation of the paramagnetic species can be calculated from observed ESR spectra and used to deduce indirectly information about the structure and mobility of the matrix. Such studies have been especially valuable for macromolecules and will be discussed in more detail below.

5.3.2. ESR Spectroscopy—Instrumentation

There are four basic components to a typical commercial ESR spectrometer: a source of microwave radiation, a microwave-resonant cavity that contains the sample of interest, a modulation and detection system, and a static magnet system. A schematic diagram of a typical modern ESR spectrometer system is shown in Fig. 4; each of the four components will be briefly discussed.

5.3.2.1. Source System. Theoretically an ESR experiment can be performed at any frequency of electromagnetic radiation as shown by the conditions for resonance described in Eq. (5.3.9). In order to achieve maximum sensitivity, however, it is important to use as high a frequency

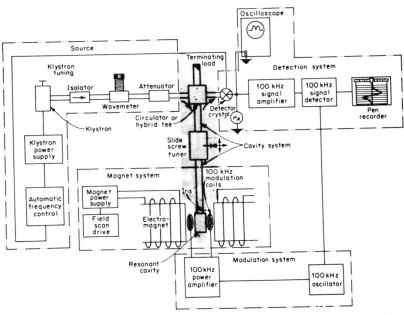

FIG. 4. Block diagram of a typical X-band ESR spectrometer employing 100 kHz phase-sensitive detection.[1]

as is feasible; the sensitivity of an ESR experiment is approximately a function of ν^2. A number of factors (including sample size, strength of the required magnetic field for the resonance condition, possible dielectric absorption by aqueous systems at high frequencies, and required homogeneity of the magnetic field) do, however, place practical limits on the frequency of the source. Most commercial ESR spectrometers operate in the microwave region using X-band ($\nu \sim 9.5$ GHz), K-band ($\nu \sim 24$ GHz), or Q-band ($\nu \sim 35$ GHz) radiation sources. X-band irradiation is the most common and can be propagated and transmitted by waveguide appropriate for the particular wavelength (for $\nu = 9.5$ GHz, $\lambda \approx 3$ cm).

A vacuum tube that produces microwave oscillations centered in a narrow frequency range (a klystron) is the normal source of monochromatic irradiation, although solid-state microwave sources are becoming more common. The frequency of the microwave radiation can be tuned (coarsely) by variation of the applied voltage to the klystron, while fine tuning (± 20 MHz) can be accomplished by a tuning stub on the klystron. Stabilization of the microwave frequency is often controlled by an automatic frequency control system and the klystron frequency can be "locked" to the resonant frequency of the sample cavity.

An *isolator,* which allows passage of microwaves in one direction only, strongly attenuates any backward reflections from the rest of the system that would induce serious variations in the klystron frequency.

The wavelength or frequency of the microwave radiation is normally determined by use of either a cylindrical resonant cavity adjusted by a micrometer (a so-called *wavemeter*), or more accurately by a frequency counter.

For many ESR experiments it is important to be able to regulate the intensity or power of the microwave radiation, and this is normally accomplished by an *attenuator,* which contains a microwave absorptive element and can be continuously varied.

5.3.2.2. Cavity System. The microwave power resulting from the source system is directed into the cavity, and from the cavity to the detector system (after reflection) by a *circulator;* this device, like the isolator, is nonreciprocal. This same function is accomplished in older instruments by a hybrid T. Microwave energy is coupled into and out of the cavity by means of an *iris,* an adjustable screw at the cavity entrance, which allows an optimum match of the waveguide impedance to that of the cavity. In many spectrometers additional matching may be accomplished by means of a *slide-screw tuner,* which is a small adjustable metallic probe inserted into the waveguide.

The heart of an ESR spectrometer is the sample container or *cavity,* which also functions as a power-collecting device. The cavity will magnify the intensity of the microwaves in a manner similar to that of an acoustic resonator with sound waves (e.g., an organ pipe). The shape and dimensions of the cavity are therefore critical since amplification is at a maximum when one dimension of the cavity is equal to some multiple of half-wavelengths of the microwave frequency. Typical ESR cavities are either cylindrical or rectangular parallelepipeds. Extensive discussions of types, dimensions, etc. of cavities may be found in a number of references[1,4] but a few points are worthy of comment at this time. The extent of microwave magnification brought about by a cavity is normally expressed as a factor of merit, called Q. This quantity can be defined by either of two alternative definitions:

$$\left\{ \begin{array}{ll} Q = \dfrac{2\pi(\text{max. microwave energy stored in the cavity})}{(\text{energy lost per cycle})}, & (5.3.19a) \\ \nu_r/\Delta\nu, & (5.3.19b) \end{array} \right.$$

where ν_r is the resonant frequency of the cavity and $\Delta\nu$ is the frequency difference between the points in the cavity at which the microwave power has dropped to one-half its maximum value. For spectrometers operating at the X-band, Q values for rectangular cavities are typically 5000–10,000, whereas significantly higher Q values ($\sim 20,000$) are realized for cylindrical cavities at the same frequency. It is thus of some advantage to utilize cylindrical cavities for aqueous samples (or other solvents with high dielectric constants) because the high dielectric constant will cause a decrease in the Q value if the sample extends into regions of appreciable electrical field from the microwaves.

5.3.2.3. Modulation and Detection System. The detection system in an ESR spectrometer is some type of microwave rectifier that converts the microwaves to a direct current. At resonance the sample absorbs microwave energy, causing a decrease in the Q of the cavity, and a resultant decrease in the intensity of the reflected microwaves. Thus resonance absorption will show up as a change in the detector current. Direct detection of this change in dc current is accompanied by considerable noise, which can be minimized by utilizing *phase-sensitive detection*. A small-amplitude alternating magnetic field is superimposed on the dc field by Helmholtz coils (normally operating at 100 kHz); thus, the rectified signal at the detector is amplitude-modulated at 100 kHz. The modulated signal is further amplified in a narrow-band amplifier and further improvement in the signal to noise ratio of the signal is accomplished by a phase-sensitive detector. One input into the phase-sensitive detector is the modulated and amplified signal and the other input is a portion of the 100 kHz oscillator signal from the modulation coils. The output signal from the phase-sensitive detector is filtered (a resistor–capacitor combination that is adjustable) and transmitted to a chart recorder or oscillo-scope. It should be noted that the output from the phase-sensitive detector is proportional to both the phase and amplitude difference of the two input signals.

If the amplitude of the 100 kHz field modulation is significantly smaller than the linewidth of the ESR signal, then the output signal will closely approximate the first derivative of the absorption curve and is normally displayed as such; some spectrometers can also display the second derivative of the absorption curve. The cause of the displayed signal approximating the first derivative of the absorption curve can be appreciated by consideration of Fig. 5. As the applied field varies from H_a to H_b, the detector current varies sinusoidally between i_a and i_b; if the modulation amplitude is small, then $i_a - i_b$ will be proportional to the slope of the absorption curve. High modulation amplitudes will lead to distorted line shapes.

A number of different types of detectors are in common use and include silicon crystal diodes, "backward diodes," bolometers, and superheterodyne oscillator systems. Each of these detector systems possess inherent advantages and disadvantages but a discussion of these is beyond the scope and intent of the present survey.

5.3.2.4. Magnet System. The magnetic field at the sample must be stable and homogeneous. For most organic free radicals in solution the field strength should vary less than ± 10 mG but variations of ± 1 G are

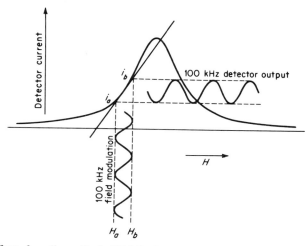

FIG. 5. Effect of small-amplitude 100 kHz field modulation on the crystal detector output current. The static magnetic field is modulated between the limits H_a and H_b; the corresponding crystal current varies between the limits i_a and i_b.[1]

usually acceptable for most inorganic systems or other systems possessing inherently broad lines.

The desired field stability is normally achieved by a highly regulated power supply coupled with a feedback control system using a Hall-effect probe to detect and correct small variations in field strength. A scanning system connected to the power supply allows for field variation in a programmed, linear fashion. For an X-band spectrometer the required field strength is of the order of 3400 G.

Exact measurement of the magnetic field strength is typically accomplished by incorporating an NMR probe in the immediate vicinity (sometimes within) of the cavity. Detection of the NMR signal and exact measurement of the frequency (via a frequency counter) allows measurement of the magnetic field strength to an accuracy of ~ 1 part in 10^5, which is often needed to determine g values with a high degree of accuracy. If, however, only field differences are required they may be accurately measured by use of a double cavity with one of the two cavities containing a suitable reference material.

5.3.2.5. Lineshapes and Linewidths. Lineshapes of ESR signals are often symmetric and fall into one of two categories—Lorentzian or Gaussian. For either of these two cases the line shapes can be described and treated mathematically in a straightforward manner; the analytical expressions describing these lines are as follows:

$$y = \begin{cases} a/(1 + bx^2) & \text{(Lorentzian)}, \quad (5.3.20a) \\ a \exp(-bx^2) & \text{(Gaussian)}. \quad (5.3.20b) \end{cases}$$

Various properties of these lines (equation of the normalized lines, peak amplitude, equations for the first and second derivative lines, peak-to-peak amplitudes, peak-to-peak widths, etc.) can be described by simple mathematical expressions.[1]

Lorentzian line shapes are often observed for ESR lines subject only to homogeneous broadening (such as result from low concentrations of a single paramagnetic molecule in fluid solution of low viscosity) whereas Gaussian lineshapes often result from inhomogeneous broadening or if the lineshape is, in fact, an envelope of unresolved hyperfine components with narrow intrinsic linewidths.

Paramagnetic centers incorporated into large biomolecules or other macromolecular matrices often exhibit asymmetric line shapes because of decreased tumbling frequency or anisotropic motion. Treatment of such lineshapes, though mathematically much more difficult, has received increasing attention and becomes important for studying molecular motion of paramagnetic species in high-molecular-weight polymers. Further information is considered below.

5.3.3. Paramagnetic Probes

The ESR phenomenon was first detected by Zavoisky[8] in 1944; the first application of ESR to macromolecular material was in 1954 when Commoner and co-workers[9] detected free radicals in various biological samples (seeds, leaves, etc.). Since that time ESR techniques have been applied extensively to biological systems and have been the subject of a number of review articles.[10-14] Because there are many naturally occurring paramagnetic species associated with specific biochemical phenomena, it is not surprising that ESR has received such extensive attention in the field of biology. Typical of the biological phenomena that have been studied by ESR are detection of reactive free radicals in enzymatic reactions,[15] study of metalloenzymes containing paramagnetic transition metals,[16] studies of carcinogenesis,[17] studies of aging,[18] studies in radiation biology,[19] studies of the photosynthetic process,[14] studies of natural and synthetic model membranes,[20] studies of biological clathrates,[21] and ESR as a tool for surveying narcotic addiction.[22]

More recently, the concept of using a paramagnetic probe to study structures that are not normally paramagnetic has come into extensive use, especially to study subtle changes in conformation, etc., of biological macromolecules and aggregates such as membranes. Such techniques were developed by McConnell in the early 1960s and have been re-

[8] E. Zavoisky, *J. Phys. (Moscow)* **9,** 211 (1945).

[9] B. Commoner, J. Townsend, and G. E. Pake, *Nature (London)* **174,** 689 (1954).

[10] G. Schoffa, "Electronspinresonanz in der Biologie." Verlag G. Braun, Karlsruhe, 1964.

[11] O. H. Griffith and A. S. Waggoner, *Acc. Chem. Res.* **2,** 17 (1969).

[12] R. C. Bray, *FEBS Lett.* **5,** 1 (1969).

[13] S. E. Lasker and P. Milvy, eds., "Electron Spin Resonance and Nuclear Magnetic Resonance in Biology and Medicine and Magnetic Resonance in Biological Systems," Ann. N.Y. Acad. Sci. No. 222. N.Y. Acad. Sci., New York, 1973.

[14] G. A. Corker, *Photochem. Photobiol.* **24,** 617 (1976).

[15] H. Beinert and G. Palmer, *Adv. Enzymol.* **27,** 105 (1965).

[16] H. Beinert, *in* "Biological Applications of Electron Spin Resonance" (H. M. Swartz, J. R. Bolton, and D. C. Berg, eds.), Chapter 8. Wiley (Interscience), New York, 1972.

[17] J. Mallard and M. Kent, *Nature (London)* **210,** 588 (1966).

[18] D. Harman and L. Piette, *J. Gerontol.* **21,** 560 (1966).

[19] E. S. Copeland, *in* "Biological Applications of Electron Spin Resonance" (H. M. Swartz, J. R. Bolton, and D. C. Borg, eds.), Chapter 10. Wiley (Interscience), New York, 1972.

[20] P. Rey and H. M. McConnell, *J. Am. Chem. Soc.* **99,** 1637 (1977), and references therein.

[21] W. Smith and L. D. Kespert, *J. Chem. Soc., Faraday Trans. 2* **73,** 152 (1977), and references therein.

[22] R. Leute, *Ann. N.Y. Acad. Sci.* **222,** 1087 (1973).

viewed.[23] The application of such techniques, either spin-label or spin-probe experiments (see below for the distinction), to studies of nonbiological macromolecules has been much more limited and will be the primary subject of the rest of this treatment.

5.3.3.1. Requirements for Paramagnetic Probes. There are a number of requirements that must be satisfied for a suitable paramagnetic probe. Ideally the probe should (a) be readily available or easily synthesized, (b) be stable in a wide variety of environments, (c) be easily incorporated into the system under study, (d) exhibit a relatively simple ESR spectrum, (e) have its chemical behavior well documented, and (f) have a well-understood and extensively characterized ESR spectrum.

The first molecule used as a paramagnetic probe was the radical–cation of chlorpromazine; this species was used by McConnell in 1965 to study the geometry of DNA intercalation.[24] Some of the other types of paramagnetic probes that have been utilized include nitric oxide, paramagnetic transition metal ions, lanthanide ions, various organic free radicals, and (most recently) charge-transfer complexes.[25] The class of molecules that has found the most extensive use as paramagnetic probes is the nitroxides; they satisfy most of the requirements mentioned above.

It is possible to incorporate the paramagnetic species into the system under study in two distinct fashions. The probe may be attached to the system by chemical bonding (normally of the covalent type) or may simply be mechanically incorporated into the system; experiments of the former type are referred to as *spin-label* experiments while those of the latter type are *spin-probe* experiments. Results of spin-label and spin-probe experiments are often complementary.

5.3.3.2. Nitroxides as Spin-Probes or Spin-Labels. Most spin-probes† are nitroxide free radicals of the general formula

$$(CH_3)_2C \overset{\displaystyle \overset{R}{|}}{} - \overset{\displaystyle \overset{}{\underset{\displaystyle \underset{O}{|}}{N}}}{} - \overset{\displaystyle \overset{R}{|}}{} C(CH_3)_2.$$

[23] C. Hamilton and H. McConnell, *in* "Structural Chemistry and Molecular Biology" (A. Rich and N. Davidson, eds.), p. 115. Freeman, San Francisco, California, 1968.

[24] S. Ohnishi and H. McConnell, *J. Am. Chem. Soc.* **87**, 2293 (1965).

[25] J. K. Jeszka and M. Kryszewski, *Rocz. Chem.* **50**, 1593 (1976).

---

† The terms spin-probe and spin-label will be used interchangeably unless it is specified or is clear from the context that one specific term is meant.

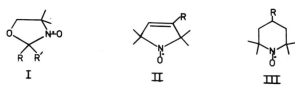

FIG. 6. Some common types of nitroxide spin-probes or spin-labels. I, 2,2-dimethyloxazoline nitroxide (R=R'=methyl); II, 2,2,5,5-tetramethylpyrroline nitroxide (R=H); III, 2,2,6,6-tetramethylpiperidine nitroxide (R=H).

Radicals of this general structure are either commercially available or easily synthesized (usually from the corresponding secondary amines).[26–29] Examples of the most commonly utilized structural types are shown in Fig. 6. For spin-probe studies, R (or R') will normally be a nonreactive group (hydrogen, alkyl, aryl, ketone, amide, ester, ether, etc.) whereas for spin-label experiments the R (or R') group may incorporate a reactive functional group (hydroxyl, amino, carboxyl, chloroformyl, etc.) to facilitate its covalent attachment to the system of interest. Some preliminary evidence has suggested that nitrogen dioxide, NO_2, may also be a useful spin-probe for polymeric materials.[30]

5.3.3.3. ESR Spectra of Nitroxides. At low concentrations in a non-viscous medium, nitroxides exhibit a symmetrical three-line spectrum as shown in Fig. 7. This is the spectrum to be expected for a radical interacting with one nitrogen nucleus ($I = 1$) and experiencing rapid enough molecular tumbling to average out the anisotropic contributions to both the g tensor and the hyperfine tensor.

The ESR spectrum actually observed for a nitroxide is, however, very dependent on two environmentally induced phenomena—the relative ease with which the nitroxide can undergo molecular motion and the polarity of the immediate environment. The former allows an evaluation of the extent of mobility allowed by the matrix in which the probe is incorporated, while the latter allows insight into the hydrophilic or hydrophobic nature of the immediate environment. Both phenomena have been used to probe details of the structure of macromolecules. As an example of

[26] A. R. Forrester, J. M. Hay, and R. H. Thomson, "Organic Chemistry of Stable Free Radicals." Academic Press, New York, 1968.

[27] S. F. Nelson, in "Free Radicals" (J. K. Kochi, ed.), Vol. II, Chapter 21, p. 527. Wiley, New York, 1973.

[28] E. G. Rozantsev, "Free Nitroxyl Radicals." Plenum, New York, 1970.

[29] E. G. Rozantsev and V. D. Scholle, *Synthesis* **190**, 202 (1971).

[30] A. M. Wasserman, L. I. Antsiferova, E. S. Osipova, and A. L. Buchachenko, *Dokl. Akad. Nauk SSSR* **222**, 384 (1975).

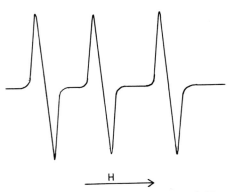

FIG. 7. Typical spectrum expected for a nitroxide of type I, II, or III (Fig. 6), in a low-viscosity fluid solution.

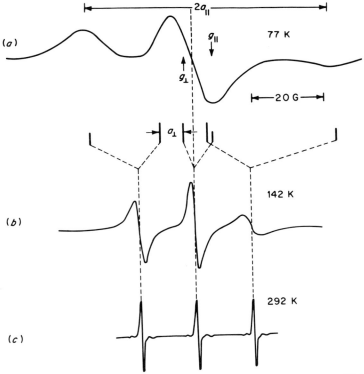

FIG. 8. First derivative spectra of the di-t-butyl nitroxide radical.   (a) At 77 K (solid). (b) At 142 K (viscous ethanol solution).   (c) At 292 K (low viscosity).   The values of $a_\parallel$, $a_\perp$, $g_\parallel$, and $g_\perp$ are taken from data on single crystals.[31]

the types of perturbation induced, Fig. 8 shows the ESR spectrum of di-*t*-butyl nitroxide in three different environments. A number of features of Fig. 8, characteristic of nitroxides in general, are of importance for the study of time-dependent phenomena involving nitroxides. Information present in nitroxide spectra includes the g tensor, the nitrogen hyperfine coupling tensor, and the individual linewidths of the ESR lines. Utilizing the formalism of quantum mechanics, the spin Hamiltonian ($\mathcal{H}$) for this system omitting the nuclear Zeeman term ($-g_N\beta_N\mathbf{H}\cdot\hat{\mathbf{I}}$) includes an electron Zeeman term ($\beta\hat{\mathbf{S}}\cdot\mathbf{g}\cdot\mathbf{H}$), and an isotropic hyperfine term ($hA_0\hat{\mathbf{S}}\cdot\hat{\mathbf{I}}$):

$$\mathcal{H} = \beta\hat{\mathbf{S}}\cdot\mathbf{g}\cdot\mathbf{H} + h\hat{\mathbf{S}}\cdot\mathbf{A}\cdot\hat{\mathbf{I}}, \tag{5.3.21}$$

where $\hat{\mathbf{S}}$ is the operator for the z component of the angular momentum or spin, $\mathbf{g}$ and $\mathbf{A}$ are the g and hyperfine coupling tensors, respectively, $\hat{\mathbf{I}}$ is the nuclear spin operator, and the other symbols have been previously defined (the g_N and β_N in the omitted nuclear Zeeman term are the nuclear g factor and nuclear magneton, respectively). In considering nitroxides, the conventional axis system places the N–O bond in the XY plane and the unpaired electron in a molecular orbital (composed of the pπ orbitals of nitrogen and oxygen) collinear with the Z axis as shown:

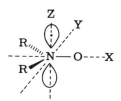

Both $\mathbf{g}$ and $\mathbf{A}$ are expected to be anisotropic but $\mathbf{A}$ approaches axial symmetry ($g_{XX} \neq g_{YY} \neq g_{ZZ}$ and $A_{XX} \simeq A_{YY} \neq A_{ZZ}$). Thus, both the observed g values and the hyperfine splitting constants A will be a function of the orientation of the nitroxide in the applied magnetic field. Accurate g values and A values are normally obtained by doping a host crystal with the nitroxide and studying the dependence of the observed ESR spectrum on the angle between the external field and the symmetry axes of the nitroxide; such studies have been carried out by Griffith[31] on a series of nitroxides. Isotropic parameters (g_0 and A_0) are normally measured in dilute solutions.

Although the actual parameters are dependent on the particular ni-

[31] O. Griffith, D. Cornell, and H. McConnell, *J. Chem. Phys.* **43**, 2909 (1965).

troxide utilized, the following approximations are often useful: $A_{ZZ} \simeq A_{YY} \simeq 5$ G, and $g_{XX} > g_{YY} > g_{ZZ}$. Also, it can be generalized that as solvent polarity increases A_0 will increase and g_0 will decrease.[32]

Let us now consider the spectra of di-t-butyl nitroxide in three different environments as shown in Fig. 8. Spectrum 8a was taken in the solid state at 77 K, 8b at 142 K in viscous ethanol solution, and 8c in a solution of low viscosity. The values of $a_{\parallel}$, $a_{\perp}$, $g_{\parallel}$, and $g_{\perp}$ are from data on single crystals. Spectrum 8a is for a randomly oriented solid and the separation between the outermost lines of the three-line spectrum is $2A_{ZZ}$ or $2A_{\parallel}$. In spectrum 8b the tumbling rate is sufficiently rapid that the line positions correspond to those in the completely averaged spectrum shown in 8c. Note, however, that the linewidths in 8b are different for each of the three lines, and that the high-field line is broader than the low-field line.

5.3.3.4. Rotational Correlation Times for Nitroxides. The ESR spectrum of a nitroxide spin-probe is very sensitive to its molecular tumbling frequency, often expressed as the rotational correlation time τ_c. This correlation time is the time required for a nitroxide radical to make one complete rotation about its axis and can be considered as the time required for the nitroxide to forget what its previous spatial orientation was. Such correlation times can be calculated from the observed spectra using various theories that have been developed for nitroxide motion. Details of the various theories have been summarized in a number of monographs,[1,2,4] to which the interested reader is referred for specific details (including assumptions and limitations of the various theories). It should, however, be noted that the rotational correlation times of interest for spin-probe or spin-label studies range from $\sim 10^{-11}$ sec (a rapidly tumbling nitroxide exhibiting a symmetric spectra due to averaging of anisotropic contributions; Fig. 8c) to 10^{-5} sec (an "immobilized" nitroxide that exhibits a spectrum approximating that of a rigid-limit or powder spectrum with very broad asymmetric peaks; Fig. 8a). No single theoretical treatment applies to the total range of interest, and thus different theories may have to be utilized depending on the actual tumbling frequency exhibited by the nitroxide.

5.3.3.5. $T_{50\,\mathrm{G}} - T_g$ Correlation Methods. An approach different from correlating nitroxide motion with characteristics of the matrix in which it is incorporated was pioneered by both Rabold[33,33a] and Russian workers[33b] and has been developed by Boyer and Kumler.[34-44] This tech-

[32] R. Briere, H. LeMaire, and A. Rasat, *Bull. Soc. Chim. Fr.* **11**, 3273 (1965).

[33] G. P. Rabold, *J. Polym. Sci., Part A-1* **7**, 1203 (1969).

[33a] G. P. Rabold, *J. Polym. Sci., Part A-1* **7**, 1187 (1969).

[33b] V. B. Stryukov and E. G. Rozantsev, *Vysokomol. Soedin., Ser. A* **10**, 626 (1968).

[34] P. L. Kumler and R. F. Boyer, *Midl. Macromol. Monogr.* **4**, 189 (1977).

nique depends upon the determination of an empirical ESR parameter $T_{50\,G}$ that is determined from the temperature-dependent ESR spectra of nitroxide-doped polymers. At low temperatures, using BzONO (Fig. 6, structure III, R = O–CO–Ph) as the paramagnetic probe, doped polymers exhibit ESR spectra in the solid-state characteristic of a highly immobilized nitroxide; such spectra typically show a very broad, asymmetric three-line spectrum with an extrema separation (the distance between the outermost lines, expressed in gauss) of 60–65 G. At high temperatures the observed spectrum approximates closely that of a freely tumbling nitroxide (symmetrical, narrow, three-line spectrum) with an extrema separation of 30–35 G. If extrema separation is plotted vs. temperature, a sigmoidal-shaped curve results, and $T_{50\,G}$ (the temperature at which the extrema separation equals 50 G) can be easily determined from the graph. In some cases the extrema separation changes very rapidly over a narrow temperature range (A of Fig. 9) while in other cases the extrema separation changes slowly over a much wider temperature range (B of Fig. 9). A schematic representation of such extrema separation vs. temperature plots is shown in Fig. 9, and typical experimental plots, of both the narrow and broad extremes, are shown in Figs. 10 and 11. The inset to Fig. 10 also shows actual experimental spectra observed at both high and low temperature extremes.

Expanding upon the earlier work of Rabold,[33,33a] Boyer and Kumler[34,35,40] also demonstrated that there is a correlation between $T_{50\,G}$ and the glass transition temperature T_g; they studied 19 different polymers/copolymers (of unambiguous T_g) and constructed a correlation curve relating $T_{50\,G}$ to T_g, which is shown in Fig. 12. A theoretical basis for the observed correlation (and the nonlinear nature of the same) was presented at the same time. Utilizing this empirical relationship addi-

[35] P. L. Kumler and R. F. Boyer, *Polym. Prepr., Am. Chem. Soc., Div. Polym. Chem.* **16** (1), 572 (1975).

[36] S. E. Keinath, P. L. Kumler, and R. F. Boyer, *Polym. Prepr., Am. Chem. Soc., Div. Polym. Chem.* **16** (2), 120 (1975).

[37] P. L. Kumler, S. E. Keinath, and R. F. Boyer, *Polym. Prepr., Am. Chem. Soc., Div. Polym. Chem.* **17** (2), 28 (1976).

[38] P. L. Kumler, S. E. Keinath, and R. F. Boyer, *Polym. Prepr., Am. Chem. Soc., Div. Polym. Chem.* **18** (1), 313 (1977).

[39] S. E. Keinath, P. L. Kumler, and R. F. Boyer, *Polym. Prepr., Am. Chem. Soc., Div. Polym. Chem.* **18** (2), 456 (1977).

[40] P. L. Kumler and R. F. Boyer, *Macromolecules* **9**, 903 (1976).

[41] R. F. Boyer and P. L. Kumler, *Macromolecules* **10**, 461 (1977).

[42] P. L. Kumler, S. E. Keinath, and R. F. Boyer, *J. Macromol. Sci., Phys.* **13**, 631 (1977).

[43] P. L. Kumler, S. E. Keinath, and R. F. Boyer, *Polym. Eng. Sci.* **17**, 613 (1977).

[44] P. L. Kumler and R. F. Boyer, unpublished data.

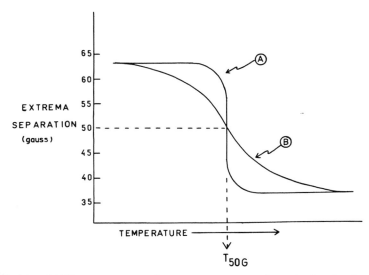

FIG. 9. Plot of ESR extrema separation vs. temperature; schematic plot showing the extreme types of behavior observed. In early work, the Russian workers[45] used an empirical parameter, T_u, defined as the temperature at which changes begin to be observed in the ESR spectrum of immobilized nitroxides. This temperature is very difficult to determine exactly from experimental data as can be appreciated by inspection of the two types of curves shown in the figure.[40] Reprinted with permission from *Macromolecules*. Copyright by the American Chemical Society.

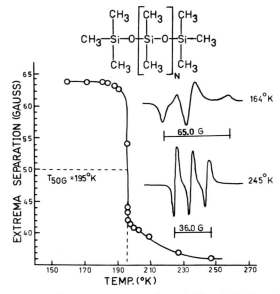

FIG. 10. Extrema separation vs. temperature for poly(dimethyl siloxane); the inset show typical spectra at both high and low temperatures.[40] Reprinted with permission from *Macromolecules*. Copyright by the American Chemical Society.

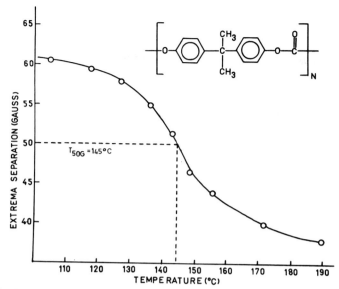

FIG. 11. Extrema separation vs. temperature for polycarbonate.[40] Reprinted with permission from *Macromolecules*. Copyright by the American Chemical Society.

tional studies on various aspects of molecular motion in polymeric materials have been carried out (see below).

5.3.4. Spin-Probe and Spin-Label Applications to Synthetic Polymers

In order to demonstrate the versatility and applicability of spin-probe and spin-label studies in the field of macromolecules, an attempt will be made to briefly summarize all of the spin-probe and/or spin-label studies on nonbiological macromolecules. Literature coverage should be complete through 1976 with many published 1977 articles summarized as well. Applications to biological macromolecules will not be discussed as there are many reviews available[10–14]; some reviews have appeared concerning such studies on synthetic polymers.[45–48] Extensive ESR work by Szwarc

[45] A. L. Buchachenko, A. L. Kovarskii, and A. M. Vasserman, *in* "Advances in Polymer Science" (Z. A. Rogovin, ed.), p. 26. Wiley, New York, 1974.

[46] W. G. Miller and Z. Veksli, *Rubber Chem. Technol.* **48,** 1078 (1975).

[47] P. Törmälä and J. J. Lindberg, *in* "Structural Studies of Macromolecules by Spectroscopic Methods" (K. J. Ivin, ed.), p. 255. Wiley, New York, 1976.

[48] A. T. Bullock and G. G. Cameron, *in* "Structural Studies of Macromolecules by Spectroscopic Methods" (K. J. Ivin, ed.), p. 273. Wiley, New York, 1976.

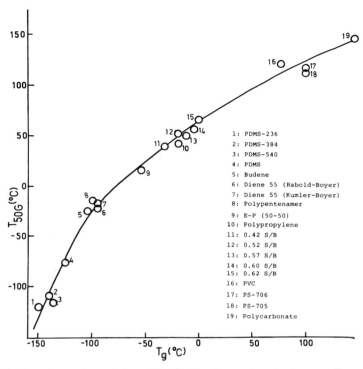

FIG. 12. Experimental correlation of T_g with $T_{50\,G}$ for a series of polymers. Experimental samples are as follows. (1–3) Low molecular weight siloxane oligomers of the formula $(CH_3)_3Si-O-[(CH_3)_2Si-O]_n-Si(CH_3)_3$; (1) $n = 1$; (2) $n = 3$; (3) $n = 4-5$; (4) poly(dimethyl siloxane), $\overline{M}_w = 305,000$; (5) Budene®, cis-1-4-polybutadiene; (6–7) Diene 55, Dow Chemical Co., poly(1,4-butadiene); (6) from Rabold[33]; (7) from Kumler and Boyer[40]; (8) poly-pentenamer; Goodyear Tire and Rubber Co.; (9) ethylene–propylene copolymer, 50:50; Exxon; (10) atactic polypropylene; (11–15) styrene–butadiene copolymers, Dow Chemical Co.; (11) 42 wt% styrene; (12) 52 wt% styrene; (13) 57 wt% styrene; (14) 60 wt% styrene; (15) 62 wt% styrene; (16) poly(vinyl chloride); Dow Chemical Co.; (17) NBS reference polystyrene #706; $\overline{M}_w = 257,800$, $\overline{M}_n = 136,500$; (18) NBS reference polystyrene #705; $\overline{M}_w = 179,300$, $\overline{M}_n = 170,900$; (19) polycarbonate of Bisphenol A, poly(oxycarbonyloxy-1,4-phenyleneisopropylidene-1,4-phenylene); General Electric Co.

and co-workers on the dynamic flexibility of molecular chains[49] does not rely on nitroxide spin-probes or spin-labels and thus will not be treated here.

At the present time (1977) the most active workers (and their nationalities) in this field are Buchachenko, Kovarskii, Rozantsev, Stryukov,

[49] K. Shimada and M. Szwarc, in "Structural Studies of Macromolecules by Spectroscopic Methods" (K. J. Ivin, ed.), p. 317. Wiley, New York, 1976.

Wasserman, and others in Russia[30,33b,50,53−78]; Brotherus, Bulla, Lindberg, Törmälä, and co-workers in Finland[79−96]; Bullock, Cameron, and co-workers in Scotland[97−108]; Boyer, Kumler, and co-workers in the United States[34−44]; Miller and co-workers in the United States[109−111b]; Regen and co-workers in the United States.[112−114] A number of other workers in a variety of countries have also reported spin-probe or spin-label studies on synthetic macromolecules.[115−142]

[50] T. A. Aleksandrova, Thesis, IKhF AN SSSR, Moscow (1972); ref. 82 in Buchachenko et al.[45]

[51] I. Kawai, Y. Yoshimi, and A. Hirai, J. Phys. Soc. Jpn. 16, 2356 (1966).

[52] W. P. Slichter, J. Polym. Sci., Part C 14 (1966).

[53] A. L. Buchachenko and A. S. Troitskaya, Izv. Akad. Nauk SSSR, Ser. Khim. p. 602 (1966).

[54] A. L. Buchachenko, A. M. Wasserman, and A. L. Kovarskii, Int. J. Chem. Kinet. 1, 361 (1969).

[55] L. N. Kaplunova, A. L. Kovarskii, T. A. Fedoseeva, A. M. Vasserman, and S. S. Voyutskii, Dokl. Akad. Nauk SSSR 204, 1151 (1972).

[56] S. M. Kireeva, A. L. Kovarskii, Y. M. Sivergin, A. M. Vasserman, and A. A. Berlin, Vysokomol. Soedin., Ser. B 14, 750 (1972).

[57] A. I. Kokorin, Y. E. Kirsh, and K. I. Zamarayev, Vysokomol. Soedin., Ser. A 17, 1618 (1975).

[58] A. L. Kovarskii, Thesis, IKhF AN SSSR, Moscow (1972); ref. 45 in Buchachenko et al.[45]

[59] A. L. Kovarskii, S. N. Arkina, and A. M. Vasserman, Vysokomol. Soedin., Ser. B 12, 38 (1970).

[60] A. L. Kovarskii, S. G. Burkova, A. M. Vasserman, and Y. L. Morozov, Dokl. Akad. Nauk SSSR 196, 383 (1971).

[61] A. L. Kovarskii, A. M. Vasserman, and A. L. Buchachenko, Dokl. Akad. Nauk SSSR 193, 132 (1970).

[62] A. L. Kovarskii, A. M. Vasserman, and A. L. Buchachenko, Vysokomol. Soedin., Ser. B 12, 211 (1970).

[63] A. L. Kovarskii, A. M. Vasserman, and A. L. Buchachenko, Dokl. Akad. Nauk SSSR 201, 1385 (1971).

[64] A. L. Kovarskii, A. M. Vasserman, and A. L. Buchachenko, Vysokomol. Soedin., Ser. A 13, 1647 (1971).

[65] A. L. Kovarskii, A. M. Vasserman, and A. L. Buchachenko, J. Magn. Reson. 7, 225 (1972).

[66] A. V. Lazarev and V. B. Stryukov, Dokl. Akad. Nauk SSSR 197, 627 (1971).

[67] G. P. Safonov, Y. A. Ol'khov, and S. G. Entelis, Vysokomol. Soedin., Ser. A, 11, 1722 (1969).

[68] A. Y. Shaulov, Thesis, IKhF AN SSSR, Moscow (1972); ref. 84 in Buchachenko et al.[45]

[69] V. B. Stryukov, A. V. Dubovitskii, B. A. Rozenberg, and N. S. Enikolopyan, Dokl. Akad. Nauk SSSR 190, 642 (1970).

[70] V. B. Stryukov, Y. S. Karimov, and E. G. Rozantsev, Vysokomol. Soedin., Ser. B, 7, 493 (1967).

[71] V. B. Stryukov and G. V. Korolev, Vysokomol. Soedin., Ser. A 11, 419 (1969).

[72] V. B. Stryukov, E. G. Rozantsev, A. I. Kashlinskii, N. G. Mal'tseva, and I. F. Tibanov, *Dokl. Akad. Nauk SSSR* **190**, 895 (1970).

[73] A. M. Wasserman, T. A. Alexandrova, and A. L. Buchachenko, *Eur. Polym. J.* **12**, 691 (1976).

[74] A. M. Vasserman, A. L. Buchachenko, A. L. Kovarskii, and M. B. Neiman, *Vysokomol. Soedin., Ser. A* **10**, 1930 (1968).

[75] A. M. Wasserman, A. L. Buchachenko, and A. L. Kovarskii, *Eur. Polym. J.* **5**, 473 (1969).

[76] A. M. Vasserman, G. V. Korolev, A. L. Kovarskii, and A. I. Malakhov, *Izv. Akad. Nauk SSSR, Ser. Khim.* p. 332 (1973).

[77] A. M. Vasserman, A. M. Kuznetsov, A. L. Kovarskii, and A. L. Buchachenko, *Zh. Strukt. Khim.* **12**, 609 (1971).

[78] O. A. Zaporozhskaya, A. L. Kovarskii, A. M. Vasserman, V. S. Pudov, and A. L. Buchachenko, *Vysokomol. Soedin., Ser. B* **12**, 702 (1970).

[79] P. Törmälä and J. Brotherus, *Finn. Chem. Lett.* p. 127 (1974).

[80] I. Bulla, P. Törmälä, and J. J. Lindberg, *Finn. Chem. Lett.* p. 129 (1974).

[81] J. Brotherus and P. Törmälä, *Kolloid-Z-Z. Polym.* **251**, 774 (1973).

[82] I. Bulla, P. Törmälä, and J. J. Lindberg, *Acta Chem. Scand., Sect A* **29**, 89 (1975).

[83] A. Savolainen and P. Törmälä, *J. Polym. Sci., Polym. Phys. Ed.* **12**, 1251 (1974).

[84] P. Törmälä, T. Pentillä, J. J. Lindberg, and J. Sundquist, *Finn. Chem. Lett.* p. 170 (1976).

[85] P. Törmälä, *Polymer* **15**, 124 (1974).

[86] P. Törmälä, H. Lättilä, and J. J. Lindberg, *Polymer* **14**, 481 (1973).

[87] P. Törmälä, *Eur. Polym. J.* **10**, 519 (1974).

[88] P. Törmälä, *Angew. Makromol. Chem.* **37**, 135 (1974).

[89] P. Törmälä and J. Tulikoura, *Polymer* **15**, 248 (1974).

[90] J. J. Lindberg, I. Bulla, and P. Törmälä, *J. Polym. Sci., Polym. Symp.* **53**, 167 (1975).

[91] P. Törmälä, K. Silvennoinen, and J. J. Lindberg, *Acta Chem. Scand.* **25**, 2659 (1971).

[92] P. Törmälä, J. Martinmaa, K. Silvennoinen, and K. Vaahtera, *Acta Chem. Scand.* **24**, 3066 (1970).

[93] P. Törmälä, Thesis, Helsinki (1973); ref. 40 in Törmälä and Lindberg[47].

[94] J. Brotherus and P. Törmälä, *Suom. Kemistseuran Tied.* **81**, 49 (1972).

[95] P. Törmälä, J. Brotherus, and J. J. Lindberg, *Suom. Kemistseuran Tied.* **81**, 11 (1972).

[96] P. Törmälä, J. J. Lindberg, and L. Koivu, *Pap. Puu* **54**, 159 (1972).

[97] A. T. Bullock, G. G. Cameron, and J. M. Elsom, ref. 16 in Bullock and Cameron.[48]

[98] A. T. Bullock, G. G. Cameron, and P. M. Smith, ref. 18 in Bullock and Cameron.[48]

[99] A. T. Bullock, G. G. Cameron, and V. Krajewski, *J. Phys. Chem.* **80**, 1792 (1976).

[100] A. T. Bullock, G. G. Cameron, and P. M. Smith, *Macromolecules* **9**, 650 (1976).

[101] A. T. Bullock, G. G. Cameron, and P. M. Smith, *Eur. Polym. J.* **11**, 617 (1975).

[102] A. T. Bullock, G. G. Cameron, and P. M. Smith, *J. Chem. Soc., Faraday Trans. 2* **70**, 1202 (1974).

[103] A. T. Bullock, G. G. Cameron, and J. M. Elsom, *Polymer* **15**, 74 (1974).

[104] A. T. Bullock, G. G. Cameron, and P. M. Smith, *Polymer* **14**, 525 (1973).

[105] A. T. Bullock, G. G. Cameron, and P. M. Smith, *J. Polym. Sci., Polym. Phys. Ed.* **11**, 1263 (1973).

[106] A. T. Bullock, G. G. Cameron, and P. M. Smith, *J. Phys. Chem.* **77**, 1635 (1973).

[107] A. T. Bullock, G. G. Cameron, and P. M. Smith, *Polymer* **13**, 89 (1972).

[108] A. T. Bullock, J. H. Butterworth, and G. G. Cameron, *Eur. Polym. J.* **7**, 445 (1971).

[109] W. G. Miller and Z. Veksli, *Rubber Chem. Technol.* **48**, 1078 (1975).

[110] Z. Veksli and W. G. Miller, *Macromolecules* **8**, 248 (1975).

[111] E. L. Wee and W. G. Miller, *J. Phys. Chem.* **77**, 182 (1973).

[111a] Z. Veksli, W. G. Miller, and E. L. Thomas, *J. Polym. Sci., Polym. Symp.* **54**, 299 (1976).

[111b] Z. Veksli and W. G. Miller, *Macromolecules* **10**, 686 (1977).

[112] S. L. Regen, *J. Am. Chem. Soc.* **96**, 5275 (1974).

[113] S. L. Regen, *J. Am. Chem. Soc.* **97**, 3108 (1975).

[114] S. L. Regen, *Macromolecules* **8**, 689 (1975).

[115] D. Braun, *Makromol. Chem.* **30**, 85 (1959).

[116] D. Braun and S. Hauge, *Makromol. Chem.* **150**, 57 (1971).

[117] B. Catoire, and R. Hagege, *Bull. Sci. Inst. Text. Fr.* **2**, 209 (1973); *Chem. Abstr.* **80**, 4728t (1974).

[118] D. B. Chesnut and J. F. Hower, *J. Phys. Chem.* **75**, 907 (1971).

[119] G. Drefahl, H.-H. Hörhold, and K. D. Hofmann, *J. Prakt. Chem.* **37**, 137 (1968).

[120] A. R. Forrester and S. P. Hepburn, *J. Chem. Soc. C* p. 701 (1971).

[121] A. R. Forrester and S. P. Hepburn, *J. Chem. Soc. C* p. 3322 (1971).

[122] O. H. Griffith, J. F. W. Keana, S. Rottschaefer and T. A. Warlick, *J. Am. Chem. Soc.* **89**, 5072 (1967).

[123] S. C. Gross, *J. Polym. Sci., Part A-1* **9**, 3327 (1971).

[124] T. Kurosaki, K. W. Lee, and M. Okawara, *J. Polym. Sci., Polym. Chem. Ed.* **10**, 3295 (1972).

[125] N. Kusumoto, H. Uchida, and M. Takayanagi, *Rep. Prog. Polym. Phys. Jpn.* **14**, 219 (1971).

[126] N. Kusumoto, M. Yonezawa, and Y. Motozato, *Polymer* **15**, 793 (1974).

[127] C. Lagercrantz and M. Setaka, *J. Am. Chem. Soc.* **96**, 5652 (1974).

[128] Y. I. Mitchenko, S. A. Gribanov, A. N. D'yachkov, and E. M. Aizenshtein, *Vysokomol. Soedin., Ser. B* **17**, 547 (1975).

[129] T. Nagamura and A. E. Woodward, *J. Polym. Sci., Polym. Phys. Ed.* **14**, 275 (1976).

[130] A. M. North, *Q. Rev., Chem. Soc.* **20**, 421 (1966); *Chem. Soc. Rev.* **1**, 49 (1972); *J. Chem. Soc., Faraday Trans.* **68**, 1101 (1972).

[131] A. Rousseau and R. Lenk, *Mol. Phys.* **15**, 425 (1968).

[132] A. Sanson and M. Ptak, *C. R. Hebd. Seances Acad. Sci., Ser. D* **271**, 1319 (1970).

[133] A. Tsutsumi, K. Hikichi, and M. Kaneko, *Polym. J.* **8**, 511 (1976).

[134] T. C. Ward and J. T. Books, *Macromolecules* **7**, 207 (1974).

[135] R. F. Boyer, *Macromolecules* **6**, 288 (1973).

[136] A. S. Waggoner, O. H. Griffith, and C. R. Christensen, *Proc. Natl. Acad. Sci. U.S.A.* **57**, 1198 (1967).

[137] R. F. Boyer, *Polymer* **17**, 996 (1976).

[138] I. C. P. Smith, G. W. Stockton, A. P. Tulloch, C. F. Polnaszek, and K. G. Johnson, *J. Colloid Interface Sci.* **58**, 439 (1977).

[139] I. D. Robb and R. Smith, *Polymer* **18**, 500 (1977).

[140] M. Rinaudo, M. Milas, and F. Sundholm, *Acta Chem. Scand., Ser. B* **31**, 102 (1977).

[141] T. Nagamura and A. E. Woodward, *Biopolymers* **16**, 907 (1977).

[142] J. D. Aplin and L. D. Hall, *J. Am. Chem. Soc.* **99**, 4162 (1977).

5.3.4.1. Types of Polymers Studied. There are very few polymer types that have not been subjected to study by either spin-probe or spin-label methods. A summary of the major types of polymers that have been studied is presented in Table III.

5.3.4.2. Applications to Synthetic Polymers. Brief summaries of all of the spin-probe and spin-label studies reported thus far are presented

TABLE III.   Polymers Studied by ESR Spin-Probe or Spin-Label Methods

Polymer	Reference
Polyethylene	33,33b,35,40,51,58,66,70,100,101,121,125, 126
Polypropylene	33,39,51,58,61,64,72,74,75,78
Poly(α-olefins)	41,51,52,70,71,74,75
Polydienes	
Polybutadiene	35,40,43,60,74,129
Polyisoprene	38,43,131
Natural Rubber	44,61,64,74,75,109
Polyacrylics and polymethacrylics	
Poly(acrylic acid) and derivatives	56,97,99,110,116
Poly(methacrylic acid) and derivatives	56,58,70,71,97,99,103,109,110,111a,111b,122
Poly(vinyl halides) and poly(vinyl esters)	
Poly(vinyl chloride)	33,40,88,117
Poly(vinyl fluoride)	40,44
Poly(vinylidene fluoride)	35,37,40
Poly(vinyl acetate)	73
Polystyrenes	
Linear polystyrene	30,33,33b,35,37,40,42,50,70,71,74,75,98,102, 104–110,111a,111b,115,119,120
Crosslinked polystyrene	110,112–114,127
Poly(p-t-butyl styrene)	42–44
Poly(α-methyl styrene)	98,102
Poly(vinyl pyridine)	57
Polyoxides	
Poly(oxymethylene)	40,69
Poly(oxyethylene)	35,40,85–89,92,93
Poly(oxyphenylenes)	36,37,41,44,83
Polycarbonates	40,41,123
Polyesters	30,62,72,91,117,123,128
Polyamides	72,84,91,93,110
Polyurethanes	117,134
Polysiloxanes	35,37,40,41,58,61,64
Cellulosics	30,90,96
Miscellaneous homopolymers	
Poly(benzyl glutamate)	111,133
Random and block copolymers	33,33a,35,36,38,40,43,60,67,97,99–101,109

below. The categorization is somewhat arbitrary and there may be some overlap between the various divisions.

5.3.4.2.1. STUDIES OF MOTION IN SOLID POLYMERS. In addition to studies of the glass transition and other relaxation phenomena in solid polymers (see below), a number of studies have concerned themselves with both rotational and translational diffusion in solid macromolecules. Kovarskii and co-workers[55] used the spin-probe technique to detect and study the diffusion layer at the interface of two thermodynamically incompatible polymers. In a study of various spin-probes in poly (vinyl acetate), Wasserman *et al.*[73] concluded that the rotational diffusion of nitroxides could only be satisfactorily interpreted by adopting various models for the diffusion with different nitroxides. Törmälä[88] studied the translational diffusion of plasticizers into poly(vinyl chloride) gels and demonstrated a high degree of lateral motion possible for the plasticizer molecules. As part of a study on spin-labeled polystyrene, Bullock and co-workers[105] showed that rotational diffusion could best be interpreted in terms of the moderate-jump diffusion model. Veksli and Miller[109] have used ESR techniques to study both the translational diffusion and the effect of motion of small molecules in various rubbers.

5.3.4.2.2. DETERMINATION OF GLASS TRANSITION TEMPERATURES. A possible correlation between ESR parameters and T_g or T_m values was first suggested by Rabold[33]; such a correlation of $T_{50\,G}$ values with T_g values was extended by Boyer[135] and greatly expanded by Kumler and Boyer.[35,40] This technique, utilizing a correlation curve of $T_{50\,G}$ with T_g has since been applied to a wide variety of polymer types including random and block copolymers.[36-38,40,42-44]

A different approach to utilizing ESR data for detecting T_g has been adopted by a number of other workers. A distinct break in the ln τ_c vs. $1/T$ plot is often observed at a temperature that corresponds to the T_g determined by other methods. This technique has been applied in a number of polymer types including polyoxides,[83,86,87,89] polyamides,[84,93] polyesters,[123] and polyethylene.[101]

5.3.4.2.3. POLYMERS EXHIBITING MULTIPLE RELAXATIONS. Both spin-probe and spin-label methods have been used to study polymers known to exhibit multiple relaxation phenomena. In general, the particular relaxation detected by the ESR methods is dependent on the size of the nitroxide; in some cases the ESR technique "sees" the glass transition but in other cases subglass transitions are detected. Polymers of this type that have been studied are polyethylene,[33,35,40,100,101] poly(2,6-dimethyl-1,4-phenylene oxide),[36,37,83] polystyrene,[37] polyamides,[84,93] poly(ethylene glycols),[86,89] polyesters,[123] and poly(benzyl glutamate).[111,133]

5.3.4.2.4. EFFECT OF PROBE SIZE. ESR spin-probe and spin-label methods are based upon observing the motion of a "guest" paramagnetic molecule in a "host" polymer; the observed ESR spectra reflect the magnitude and type of motion allowed by the polymeric matrix. These considerations suggest that probes of varying size and/or functionality may selectively respond to different relaxations in a given polymer and/or respond differently to the same morphological transitions. Such variations in the observed ESR spectra have, in fact, been detected and studied by a number of different workers[33,33a,36,37,39,41,51,64,73,74,80,106,123]; these studies have all suggested that there is a correlation between the size of the paramagnetic probe and the size of the polymer segment experiencing increased motion upon traversing a particular transition. It has recently been shown[41] that log activation volumes at T_g increase linearly as a function of T_g and that the tumbling behavior of nitroxide probes is semiquantitatively related to these activation volumes.

5.3.4.2.5. MOLECULAR WEIGHT DEPENDENCE. Glass temperatures determined by ESR spin-probe methods have been found to be very sensitive to molecular weight in polystyrenes[37,39,42]; the effect of dispersity was also investigated and it was shown that the spin-probe technique is sensitive to the number-average molecular weight. Rotational correlation times have also been determined for a series of polystyrenes of varying molecular weight by Bullock and co-workers.[102,105,106] In a spin-labeling study these authors reported no molecular-weight dependence but were observing a subglass transition (due to phenyl group rotation). They have, however, shown that in low-molecular-weight polystyrenes τ_c is a function of molecular weight and have suggested that end-over-end motion (in addition to the molecular-weight independent local-mode motion) becomes important at low molecular weights.

As part of a spin-probe study on anionic polystyrenes[37,39,42] it was suggested that free-volume effects play an important role in ESR studies of low-molecular-weight polystyrenes. It was shown that the shape of the extrema separation vs. temperature plot is a function of molecular weight; the sigmoidal shaped curve is quite narrow at low molecular weight (i.e., curve A in Fig. 9) and broadens significantly (i.e., curve B in Fig. 9) as molecular weight increases. It was also noted that the upper limit of the extrema separation (i.e., plateau at left side of Fig. 9) increased significantly as the molecular weight decreased (for example, styrene monomer and dimer exhibited upper-limit extrema separations of ~75 and ~70 G, respectively, in contrast to the upper limit of 60–65 G for high-molecular-weight polystyrene samples). Both of these observations were ascribed to the effects of free volume and its distribution and suggested the potential utility of the ESR spin-probe method for studying free volume effects in high polymers.

Studies on the effect of molecular weight have also been reported for poly(ethylene glycol)[86,87,89,93] and poly(methyl methacrylate).[99]

5.3.4.2.6. NETWORK POLYMERS. Crosslinked unsaturated polyesters have been studied by a number of workers[30,56]; τ_c was shown to be independent of percentage cross-linking but very dependent on the flexibility of the crosslinks. Various crosslinked elastomers have been studied by Russian workers[58,59,67] and the effect of crosslinking via vulcanization was studied. Törmälä[88] has studied the effect of plasticizers on PVC plastisols and gels; there was only slight variation in τ_c with the percentage concentration of PVC in plastisols, but gelation was shown to facilitate the penetration of plasticizer into the polymeric matrix. In a study of thioglycolic acid lignin[90] (TGL, a constituent of wood) the effects of solvation were examined; it was shown that spin-labeled TGL experienced two different environments, which was explained by differences between nitroxides on the surface and nitroxides located in the interior of the inner stiff network.

Network polymers of polystyrene have been studied by both Regen[112-114] and Miller.[110,111a,111b] Regen showed that the rotational freedom of the nitroxide was dependent on both the extent of crosslinking and the extent of swelling and suggested that the extent of swelling had a significant effect on the motion of nitroxide *imbibed* in the solvent channels of the polymer. Veksli and Miller[110,111b] have suggested that surface motion of nitroxides is virtually identical with internal motion and indicated a strong dependence of τ_c on the nature of the solvent, in direct contrast to the results of Regen.[112]

Polyurethanes have been the subject of two studies. Solvent penetration into polyurethane fibers was studied by Catoire and Hagege[117] using the stable free radical diphenylpicrylhydrazyl (DPPH) as the spin probe. In an extensive study of crosslinked polyurethane elastomers, Ward and Books[134] observed a number of differences in nitroxide motion and related these to composition, temperature, state of swelling, and effect of strain.

Ion-exchange resins have also been the subject of two studies. In 1971, Chesnut and Hower[118] studied cation-exchange resins by utilizing nitroxide spin-probes; they found minor variations in the observed ESR spectra as the cation was varied but large variations as the extent of crosslinking was varied. Lagercrantz and Setaka[127] studied both anion- and cation-exchange resins by spin-labeling techniques. Anion exchange resins dramatically slowed the motion of nitroxides in aqueous solution and there was a considerable effect upon varying the pH of the solutions.

A recent study of ion-exchange membranes has used a series of nitroxides (including neutral, anionic, and cationic nitroxides) to study the interaction inside of high-polymer membranes, both neutral (cellulose) and cationic [poly(carboxylic acid) and sulfonated poly(sulfonic acid)] membranes were included in the study.[140]   Data were reported relevant to the degree of swelling of the membrane, the nature of the nitroxide probe, the effects of hydrogen bonding, and the nature of the electrostatic and stereochemical interaction between nitroxide and the membrane.

Two recent studies, although dealing primarily with biopolymers, are of interest.   Nagamura and Woodward[141] have used the spin-probe technique to study molecular motion in collagen and the effect of water content on the observed motion.   Aplin and Hall[142] studied agarose, a commonly used support matrix for affinity chromatography, by the spin-label technique.

5.3.4.2.7. SURFACE AND INTERFACIAL PHENOMENA. Both spin-probe and spin-label techniques have been used to study surface and interfacial phenomena.   Rabold[33a] determined the effective surface area (per soap molecule) occupied on the surface of styrene–butadiene latexes.   The diffusion of nitroxides into micelles[81] and the association of nitroxides at the surface of micelles[79] have been studied.   Veksli and Miller[109,110] have also shown that ESR methods are useful in the study of both adsorption and motion of oligomers on polymeric surfaces.   Kumler, Keinath, and Boyer[38,43] have presented evidence suggesting that the ESR spin-probe method can be used to study interfacial material in multiphasic block copolymers.   Robb and Smith[139] have studied the adsorption of spin-labeled poly(vinyl pyrrolidone) and poly(methyl methacrylate) onto carbon and silica surfaces using both ESR and IR techniques.   The experimental results were also compared with NMR results.

5.3.4.2.8. MORPHOLOGY AND ORDER IN CRYSTALLINE POLYMERS. Very recent spin-probe studies on isotactic polypropylene[39] give evidence for the existence of two environments.   Possible explanations of the two distinct observed environments were presented and included (a) partitioning of the probe between atactic and isotactic regions of the polymer prepared by Ziegler–Natta catalysis, (b) probe present in the interface between atactic and isotactic blocks, (c) probe present in loose chain folds, (d) adsorbed atactic material on crystal faces, and (e) some probe molecules present within the crystalline phase, perhaps as lattice defects. Although the experiments did not allow a distinction among these interpretations, it was reported that atactic polypropylene, polyethylene, and isotactic polybutene-1 did not give evidence of probe in two environments.

A number of studies have demonstrated the effect of orientation on crystalline or partially crystalline polymers. Kovarskii[58] examined oriented polyethylene, polypropylene, and natural rubber by the spin-probe technique; even though the nitroxide probes were not oriented the probe rotation frequencies decreased upon stretching. Stryukov and co-workers[72] studied oriented polypropylene, poly(ethylene terephthalate), and Nylon 6; in both the polyester and the polyamide they observed that the probe was present in two environments and they related this observation to the volume of microvoids present in interfibrillar regions. Mitchenko and co-workers[128] have discussed the molecular mobility in oriented poly(ethylene terephthalate) in terms of the packing density of domains and the appearance of holes. Finally, there have been two studies of orientation in synthetic polypeptides; Wee and Miller[111] discussed orientation effects in solutions of poly(benzyl glutamate), and Tsutsumi and co-workers[133] studied the same polypeptide in the solid state.

Kovarskii[58] has studied the rotational correlation times of nitroxides in polypropylene films containing varying sizes of spherulites but the same level of crystallinity; the τ_c were independent of spherulitic size.

The crystallization of oriented and nonoriented poly(ethylene terephthalate) has been subjected to investigation.[62] The rotational correlation times of nitroxide probes varied as a function of the density of the polymer; the kinetics of the τ_c variation were consistent with Avrami's equation for crystallization kinetics.

Stryukov and co-workers[69] have studied the supermolecular structure in polyformaldehyde by the spin-probe method; the rotational frequencies of nitroxide probes were higher in polymer produced by anionic polymerization than in polymer produced by cationic polymerization, leading to the conclusion that the anionically prepared polymer has a looser packing of amorphous regions.

Törmälä and co-workers[86,89] have extensively studied partially crystalline poly(ethylene glycols). Four different rotational relaxation regions were detected, each possessing different activation energies.

Based upon an extensive study of the relaxation behavior of spin-labeled low-density polyethylene, Bullock and co-workers[101] have shown that the processes governing relaxation in the melt and in dilute solution are essentially the same.

Three different studies have appeared concerning the application of spin-probe and spin-label studies to polymer single crystals. In a study of polyethylene single crystals, Kusumoto and co-workers[125] suggested that the fold structure at the surface became tighter upon annealing. This work was expanded in 1974[126] and it was suggested that the mobility at the surface of annealed crystals was uniquely determined by the increased thickness of the lamellae upon annealing. Nagamura and Woodward[129] studied the surface regions of poly(*trans*-1,4-butadiene) by the spin-probe method and concluded that the amount of amorphous material at the crystal surfaces depends on the solvent and temperature used for crystal growth.

5.3.4.2.9. COPOLYMERS. Many studies have been carried out on random copolymers[33,33a,35,40,60,97,99–101,109] and the ESR techniques normally give results that are consistent with the average predicted properties. Fewer studies have been reported on multiphasic block copolymers[36,38,43,60]; in such cases there is normally a partitioning of the probe between the component phases, and the motion of the probe can be followed independently in each phase. Evidence has also been presented[36,38,43] that suggests the ESR spin-probe method may be unique in being able to detect interfacial material in multiphasic blocks, which would allow independent estimates of the composition of such interphases.†

5.3.4.2.10. EMULSION POLYMERIZATION: DETERGENTS, MICELLES, LATEXES. In view of the importance of emulsion polymerization, it is not surprising that a number of studies relate to this area. Several studies have focused on details of the structure of micelles and changes occurring at the critical micelle concentration[79,81,94,95,136]; one study focused on the interaction of detergents with styrene–butadiene latexes.[33a] Veksli and Miller[109,110,111a,111b] have studied both surface-labeled and homogeneously labeled latexes and their interaction with solvents and nonsolvents; differences in behavior were correlated with the equilibrium solvent uptake from the saturated vapor.

5.3.4.2.11. STUDIES OF POLYMERS IN SOLUTION. A variety of studies have been carried out using spin-probes or spin-labels on polymers in solution. In a series of early studies, concentrated solutions of various polymers were examined in both poor and good solvents[50,53,76]; the general conclusions from these studies were that the ESR technique could be useful in studying the microheterogeneous nature of concentrated polymer solutions. A large number of studies have examined the dependence of rotational correlation times on the properties of the solvent,[99–103,106,108–110,111a,111b] and have found significant variation dependent upon both the viscosity of the solvent and the concentration of the polymer. Such variations have also been related to the critical concentration for entanglement. It thus appears that the ESR method is a valuable technique for studying the dynamics of polymer solutions.

The extensive studies of Szwarc and co-workers[49] should also be noted.

† A detailed discussion of block copolymers is given in Part 16 (this volume, Part C).

These authors have prepared varying length hydrocarbon chains containing an α-naphthyl moiety at each end of the chain. By appropriate reduction techniques one end of the chain is converted to a paramagnetic radical anion and ESR techniques are used to study intramolecular electron transfers. This technique offers an extremely powerful method for studying the intrinsic dynamic flexibility of molecular chains in solution.

5.3.4.2.12. THE EFFECT OF PLASTICIZERS. It was first suggested by Rabold[33] that ESR methods would be valuable for the study of plasticized polymers. In a study of rigid-chain, dense network polymers of methacrylates,[56] it was shown that the principal chains of the polymer have a low mobility. This mobility is not reduced by crosslinking with more flexible oligomer blocks, provided that the crosslinking bridge is not so large that it leads to plasticization of the polymer. In a study more directly designed to investigate polymer–plasticizer interactions, Törmälä[88] found that the rotational relaxation times of probe radicals in plasticized poly(vinyl chloride) decreased as the amount of plasticizer increased. A linear correlation between τ_c and Young's modulus indicated that the microstate considerably influenced the mechanical behavior of the PVC gel.

5.3.4.2.13. THERMAL OXIDATIVE DEGRADATION. Because the processes that accompany the oxidation of polymers (primarily degradation and crosslinking) will modify the molecular mobility, it is not surprising that ESR techniques have been used to study such phenomena. Such studies have been carried out on polypropylene[78] and methylvinylsiloxane rubber.[58]

5.3.4.2.14. POLYMERIZATION MECHANISMS. Some spin-label and spin-probe studies have yielded information directly related to the mechanisms of polymerization reactions. Bullock and co-workers[99,103] observed nonlinear Arrhenius plots of log τ_c vs. $1/T$, which reflected the temperature dependence of the termination rate coefficient in the free-radical polymerization of methyl methacrylate. Also, Shaulov[68] used spin-labeling to model the recombination of polymer radicals.

5.3.4.2.15. COMPARISON OF SPIN-LABEL AND SPIN-PROBE EXPERIMENTS. Only in a few studies have attempts been made to investigate both spin-probe and spin-label experiments on the same polymer. Wasserman et al.[73] studied poly(vinyl acetate) by both spin-probe and spin-label methods, in both solution and the solid state, using two different sizes of nitroxide probes; these authors concluded that the rotational characteristics of the nitroxide radicals were very dependent on the probe sizes.

Törmälä and co-workers[85,86] have applied both methods to poly(ethylene glycols). They detected little difference in the rotational motion of the nitroxides in solid polymers, but got considerable variation in solution studies.

Bullock and co-workers[101] have studied polyethylene by both spin-probe and spin-label methods; the spin-label experiments resulted in detection of both the α and β relaxation processes in solid polyethylene, while the spin-probe experiments gave a transition temperature attributed to the γ relaxation. A different interpretation of this probe data has, however, been suggested by Boyer.[137]

5.3.4.3. Comparison of Spin-Probe and Spin-Label Results with Other Experimental Methods. All of the above results suggest that spin-probe and spin-label techniques are valuable additions to the arsenal of methods available for the study of both structure and dynamics in polymeric systems. A number of studies reported above have included comparison of ESR results with results gathered by other methods. A brief summary of these comparisons follows.

5.3.4.3.1. DILATOMETRY. Molecular-weight dependence of T_g as determined by ESR has been compared with values by dilatometry[37,42]; the agreement is quite good.

5.3.4.3.2. THERMAL METHODS. Molecular-weight dependence of T_g as determined by ESR has been shown to compare favorably with glass temperatures determined by differential scanning calorimetry.[37,42]

5.3.4.3.3. DYNAMIC MECHANICAL METHODS. A number of workers have looked for correlation of ESR-derived parameters with data derived from dynamic mechanical methods. Boyer and Kumler[38,43] have shown that there are distinct differences between dynamic mechanical and ESR methods as applied to block copolymers. If the two phases of the block copolymer are quite distinct, dynamic mechanical methods do, in fact, detect both phases. However, when there is a low content of hard phase and/or the molecular weight of the polystyrene hard phase is low, there are significant amounts of interfacial material; in such circumstances the ESR spin-probe method detects both the soft phase and the interphase, whereas dynamic mechanical methods tend to give diffuse results, which probably represent observation of a composite phase. Many of the ESR results of the Finnish workers have been compared directly to data derived from mechanical spectroscopy.[83-86] Bullock and co-workers reported that the fit of the correlation between ESR parameters and dynamic mechanical parameters was dependent on the particular diffusion model adopted to treat the nitroxide motion.[105] It should also be noted that two studies have reported a direct comparison of ESR-derived parameters with Young's moduli.[84,88]

5.3.4.3.4. DIELECTRIC RELAXATION. A number of studies have shown a correlation between various ESR derived parameters (E_a, τ_cs) and dielectric relaxation data.[83-86,99,102,105,106,108,109] In at least one case[87] a glass transition temperature derived by ESR studies agreed very well with the T_g derived from dielectric relaxation data.

5.3.4.3.5. FLUORESCENCE DEPOLARIZATION. In a study of end-labeled PMMA, Bullock and co-workers[103] found τ_c values from ESR to be significantly lower than τ_c values from fluorescence depolarization but suggested the difference was due to different solvents being used in the two cases. In a later study[99] these same workers showed excellent agreement between τ_c values for both segmental and end-group mobilities of methacrylates compared to data from fluorescence depolarization measurements when the relative viscosities of the solvents were taken into account.

5.3.4.3.6. NUCLEAR MAGNETIC RESONANCE. Not surprisingly, the most extensively investigated correlations are those of ESR data with NMR data. Many studies have shown correlations of either activation energies or rotational frequencies with similar data from NMR experiments; such correlations have been shown for ethylene/propylene copolymers,[33,40] polyethylene,[51,101] natural rubber,[64,75] polystyrene,[71,74,105,106,108] polymethacrylates,[71,109] polyisobutylene,[71,74] polybutadiene,[36,74] polypropylene,[74] poly(vinyl acetate),[73] and poly(ethylene glycols).[86,87,93] In most of these studies there was very good agreement between activation energies determined by the two methods, but there were often discrepancies between frequencies of rotation[51,52,64,71,74,75] that have been ascribed to differences in size between the tumbling radicals seen by ESR and the main chain segmental motion observed in NMR experiments.

A very recent study[39] has shown evidence for two environments in isotactic polypropylene, a result that agrees very well with comparable NMR data.

Recently, workers at the National Research Council of Canada have carried out an extensive comparison of the deuterium NMR and the ESR spin-label techniques for characterizing organization in both model and natural biological membranes[138]; each technique is shown to possess both distinct advantages and disadvantages and these results should be considered in any study of synthetic high polymers.

A recent study by Robb and Smith[139] has also compared the ESR spin-label technique to both IR and NMR methods as a tool for studying adsorption of polymers at solid–liquid interfaces.

5.4. Small-Angle Neutron Scattering †

By J. S. King

5.4.1. Introduction

The use of small-angle neutron scattering began in the late 1960s as a technique to observe the structures of large defect clusters in alloys and magnetic superlattices.[1] In 1971–1972 it became apparent, as suggested by Allen and Edwards[2] and others, that the technique had great potential for the study of the size, conformation, and shape of large polymer molecules in solution and solid, where the characteristic chain dimensions may be between 30 and 1000 Å. The potential exists for two main reasons. First, there is a large difference in the neutron–nuclear scattering intensities between deuterated and protonated monomers. This means that when a few fully deuterated chains are embedded in a protonated host, there is a strong contrast factor by which isolated but chemically undisturbed chains may be "tagged" for observation. Second, the range of momentum transfer, and hence range of shape dimensions, is much greater than for light scattering. These two factors make it possible to extend conformation measurements into concentrations and dimensional ranges not heretofore possible.

Initial experiments were performed at Saclay[4] and Jülich[5] that demonstrated that the potential could indeed be exploited. Since that time a number of small-angle experiments on polymer systems have been successfully completed at Jülich[6-9] and particularly at Grenoble.[3,10-13] Cur-

[1] W. Schmatz, T. Springer, J. Schelten, and K. Ibel, *J. Appl. Crystallogr.* **7**, 96 (1974).

[2] G. Allen and S. F. Edwards, private communication, in Benoit *et al.*[3]

[3] H. Benoit, D. Decker, J. S. Higgins, C. Picot, J. P. Cotton, B. Farnoux, G. Jannink, and R. Ober, *Nature* (*London*), *Phys. Sci.* **245**, 13 (1973).

[4] J. P. Cotton, B. Farnoux, and G. Jannink, *J. Chem. Phys.* **57**, No. 1, 290 (1972).

[5] R. G. Kirste, W. A. Kruse, and J. Schelten, *Makromol. Chem.* **162**, 299 (1973).

[6] D. G. H. Ballard, G. D. Wignall, and J. Schelten, *Eur. Polym. J.* **9**, 965 (1973).

[7] G. D. Wignall, D. G. H. Ballard, and J. Schelten, *Eur. Polym. J.* **10**, 861 (1974).

[8] J. Schelten, D. G. H. Ballard, J. Schelten, G. D. Wignall, G. W. Longman, and W. Schmatz, *Polymer* **17**, 751 (1976).

[9] D. G. H. Ballard, A. Cunningham, and J. Schelten, *Polymer* **18**, 259 (1977).

† See also Volume 5A of this series, Sections 2.1.2.2.1.4; 2.2.2.2.2.3; 2.2.2.2.2.4 as well as Volume 5B, Section 2.7.4.

METHODS OF EXPERIMENTAL PHYSICS, VOL. 16A

rently, small-angle spectrometers are in various stages of design or development at MURR (Missouri), Brookhaven, Oak Ridge, and NBS in this country,[14] and a significant fraction of the programs planned for them are experiments on polymer systems.

5.4.2. Elastic Neutron Scattering Cross Section at Small Angles

At the outset, we must develop expressions for the neutron–nuclear scattering intensity from a collection of monomers. The intensity is a function of angle and therefore proportional to the differential scattering cross section $d\sigma/d\Omega$ for elastic, coherent neutron scattering[15] from a collection of nuclei. The momentum change Q of the neutron after scattering from such a collection of nuclei obeys the Bragg law,

$$Q = |\mathbf{K}| = 2k \sin \frac{\theta_s}{2}, \qquad (5.4.1)$$

where θ_s is the scattering angle for neutrons with incident and scattering wave vectors $k = 2\pi/\lambda$. We are interested in a region of Q around the origin of reciprocal space such that $Q/k \ll 1$ and $\sin \theta_s/2 \approx \theta_s/2$. For this region Q is perpendicular to k and simply given by

$$Q = k\theta_s. \qquad (5.4.2)$$

For practical reasons λ normally will be limited to a range of about 2.0–12 Å (for a lower limit of Q of, say, 0.001 Å^{-1} this requires observation at scattering angles down to 2.0–6.0 minutes of arc).

The neutron differential cross section expresses the normalized scattering intensity per unit solid angle for a plane wave (neutron) incident on a collection of scattering centers (nuclei) from which the outgoing spherical waves interfere. The sum of these waves each multiplied by a scattering probability for each scattering center gives an angularly dependent scattering amplitude whose absolute square is, in fact, just the structure

---

[10] H. Benoit, J. P. Cotton, D. Decker, H. Benoit, B. Farnoux, J. Higgins, G. Jannink, R. Ober, C. Picot, and J. des Cloizeaux, *Macromolecules* **7**, No. 6, 863 (1974).

[11] R. Ober, J. P. Cotton, B. Farnoux, and J. S. Higgins, *Macromolecules* **7**, No. 5, 634 (1974).

[12] G. Jannink, M. Dauod, J. P. Cotton, B. Farnoux, G. Jannink, G. Sarma, H. Benoit, M. Duplessix, C. Picot, and P. G. de Gennes, *Macromolecules* **8**, 804 (1975).

[13] H. Benoit, D. Decker, R. Duplessix, C. Picot, P. Rempp, J. P. Cotton, B. Farnoux, G. Jannink, and R. Ober, *J. Polym. Sci., Polym. Phys. Ed.* **14**, 2119 (1976).

[14] See, for example, R. M. Brugger *et al.*, "A Small Angle Neutron Scattering Spectrometer at MURR." Missouri Univ. Res. Reactor Facility, University of Missouri, Columbia, 1976.

[15] See, for example, G. E. Bacon, "Neutron Diffraction," 3rd ed., Oxford Univ. Press, London and New York, 1975.

factor for the collection of scattering centers. The scattering probability is given by a_i, the "bound atom" coherent scattering length for the ith nucleus.[15] The differential cross section is

$$d\sigma/d\Omega = S(Q) = \left\langle \left| \sum_i a_i \exp[i\mathbf{Q} \cdot \mathbf{r}_i] \right|^2 \right\rangle. \qquad (5.4.3)$$

The brackets denote that an ensemble average must be made over all space configurations possible for the polymer system.[16]

We can examine, initially, the ideal case of a two-component system of a small concentration of long-chain polymer molecules in a host solvent. That is, we wish to observe the structure factor for an isolated chain in a host, which may be a liquid solvent or the same polymer itself in either a liquid or solid state. Q is small enough so that phase differences for atomic components of a single monomer or a chain segment of several monomers are vanishingly small. We may then treat the monomer (or segment) as a rigid unit and simply add together the scattering lengths of atoms in each monomer. Call the scattering length of these monomer sums a_m for the chain elements. Divide the solvent artifically into volume elements of a size equal to the volume per chain monomer and call the scattering length of the equivalent solvent elements a_s. Then Eq. (5.4.3) is a double sum

$$S(Q) = \left\langle \left| a_m \sum_j^{\text{chain}} \exp[i\mathbf{Q} \cdot \mathbf{r}_j] + a_s \sum_k^{\text{solvent}} \exp[i\mathbf{Q} \cdot \mathbf{r}_k] \right|^2 \right\rangle, \quad (5.4.4)$$

where $\mathbf{r}_j$ is the jth chain monomer position and $\mathbf{r}_k$ is the solvent "monomer" position. If we add an a_s term to the last sum for each chain element at $\mathbf{r}_j$, and subtract the same term from the first sum, we transform (5.4.4) to a sum over chain elements with a new amplitude factor $a_m - a_s$, plus a sum over all elements of a "perfect" solvent with no chains dissolved[17]:

$$S(Q) = \left\langle \left| (a_m - a_s) \sum_j^{\text{chain}} \exp[i\mathbf{Q} \cdot \mathbf{r}_j] \right.\right.$$
$$\left.\left. + a_s \sum_k^{\substack{\text{all} \\ \text{space}}} \exp[i\mathbf{Q} \cdot \mathbf{r}_k] \right|^2 \right\rangle. \qquad (5.4.5)$$

For an ideal sample of uniform density, the second term contributes nothing to the Q-dependent coherent scattering since this can only result

---

[16] A brief general review of scattering theory may be found in "Molecular Spectroscopy with Neutrons," by H. Boutin and S. Yip. MIT Press, Cambridge, Massachusetts, 1968.

[17] G. C. Summerfield, "Spectroscopy in Biology and Chemistry—Neutron, X-Ray, Laser," Chapter 10. Academic Press, New York, 1974.

from the spatial fluctuations of the scattering lengths. We are consequently left with only the first term. Now, finally, assume that the chain is composed of fully deuterated monomers and the solvent of only protonated elements. To denote this system, we replace a_m by a_D and a_s by a_H, and Eq. (5.4.3) reduces to

$$d\sigma/d\Omega = S(Q) = |a_D - a_H|^2 \left\langle \sum_{i,j} \exp[i\mathbf{Q} \cdot \mathbf{r}_{ij}] \right\rangle. \qquad (5.4.6)$$

The bracketed term of Eq. (5.4.6) is common to all spectroscopic structure expressions, but the contrast term $|a_D - a_H|^2$ is unique for neutron scattering because of the large difference in coherent scattering lengths between protonated and deuterated monomers. To the extent that complete deuteration of chains does not perturb the chemistry of the solution, we have a "tagging" method, which allows observation of an isolated chain embedded in its typical environment. Such tagging becomes particularly dramatic when the solvent is a solid composed of the same polymer chains as the chain under observation.

While Eq. (5.4.6) is adequate for the present discussion, a general derivation, which accounts for concentration and density fluctuations in both solute and solvent, may be found in Summerfield et al.,[18] which makes use of correlation statistical arguments.

Both deuterons and protons have incoherent scattering lengths as well, and the value for protons is particularly large. These, however, produce only a constant background independent of Q. Table I lists the scattering lengths for D, H, C, and S as recently compiled by Bacon.[19] The scattering length is also given for several monomers. Table II lists approximate contrast factors for common two-component polymer solutions.

Equation (5.4.6) is the differential cross section per single tagged chain. In the limit of $Q = 0$, the bracketed term must equal n^2, the square of the number of monomers per chain. Hence, $S(Q) = |a_D - a_H|^2 n^2$ multiplied by a function of Q, which we call $f(Q)$. $f(Q)$ is unity at $Q = 0$, tends to zero at large Q, and is, in fact, the square of the normalized Fourier transform of the pair correlation function for elements of the chain. In dilute solutions the correlation is between segments of a single tagged chain. In concentrated solutions the correlation encompasses other chains as well unless the doping method specifically avoids it.

If we designate the experimental Q measured at a given detection angle as Q_0, the observed SANS count rate is the integral of the cross section over the resolution multiplied by constants for the beam, sample, and de-

[18] G. C. Summerfield, J. S. King, and R. Ullman, *Macromolecules* **11**, 218 (1978).
[19] G. E. Bacon, *in* "Neutron Diffraction Newsletter" (W. Yelon, ed.). Neutron Diffraction Commission of the International Union of Crystallography, 1977.

TABLE I.   Coherent Scattering Lengths of Some
Common Monomers and Solvents

Nucleus	Chemical repeat unit	ρ/m (gm/cm^3-amu)	a (10^{-12} cm)
^{1}H			−0.374
^{2}D			+0.667
^{12}C			+0.665
^{32}S			+0.280
	Polystyrene (C$_8$H$_8$)	0.010	+2.328
	d-Polystyrene (C$_8$D$_8$)	0.010	+10.656
	Polyethylene (CH$_2$)	0.067	−0.083
	d-Polyethylene (CD$_2$)	0.067	+1.999
	Carbon disulfide (CS$_2$)	0.017	+1.225
	Benzene (C$_6$H$_6$)	0.011	+1.746
	d-Benzene (C$_6$D$_6$)	0.011	+7.992
	Cyclohexane (C$_6$H$_{12}$)	0.0092	−0.498
	d-Cyclohexane (C$_6$D$_{12}$)	0.0092	+11.994
	Xylene (C$_8$H$_{10}$)	0.0082	+1.58
	Toluene (C$_7$H$_8$)	0.0094	+1.663
	d-Toluene (C$_7$D$_8$)	0.0094	+9.991

TABLE II.   Some Approximate Contrast Factors

Polymer	Solvent	Contrast factor (barns/monomer)
PS	CS$_2$	0.09
PSD	CS$_2$	74.3
PSD	PS	69.3
PS	C$_6$H$_{12}$	7.76
PSD	C$_6$H$_{12}$	123.4
PSD	C$_7$H$_8$	82.7
PSD	C$_7$D$_8$	1.6
PS	C$_6$H$_6$	0.14
PS	C$_6$D$_6$	43.9
PSD	C$_6$H$_6$	75.5
PE	C$_8$H$_{10}$	0.08
PED	C$_8$H$_{10}$	3.26
PE	C$_6$H$_{12}$	0.00
PED	C$_6$H$_{12}$	4.27
PED	PE	4.33

tector conditions:

$$I(Q_0) =$$
$$I_0 \left(\frac{c\rho t A}{M}\right) |a_H - a_D|^2 n^2 \epsilon(Q_0) \exp[-\Sigma t] \int_{-\infty}^{\infty} R(Q_0 - Q) f(Q) \, dQ, \quad (5.4.7)$$

where I_0 is the neutron flux on the sample, ρ the solution density, c the weight fraction concentration of tagged chains, t the sample thickness, M the chain molecular weight, $R(Q_0 - Q)$ the spectrometer resolution, $\epsilon(Q_0)$ the detector efficiency, and A Avogadro's number. $\exp[-\Sigma t]$ is the sample transmission factor and Σ is the total macroscopic neutron cross-section. For polymer experiments Σ is dominated by the incoherent scattering cross section of hydrogen. If we lump all constants in Eq. (5.4.7) except c and M, and take $R(Q_0 - Q)$ to be a δ function, we obtain the simple law

$$I(Q) = KcMf(Q). \quad (5.4.7a)$$

5.4.3. SANS Spectrometer Design

In principle, the requirements for SANS spectrometers are simple by comparison with other neutron instrumentation: a well-collimated, moderately monochromatic thermal neutron beam is extracted from the neutron source (today universally a thermal research reactor) by means of a mechanical velocity selector or Bragg crystal system, followed by a multiple-aperture system. The beam elastically scatters through small angles and the intensity as a function of scattering angle is detected and recorded by neutron detectors and data acquisition instrumentation downstream. It is assumed (thus far) that small-Q inelastic scattering can be ignored.

In reality, and in common with most neutron spectrometry, the neutron source intensity is a primary limitation. This dictates the need to find an optimal compromise between intensity and resolution and to seek extremely large flight paths to accommodate large samples. We briefly review the spectrometer components and their requirements. It will be evident that the success of SANS is due in large part to the exploitation of large area two-dimensional position-sensitive neutron detectors. The performance of the current premier device—the D11 camera at the Institut-Laue-Langevin at Grenoble—will be discussed for illustration and compared with other designs.

Figure 1 gives a schematic representation of the essential features of a SANS system. Thermal neutrons escape the source volume with a nearly Maxwellian spectrum whose characteristic temperature is a few

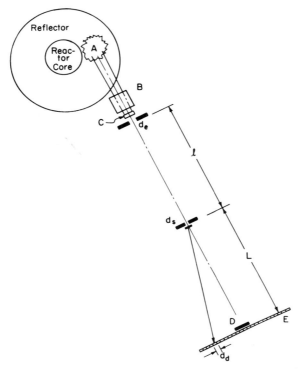

FIG. 1. Sketch of a SANS spectrometer. (A) Effective neutron source volume; (B) wavelength or velocity monochromator; (C) source intensity monitor; d_e, d_s, source and sample irises; l, L, evacuated flight paths; (D) beam stop; (E) area detector; d_d individual detector element.

degrees above the physical temperature of the source volume moderator. A thin slice of this distribution is transmitted by the monochromating system and enters an evacuated flight path l with angular limits defined by the diameters d_e and d_s, which are the source and sample apertures, respectively. The direct and scattered beams pass through a second evacuated flight tube of length L and impinge on the two-dimensional detector plane. The flight tubes are evacuated ($\leqq 10$ μm) to eliminate air scattering, which would produce an effective degradation of up to 8% per meter. A carefully dimensioned beam stop masks the detector from the total direct beam.

5.4.3.1. *Q Regime and Flight Path.* Ideally, SANS should cover the Q region from 0 to the lower limit of conventional diffractometers, which is perhaps 0.2 Å^{-1}. Certainly it is desirable that the lower limit overlap the upper light scattering region of about 0.002 Å^{-1}. The practical limits of the present D11 camera are $0.002 < Q < 0.2$ Å^{-1}, although in principle D11 is designed, at very reduced intensity (see Fig. 5), to go below 10^{-3} Å^{-1}. Polymer chain properties such as the radius of gyration R_g, the molecular weight M, and the virial coefficients, may require extrapolation of measured intensities to $Q = 0$. A practical lower bound can only approach 0, and we may adopt the criterion for the lower limit to be

$$R_g \check{Q} \leq 1.0.$$

As an example, the measured R_g^2 for polystyrene samples in poor solvents is proportional to the weight average molecular weight M_w and is approximately

$$R_g^2(\text{Å}^2) \approx \tfrac{1}{3} M_w \quad (\text{amu}).$$

Polystyrene chains of $M_w \leq 0.75 \times 10^6$ and $R_g \leq 500$ Å could thus be observed adequately using $\check{Q} = 0.002$ A^{-1}. According to Eq. (5.4.2) this requires for 10 Å neutrons a minimum angle of observation of 11 minutes.

This limit puts severe demands on the absence of a halo around the direct beam, on the magnitude of final flight path, and on the aperture sizes. Typically for the D11 source $d_e \approx 3.0$ cm, $d_s = 1.0$ cm, and with $l = L$ the direct beam has an outer radius at the detector of ≈ 3.0 cm [see Eq. (5.4.12)]. Undistorted data can begin at twice this radius, or 6.0 cm. To obtain data at an angle of 11 minutes requires, therefore, $l = L = 18.8$ m, giving a total flight path of 37.7 m. In fact, the normal L detector station for this Q regime is 20 m at D11.

To meet, on the same instrument, the high-Q limit of 0.2 Å^{-1} requires that the magnitudes of L and l be shortened so that the outer detector elements subtend a maximum angle $\hat{\theta} = \check{Q}/k$. For the D11 detector, using λ shortened to, say, 6 Å requires $l = L = 2.4$ m. Thus, the spectrometer must provide a large adjustment in source to sample, and sample to detector flight paths to cover the extremes of Q we have chosen.

5.4.3.2. Resolution. The resolution function in Eq. (5.4.7) is due to the limits on both the angular collimation and the source wavelength interval. From the simple law of Eq. (5.4.2) this can be written as an rms deviation in Q about the nominal measured $Q = Q_0$. It is the fractional uncertainty in Q that is important in SANS. Hence we write

$$\left(\frac{\Delta Q}{Q}\right)^2_{\text{FWHM}} = \left(\frac{k}{Q}\right)^2 \Delta\theta^2 + \left(\frac{\Delta\lambda}{\lambda}\right)^2, \tag{5.4.8}$$

where $\Delta\theta$ and $\Delta\lambda$ are the effective FWHM (full-width at half-maximum) limits of each variable. From the rms variance of a Gaussian distribution it can be shown[20] using the dimensional symbols of Fig. 1, that

$$\left(\frac{\Delta Q}{Q}\right)_{\text{FWHM}} = \left(\frac{2\ln 2}{3}\right)^{1/2} \left\{\frac{k^2}{Q^2}\left[\frac{d_e^2}{l^2} + \frac{d_d^2}{L^2} + d_s^2\left(\frac{1}{l} + \frac{1}{L}\right)^2\right] + 2\left(\frac{\Delta\lambda}{\lambda}\right)^2\right\}^{1/2}.$$

(5.4.9)

The $\Delta\lambda$ term is independent of Q while the $\Delta\theta$ term grows as Q^{-1}. The latter dominates the resolution in the low-Q regime but rapidly becomes negligible as Q^2 becomes large. $\Delta\lambda$ is the FWHM interval in λ transmitted by the monochromator.

Both $\Delta\theta$ and $\Delta\lambda$ affect the intensity [see Eq. (5.4.13)] and it has been argued[1] that intensity is optimized for a given resolution at some preferred Q region when all terms of Eq. (5.4.9) are roughly equal. This leads to a symmetric geometry where $l = L$ and $d_e = d_d = 2d_s$. In reality there may be restrictions either in design (such as the upper limit of $\Delta\lambda$ in some monochromators, or the available flight paths), or in operation (varying L without varying l) that obviate this symmetry. The consequences are undesirable but not severe. For example, it is often possible to obtain adequate intensity, when reducing L to reach a higher Q regime, to leave l fixed. However, Eq. (5.4.9) dictates obvious limits, for example, on the necessary resolution required of the detector elements (d_d) for given source and sample diameters and Q regime.

The $\Delta\lambda/\lambda$ contribution is determined by the monochromator design. D11 and other European spectrometers employ a rotating helical slot collimator, which is a mechanical velocity selector. Such devices are not new and are well understood.[21] They are particularly effective at wavelengths greater than 2 Å but require long horizontal beam paths not always available in a reactor facility. The high-resolution rotor at D11 ("Brunhilde") is designed to transmit a $\Delta\lambda/\lambda$ of 8.9% FWHM with an efficiency of about 80%. Alternatively, designs are now in progress that consist of stacked pyrolytic graphite crystals or multilayered artificial crystals,[22] which monochromate by Bragg scattering from a mosaic that is several degrees broad in the Bragg scattering plane but narrow in non-Bragg directions. These crystals can have a reflecting efficiency of 95%[23] and by slightly misaligned stacking can reach $\Delta\lambda/\lambda$ up to 4 or 5%.† Fig-

[20] This expression differs from that of Eq. 10, reference 1, in the choice of a Gaussian rather than rectangular wavelength interval, $\Delta\lambda$.

[21] R. M. Brugger, in "Thermal Neutron Scattering" (P. A. Eglestaff, ed.), Chapter 2. Academic Press, New York, 1965.

[22] B. Schoenborn, private communication.

[23] T. Riste and K. Otnes, *Nucl. Instrum. & Methods* **75**, 197 (1969).

---

† Data taken at 5 Å at MURR using a UCAR-ZYH graphite monochromator show peak reflectivity of 93 ± 2% and a FWHM double crystal rocking width of 2.45°.

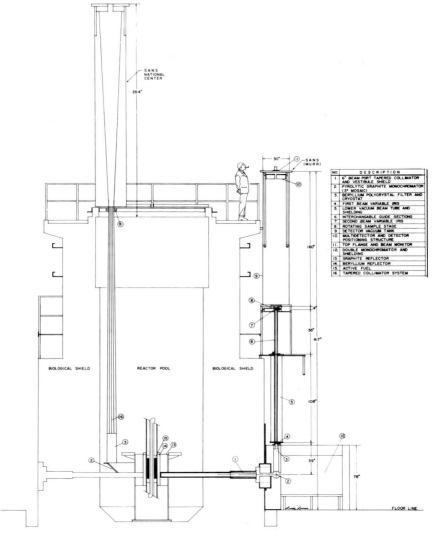

NO.	DESCRIPTION
1.	6" BEAM PORT TAPERED COLLIMATOR AND VESTIBULE SHIELD
2.	PYROLYTIC GRAPHITE MONOCHROMATOR (3° MOSAIC)
3.	BERYLLIUM POLYCRYSTAL FILTER AND CRYOSTAT
4.	FIRST BEAM VARIABLE IRIS
5.	LOWER VACUUM BEAM TUBE AND SHIELDING
6.	INTERCHANGABLE GUIDE SECTIONS
7.	SECOND BEAM VARIABLE IRIS
8.	ROTATING SAMPLE STAGE
9.	DETECTOR VACUUM TANK
10.	MULTIDETECTOR AND DETECTOR POSITIONING STRUCTURE
11.	TOP FLANGE AND BEAM MONITOR
12.	DOUBLE MONOCHROMATOR AND SHIELDING
13.	GRAPHITE REFLECTOR
14.	BERYLLIUM REFLECTOR
15.	ACTIVE FUEL
16.	TAPERED COLLIMATOR SYSTEM

FIG. 2. Schematic layout of two spectrometer designs at the 10 MW Missouri Research Reactor (MURR).[14]

ure 2 depicts two spectrometers designed at the University of Missouri (Columbia) using pyrolytic crystals.

The major resolution problem is due to the angular spread set by the beam aperture system, and this is most critical as Q approaches its minimum value. A common and useful alternative to Eq. (5.4.9) is just the product of k times the sum of extreme angles for the beam optics

$$\Delta Q = \frac{2\pi}{\lambda} \left(\frac{d_e + d_s}{2l} + \frac{d_d + d_s}{2L} \right). \tag{5.4.10}$$

If we use this expression with $d_d = 0$ we see it is just

$$\Delta Q = \frac{2\pi}{\lambda} \frac{\check{R}}{L}, \tag{5.4.11}$$

where $\check{R}$ is the extreme radius of the direct beam profile at the detector plane,

$$\check{R} = \frac{d_s}{2} + \frac{L}{l} \left(\frac{d_s + d_e}{2} \right). \tag{5.4.12}$$

Since $\check{R}$ can be directly observed and since there is some arbitrariness in the value of d_d made available, Eq. (5.4.11) is a useful figure of merit. It is possible to obtain an extremely sharp definition of $\check{R}$, using only cadmium apertures. Figure 3 is the profile of a small-angle cadmium aperture system at Münich.[24] This shows an absence of halo for well over four decades of the direct beam. It must be added, however, that the direct beam stop must be exactly located and totally absorbing to avoid scattering of the relatively intense direct beam into the low-Q detector region.

5.4.4.3. Intensity. The most critical feature of the SANS spectrometer is the magnitude of the product of the source intensity on the sample and the sample area, for adequate resolution. This can be stated as

$$I_0 = \left(\frac{d^2\phi}{d\lambda \, d\Omega} \right) \Delta\lambda \, \Delta\Omega A_e,$$

or in terms of the geometry of Fig. 1 as

$$I_0 = \left(\frac{d^2\phi}{d\lambda \, d\Omega} \right) \Delta\lambda \frac{d_s^2}{l^2} d_e^2. \tag{5.4.13}$$

The differential is the flux of neutrons per second per cm² per wavelength per steradian at the effective source plane of the reactor. $\Delta\lambda$ is the wavelength interval transmitted by the monochromator and d_e^2 is the effective

[24] P. Herget, Ph. D. Dissertation, Technische Hochschule, München (1968).

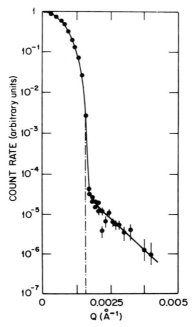

FIG. 3. Direct beam profile through cadmium irises, Münich experiment.[24]

source plane area. The source spectrum is approximated by a Maxwell–Boltzmann energy spectrum

$$\frac{d\phi}{dE} = \phi_0 \frac{E}{E_0^2} \exp[-E/E_0], \qquad (5.4.14)$$

where E_0 is the most probable energy,

$$E_0 = \frac{h^2}{2m_n\lambda_0^2} = k_B T, \qquad (5.4.15)$$

determined by the neutron temperature characteristic of the reactor moderator volume immediately surrounding the source plane. ϕ_0 is the integral of the Maxwellian over all E (or λ). The most common reactor sources have a neutron temperature a few degrees above the core moderator temperature, $E_0 \approx 0.028$ eV, but a "cold source" moderator volume may be inserted outside the reactor core periphery to enhance the long-wavelength flux. Here T may be as low as 21 K ($E = 0.0019$ eV). While ϕ_0 is still the thermal integral about the lower temperature, its magnitude will be reduced by the decoupling effect between core and cold modera-

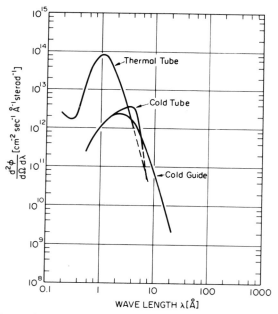

FIG. 4. Comparison of $d^2\phi/d\lambda\, d\Omega$ for three different source planes at Ill.[25]

tor. Figure 4 gives the first term of Eq. (5.4.13) at the exit plane of three different beam collimators at ILL; e.g., at the collimator tube from the normal thermal source in the reactor reflector, at the collimator tube from a cold source (25 liters of liquid hydrogen) and from the same cold source after traversing a 60 m long totally reflecting neutron guide.[25] The first two have equivalent collimator geometry and, if it is assumed both are Maxwellian, provide an opportunity to compare the intensities obtained from cold and thermal moderators. The exit plane of the neutron guide is the D11 source.

To compare intensities it is instructive to express the double differential of Eq. (5.4.13) in terms of the effective thermal integrals at the reactor moderator surfaces themselves. If these are assumed isotropic then using (5.4.14) and (5.4.15)

$$\frac{d^2\phi}{d\lambda\, d\Omega} = \frac{1}{4\pi}\frac{d\phi}{dE}\frac{dE}{d\lambda} = \frac{\phi_0}{4\pi}\left[2\left(\frac{\lambda_0}{\lambda}\right)^4 \exp[-\lambda_0{}^2/\lambda^2]\right]\frac{1}{\lambda}, \quad (5.4.16)$$

[25] "Neutron Research Facilities at the High Flux Reactor of the Ill." Institut Max Von Laue-Paul Langevin, Grenoble, France, 1975.

and

$$I_0 = \frac{\phi_0}{4\pi} F(\lambda, T) \frac{\Delta\lambda}{\lambda} \frac{d_e^2 d_s^2}{l^2},$$

where

$$F(\lambda, T) = \left[2 \left(\frac{\lambda_0}{\lambda}\right)^4 \exp[-\lambda_0^2/\lambda^2] \right]. \tag{5.4.17}$$

ϕ_0 is the effective Maxwellian integral. For the usual warm reflector source it corresponds to the commonly quoted available flux characteristic of the research reactor (1.2×10^{15} neutrons/cm²/sec for the ILL). $F(\lambda, T)$ is then very small if we seek large λ [e.g., for $\lambda = 4.75$ Å, $F(\lambda, T)$ is 0.029]. It will be found that Eq. (5.4.16) with $E_0 = 0.028$ and $\phi_0 = 1.2 \times 10^{15}$ does indeed reproduce the curve of Fig. 4 for the thermal tube. If we repeat the process for the cold tube, recognizing the approximate nature of the Maxwellian, at the same λ, $E_0 \approx 0.0019$ eV, $F(\lambda, T) \approx 1.06$, but ϕ_0 must now be reduced to about 1.4×10^{14} to reproduce Fig. 4. The effective ϕ_0 for the cold guide is reduced further at this λ, presumably due to guide losses.

For the same beam geometry and λ, relative intensities can of course be read directly from curves such as those of Fig. 4. In attempting to compare the performance of different spectrometers this is not valid since all the quantities of Eqs. (5.4.9) and (5.4.13) must be considered. A reasonable comparison can be made as follows: (a) select a minimum undistorted spectrometer Q for a matched-geometry condition; (b) select the optimum available wavelength for each spectrometer; (c) adjust l so that $\Delta Q(\check{Q})$ is the same in each case; and (d) compare intensities. These conditions are expressed by

$$\check{Q} = \frac{2k\check{R}}{l} = \frac{8\pi}{\lambda} \frac{d_s}{l},$$

$$\Delta Q(\check{Q}) = \frac{k\check{R}}{L} = \frac{4\pi}{\lambda} \frac{d_s}{l} = 0.5\check{Q},$$

$$I_0 = \frac{d^2\phi}{d\lambda\, d\Omega} \Delta\lambda \frac{4d_s^4}{l^2} = \left(\frac{\phi_0}{4\pi} \frac{F(\lambda, T)}{\lambda}\right) \Delta\lambda \frac{4d_s^4}{l^2}.$$

Such a comparison is shown in Fig. 5 for D11 at ILL and for the MURR-SANS intermediate flux spectrometer under construction at Columbia, Missouri.[14] The D11 curves were prepared by Ibel[26] and have

[26] K. Ibel, *J. Appl. Crystallogr.* **9**, 296 (1976).

TABLE III.   Calculated Comparison of D11 and MURR–SANS Intensities

	D11	MURR–SANS
λ	6.3 Å	4.75 Å (fixed)
k	0.99 Å^{-1}	1.32 Å^{-1}
d_s	$(15/4)^{1/2} = 1.94$ cm	$(\pi/4)^{1/2} = 0.88$ cm
d_e	$2d_s = 3.88$ cm	$2d_s = 1.77$ cm
l	700 cm	425 cm
$\check{Q}$	0.011 Å^{-1}	0.011 Å^{-1}
$\Delta Q(\check{Q})$	0.0055 Å^{-1}	0.0055 Å^{-1}
$\Delta\lambda/\lambda$	8.9%	4.2%
$\Delta\lambda$	0.56 Å	0.20 Å
$(d^2\phi/d\lambda\ d\Omega)$	4.0×10^{11}/cm^2 S Å ster	5.4×10^{10}/cm^2 S Å ster
I_o (calculated)	2.6×10^7/sec	1.5×10^5/sec
I_o (observed)	1.1×10^7/sec	

been normalized by him to measurement at 10 Å.   Table III lists the parameters and intensities calculated by the author for a selected value of $\check{Q} = 0.011$ Å^{-1}.   The (observed) D11 curves of Fig. 5 appear to be lower than Table III by a factor of about 2.4.   This is not unreasonable in view of omitted detail, especially the effective size of d_e and the detector efficiency.   The ratio of calculated intensities for D11 and MURR-SANS is 174, which is equal to the product of the ratios of source terms (7.4), of $\Delta\lambda$

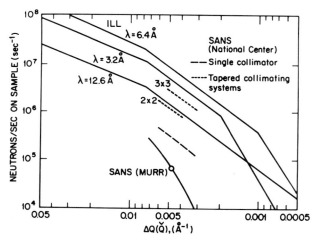

FIG. 5. Beam strength in neutrons per second on sample area vs. geometric resolution. Top solid lines apply to D11 for a sample area of 15/4 cm^2.[26]   Bottom solid curve applies to SANS–MURR for a sample area of $\pi/4$ cm^2.[14]   Dashed curves are estimates for a large SANS–MURR using multiple tapered flight tubes.

terms (2.8), and of beam geometry terms (8.4). The effect of the cold source is embedded in all three ratio terms. The fact that the core thermal flux at ILL is 10 times that of MURR, due to a higher reactor power level, must be kept in mind when considering this ratio of 174.

Several observations, based on resolution and intensity estimates such as discussed above, can be made:

(1) It is common in SANS (including D11) to run polymer samples closer to the 1.0 cm diameter listed for MURR-SANS in Table III. Such samples would lower the D11 I_0 by a factor of 4.8 without significantly altering the $\check{Q}$ or ΔQ (an "unmatched" case). Such practice would make the ratio above nearer 36 than 174.

(2) The intensity-resolution regime available at D11 is quite comparable with the best small-angle X-ray spectrometers. The 10 m system at ORNL for instance is capable of delivering on target 10^6 photons per second of wavelength 1.57 Å, and a minimum Q of about 0.002 Å^{-1} can be reached.[27] The target area is typically 100 times smaller, however.

(3) As can be shown by eliminating l between Eqs. (5.4.10) and (5.4.13), I_0 is proportional to $(\check{Q})^2$, over the intermediate range where l can be adjusted. This quadratic cost in intensity is evident in the middle region of the D11 curves in Fig. 5.

(4) The controversial question of use of a cold source is shrouded in cost of operation, questions of reactor safety, and reactor design. Our conclusions, based on comparisons made above for D11 are (a) for wavelengths shorter than 10.0 Å the intensity improvement is ≤ 4.0, (b) the $\check{Q}$ reachable is improved by about 2, (c) operation at 4.75 Å as opposed to 10 Å imposes neither detector problems nor interference from multiple Bragg scattering.

(5) The intensity is proportional to $\Delta\lambda/\lambda$ and for all systems today there is no reason not to raise this to the level of about 10%. Consequently, the rotating helical slot velocity selector has an advantage of perhaps 2.0 over potential crystal systems. The rotor does require, however, a large ($\approx 2l$) horizontal flight path.

5.4.3.4. Area Detectors The use of a large two-dimensional detector has become a common trend in many fields of spectroscopy today. It provides simultaneous acquisition of events over a scattering surface, rather than at a point. It eliminates the need for data corrections due to slit source geometry, so extensively developed heretofore in X-ray small-angle scattering.† It is particularly powerful in SANS on polymers

---

[27] R. W. Hendricks, "The ONRL 10-Meter Small-Angle X-Ray Scattering Camera." Oak Ridge Natl. Lab., Oak Ridge, Tennessee, 1977.

---

† See Chapter 6.2 (this volume, Part B)

because the scattering function $S(Q)$ tends to vary monotonically as Q^{-n}, where n is 1.0 or greater. Data collected over an area with $Q = 0$ at the center tend to retain constant statistics as Q increases toward its maximum value, where counting rates are low.

Space does not permit a detailed review of SANS area detectors now in use or under construction. The reader is referred to several reviews in the literature.[28–33] Detectors have been developed and built at CENG (Grenoble), and at ONRL and BNL in the U.S. The D11 detector (CENG) has an active area of 64×64 cm^2 with a resolution of 1.0 cm. Larger area detectors that at the same time have better resolution are under design or construction. All have a common feature that the scattered beam passes perpendicularly through a sandwich of grids—the x coordinate cathode, the anode, then the y coordinate cathode. Neutron capture reactions produce energetic ions in the gas volume and the subsequent gas multiplication creates many electron–ion pairs. The ions induce a charge pulse on both the x and y cathodes near the position of the neutron reaction. The methods of identifying which x and which y positions fired vary. The BNL system, for example, is shown in Fig. 6. Both cathodes are continuous resistive wires, and the position of an event is determined by comparing the charge collected at one end to the sum of charges collected at both ends of a cathode.

There are significant difficulties with all designs such as nonuniform position sensitivity, low efficiency (now 40–70%), long-term high-pressure gas stability, scattering from thick front wall high-pressure (10 atm) gas containment materials, use of Cartesian rather than the more desirable cylindrical cell geometry. The associated electronic amplification, ADC, and microprocessing logic is extensive, but thanks to much development in the high-energy physics field is becoming relatively cheap. It is reasonable to expect that by 1981 area detectors up to 80×80 cm^2 with efficiencies greater than 80%, and resolution down to 2–3 mm, will be operating reliably.

It may be added that a very economic alternative is to assemble a square array of one-dimensional position-sensitive tubular detectors. These are almost "off the shelf" items and the signal processing is not at

[28] R. B. Owen and M. L. Awcock, *IEEE Trans. Nucl. Sci.* **15**, 290 (1968).

[29] C. J. Borkowski and M. K. Kopp, *IEEE Trans. Nucl. Sci.* **19**, 161 (1972).

[30] J. Alberi, J. Fischer, V. Radeka, L. C. Rodgers, and B. Schoenborn, *Nucl. Instrum. & Methods* **127**, 507 (1975).

[31] R. Allemand, J. Bourdel, E. Roudant, P. Convert, K. Ibel, J. Jacobe, J. B. Cotton, and B. Farnoux, *Nucl. Instrum. & Methods* **126**, 29 (1975).

[32] J. Fischer, J. Fuhrmann, S. Iwata, R. Palmer, and V. Radeka, *Nucl. Instrum. & Methods* **136**, 19 (1976).

[33] B. E. Fischer, *Nucl. Instrum. & Methods* **141**, 173 (1977).

[34] J. Schelten, *Kerntechnik* **14**, No. 2, 86 (1972).

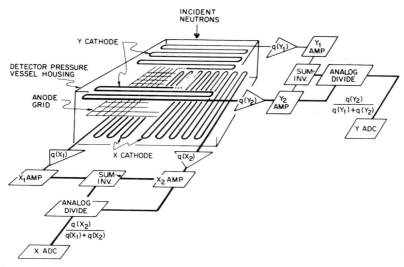

FIG. 6. Area detector employing two continuous resistive wire cathodes, after the BNL design.[30]

all prohibitive. This is the scheme used originally at Jülich[34] and being refined at Missouri.

5.4.4. Experimental Results

Because SANS is in its infancy, there have not been a large number of experiments reported. We briefly review results in those areas where significant effort has occurred. Despite the small number there have already been some notable answers to classic polymer questions. To date data treatment has been somewhat simply presented: scattering from "control" blanks have been subtracted from the tagged sample results, after the incoherent ("flat") background has been removed from both. Corrections for resolution in Q have been ignored, and multiple coherent scattering assumed negligible. Inelastic coherent contributions are also assumed to be negligibly small. Normalization to absolute scattered intensities has often been found important and can be done, for example, by measurement on known liquid incoherent (nondirectional) scattering samples.

The simplest linear polymer system imaginable is an isolated random coil dissolved in a solvent that does not support long-range monomer–monomer interactions. Such a solvent is called a theta (θ) solvent and the coil obeys Gaussian statistics. This system serves as an ideal limiting case against which SANS experiments on real systems can be compared. The scattering from a Gaussian coil can be characterized by the Debye function,[35] which in the notation of Eq. (5.4.7a) is

$$I(Q) = KcM \frac{2}{x^2} (x - 1 + e^{-x}), \qquad x = R_g^2 Q^2, \qquad (5.4.18)$$

where R_g^2 is the mean square radius of gyration of the coil with respect to its mass center. R_g^2 can be expressed in terms of the characteristic length b of the statistical chain elements (one or a few monomers) comprising the random walk steps of the free chain

$$R_g^2 = \frac{b^2}{6} \frac{M}{m}, \qquad (5.4.19)$$

m being the mass of the element. An expansion of (5.4.18) allows us to concentrate on two regions in Q space. The "low-Q" region, for which $x < 1$, can be written

$$\frac{Kc}{I(Q)} = \frac{1}{M} \left(1 + \frac{R_g^2 Q^2}{3} + O(Q^4) + \cdots \right), \qquad Q^2 < \frac{1}{R_g^2}, \quad (5.4.20)$$

and the "intermediate-Q" region, which can be written

$$\frac{I(Q)}{KcM} = \frac{12}{Q^2 b^2 M/m}, \qquad \frac{1}{R_g^2} < Q^2 < \frac{1}{b^2}. \qquad (5.4.21)$$

There are two distinguishing features of this simple system: (a) there is no concentration dependence of the right-hand side of (5.4.20) or (5.4.21); (b) R_g^2 is linearly dependent on molecular weight as seen in (5.4.19).

The low-Q and intermediate-Q regions in Q space have been studied for several real polymer systems by SANS. The governing equations are in general more complex than (5.4.19), (5.4.20), and (5.4.21) because of the presence of concentration-dependent terms. Nevertheless, the low-Q regime provides a measure of R_g^2, while the intermediate region, when compared with conformation models, reveals the chain shape.

5.4.4.1. Chain Conformation in Amorphous Bulk Polymer Light-scattering experiments have established basic properties of polymer solutions in good and bad solvents, where chain size and solute–solvent contrast factors permit. Gaussian behavior [Eqs. (5.4.19) and (5.4.20)] has been demonstrated for θ solvents and swelling in good solvents has been shown. One classic prediction, due to Flory[36] and others, but heretofore not observable, has been that chains in their own bulk return to Gaussian behavior, i.e., see the bulk liquid or solid, as a θ solvent. Neutron scat-

---

[35] P. Debye and B. Bueche, *J. Chem. Phys.* **20**, 1337 (1952).

[36] P. J. Flory, "Principles of Polymer Chemistry," Chapter XII. Cornell Univ. Press, Ithaca, New York, 1953.

tering on polystyrene,[6,10] on poly(methyl methacrylate),[4] and on poly(di-methyl siloxane)[37] have now given evidence to support the prediction. The polystyrene experiments by Benoit *et al.*[10] are particularly complete. Neutron experiments were sequentially performed (a) on deuterated PS chains (PSD) in a good solvent, CS_2, (b) on PSD in a θ solvent, C_6H_{12} at 36°C, and (c) on room temperature bulk PS doped with small concentrations (2, 1, $\frac{1}{2}\%$) of PSD chains. For general solvents a concentration-dependent term must be added to the low-Q equation (5.4.20). To first order in Q^2 and c, this is the so-called Zimm formula,[38]

$$\frac{Kc}{I(Q)} = \frac{1}{M}\left(1 + \frac{R_g^2 Q^2}{3} + \cdots\right) + 2A_2(Q)c. \qquad (5.4.20a)$$

Scattering data presented as $Kc/I(Q)$ vs. $Q^2 + 100c$ comprise what is known as a Zimm plot and this is commonly used for the low-Q data. $A_2(0)$ is the second virial coefficient. The limiting, extrapolated curves $Kc/I(Q)$ for $c = 0$ and $Kc/I(Q)$ for $Q = 0$ can be used to measure directly $A_2(0)$ and R_g^2. Similarly, for general solvents, Eq. (5.4.19) must be replaced by

$$R_g^2 = KM^{2\nu}, \qquad (5.4.19a)$$

where ν is a critical exponent determined by the excluded volume in solutions of real, nonintersecting chain segments. ν should equal $\frac{1}{2}$ only for chains in a θ solvent.

The Zimm plot for PSD in solid PS shows no concentration effect ($A_2 = 0$), as is evident in Fig. 7, taken from Benoit *et al.*[10] Further, a plot of R_g^2 vs. M_w,* from Zimm plots for samples with different M_w as seen in Fig. 8, shows that, within good error limits, not only does $\nu = \frac{1}{2}$ but the constant of Eq. (5.4.19a) is the same for the bulk as for a known θ solvent.† Finally, the Debye approximation, Eq. (5.4.19), is fulfilled extremely closely well beyond the low-Q region, as may be seen by the Kratky plot of Fig. 9 for PSD chains of $M_w = 98,000$.[6] Wignall *et al.*[6] have further demonstrated that these results are not altered by variation in temperature from 20 to 155°C, which includes the glass transition. These results, taken together, leave little doubt about the Flory prediction.

[37] R. G. Kirste and B. R. Lehnen, *Makromol. Chem.* **177**, 1137 (1976).
[38] B. H. Zimm, *J. Chem. Phys.* **16**, 1093 (1948).

* For polystyrene chains the measured $[R_g^2]^{1/2}$ must be multiplied by a known constant $\approx [M_z/M_w]^{1/2}$ to give the correct proportionality with M_w. R_g is then identified as R_w and is so shown in Fig. 8.[10] For monodisperse chains $R_g = R_w$.

† For CS_2 ν was observed to be $\frac{3}{5}$ in accord with prediction for good solvents.

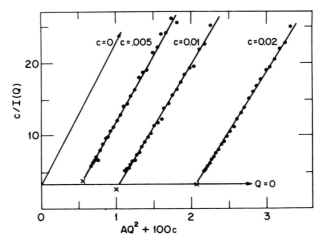

FIG. 7. Zimm plot for bulk solid polystyrene doped with three concentrations of deuterated polystyrene. The slope of the $Q = 0$ extrapolated curve measures the second virial coefficient and here is effectively 0. The slope and intercept of the extrapolated curve $c = 0$ measures the radius of gyration and molecular weight. Reprinted with permission from H. Benoit, *Macromolecules* **7** (6), 863 (1974). Copyright by the American Chemical Society.

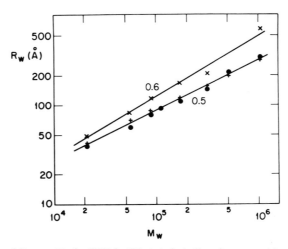

FIG. 8. Plot of R_w vs. M_w for PSD in CS_2 ($\times$), in bulk polystyrene ($\bullet$), and in C_6H_{12} (+). For CS_2 $\nu = \frac{3}{5}$, and for both C_6H_{12} and PS $\nu = \frac{1}{2}$. Reprinted with permission from H. Benoit, *Macromolecules* **7** (6), 863 (1974). Copyright by the American Chemical Society.

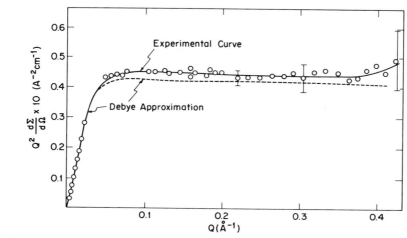

FIG. 9. Kratky plot of 5% PSD–95% PS atactic polystyrene covering the intermediate Q region, compared to the Debye approximation.[6]

5.4.4.2. Concentration Effects in Semidilute Polymer Solutions. A second classic problem has been a realistic description of the changes in chain conformation in concentrated solutions. The onset of significant chain overlap should occur near

$$c > c^* \approx M/AR_g^3. \qquad (5.4.22)$$

Changes, perhaps dramatic changes, in size and shape might be expected to occur to accommodate this interpenetration. The prevailing theory has been the mean-field theory of Edwards[39] and Flory,[40] but an unambiguous interpretation of light scattering has been confined[41] to the limiting region of $c \ll c^*$ and $Q \to 0$, where the asymptotic Eq. (5.4.20a) is always fulfilled. The reasons are twofold: first, light scattering cannot reach the interesting intermediate-Q regime. In this regime, with concentration effects present, the Edwards–Flory theory predicts a Lorentzian form of Eq. (5.4.21)[12,39]:

$$\frac{I(Q)}{KcM} = \frac{1}{Q^2 + \zeta^{-2}}, \qquad \zeta^{-2} = \frac{12cA}{mb^2}\, v, \qquad (5.4.21a)$$

where ζ is the "screening length" and v the excluded volume per statistical segment of true nonintersecting chains.

[39] S. F. Edwards, *Proc. Phys. Soc., London* **88,** 265 (1966).
[40] P. J. Flory, "Principles of Polymer Chemistry," Chapter XI. Cornell Univ. Press, Ithaca, New York, 1953.
[41] H. Benoit and C. Picot, *Pure Appl. Chem.* **12,** 548 (1966).

The second reason is that while we wish to observe changes in conformation of a single chain due to a changing concentration environment, when $c \geq c^*$ light scattering makes no distinction between inter- and intramolecular chain segment correlations. SANS circumvents this problem by retaining a very low concentration of tagged chains (PSD) in an increasing concentration of untagged chains (PS). Experiments using this technique for PSD + PS in CS_2 have been successfully completed.[12] The concentration dependence of $R_g(c)$, of $[R_g^2(c)/M_w]$, and of $A_2(c)$ have been measured in the low-Q limit, and the dependence of ζ on c found in the intermediate-Q (shorter correlation distance) regime.

The quantitative results show that near c^*, $R_g^2(c) \sim c^{-x}$, where $x = 0.025 \pm 0.02$, $R_g^2(c) \sim M_w^y$, where $y = 1.0$, and $\zeta(c) \sim c^{-z}$, where $z = 0.72 \pm 0.06$. These values are in disagreement with the Edwards–Flory predictions, which are 0.0, 1.0, and 0.5, respectively. It appears that the theory must be refined to fit these new results.

5.4.4.3. Polymer Networks and Rubber Elasticity Crosslinked poly(styrene-b-divinyl benzene) rubbers have been studied by SANS to test the Gaussian theory of rubber elasticity.[13] Model networks were prepared by block copolymerization of styrene and DVB. Two types of measurements can be made. In one the nodules representing the linkage points between styrene and DVB can be "tagged" by deuterating the ends of the PS chains. In the second, a few of the PS chains between linkage points are fully deuterated. Displacement of the linkage points after swelling or uniaxial deformation can be observed from the first experiment. Changes in the conformation of the chains connecting linkage points after uniaxial deformation can be inferred from the second experiment.

The data apparently show that the motions of the linkage points are approximately affine under swelling or stretching. The scattering profile, however, does not show monotonic intensity drop with increasing Q expected from a Gaussian distribution of elastic chains, but rather a well-defined peak, more characteristic of a quasi-lattice. The peak moves to lower Q when the network swells. The change in conformation of the fully deuterated tie chains is not affine. Questions of how detailed and over what distortion region the affine behavior should be expected to hold remain to be answered. Similar questions have been raised for the interpretation of PSD-doped PS samples that have been hot-stretched and subsequently quenched.[42]

It is fair to conclude that these experiments have shown that rubber net-

[42] C. Picot, R. Dupplessix, D. Decker, H. Benoit, F. Boue, J. P. Cotton, M. Daoud, B. Farnoux, G. Jannink, M. Nierlich, A. J. de Vries, and P. Pincus, *Macromolecules* **10**, 436 (1977).

works behave in a more complex pattern than is described by the simple Gaussian theory. This is a subject of fundamental importance for which further SANS experiments can and hopefully will bring major clarifications.

5.4.4.4. Structure of Block—Copolymer Chains.

Two noncompatible block copolymers may exhibit segregated or integrated chain structure. Expansion of the chains over their homopolymer states, in dilute solution, is affected by the degree of integration. By deuterating one of the polymers the SANS technique should provide data on the degree of segregation.[43,44] Han and Mozer[43] have reported data on a PSD–PMMA diblock-copolymer in deuterated toluene, and compared the SANS results with light-scattering data from the same copolymer in protonated toluene. The technique is an interesting study in contrast factors for both methods; only the PMMA is dominantly observed in SANS, while only PSD is observed in light-scattering photometry. The contrast for PSD alone in d-toluene was sufficient to demonstrate the degree of agreement obtainable from the two methods [R_g(SANS) = 210 ± 40 Å, R_g(LS) = 215 ± 18 Å]. (Excluded volume exponents were possible only from the neutron data and are included in the paper.)

Comparison of the measured R_g showed the PSD expanded slightly (7%) over its homopolymer state while PMMA, even though in a good solvent, showed a surprising compaction (15%) over its homopolymer θ solvent free dimension, and also exhibited an $A_2 = 0$.

Interpretation of these results are not entirely unambiguous and several structural models were examined. The best apparent model is that of a partially segregated system in which the PMMA block forms the core with PS as a surrounding but also penetrating shell.

5.4.4.5. Chain Configuration in Partially Crystalline Polymers—Polyethylene.

Neutron experiments on partially crystalline polystyrene and particularly on polyethylene (PE) show that measuring chain properties in these systems is much more difficult, and more complex than for the amorphous solids (Section 5.4.4.1). Extensive SANS data have been obtained on polyethylene samples melt-crystallized,[8,45,46] melt-crystallized under high pressure,[47] and precipitation-crystallized from dilute solutions.[46,48]

[43] C. C. Han and B. Mozer, *Macromolecules* **10**, 44 (1977).

[44] M. Duval *et al., Polym. Lett.* **14**, 585 (1976).

[45] J. Schelten, *Polymer* **15**, 682 (1974).

[46] G. C. Summerfield, J. S. King, and R. Ullman, *J. Appl. Cryst.* **11**, 548 (1978).

[47] D. G. H. Ballard, A. Cunningham, and J. Schelten, *Polymer* **18**, 259 (1977); D. G. H. Ballard, J. Schelten, and G. W. Longman, *Polym. Prepr., Am. Chem. Soc., Div. Polym. Chem.* **18**, No. 2, 167 (1977).

[48] D. M. Sadler and A. Keller, *Polymer* **17**, 37 (1976).

The primary difficulty is preparing reproducible samples that have randomly dispersed tagging molecules (PED) in the host (PEH). This arises for the melt-crystallized samples because the protonated molecules have a crystallization temperature approximately 6°C higher than the deuterated molecules. A similar differential exists for precipitation crystallization from solutions, for example, from hot xylene. This tends to segregate the PED and PEH isotopes, as first pointed out by Mandelkern,[49] and for SANS by Schelten.[45] Such a condition defeats the SANS method and special preparative procedures have been used to minimize the segregation. These include the self-seeding technique used by Keller[50] to ensure highly dispersed nucleation in solution, and especially a rapid quench of the polymer blend from the molten state or from dilute solution. It appears possible, but not yet on a satisfactorily consistent basis, to avoid segregation in quenched samples. Segregation appears most obviously in the Zimm plot extrapolation to $Q \leqq 0$, which gives an apparent molecular weight M_{app}, higher than the weight obtained from gel permeation chromatography (GPC) or other external measurements. A means of avoiding segregated samples is to accept only those for which the Zimm plots show the ratio (M_{app}/M_{GPC}) to be close to unity. Because of differences in viscosity, in temperature, and in the method of quench, segregation may be quite different for melt-crystallization than for precipitation. Some authors[8] interpret the effects of segregation as a clustering of PED chains, while others[18] require only a natural dispersion in the concentration of PED from one crystallite to another.

There are other difficulties. One is the scattering from voids, which in the low-Q region can be a very large additional scattering component.† This cannot be removed by simple subtraction of the scattering from the undoped control dummy because the presence of PED changes the void effect.[18] Another problem is the possible concentration of PED chains in the amorphous volumes, which for quenched melts can constitute 35% or more of sample volume. Another is the effect on mixing of both the average molecular weight and its dispersity for the host matrix as compared to that of the tagged chains. Finally, the dependence on concentration has not thus far been firmly established.

Despite these difficulties, Schelten and co-workers have determined from Zimm plots that there is no significant difference between molten

[49] F. S. Stehling, E. Ergos, and L. Mandelkern, *Macromolecules* **4**, 672 (1971).
[50] D. J. Blundell and A. Keller, *J. Macromol. Sci.* **132**, 301 (1968).

† As an extreme example the scattering cross section from a void in a 1% PED–99% PEH matrix where the void radius is equal to the R_g of a PED chain in PEH is 5×10^3 bigger for the void than for the chain.

and melt-crystallized PED–PEH samples.[8] They report for both sample types a Gaussian behavior for the radius of gyration. When corrected for polydispersity they report

$$\overline{(R_w^2)}^{1/2} = (0.46 \pm 0.05)M_w^{1/2}.$$

Since the melt should exhibit, again, all the properties of a Gaussian coil, this agreement is remarkable because it indicates that the long-range crystalline order, comprising at least 65% of the sample, has no observable effect on the size of the chain.

There is, however, less agreement between the two sample types in the intermediate-Q regime, as revealed in Kratky plots. It is here that the data are most sensitive to physical models of shape. Yoon and Flory[51] have developed an adjacent reentry model for melt-crystallized PEH, and have compared it with Schelten's data. A major interest is to determine whether the Kratky data confirm or deny the existence of adjacent chain folding along a preferred direction in the lamellae, as opposed to a random or "switchboard" model. The subject has been extensively debated since a series of infrared papers were presented by Krimm[52] demonstrating adjacent reentry along ⟨020⟩ for melt solid and along ⟨110⟩ for crystal lamellae precipitated from dilute solution. The Schelten data agree most closely with the Yoon–Flory curve for random reentry ("single-stem" reentry) and hence apparently disagree with the Krimm model.

An alternative SANS study has been made that attempts to examine the differences in scattering between melt-crystallized and precipitated PED–PEH crystals.[46] Intuitively, the precipitate-crystallized solid should be a simpler matrix for tagging, because a given PED chain tends to be confined to a single lamella. At the same time, greater interference from void scattering should occur because the samples are powders. These may be subsequently cold-pressed to remove most large voids.

The results of these experiments show consistent differences between melt- and precipitate-crystallized samples. For molecular weight 97,000 the Zimm plot for melt-crystallized samples shows an $\overline{(R_g^2)}^{1/2}$ of 152 ± 8 Å, which is in reasonable agreement with the Schelten formula above. The $\overline{(R_g^2)}^{1/2}$ for precipitation-crystallized samples is substantially smaller, Zimm plots giving values of 98 ± 15 Å for unpressed powders and 122 ± 8 Å for the same powders after cold-pressing at 20,000 psi.

Of greater importance is a comparison of the Kratky plots between

[51] D. Y. Yoon and P. J. Flory, *Polymer* **18**, 509 (1977).
[52] M. I. Bank and S. Krimm, *J. Appl. Phys.* **39**, 4951 (1968); *J. Polym. Sci., Part A-1* **7**, 1789 (1969); S. Krimm and J. H. C. Ching, *Macromolecules* **5**, 209 (1972); J. H. C. Ching and S. Krimm, *ibid.* **8**, 894 (1975).

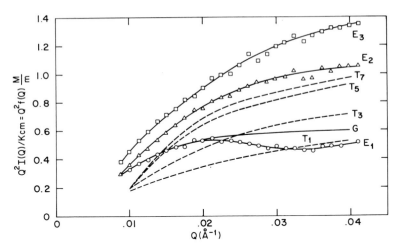

FIG. 10. Comparison of Yoon–Flory function $F_n(\mu)$: T_1, T_3, T_5, $T_7 = 1, 3, 5, 7$ adjacent reentry stems, with experimental curves; E_1, melt-crystallized 1% PED–99% PEH sample; E_3, precipitate-crystallized 1% PED–99% PEH sample; $E_2 = E_3$ after cold press; G, Gaussian coil.

samples, and with the Yoon–Flory model. These plots are shown in Fig. 10. The ordinate is the Yoon–Flory function $F_n(\mu)$, which is different by a constant M/mc from the usual Kratky ordinate. A calculated Gaussian coil for $R_g = 140$ Å is included for comparison. All the curves have the same ordinate, which for the experimental curves is $Q^2 I(Q)/cKm$ in our notation. There is no normalization between any curves—they are absolute values, T_1, T_3, T_5, and T_7 are the Yoon-Flory plots for no adjacent reentry, and for 3, 5, or 7 adjacent reentry stems in a lamella (the curve for 28 adjacent stems is only slightly above T_7). E_1 is a 1% PED melt-crystallized sample, E_3 a 1% PED precipitation-crystallized sample, and E_2 the sample E_3 after cold-pressing. The model is not for the identical parameters as the experiment and because of molecular-weight difference the experimental curves should diverge from the model at the lower-Q limit but not about $Q = 0.04$ Å^{-1}.

The comparisons in Fig. 10 seem to support strongly adjacent reentry for the precipitated samples as opposed to no adjacent reentry for the melt solid. The agreement can only be said to be qualitative at this stage and closer comparison of model and sample parameters are clearly needed. The effects of pressing on lamellar structures and the proper treatment of amorphous volumes in the precipitate-crystallized samples are as yet unresolved.

AUTHOR INDEX FOR PART A

Numbers in parentheses are footnote reference numbers and indicate that an author's work is referred to although the name is not cited in the text.

A

Abbate, S., 106
Abbe, J. C., 400
Abbott, F., 437
Abbott, S. D., 45
Abe, M., 59
Abe, Y., 106
Abragam, A., 244, 245(6), 246, 247(11), 248(11), 250(11), 253(11), 266(6), 267(6), 332, 334, 361
Abraham, R. J., 266
Abramovitz, M., 416
Ache, H. J., 372
Adams, Jr., E. T., 66
Adams, H. E., 64
Adams, J. Q., 285
Adcock, J. I., 44
Adler, J. G., 153, 158, 167, 168(17), 169
Afifi-Effat, A. M., 34
Aizenshtein, E. M., 467(128), 469, 470(128), 475(128)
Akhiezer, A. I., 372
Alderman, D. W., 326, 340(142), 343(142)
Ailion, D., 256
Aithal, H. N., 410
Alberi, J., 496, 497(30)
Alefeld, B., 228
Alexandrova, T. A., 467(73), 468, 470(73), 472(73), 476(50), 477(73)
Alfonso, G., 430
Alla, M., 358, 360
Allemand, R., 496
Allen, G., 208, 210, 229(19), 236(13), 237, 238(70), 302, 311, 312(91), 480
Allen, L. C., 11
Allen, P. E. M., 4
Allen, P. W., 31
Allerhand, A. A., 275, 303
Allison, S., 350, 351(181)
Almin, K. E., 48
Alms, G. R., 180
Alpert, N. L., 241

Altgelt, K. H., 44, 46, 50
Ambler, M. R., 47, 57, 58, 62, 63
Ambrose, E. J., 74
Anderson, C. D., 372
Anderson, F. R., 37
Anderson, W. A., 270
Andrew, E. R., 257, 258(26), 323(26), 326(26), 329(26), 340, 341
Andrews, R. D., 194, 195(23), 201(23), 202
Andrich, M. P., 411, 436
Antle, P. E., 53
Antsiferova, L. I., 459, 467(30), 470(30), 473(30)
Anufrieva, E. V., 435
Aplin, J. D., 467(142), 469, 474
Archibald, W. J., 68
Arkina, S. N., 467, 473(59)
Armstrong, J. L., 25
Arrington, M. C., 46, 47
Arven, J. A., 4
Asahina, M., 93
Asai, T., 432
Ashcraft, R. W., 59
Atkinson, C. M. L., 52, 57(192)
Atsumi, M., 422
Attebery, J. M., 45
Awcock, M. L., 496
Axelson, D. E., 288, 306
Azumi, T., 422

B

Bacon, G. E., 205, 208(1), 481, 482(15), 483
Badley, R. A., 411, 436(32)
Baekeland, L. H., 4
Bailey, G. F., 88, 97
Baker, C. A., 37, 39
Baker, D. R., 51, 53
Bakos, D., 48
Baldwin, R. L., 66
Balke, S. T., 55, 59, 60(254)
Ballard, D. G. H., 480, 499(6), 501(6), 503

507

SUBJECT INDEX FOR PART A

AUTHOR INDEX FOR PART B

Numbers in parentheses are footnote reference numbers and indicate that an author's work is referred to although the name is not cited in the text.

A

Adachi, K., 214
Adam, G., 324
Agboatwala, M. C., 276, 277(160)
Akahori, T., 269
Albers, J. H. M., 308
Alcock, T. C., 342
Aleksandrov, A. A., 213
Alexander, L. E., 5, 20, 67, 74, 118, 129, 133(2), 137, 138(13), 139
Alexander, W. J., 270, 273
Ali, M. S., 328
Allegra, G. E., 93, 94, 95, 96
Allen, P. W., 373
Allison, S. K., 27, 30
Allou, A. L., 325
Allou, Jr., A. L., 295, 300, 305, 306
Anderegg, J. W., 180, 181(91)
Anderson, F. R., 217, 238, 244
Anderson, H. C., 334
Andrews, E. H., 385
Aqua, E. N., 144
Arai, N., 253, 254(80)
Arakawa, T., 253, 254, 308, 350
Arbogast, J. F., 366
Arimoto, H., 82
Armeniades, C. D., 372
Arnott, S., 97, 113
Atkinson, C. M. L., 325
Averback, B. V., 147
Avrami, M., 347, 348

B

Bacon, R. C., 256, 262(91)
Baer, E., 238, 245(27), 247, 249, 250, 251,

252, 253, 254(27), 304, 308, 327(72), 362, 372, 386, 387(114), 388(114), 392(114), 393(114), 395(114), 396
Baev, A. S., 213
Bagchi, S. N., 102(65), 103, 105, 106, 109(65, 68), 112(65), 144, 147(26), 148(26), 164(26), 171(26)
Bailey, G. W., 240
Bair, H. E., 254, 300, 305(57), 314, 315, 331, 352, 364
Baker, C., 369
Baker, P. N., 208
Ballard, D. G. H., 269
Balta Calleja, F. J., 299
Balwit, J. S., 279
Baranov, V. G., 368
Barrales-Rienda, J. M., 337
Barrett, K. E. J., 332
Barton, D. H. R., 268, 269
Barton, J. M., 329, 334
Bascom, W. D., 332
Bassett, D. C., 216, 217(21), 229, 233(21), 238, 245, 248, 297, 299, 312, 315, 339, 346
Bassett, G. A., 217, 222, 237
Battista, O. A., 276
Baudisch, J., 269
Baumeister, W., 213
Beacheli, H. C., 256, 262(93)
Beck, H. N., 396
Beech, D. R., 308
Bekkedahl, N., 370, 371
Bell, A. T., 241
Bell, D., 175
Bell, J. P., 255, 274, 276, 277, 306, 307, 308

537

Benedetti, E. 93, 94(59, 61), 95(59), 96(61)

Berghmans, H., 301, 303(59)

Bergmann, K., 237

Berkelhamer, L. H., 291

Bernard, M. Y., 186

Bertein, F., 186

Bezruk, L. I., 244

Bhat, N. V., 244

Bibikov, V. V., 279

Billmeyer, F. W., 151, 325

Billmeyer, Jr., F. W., 18

Binsbergen, F. L., 396

Blackadder, D. A., 229, 325

Blaine, R. L., 325

Blais, P., 241, 242(51)

Blanchard, L. P., 336

Blazek, A., 287

Blundell, D. J., 230, 256, 260, 262(97, 100, 105), 263(105), 307, 382, 384

Blundell, D. S., 169

Bobalek, E. G., 256, 262(92)

Bodily, D. M., 279, 280, 281, 328

Bodnar, M. J., 241, 242(50)

Boersma, S. L., 291

Bohlin, L., 296

Bolz, L. H., 242, 243

Bonart, R., 144, 147(27), 148(27), 172

Boon, J., 306

Booth, A., 336

Booth, C., 307, 308

Borchardt, H. J., 291, 292(9)

Borkowski, C. J., 46, 61, 155

Bosch, W., 366

Boyer, R. F., 252, 331, 374

Bragg, W. H., 1

Bragg, W. L., 1

Bramer, R., 164

Bravais, A., 3

Brennan, W. P., 318, 319, 320, 321, 322, 324

Brenner, N., 289, 290(5), 321

Broadhurst, M. G., 325, 351

Broers, A. N., 228

Brumberger, H., 182, 330

Buben, N. Y., 279

Buchanan, D. R., 148, 149, 150

Budzol, M., 279

Buerger, M. J., 11, 71, 72, 77, 92, 93

Bullough, R. K., 92

Bunn, C. W., 15, 74, 79, 80, 81, 82, 88, 89, 90(39), 93, 95, 325, 342

Burmester, A., 163, 167(59), 299

C

Calvert, P. D., 387, 395

Canterino, P. J., 245

Carlson, G. L., 241, 242(49)

Carrlson, D. J., 241, 242(51)

Carroll, B., 332

Carter, D. R., 362

Castaing, R., 186

Ceccorulli, G., 336

Cernee, F., 332

Chabre, M., 155

Challa, G., 121, 255, 303, 304(66), 306

Chan, K. S., 330

Chandler, L. A., 306

Chang, M., 260(121), 261(121), 264, 269(121), 270, 271, 272

Chapiro, A., 278

Charlesby, A., 278, 279, 280

Chatani, Y., 93, 94(58), 95, 96, 97, 98

Chatterjee, A. M., 396

Chaudhuri, A. K., 256

Chiang, R., 237, 298, 300

Chiu, J., 334, 335

Christiansen, A. W., 304, 327(72)

Cirlin, E. H., 334

Clampitt, B. H., 305

Clark, D. T., 245

Clark, E. S., 20, 83, 85, 87, 88, 90, 113, 114

Cobbold, A. J., 238, 255, 262(31)

Cochran, W., 83, 84, 92, 113

Cogswell, F. N., 307

Cohen, J. B., 142

Coleman, M. M., 174

Collins, E. A., 306

Compton, A. H., 27, 30

Connor, T. M., 256, 262(100)

Cooper, C. W., 248

Cooper, M., 384

Cooper, S. L., 307, 335

Cooper, W., 304, 305

Corey, R. B., 90, 91

Cormia, R. L., 386, 387, 388, 392, 393

Cormier, C. M., 325

Corradini, P., 16, 17(14), 37, 78, 81, 88, 89

Cox, W. P., 330

Cox, W. W., 304, 305, 327(71)

Crescenzi, V., 332

Crick, F. H. C., 83, 84(41), 113(41)

Crist, B., 167, 171, 172(69)

Critchley, J. P., 329

AUTHOR INDEX FOR PART C

Numbers in parentheses are reference numbers and indicate that an author's work is referred to although the name is not cited in the text.

A

Abagyan, G. V., 189, 195, 201(17, 34), 202(34)
Abrams, A. I., 366
Adams, V., 493
Aggarwal, S. L., 295, 300(110), 311(110)
Aiken, W. H., 363
Ailhaud, H., 303
Akhmed-Zade, K. A., 196, 201(50)
Alexander, L. E., 175
Alfonso, G. C., 295, 297(109)
Alfrey, T., 17, 336
Alger, R. S., 186
Allen, F. G., 435
Allen, G., 283, 285
Alpert, D., 359, 370
Alston, L. L., 461, 481(60)
Altamirano, J. O., 311
Al'tzitser, V. S., 311
Amakawa, K., 477
Amborski, A. J., 490, 495(111)
Ambrose, J. F., 350
Amelin, A. V., 227
American Society for Testing Materials, 363, 365(240), 366, 371(266)
Anderson, R. L., 36
Anderson, R. S., 186
Andersson, P., 101
Andrade, E. N. da C. A., 15, 27
Andreatch, P., Jr., 61
Andrews, E. H., 186, 214, 216, 253, 254,

Andrews, R. D., 33
Angelo, R. J., 311
Anon, 372
Aozasa, M., 492
Arakawa, T., 91, 115(4)
Ar'ev, A. M., 229

Argon, A. S., 250, 251, 252, 253(48), 255, 264(48, 56), 265(48)

Arkawa, Y., 372
Arlie, J. P., 304
Armstrong, A. A., 375
Arridge, R. G. C., 429
Artbauer, J., 452, 456(23), 464(23), 465(23), 474
Artemov, P. G., 108
Asai, H., 295, 298(104)
Asay, J. R., 66, 74, 75, 77(16, 22), 78(22)
Ash, R., 327, 358, 359, 362
Ashcraft, C. R., 395, 396
Assink, R. A., 356
Ast, D. G., 244, 246, 247(32, 35), 248(32), 253(32), 256(21), 271(32)
ASTM Standards, 122
Auckland, D. W., 449, 451, 461, 470(12, 15), 478(15), 480(15), 486, 487
Austen, A. E. W., 452

B

Backman, D. K., 190(19), 191, 202, 203(19)
Bäckström, G., 428, 432, 433(31, 32), 434(32)
Backström, J., 101
Baer, E., 91, 93, 94, 100, 244, 245, 249, 250(41), 257, 260, 261(73)
Baer, M., 289, 312, 313(216)
Baessler, H., 464
Bagley, E. B., 47, 337, 350(119)
Bahder, G., 449, 462, 482, 488(75)
Bailey, C. D., 366
Bailey, W. J., 104
Baird, J. C., 186, 188(10)
Baker, R. W., 362
Baker, W. O., 388
Baker, W. P., 449
Bakr, A. M., 322

549

SUBJECT INDEX FOR PART B

SUBJECT INDEX FOR PART C